Jasmonates and Brassinosteroids in Plants

This book provides a comprehensive update on recent developments of jasmonates (JAs) and brassinosteroids (BRs) in plant signaling and biotechnological applications. Over the last few decades, an enormous amount of research data has been generated on these two signaling molecules. This valuable compilation will enhance the basic understanding of the mechanisms of action of JAs and BRs and the ensuing tolerance mechanisms of crops for sustainable agriculture and human welfare during climate change. This book covers topics regarding the occurrence of JAs and BRs in plants, biosynthesis, their role in plant growth and development, the role of these plant growth regulators in abiotic stress tolerance in plants, crosstalk of reactive oxygen species and plant stress mitigation, regulation of JA and BR signaling pathways by microRNA, and physiological and anatomical roles of JAs and BRs such as wound healing, regeneration, and cell fate decisions. In addition, this book covers the crosstalk of JAs and BRs with neurotransmitters in plant growth and development, as well as the bio-fortification of crop plants with BRs in managing human health issues in their new role in human wellbeing.

This book will be beneficial to scientists, researchers, agriculturists, horticulturists, and industries related to crop and food production

Key Features

This book reviews the global scientific literature and experimental data of the authors on the occurrence of JAs and BRs in various plants:

- Provides updated information on recent developments in JA and BR signaling and biotechnological applications in plants
- Highlights the physiological, metabolic, and molecular mechanisms of JAs and BRs in variable climates
- Addresses abiotic and biotic tolerance management by JAs and BRs
- Describes the role of JAs and BRs in sustainable agriculture and human welfare in an eco-friendly manner

Jasmonates and Brassinosteroids in Plants

Metabolism, Signaling, and Biotechnological Applications

Edited by
Ramakrishna Akula and Geetika Sirhindi

CRC Press
Taylor & Francis Group
Boca Raton London New York

CRC Press is an imprint of the
Taylor & Francis Group, an **informa** business

First edition published 2023
by CRC Press
6000 Broken Sound Parkway NW, Suite 300, Boca Raton, FL 33487-2742

and by CRC Press
2 Park Square, Milton Park, Abingdon, Oxon, OX14 4RN

Library of Congress Cataloging-in-Publication Data

Names: Akula, Ramakrishna, editor. | Sirhindi, Geetika, editor.
Title: Jasmonates and brassinosteroids in plants: metabolism, signaling, and biotechnological applications / Ramakrishna Akula, Geetika Sirhindi.
Description: First edition | Boca Raton, FL: CRC Press, [2022] | Includes bibliographical references and index.
Identifiers: LCCN 2022006708 (print) | LCCN 2022006709 (ebook) | ISBN 9780367627560 (hardback) | ISBN 9780367627584 (paperback) | ISBN 9781003110651 (ebook)
Subjects: LCSH: Plant hormones. | Brassinosteroids. | Jasmonic acid.
Classification: LCC QK898.H67 J37 2022 (print) | LCC QK898.H67 (ebook) | DDC 571.7/42--dc23/eng/20220228
LC record available at https://lccn.loc.gov/2022006708
LC ebook record available at https://lccn.loc.gov/2022006709

ISBN: 9780367627560 (hbk)
ISBN: 9780367627584 (pbk)
ISBN: 9781003110651 (ebk)

DOI: 10.1201/9781003110651

Typeset in Palatino
by Deanta Global Publishing Services, Chennai, India

Contents

Preface

The present dynamic climate has challenged all plant forms at every point of their growth and development around the world, while every country faces challenges in feeding an ever-increasing population on a regular basis. To accomplish the goal of feeding all people with nutritious food, scientists are working to explore new and emerging ways of enhancing crop productivity. For this, the most common practice in plant breeding is engineering the gene(s) in crops responsible for high productivity. But this way does not have guaranteed success as a number of environmental factors play a role in the sustainable viability of such engineered crops. Second, genetic manipulation practices are very laborious and time-consuming and also show deleterious effects on a dynamic environment. Another widespread practice to enhance crop productivity is using chemicals and fertilizers. But such practices are highly detrimental not only to humans but also to the environment and life on Earth, directly or indirectly. For tackling such problems, there is a new emerging path for enhancing crop productivity in an eco-friendly manner, which is chemical genetics technology. This technology involves the use of eco-friendly chemicals like plant growth regulators (PGRs) for sustainably enhancing plant productivity in dynamic climates. Jasmonates (JAs) and brassinosteroids (BRs) are two such young PGRs that are now being used to enhance crop productivity in changeable environmental conditions for sustainable agriculture development.

However, no blueprint is available for explaining the metabolism, signaling, and biotechnological applications of different forms of BRs and JAs in a single platform. Thus, this edited book presents a collection of vignettes written by experts in BR and JA research. This book is an up-to-date overview of the current progress in exploring the potential of BRs and JAs in sustainable agriculture in a dynamic environment, showing that these PGRs are beneficial for humans in the present scenarios of climate adversity and a pandemic.

This volume on BRs and JAs covers and compiles the latest information in one platform that will be beneficial for young researchers and scientists as well as those students who want to explore chemical genetics for sustainable agriculture development.

Dr. Ramkrishnan Akula
Dr. Geetika Sirhindi

Foreword

Brassinosteroids (BR) and jasmonates (JA) are two new classes of plant growth regulators (PGRs) that are used to manage various aspects of plant growth and development in the present global environment. BR and JA are not just hormones, managing plant growth and productivity under optimal growth conditions or when not facing extreme environments, they also assist plants in the protection and management of growth in internal or external environments. Under some conditions, BRs and JAs are amalgamated with other PGRs and are synthesized endogenously or applied exogenously to accomplish sustainable growth and development in dynamic environments. A common objective of PGRs in the field of plant physiology, particularly stress physiology, is to protect and defend sustainable plant development. Additional objectives are the enhancement of yield in unexpected climatic changes by ensuring the availability of metabolites and maintaining the homeostasis of cellular structures and biochemistry within the cell.

Environmental stresses are the most limiting factor for agricultural yield. A substantial proportion of annual crop yield is lost to abiotic and biotic constraints. To manage these losses, plants have a number of endogenous and exogenous factors that work in synchronization for the sustainable growth and productivity of crops. The intense use of chemical fertilizers is another area of concern that has to be questioned for protecting the environment both for human life as well as crop productivity. Recent advances in agriculture biotechnology and plant breeding programs have tried to find solutions to the previously mentioned challenges, which are somewhat laborious and have failed with sudden climatic encounters. The extent of crop yield loss as a result of endogenous and exogenous factors can be managed by manipulating plant metabolism and biochemistry by using plant growth regulators. BR and JA are two of the finest PGRs at present.

The challenges in plant stress management and a sustainable yield of agriculture crops by BR and JA are both difficult and an interesting line of research. People are working on them with enthusiasm, tenacity, and dedication to understanding the mechanisms and signaling pathways involved in BR and JA functioning in plants. This will open new insights and help in developing innovative methods of analysis and providing fresh solutions to keep up with the ever-changing threats to agriculture productivity. With the aberrant climate change challenge, it is necessary to provide eco-friendly and sustainable practices, with state-of-the-art knowledge on these eco-friendly PGRs frontiers in agriculture sustainability. This book is a good step in that direction.

This book *Jasmonates and Brassinosteroids in Plants: Metabolism, Signaling, and Biotechnological Applications* edited by Drs. Ramakrishna Akula and Geetika Sirhindi provides a valuable window on BRs and JAs. It covers the necessary components of BRs and JAs such as metabolism, signaling, and biotechnological applications in different crops for crop management by BR and JA in dynamic environments. The chapters, written by experts in their respective fields, cover a large array of recent topics pertaining to JA and BR occurrence, these include the following: biosynthesis; their role in plant growth and development; the role of these PGRs during various abiotic stresses in plants; crosstalk of reactive oxygen species and plant stress mitigation; regulation of the JA and BR signaling pathways by microRNA; and the physiological and anatomical role of BRs and JAs in wound healing, regeneration, and cell fate decisions. In addition, the crosstalk of JA and BR with neurotransmitters in plant growth and development is discussed as is the bio-fortification of crop plants with BRs in managing human health issues. This comprehensive analysis of different aspects of BR and JA should make this volume equally valuable to new start-up researchers in the field of plant stress physiology and plant growth regulators, and to application-orientated scientists including teachers as well as students entering the fields of plant biology and plant physiology. I am sure readers in the field of agriculture, particularly in plant stress physiology management, biotechnology, and agriculture applications of PGRs, will find this book beneficial. I extend my congratulations to the publishers for publishing this comprehensive volume of BRs and JAs.

Prof. Frantisek Baluska
Group leader
Department of Plant Cell Biology
Institute for cellular and molecular Botany
University of Bonn, Germany

Acknowledgments

We are very much thankful to our contributors who have devoted their valuable time and energy and shared their research experiences and findings of exploring different potentials of JAs and BRs for sustainable agriculture and human welfare. We are grateful to Renu Upadhyay, Assistant Commissioning Editor, Jyotsna Jangra, Editorial Assistant, and all the other staff members of CRC, Taylor & Francis, who were directly or indirectly associated with the project for their unconditional support and help, valuable suggestions to make this publication on time, and for their help in bringing out the volume in a beautiful and appealing manner. The Editors are thankful to our families who have encouraged us to take up this task and permitted us to take time away from our families. We are also thankful to our friends and academic colleagues for their support in the successful accomplishment of this book.

Dr. Ramkrishnan Akula
Dr. Geetika Sirhindi

Editor Biographies

Dr. Ramakrishna Akula is currently a scientist at Bayer Crop Science, Bangalore, India. Dr. Ramakrishna holds a master's degree from Sri Krishna Devaraya University, Anantapur, Andhra Pradesh, India. He obtained his PhD in biochemistry from the University of Mysore, CSIR-CFTRI, Mysuru. He is a Senior Research Fellow of CSIR, New Delhi. He is involved in various research fields such as plant biotechnology, biochemistry, plant secondary metabolites, food science and technology, and vegetable quality analytics. He has published several research papers in leading international journals and has published books, reviews, and book chapters. He is also serving as an editorial board member and a reviewer for reputed international journals. He has presented over 20 research papers in symposia in India and abroad, as well as invited lectures in India. He attended the Fifth International Symposium on Plant Neurobiology, held in 2009, in Florence, Italy. He also attended the Technical Community of Monsanto (TCM), held in 2016, in St. Louis, Missouri, USA. He is a member of the Indian Science Congress Association and Society for Biotechnologists (India). He is a fellow of the Society for Applied Biotechnology, India (2012), and recipient of a Global Vegetable Research Excellence Award (2017), Global Technology Recognition, Rapid Recognition Award, Special Recognition from the Monsanto Company, and Veg R&D Champion (2020) from Bayer Crop Science.

Dr. Geetika Sirhindi is currently working as an Associate Professor in Plant Physiology at the Department of Botany, Punjabi University, Patiala, India. She completed her master's in botany and her PhD in life sciences from Punjabi University, Patiala, Punjab, India. Her research interest is in chemico-genetical exploration of brassinosteroids and jasmonates in selected major agriculture crops in India for sustainable agriculture under dynamic stress environments and human welfare. She has completed three major research projects awarded by UGC, New Delhi, India (2008–2011; 2013–2016) and SERB-DST, New Delhi, India (2014–2017). She has published more than 45 papers in national and international journals and more than 13 book chapters. She has presented her research findings at various international platforms. She has been awarded the Nature Publication Group Award in 2007 and has delivered a number of invited lectures in India and abroad and visited numerous countries to present her research work in international symposia. She is a life member of the Indian Science Congress, Indian Botanical Society, Indian Society of Plant Physiology, and the Biotechnology Research Society of India.

Contributors

Ramakrishna Akula
Vegetable R&D Department
Bayer Crop Science Division
Chikkaballapur, India

Barket Ali
J&K Higher Education Department
Government Degree College, Kishtwar
Jammu and Kashmir, India

B. Asthir
Department of Biochemistry
Panjab Agricultural University
Punjab, India

Andrzej Bajguz
Faculty of Biology and Chemistry
Institute of Biology
University of Białystok
Bialystok, Poland

Laxmipreeya Behera
Odisha University of Agriculture and Technology
Odisha, India

Renu Bhardwaj
GNDU
Punjab, India

L.A. Bravo
Universidad Catolica de Temuco
Temuco, Chile

Ashmita Chhikara
Maharshi Dayanand University
Haryana, India

Magdalena Chmur
University of Bialystok
Bialystok, Poland

V. R. Devaraj
Bengaluru City University
Bengaluru, India

Neha Dogra
Punjabi University
Punjab, India

C. Figueroa
Universidad de La Frontera
Temuco, Chile

Mona Gergis Dawood
Botany Department
National Research Centre
Egypt

Ritu Gill
Centre for Biotechnology
Maharshi Dayanand University
Haryana, India

J. González-Villagra
Universidad Catolica de Temuco
Temuco, Chile

Takanori Horibe
College of Bioscience and Biotechnology
Chubu University
Aichi, Japan

C. Inostroza-Blancheteau
Universidad Catolica de Temuco
Temuco, Chile

Pooja Jha
Maharshi Dayanand University
Haryana, India

G. Kaur
Punjab Agricultural University
Punjab, India

Gurvarinder Kaur
Punjabi University
Punjab, India

Harpreet Kaur
Khalsa College
Amritsar, India

Navdeep Kaur
Department of Biotechnology
Guru Nanak Dev University
Punjab, India

Shruti Kaushik
Punjabi University
Punjab, India

Manish Kumar
Dev Samaj College for Women
Punjab, India

Pratap Kumar Pati
Department of Biotechnology
Guru Nanak Dev University
Punjab, India

Anil Kumar Singh
Punjabi University
Punjab, India

Punam Kundu
Maharshi Dayanand University
Haryana, India

José López Bucio
Instituto de Investigaciones Químico-Biológicas
Universidad Michoacana de San Nicolas de Hidalgo
Michoacán, Mexico

Isha Madaan
Punjabi University
Punjab, India

Deeksha Marothia
Guru Nanak Dev University
Punjab, India

M. Morales
University of Barcelona
Barcelona, Spain

Soumya Mukherjee
Department of Botany
Jangipur College
University of Kalyani
West Bengal, India

R. D. Myrene
Department of Biochemistry
Mount Carmel College
Bengaluru, India

M. Naeem
Aligarh Muslim University
Aligarh, India

Satya Narayan Dash
College of Agriculture
Odisha University of Agriculture and Technology
Odisha, India

Jannela Praveena
College of Agriculture
Odisha University of Agriculture and Technology
Odisha, India

Gyana Ranjan Rout
Department of Agricultural Biotechnology
Odisha University of Agriculture & Technology
Odisha, India

Javier Raya-González
Universidad Michoacana de San Nicolas de Hidalgo
Michoacán, Mexico

M. Reyes-Díaz
Faculty of Engineering and Sciences
Department of Chemical Sciences and Natural Resources
Universidad de La Frontera
Temuco, Chile

Shivani Saini
Goswami Ganesh Dutta Sanatan Dharma College
Sector 32-C
Chandigarh, India

Ritu Sharaya
Maharshi Dayanand University
Haryana, India

Anmol Sidhu
Punjabi University
Punjab, India

Sarvajeet Singh Gill
Maharshi Dayanand University
Haryana, India

Geetika Sirhindi
Department of Botany
Punjabi University
Patiala, India

A. Thilagavathy
Biochemistry Department
Bangalore University
Bangalore, India

1

Biosynthesis and Inactivation of Brassinosteroids in Plants

Andrzej Bajguz and Magdalena Chmur

CONTENTS

1.1 Introduction

Brassinosteroids (BRs) are an essential class of steroidal phytohormones consisting of more than 70 different compounds. The common part of BR biochemical structures is the skeleton of 5α-cholestane, 5α-ergostane, or 5α-stigmastane (Figure 1.1). Differentiation of BRs is an effect of the presence of various functional groups within rings A, B, and the side chain (Figure 1.2). Thus, the most active and recognizable type of BRs comprise ring A with two hydroxyl group in the C-2 and C-3 positions in the conformation of C(2α, 3α), and can create isomers of C(2α, 3β), C(2β, 3α), or C(2β, 3β). BRs possess only one hydroxyl group in C-3α or C-3β, or three hydroxyl groups in C(1β, 2α, 3α) or C(1α, 2α, 3β). Compounds with a ketone group in C-2 or an epoxide group in C(2α, 3α) or C(2β, 3β) are known too. The chemical structure of ring B includes 7-membered 7-oxalactone or 6-membered with 6-oxo, 6-deoxy, or 6-hydroxy type of substitute. The side-chain structures depend on the BR type but most BRs possess the hydroxyl group in C-22 and C-23 positions in the side chain (Kanwar et al. 2017, Zullo and Bajguz 2019).

This chapter aims to summarize BR metabolism in plants, including types of biosynthetic and catabolic reactions, levels of BRs in transgenic organisms, and effects of BR deficiency on growth.

1.2 Biosynthesis Pathways of Brassinosteroids

The endogenous level of bioactive BRs is regulated through biosynthesis and catabolism. Until now, three independent pathways of BR biosynthesis leading to the synthesis of C_{27},C_{28},C_{29}-type are known. Early steps of their synthesis are common and occur via mevalonate (MVA) or non-MVA pathways (Hunter 2007, Miziorko 2011), while later steps differentiate the BR biosynthesis pathways (cycloartenol- and cycloartanol-dependent). The C_{27}-BRs (Figure 1.3) do not possess any substitute in the C-24 position and are derived from cholesterol (CR), which is finally converted to 28-norBL. The C_{28}-BRs (Figure 1.4) are produced predominantly from campesterol and 22α-hydroxycampesterol, and contain α-methyl, β-methyl, or methylene groups in the side chain. Whereas the C_{29}-BRs (Figure 1.5) with an α -ethyl group are synthesized from sitosterol and lead to 28-homoBL (Chung and Choe 2013, Fujioka et al. 2002, Fujita et al. 2006, Kim et al. 2018, Kwon and Choe 2005, Ohnishi, Watanabe, et al. 2006, Roh et al. 2017, Rozhon et al. 2019, Sonawane et al. 2016, Wang et al. 2017).

1.2.1 Biosynthesis of C_{27}-Brassinosteroids

The C_{27}-BR biosynthesis pathway starts from the conversion of cycloartenol to cycloartanol (Figure 1.6) by the sterol side chain reductase 2 (SSR2) and proceeds through a synthesis of 31-norcycloartanol from cycloartanol by the C4-sterol methyloxidase3 (SMO3), and further biochemical changes of 31-norcycloartanol up to CR, which is synthesized from 7-dehydrocholesterol by the 7-dehydrocholesterol reductase 2 (Sonawane et al. 2016). The biosynthesis of C_{27}-BRs might occur through the late C6 oxidation pathway. First, CR is converted to cholestanol (a C_{27}-BR biosynthesis precursor) by

DOI: 10.1201/9781003110651-1

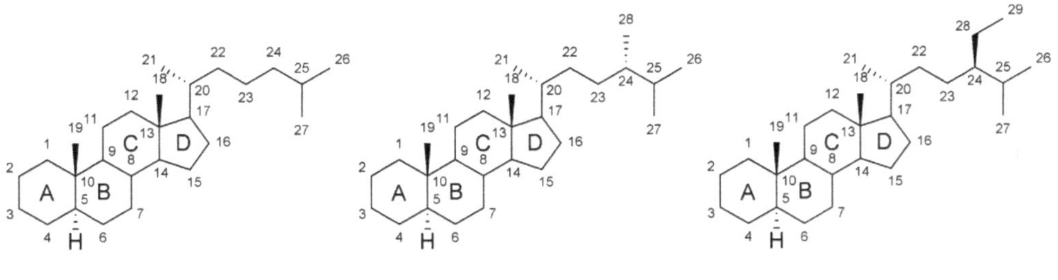

FIGURE 1.1 The structural basis of brassinosteroid skeleton.

the 5α-reductase encoded by the *DET2* gene (Figure 1.7). In the next steps of the C_{27}-BR biosynthesis pathway, 6-deoxo-28-norcathasterone, 6-deoxo-28-norteasterone, 6-deoxo-28-nor-3-dehydroteasterone, 6-deoxo-28-nortyphasterol, and 6-deoxo-28-norcastasterone are synthesized in consecutive reactions. On the other hand, the early C6 oxidation pathway is initiated through oxidation of cholestanol into 6-oxocholestanol, which is then followed by a synthesis of 28-norcathasterone, 28-norteasterone, 28-nor-3-dehydroteasterone, 28-nortyphasterol, 28-norcastasterone, and 28-norbrassinolide. Enzymatic conversions of BRs from the late C6 oxidation pathway to the early C6 oxidation counterparts have been evidenced, for example, 6-deoxo-28-nortyphasterol to 28-nortyphasterol and 6-deoxo-28-norcastasterone to 28-norcastasterone. The oxidation/hydroxylation reactions are catalyzed by cytochrome P450 enzymes, that is, CYP85A1 and CYP85A2 oxidases (Fujita et al. 2006, Joo et al. 2015, Joo et al. 2012, Kim et al. 2005, Zhao and Li 2012).

1.2.2 Biosynthesis of C_{28}-Brassinosteroids

The first metabolite in the C_{28}-BR biosynthesis is episterol, which is synthesized from 24-methylenelophenol. This reaction is catalyzed by the C-4 sterol methyl oxidase 2 (SMO2) (Figure 1.6). Then, episterol is converted to 5-dehydroepisterol by the sterol C-5(6) desaturase encoded by the *DWF7* gene, which is then converted to 24-methyleneCR (catalyzed by 7-dehydrocholesterol reductase encoded by the *DWF5* gene). Further stages of the C_{28}-BR biosynthesis may proceed through the late and early C-22 oxidation pathways (Choe 2006, Ohnishi 2018). An essential difference between the C-22 oxidation sub-pathways is the synthesis of 6-deoxocathasterone (6-deoxoCT) from (22*S*,24*R*)-22-hydroxy-5α-ergostan-3-one, without synthesis of campestanol (CN) (CN-independent pathway of BRs biosynthesis) as a result of the early C-22 oxidation (Choe 2004, Ohnishi 2018).

The reduction of 24-methyleneCR to campesterol initiates the late C-22 oxidation pathway and is catalyzed by the C-24(25)-sterol reductase (Figure 1.6). Campesterol is then transformed into (24*R*)-ergostan-4-en-3β-one, (24*R*)-5α-ergostan-3-one, and CN. In the early C-22 oxidation pathway, C-22α hydroxylation of 24-methyleneCR is synthesized to 22-hydroxy-24-methyleneCR by the C-22α hydroxylase (encoded by the *DWF4* gene). The next reactions are analogous to the late C-22 oxidation pathway and result in the synthesis of 22-hydroxy forms of the corresponding compounds (Dockter et al. 2014, Fujiyama et al. 2019, Schaeffer et al. 2001). Biochemical changes of 22-hydroxymethyleneCR can lead to the synthesis of 6-deoxodolichosterone, which may be further converted into dolichosterone (DS) and dolicholide (DL), as well as to castasterone (CS) and BL (Roh et al. 2017, Verhoef et al. 2013, Ye et al. 2011).

Campestanol may be a substrate of the BR biosynthesis in both the early and late C6 oxidation pathways (Figure 1.7) (Shimada et al. 2001). The early C6 oxidation pathway begins from hydroxylation of CN to 6α-hydroxyCN and its subsequent oxidation to 6-oxo-CN. The latter is transformed to CT by the 22α-hydroxylase. Cathasterone (CT) is converted in the consecutive reactions to teasterone (TE), 3-dehydroteasterone (3-DT), typhasterol (TY), CS, and BL, respectively (Chung and Choe 2013, Fujioka et al. 2002, Joo et al. 2015, Kim et al. 2018, Kwon and Choe 2005, Lee et al. 2010, Lee et al. 2011, Ohnishi 2018, Ohnishi, Szatmari, et al. 2006, Ohnishi, Watanabe, et al. 2006, Roh et al. 2020, Shimada et al. 2001, Zhao and Li 2012). However, the late C6 pathway begins with hydroxylation of CN into the 6-deoxoCT catalyzed by the 22α-hydroxylase. In the CN-independent pathway, 22-hydroxy-5α-ergostan-3-one synthesizes 6-deoxoCT, which is hydroxylated through the C-23 hydroxylase (encoded by the *CPD* gene) to the 6-deoxoteasterone, which is oxidized through the *CYP90D* C3-oxidase into the 3-dehydro-6-deoxoteasterone (6-deoxo-3-DT); 6-deoxo-3-DT is converted to 6-deoxoTY by the *D11* CYP724B1. Then, 6-deoxoTY is hydroxylated to 6-deoxoCS by the 2α-hydroxylase (*DDWF1* gene); 6-deoxoCS is converted to CS (BR-6-oxidasel and BR-6-oxidase2 catalyze the reaction). Then, CS is converted to BL by the BR-6-oxidase2 (CYP85A2) (Choe 2004, 2006, Nakano and Asami 2014, Ohnishi 2018, Son et al. 2012, Vriet et al. 2013, Zhao and Li 2012).

1.2.3 Biosynthesis of C_{29}-Brassinosteroids

In the C_{29}-BR biosynthesis pathway, 24-methylenelophenol is converted by sterol methyltransferase 2 (SMT2) into 24-ethylidenelophenol, which is transformed into avenasterol by the sterol methyl (Figure 1.6) oxidase 2. Isofucosterol and β-sitosterol are products of avenasterol in reactions catalyzed by the DWF7, DWF, and SSR1 (DWF1) enzymes. β-sitosterol, as a precursor of the C_{29}-BRs, is hydroxylated into 6-deoxo-28-homoTY and oxygenated into 28-homoTY by the *CYP724B2* and *CYP90B3* C-22 hydroxylase, respectively (Figure 1.7) (Ohnishi, Szatmari, et al. 2006); 28-homoTY is

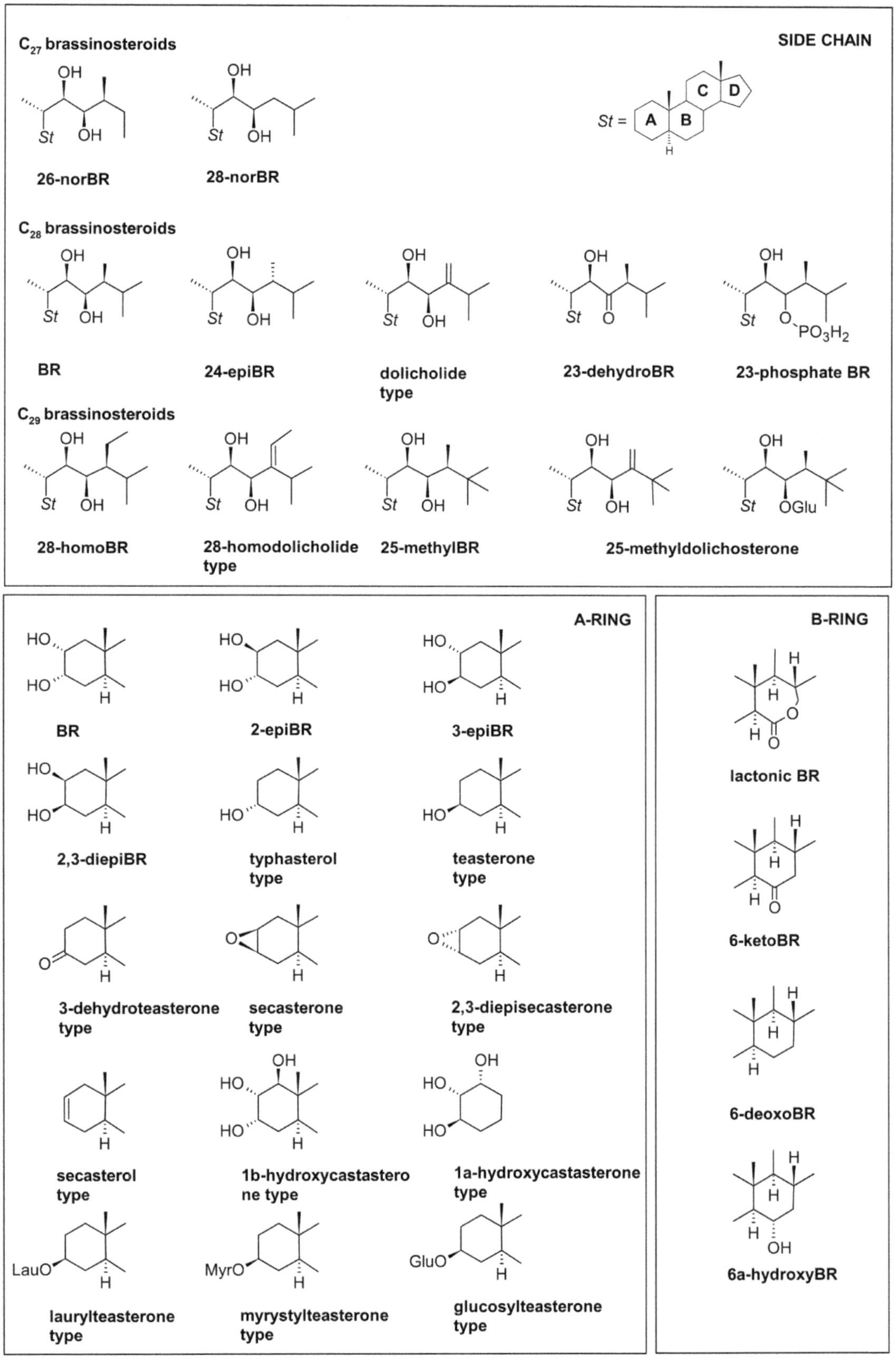

FIGURE 1.2 The structural variations of brassinosteroids in rings A and B, and side-chain.

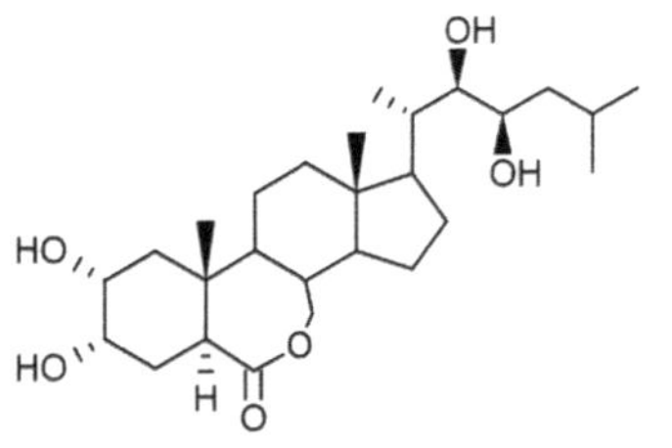

28-norbrassinolide (28-norBL)

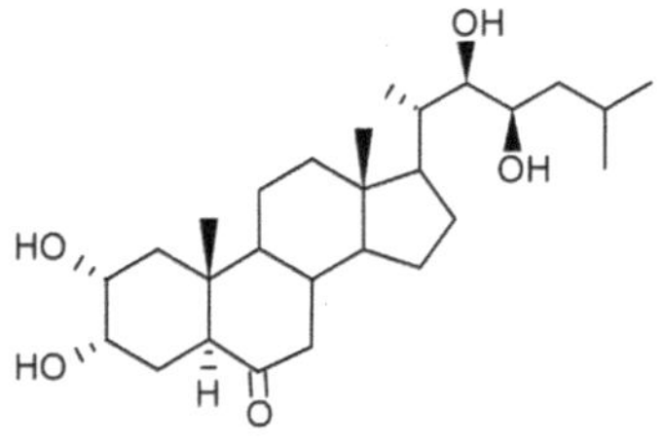

28-norcastasterone (28-norCS)

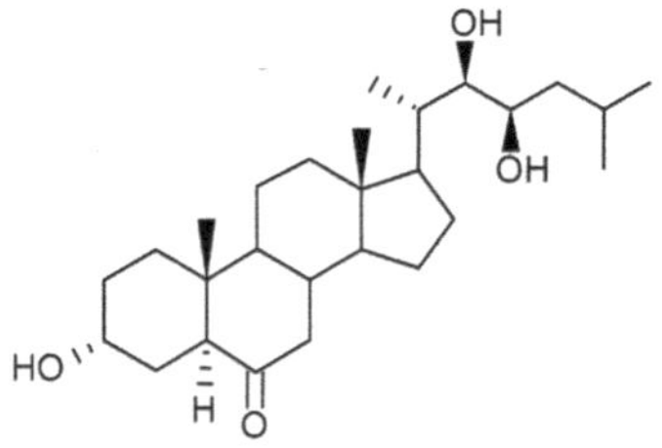

28-nortyphasterol (28-norTY)

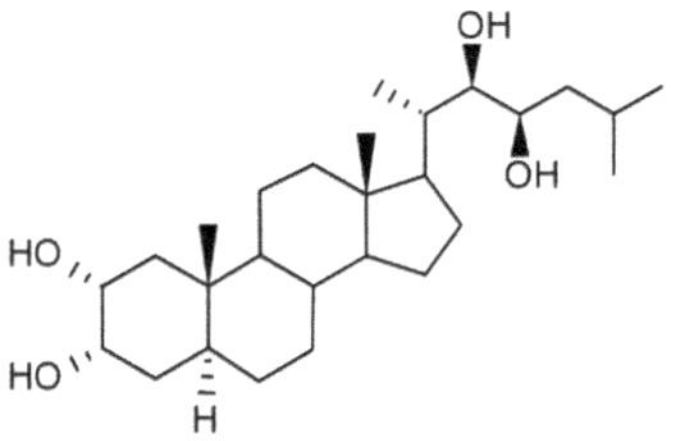

6-deoxo-28-norcastasterone (6-deoxo-28-norCS)

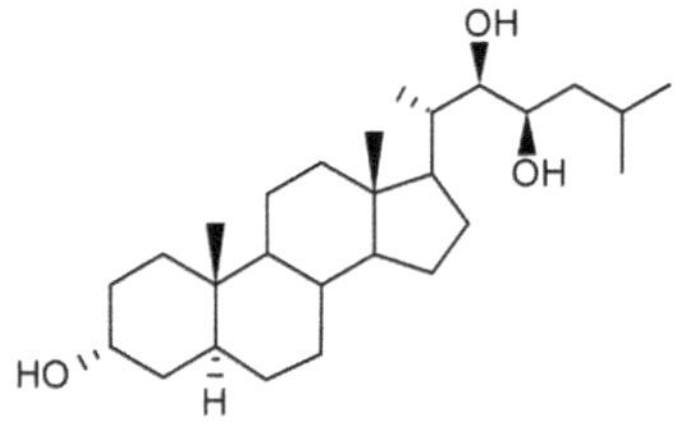

6-deoxo-28-nortyphasterol (6-deoxo-28-norTY)

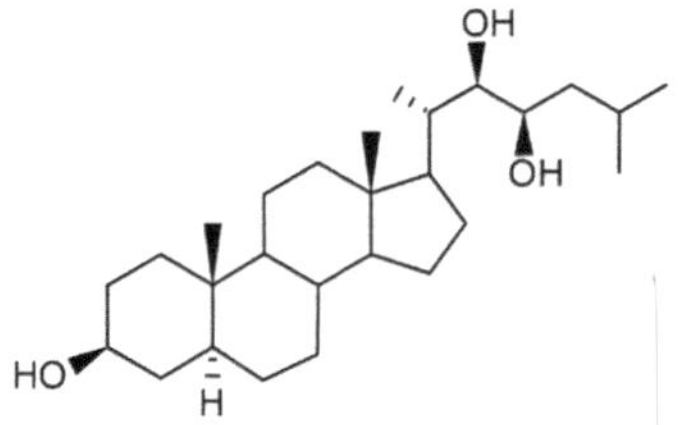

6-deoxo-28-norteasterone (6-deoxo-28-norTE)

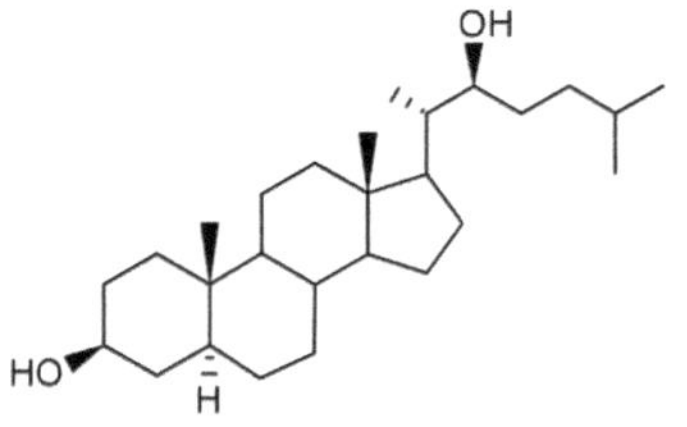

6-deoxo-28-norcathasterone (6-deoxo-28-norCT)

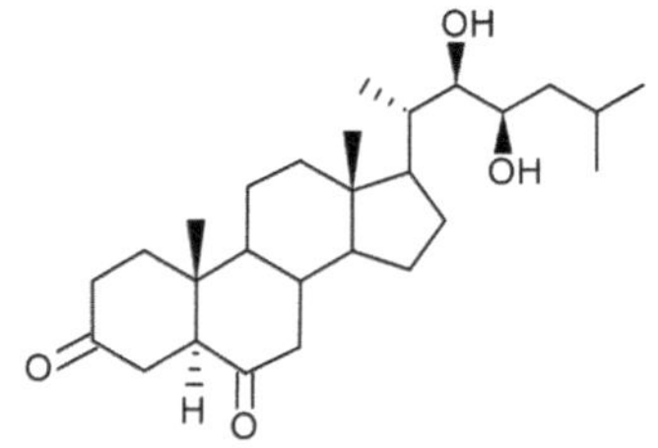

28-nor-3-dehydroteasterone (28-nor-3-DT)

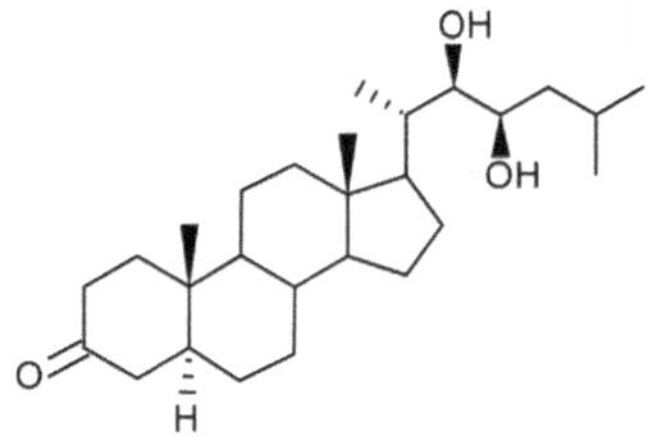

6-deoxo-28-nor-3-dehydroteasterone (6-deoxo-28-nor-3-DT)

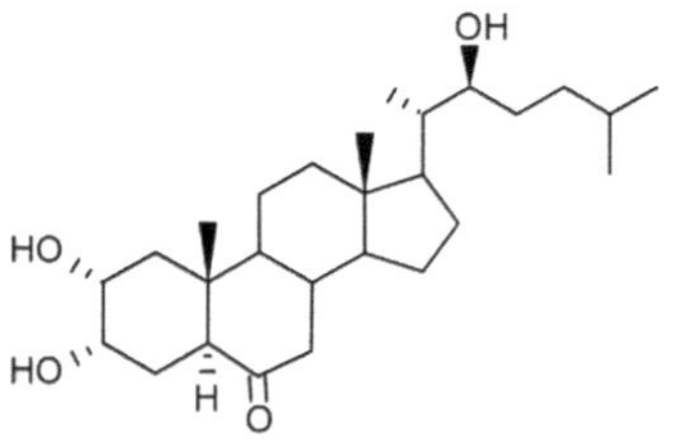

28-norcathasterone (28-norCT)

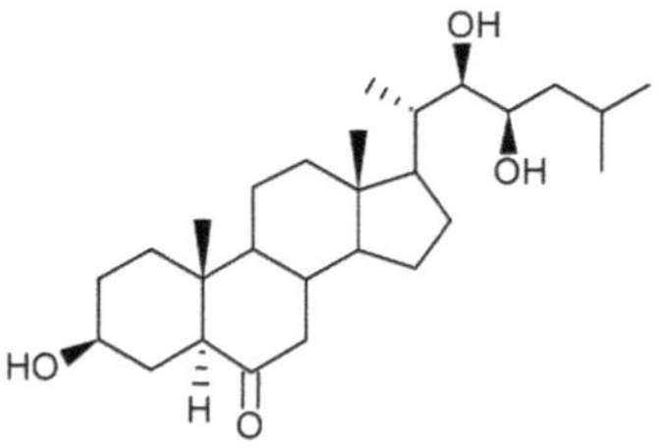

28-norteasterone (28-norTE)

FIGURE 1.3 The C_{27}-brassinosteroids involved in biosynthetic pathway.

converted to 28-homoCS and 28-homoBL via the CYP85A1/A2 oxidases, and 28-homoTY can also be formed from 28-homoTE. The synthesis of 28-homodolicholide and CS from isofucosterol via 22-hydroxyisofucosterol, 6-deoxo-28-homoDS, and 28-homoDS is suggested. Moreover, CS can be converted from β-sitosterol through the 22-homositosterol, 6-deoxohomositosterol, and 28-homoCS. Furthermore, 28-homoTE, 28-homoTY, and 28-homoCS can be converted into 26-nor-28-homoTE, 26-nor-28-homoTY, and 26-nor-28-homoCS, respectively (Joo et al. 2015, Kim et al. 2018, Lee et al. 2011, Roh et al. 2017).

1.2.4 Links between C_{27}-C_{28} and C_{28}-C_{29} Pathways

The C_{27}-BRs biosynthetic pathway links with the C_{28} pathway through the following reactions: 28-norTE → TE, 28-nor-3-DT → 3-DT, 28-norTY → TY, and 28-norCS → CS (Figure 1.7). C_{29}-BRs conversion to C_{28}-BRs occurs through the following reactions: 28-homoTE → TE, 28-homoTY → TY, 28-homoCS → CS, 28-homoDS → CS, and 28-homoDS → DS → CS. Direct substrates to synthesize CS are 28-norCS, DS, 6-deoxoCS by 6α-hydroxyCS, 28-homoDS, and 28-homoCS (Fujita et al. 2006, Joo et al. 2015, Joo et al. 2012, Kim et al. 2018).

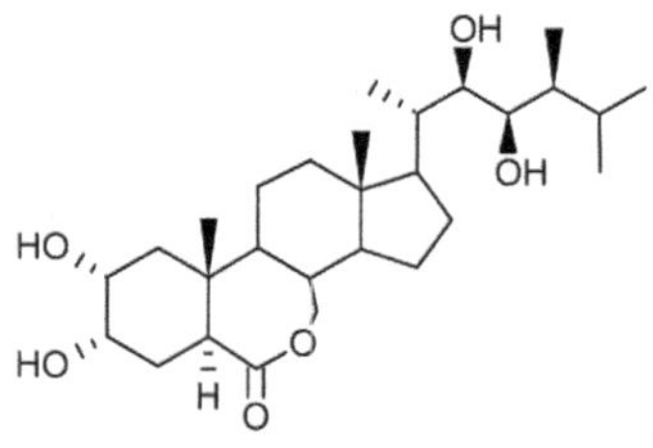

brassinolide
(BL)

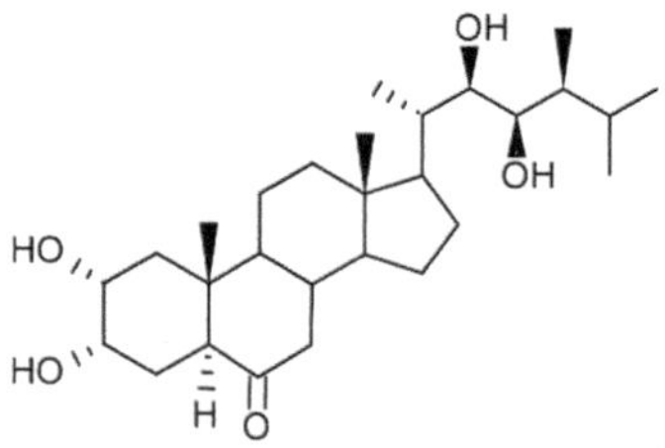

castasterone
(CS)

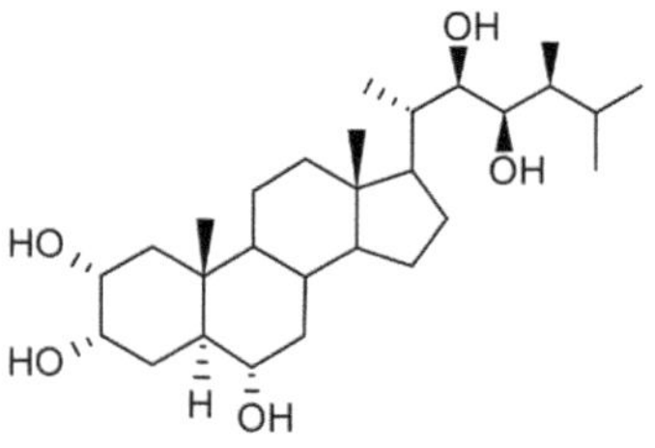

6-α-hydroxycastasterone
(6-α-OH-CS)

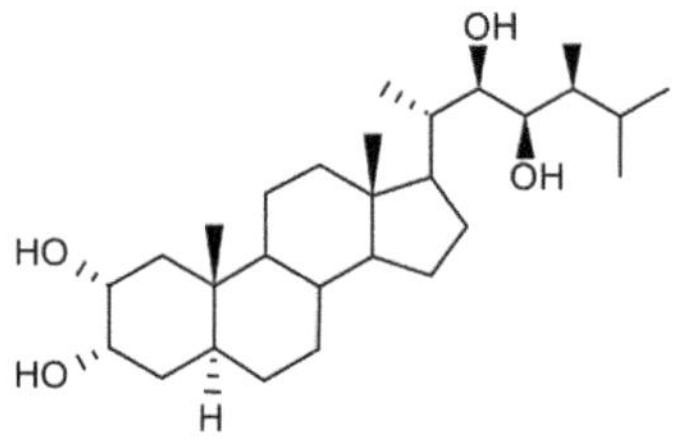

6-deoxocastasterone
(6-deoxoCS)

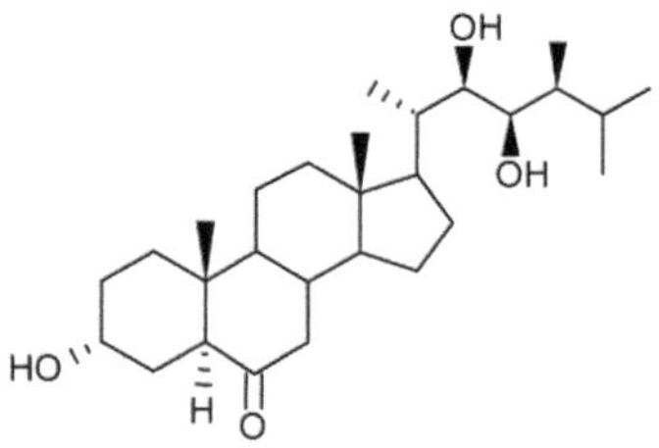

typhasterol
(TY)

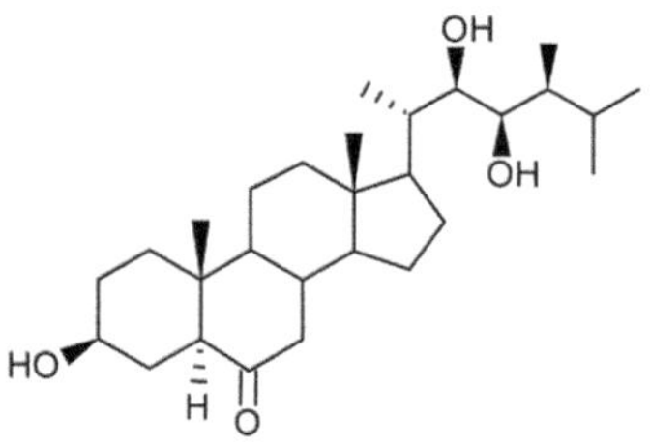

teasterone
(TE)

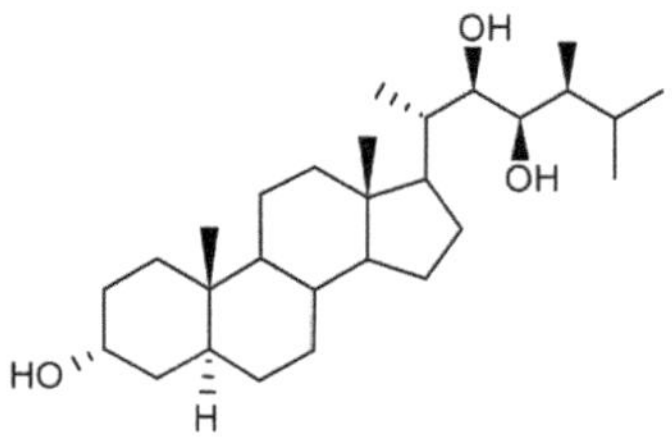

6-deoxotyphasterol
(6-deoxoTY)

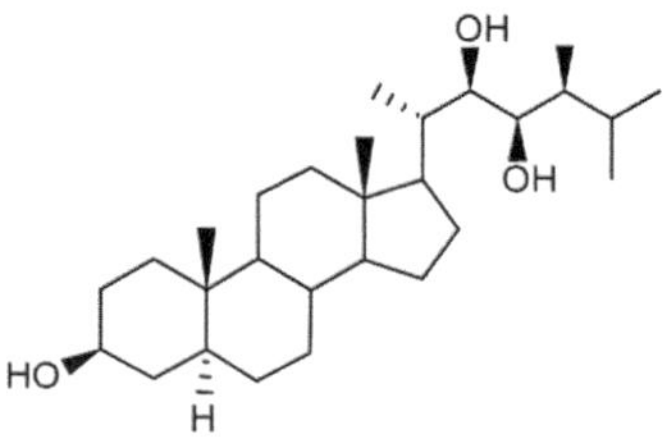

6-deoxoteasterone
(6-deoxoTE)

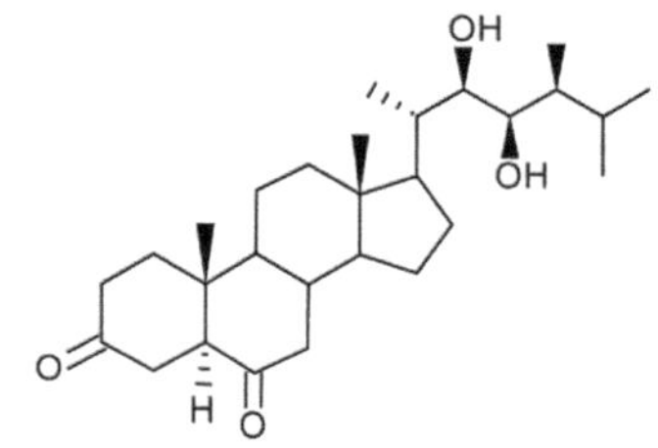

3-dehydroteasterone
(3-DT)

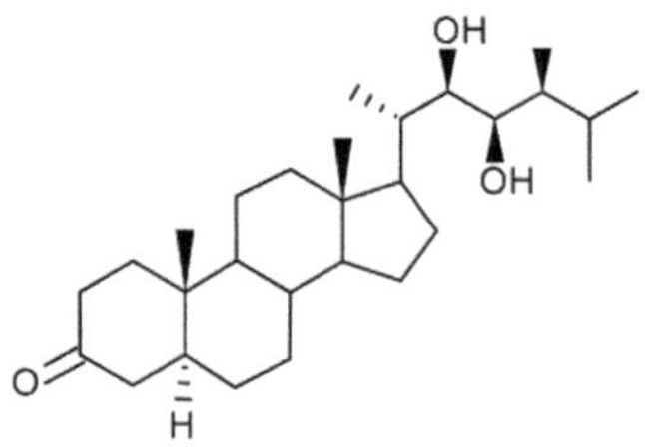

6-deoxo-3-dehydroteasterone
(6-deoxo-3-DT)

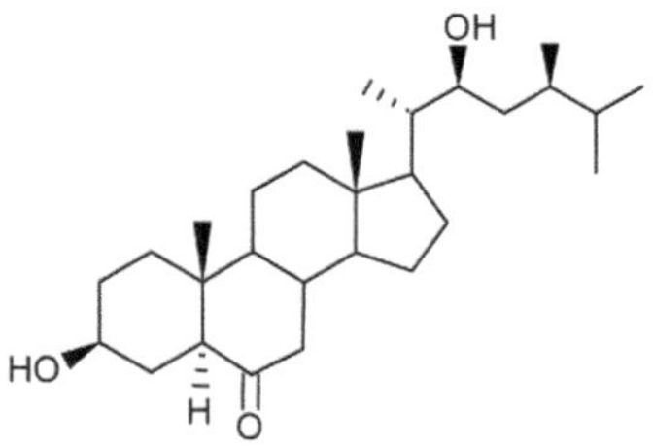

cathasterone
(CT)

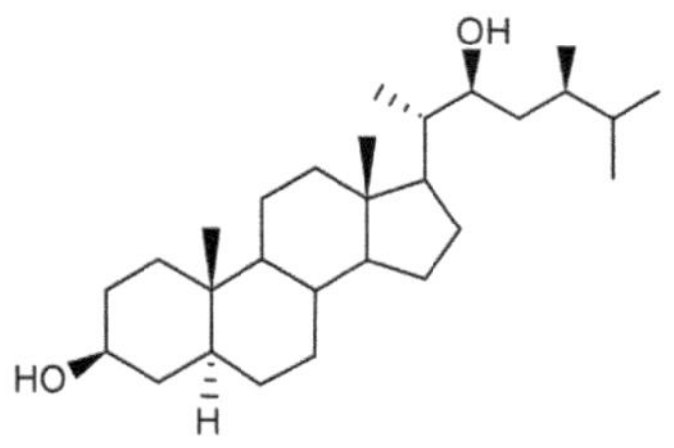

6-deoxocathasterone
(6-deoxoCT)

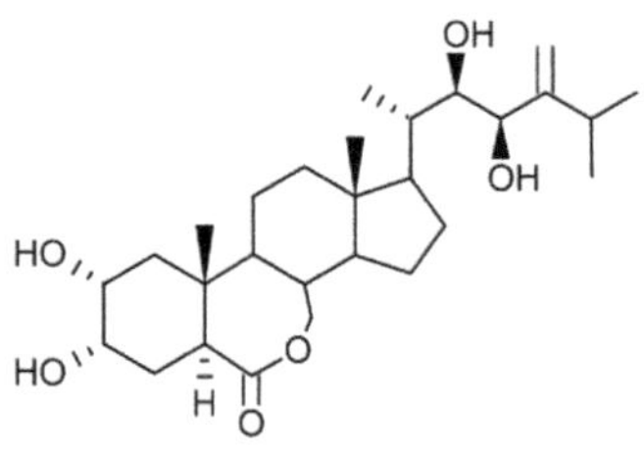

dolicholide
(DL)

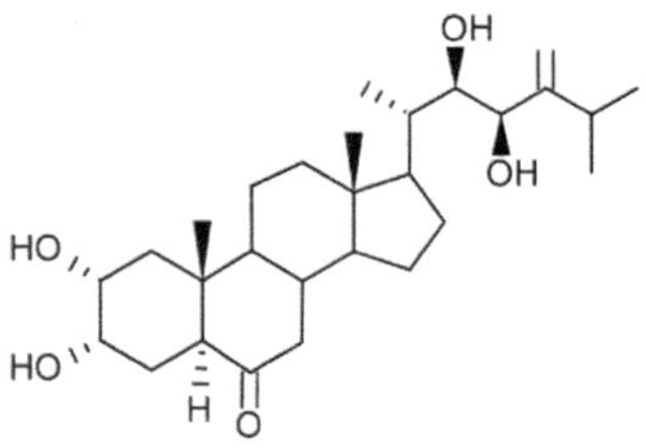

dolichosterone
(DS)

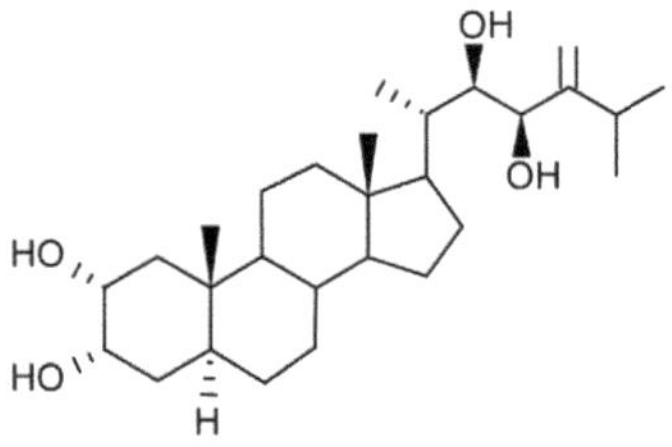

6-deoxodolichosterone
(6-deoxoDS)

FIGURE 1.4 The C_{28}-brassinosteroids involved in biosynthetic pathway.

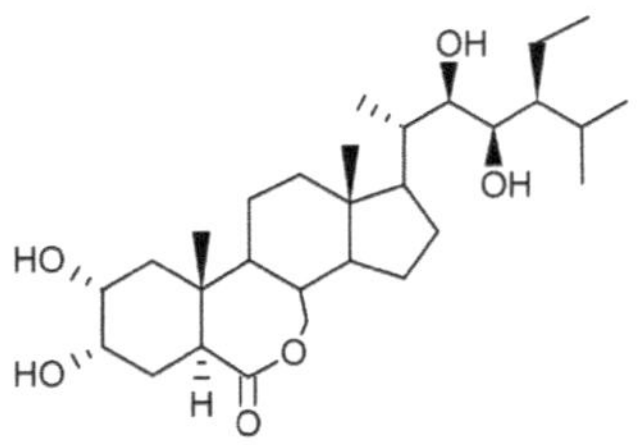

28-homobrassinolide (28-homoBL)

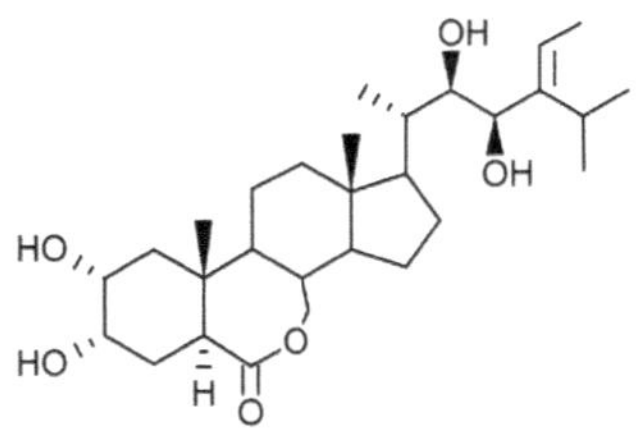

28-homodolicholide (28-homoDL)

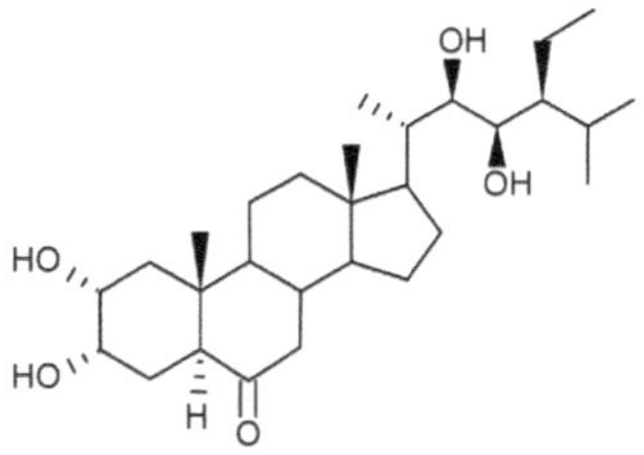

28-homocastasterone (28-homoCS)

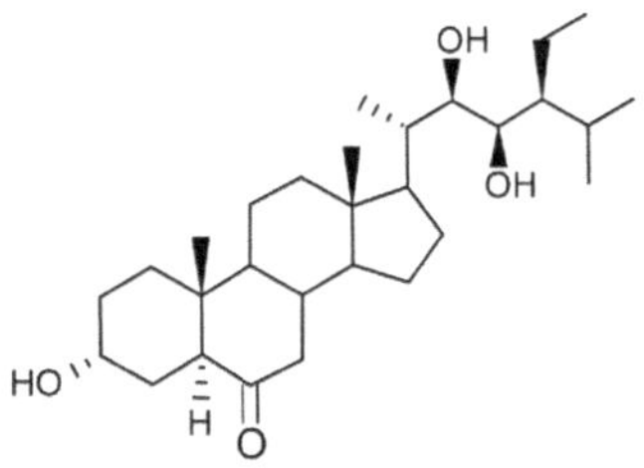

28-homotyphasterol (28-homoTY)

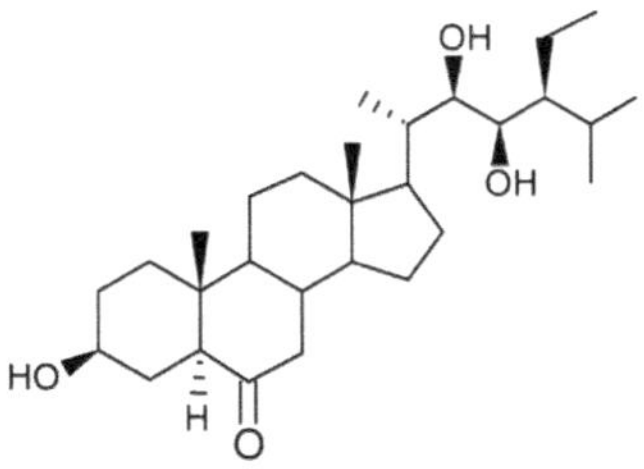

28-homoteasterone (28-homoTE)

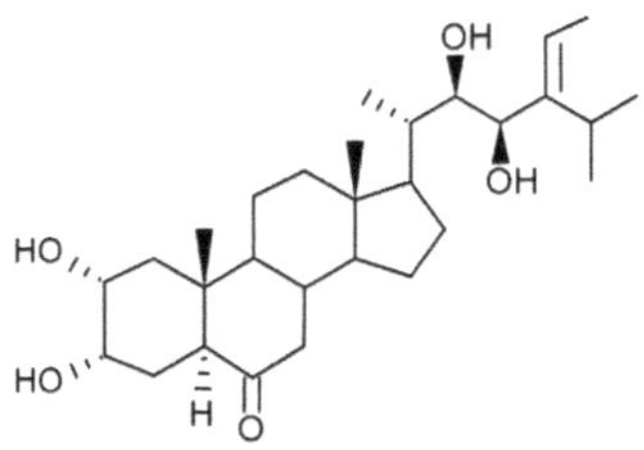

28-homodolichosterone (28-homoDS)

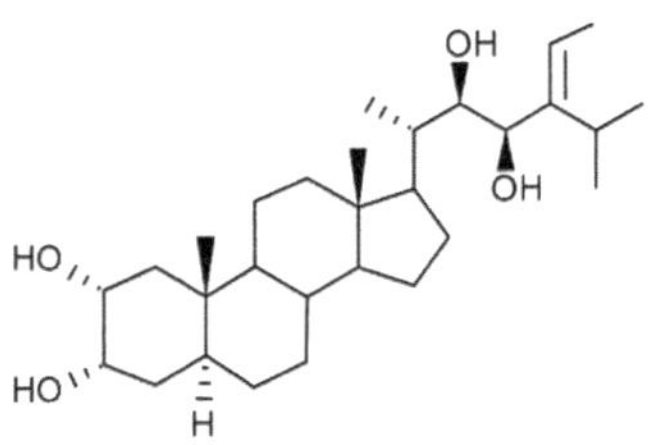

6-deoxo-28-homodolichosterone (6-deoxo-28-homoDS)

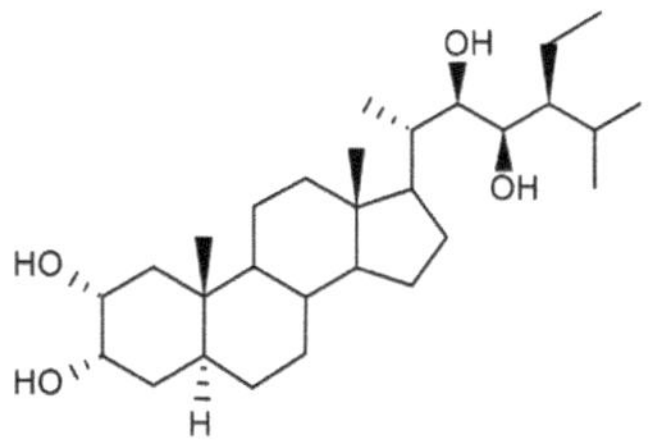

6-deoxo-28-homocastasterone (6-deoxo-28-homoCS)

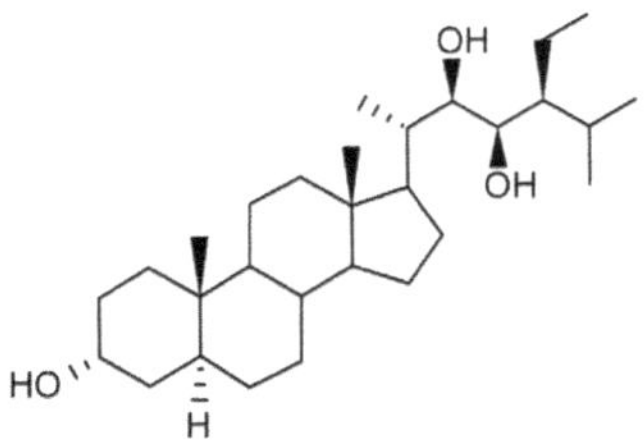

6-deoxo-28-homotyphasterol (6-deoxo-28-homoTY)

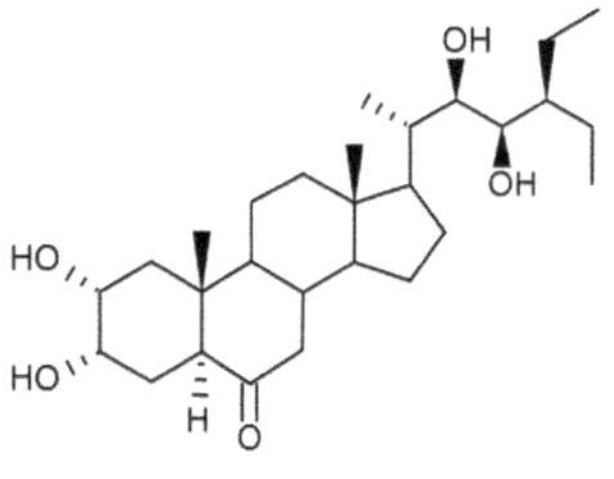

26-nor-28-homocastasterone (26-nor-28-homoCS)

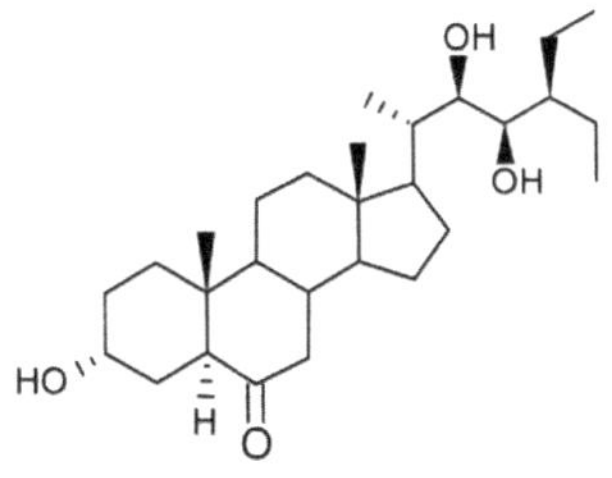

26-nor-28-homotyphasterol (26-nor-28-homoTY)

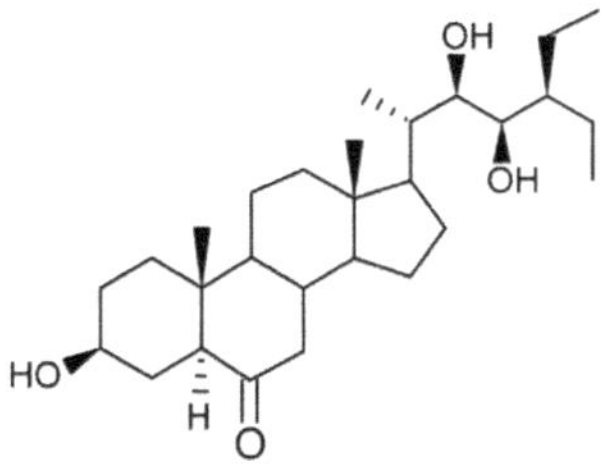

26-nor-28-homoteasterone (26-nor-28-homoTE)

FIGURE 1.5 The C_{29}-brassinosteroids involved in biosynthetic pathway.

1.2.5 Inhibitors of Brassinosteroid Biosynthesis

Inhibitors are tools useful not only for investigating biosynthetic pathways but also for manipulating the BR level in crop plants. Until now, 17 inhibitors have been discovered: KM-01, brassinazole (Brz), Brz2001, Brz220, propiconazole, YCZ-18, yucaizol, fenarimol, spironolactone, triadimefon, imazalil, 4-MA, VG106, DSMEM21, finasteride, AFA76, and brassinopride; however, the sites of action of only nine compounds are known (Table 1.1; Asami et al. 2000, Asami et al. 2001, Hartwig et al. 2012, Oh, Matsumoto, Yamagami, Hoshi, et al. 2015, Oh, Matsumoto, Yamagami, Ogawa, et al. 2015, Rozhon et al. 2019, Rozhon et al. 2014, Sasaki et al. 2013, Sekimata et al. 2008, Sekimata et al. 2002).

1.3 Catabolism of Brassinosteroids

The catabolism of BRs is a necessary process involving the regulation of bioactive BR content. During catabolic reactions, the principal effect is inactivation of biologically active BRs.

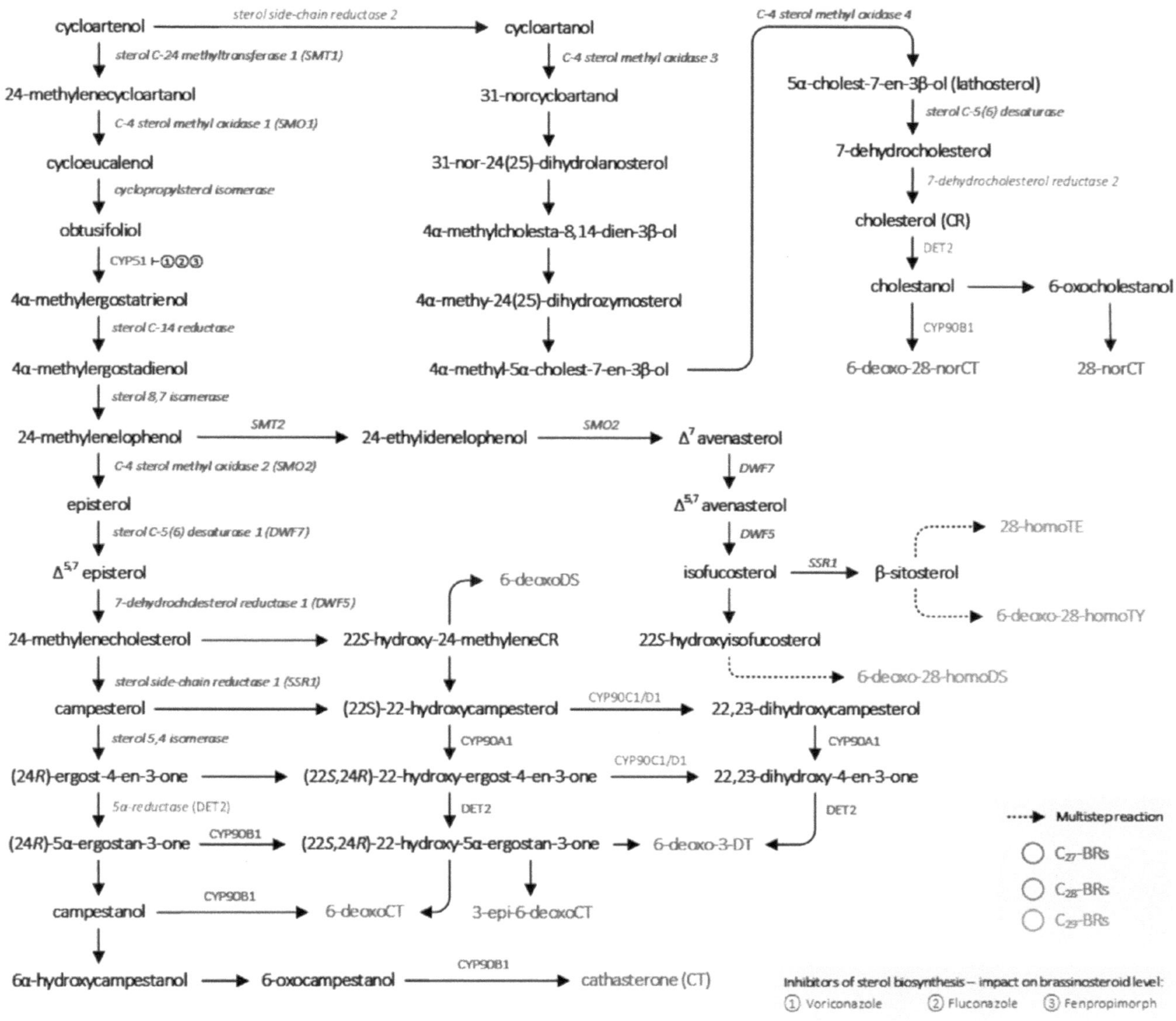

FIGURE 1.6 Biosynthesis of sterols leading to brassinosteroids.

Metabolites are a result of the following chemical reactions (Figure 1.8, Table 1.2):

- dehydrogenation (cleavage of a hydrogen atom from hydroxyl group) at C-3 or C-23;
- demethylation (removal of methyl group) at C-26 or C-28;
- epimerization (modification of hydroxyl group configuration) at C-2, C-3, or C-24;
- esterification (reaction between acids and the hydroxyl group of BRs leading to the synthesis of esters) at C-3;
- glycosylation (addition of sugar residue) at C-2, C-3, C-23, C-25, or C-26;
- hydroxylation (addition of hydroxyl group) at C-20, C-25, or C-26;
- side-chain cleavage at C-20/C-22;
  and
- sulfonation (addition of sulfone group) at C-22 (Bajguz 2007, Choe 2006).

The conversion of BRs is an essential mechanism affecting plant growth and development. BRs cannot be transported a long distance (Symons et al. 2008). Thus, their inactivation is a necessary process regulating the level of BRs in plants (Yang et al. 2014). Many studies with plant mutants confirmed that catabolic reactions cause a significant decrease in BR content (Rouleau et al. 1999, Sakamoto et al. 2011, Suzuki et al. 1993). Although most metabolic reactions lead to the creation of BR derivatives (Table 1.2), synthesis of other active BRs during metabolic transformation also occurs. For instance, CS is converted into 28-norcastasterone (28-norCS) during the 28-demethylation process (Fujioka et al. 2000). Several reports about the genetic regulation of catabolism and the influence of BRs reduction on the phenotype of mutated plants do exist (Joo et al. 2015). Figure 1.9 presents the chemical structures of BR metabolites.

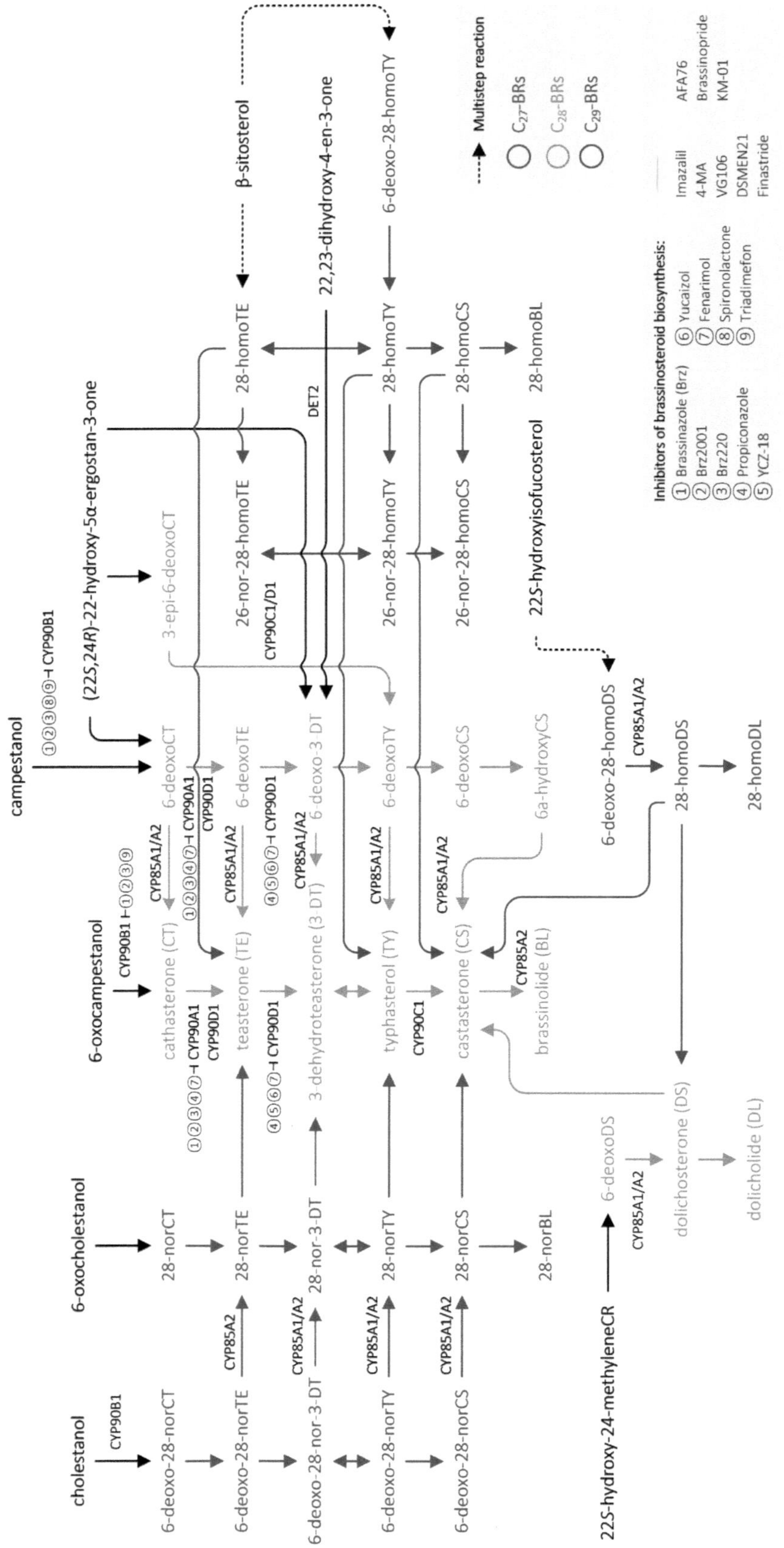

FIGURE 1.7 Biosynthesis of C_{27}-, C_{28}-, and C_{29}-brassinosteroids.

TABLE 1.1

Targets of Inhibitors Interfering with Biosynthetic Pathways for Brassinosteroids (Rozhon et al. 2019)

Inhibitor	Structure	Inhibition of enzyme	Site of action
Inhibitors of brassinosteroid biosynthesis			
KM-01	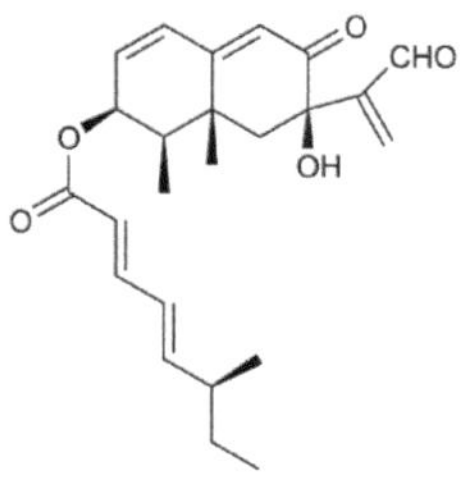	?	?
Brassinazole (Brz)	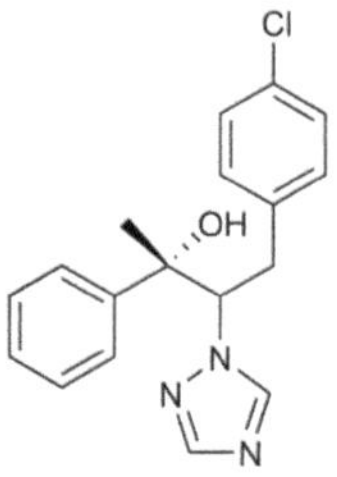	CYP90A1 CYP90B1	CT → TE 6-deoxoCT → 6-deoxoTE campestanol → 6-deoxoCT 6-oxocampestanol → CT
Brz2001	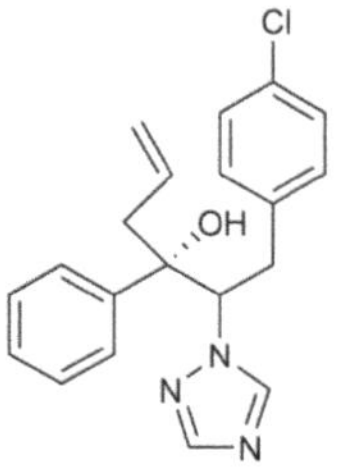	CYP90A1 CYP90B1	CT → TE 6-deoxoCT → 6-deoxoTE campestanol → 6-deoxoCT 6-oxocampestanol → CT
Brz220	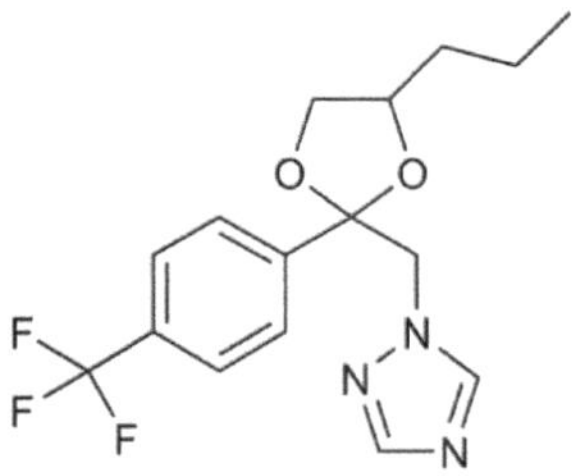	CYP90A1 CYP90B1	CT → TE 6-deoxoCT → 6-deoxoTE campestanol → 6-deoxoCT 6-oxocampestanol → CT
YCZ-18	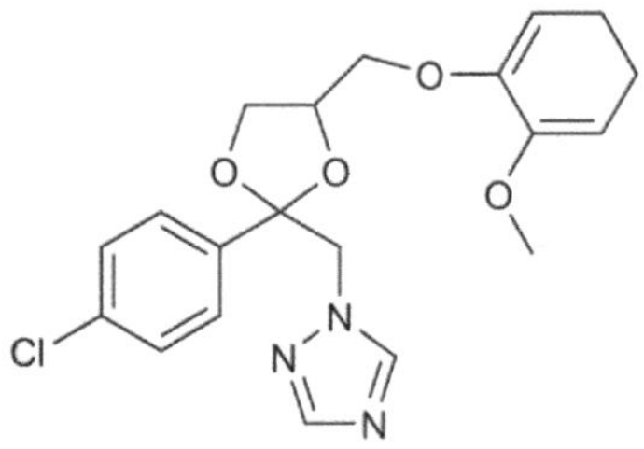	CYP90D1	6-deoxoTE → 6-deoxo-3DT TE → 3DT
Yucaizol	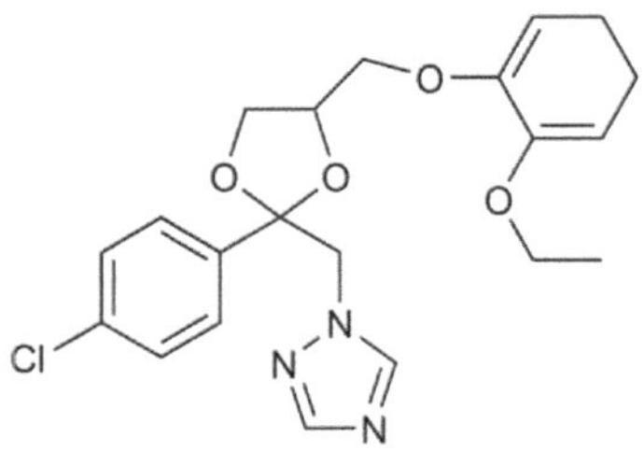	CYP90D1	6-deoxoTE → 6-deoxo-3DT TE → 3DT

(*Continued*)

TABLE 1.1 (CONTINUED)

Targets of Inhibitors Interfering with Biosynthetic Pathways for Brassinosteroids (Rozhon et al. 2019)

Inhibitor	Structure	Inhibition of enzyme	Site of action
Propiconazole	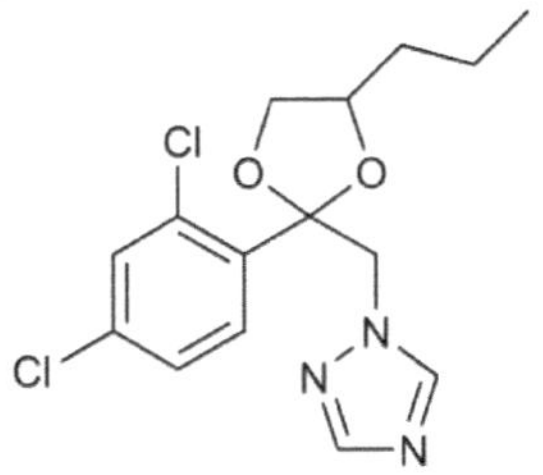	CYP90D1	CT → TE TE → 3DT 6-deoxoCT → 6-deoxoTE 6-deoxoTE → 6-deoxo-3DT
Triadimefon	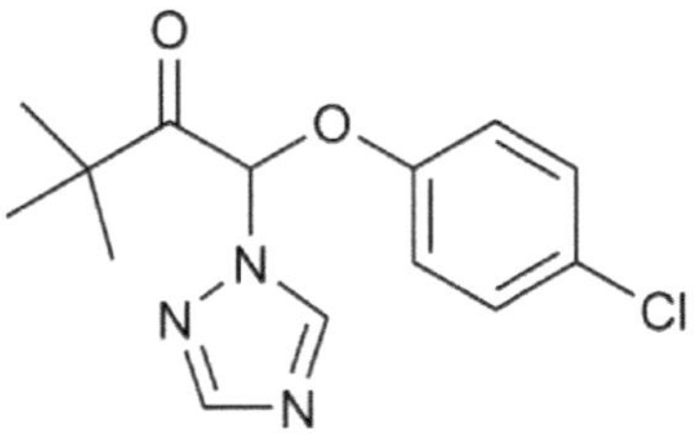	CYP90B1	campestanol → 6-deoxoCT 6-oxocampestanol → CT
Fenarimol	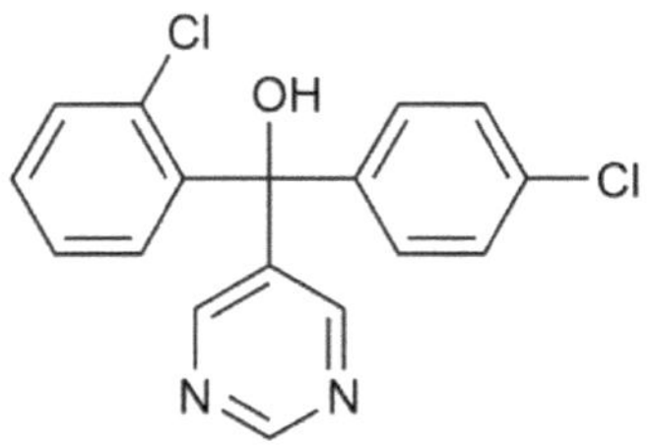	CYP90D1	CT → TE TE → 3DT 6-deoxoCT → 6-deoxoTE 6-deoxoTE → 6-deoxo-3DT
Imazalil	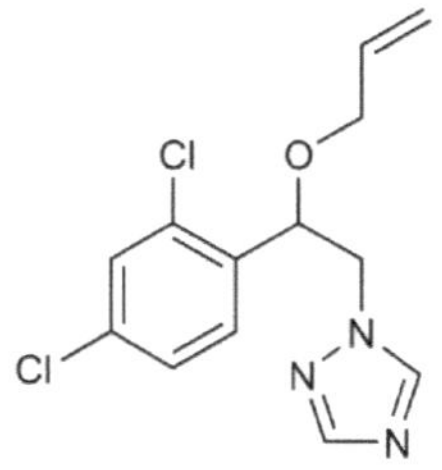	?	?
4-MA	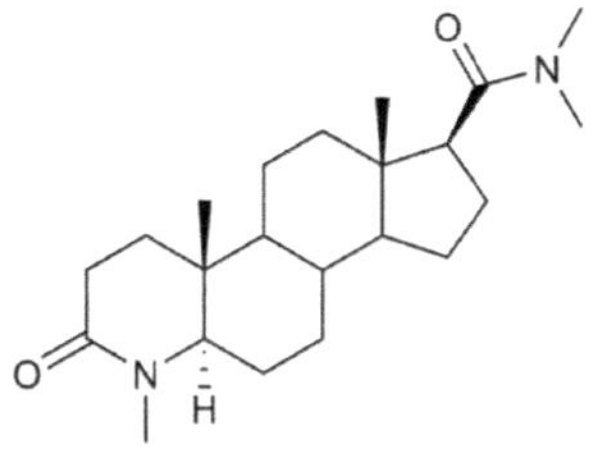	DET2	?
VG106	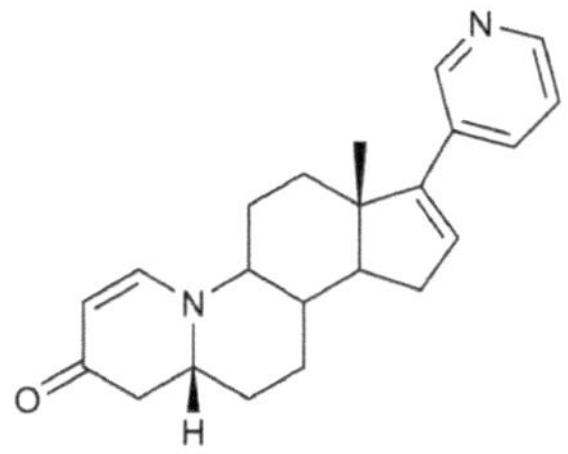	DET2	?
DSMEN21	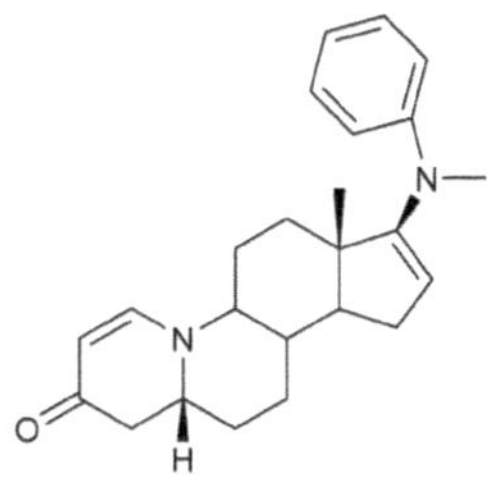	DET2	?

(*Continued*)

TABLE 1.1 (CONTINUED)

Targets of Inhibitors Interfering with Biosynthetic Pathways for Brassinosteroids (Rozhon et al. 2019)

Inhibitor	Structure	Inhibition of enzyme	Site of action
Finasteride	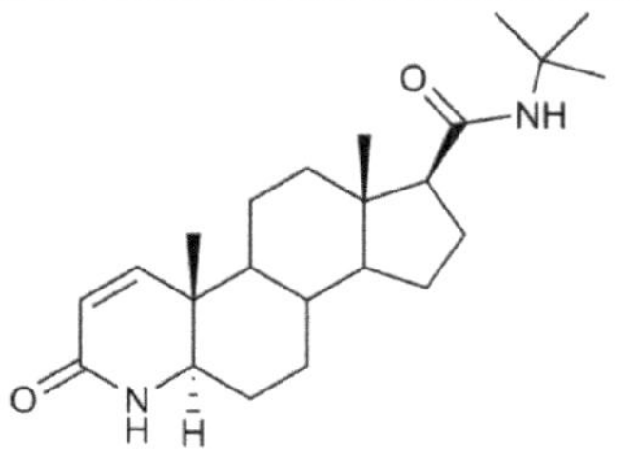	DET2	?
AFA76	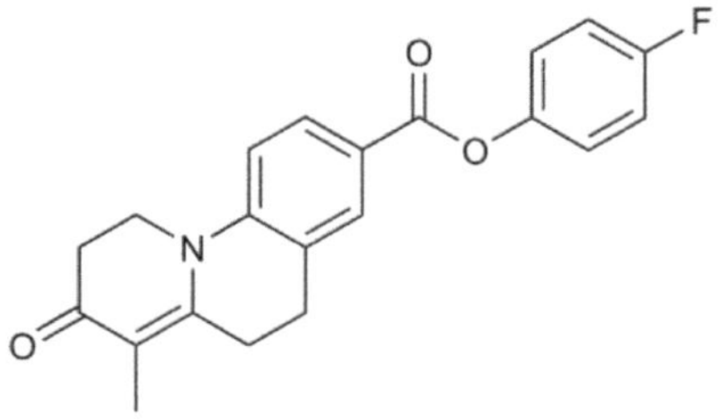	DET2	?
Spironolactone	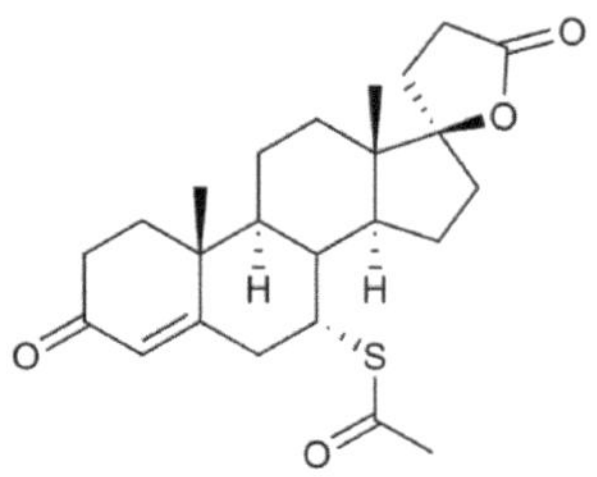	CYP90B1	campestanol → 6-deoxoCT
Brassinopride	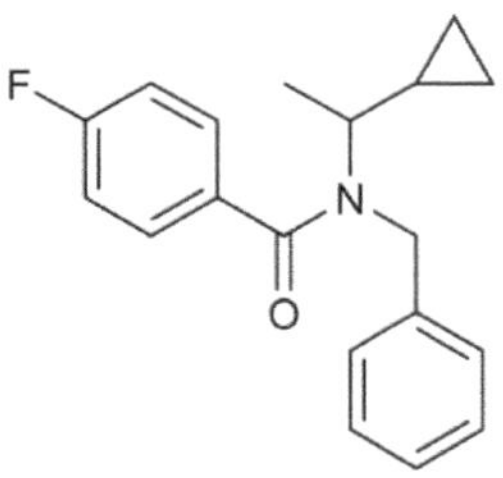	?	?
Inhibitors of sterol biosynthesis, which have an impact on the brassinosteroid level			
Voriconazole	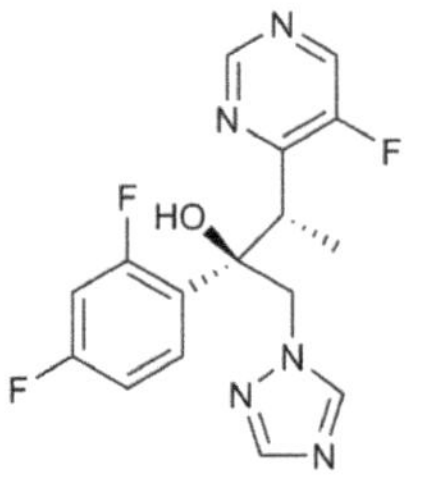	CYP51	obtusifoliol → 4α-methylergostatrienol
Fenpropimorph	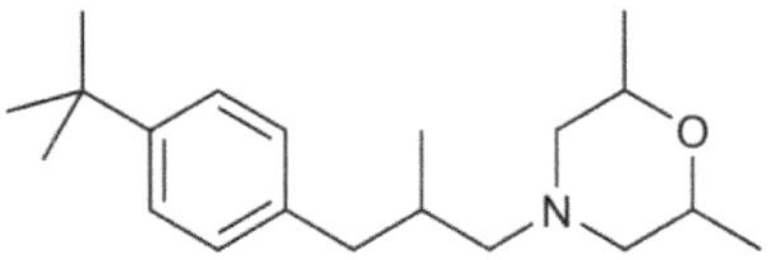	CYP51	obtusifoliol → 4α-methylergostatrienol
Fluconazole	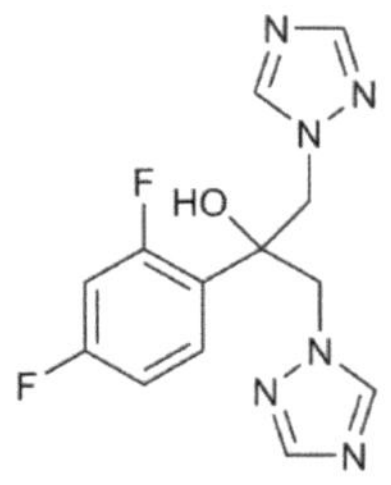	CYP51	obtusifoliol → 4α-methylergostatrienol

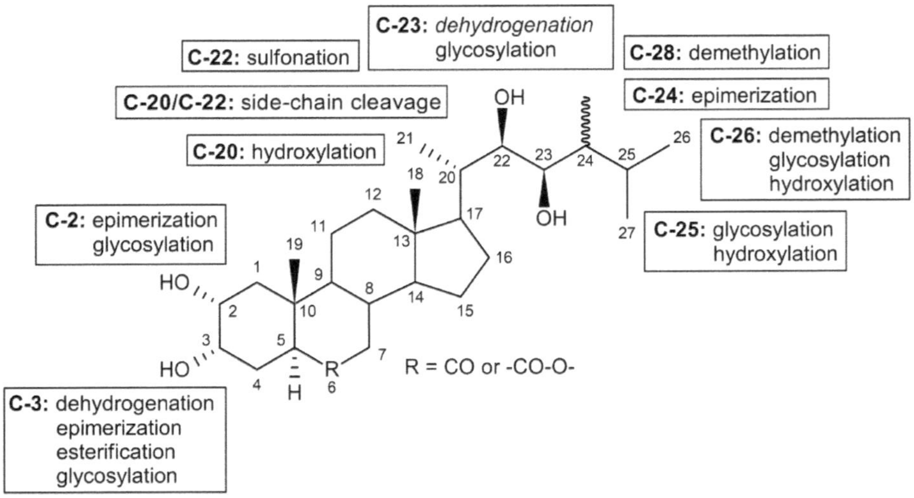

FIGURE 1.8 The sites in the brassinosteroid molecule which are modified by metabolic reactions.

1.3.1 Conversion of Brassinolide and Castasterone

Brassinolide (BL) and CS are the most active BRs essential in plant physiology (Roh et al. 2017). Their presence has been confirmed in the majority of plant species. CS is a direct precursor of BL biosynthesis. Conversion of CS into BL occurs via Baeyer–Villiger oxidation at the C-6/C-7 position of the B ring (Ohnishi 2018). Poppenberger et al. (2005) reported that BL and CS could create a C-23 glucoside via glycosylation in *Arabidopsis thaliana*, whereas in mung bean (*Vigna radiata*), glycosylation of BL leads to the production of 23-*O*-β-D-glucopyranoside (Figure 1.10) (Suzuki et al. 1993). In pollen and anthers of Japanese cedar (*Cryptomeria japonica*), the presence of a dehydrogenated form of BL viz 23-dehydroBL (cryptolide) was noted (Figure 1.10) (Watanabe et al. 2000). The gas chromatography-mass spectrometry (GC-MS) analysis indicated the demethylation of BL into 26-norBL (Kim et al. 2000). The studies of CS metabolism (Table 1.2) in rice, tomato, *Arabidopsis thaliana*, and *Catharanthus roseus* indicated that CS could be converted to 28-norCS through the demethylation process. Although 28-norCS is classified as a naturally occurring BR, affecting the intracellular level of CS (Fujioka et al. 2000). In tomato, 28-norCS was hydroxylated in 26-hydroxy-28-norCS (Table 1.2; Ohnishi, Nomura, et al. 2006). CS can also be demethylated in the C-26 position giving a 26-norCS (Kim et al. 2004). In tomato, C-26 hydroxylation is involved in 28-norCS inactivation (Ohnishi, Nomura, et al. 2006). On the other hand, in rice, the demethylation of 6dCS side chain was observed, leading to the creation of aldehyde and carboxylate forms (Sakamoto et al. 2011).

BL and CS also create an isomer form of other BRs, 24-epibrassinolide (EBL), and 24-epicastasterone (ECS) (Park et al. 2000). Various types of their catabolic reactions are known (Figure 1.11). Furthermore, hydroxylation of CS into C-26 hydroxy compounds was observed in tomato (*Lycopersicon esculentum*; Ohnishi, Nomura, et al. 2006), *Arabidopsis thaliana* (Nakamura et al. 2005, Nelson et al. 2004), cotton (*Gossypium hirsutum*; Yang et al. 2014), and rice (*Oryza sativa*). Interestingly, different genes are involved in the hydroxylation process in these species (Table 1.2).

1.3.2 Conversion of 24-Epibrassinolide and 24-Epicastasterone

The EBL and ECS metabolic pathways have been extensively examined (Figure 1.11; Hai et al. 1995, Hai et al. 1996). Structural changes in ECS and EBL skeleton take place in C-2, C-3, C-20, C-22, C-25, and C-26 (Table 1.2). In the case of EBL, the presence of the two-stage metabolic pathways was noted for both C-25 and C-26. It was converted into C-25 or C-26 hydroxylated forms, which were glycosylated. Similar metabolic changes were found for ECS. It also has the capability of dehydrogenation to 3-dehydroECS, which is epimerized to 3,24-diECS. Obtained diepimer can be hydroxylated to 25-hydroxy-3,24-diECS, as well as glycosylated to 3β-D-glucopyranosyl-3,24-diECS or 2α-D-glucopyranosyl-3,24-diECS. These processes have been confirmed in bean (*Phaseolus vulgaris*; Park et al. 2000) and serradella (*Ornithopus sativus*; Kolbe et al. 1995). Moreover, 3,24-diECS is esterified in C-3β to give fatty acids conjugates, i.e., 3,24-diECS-3β-laurate, 3,24-diECS-3β-myristate, and 3,24-diECS-3β-palmitate. In serradella, the catabolic transformation of EBL also leads to 3,24-diEBL, which is esterified to 3,24-diEBL-3β-laurate, 3,24-diEBL-3β-myristate, and 3,24-diEBL-3β-palmitate (Kolbe et al. 1995). Further, 3,24-diECS can be hydroxylated in the C-20 position to 20-hydroxy-3,24-diECS, in which cleavage of the side chain occurs. As a result of this process, complex compounds, such as 2α,3β,6β-trihydroxy-5α-pregnan-20-one and 2α,3β-dihydroxy-5α-pregnan-6,20-dione, are formed. Similar reactions are also observed for 3,24-diECS (Kolbe et al. 1994). Metabolism of EBL was studied in wheat (*Triticum aestivum*; Nishikawa, Shida, et al. 1995) and cucumber (*Cucumis sativus*; Nishikawa, Abe, et al. 1995). In these plants, the epimerization of EBL in the C-2 position, creating 2,24-diEBL, was observed; whereas sulfonation of 24-ECT, creating 24-ECT-22-sulfate, was observed in *Brassica napus* (Figure 1.12, Table 1.2; Rouleau et al. 1999).

1.3.3 Conversion of Teasterone and Its Derivatives

Metabolic processes of teasterone (TE) were reported in anthers of *Lilium longiflorum*, in which esterification of

TABLE 1.2
Metabolic Reactions of Brassinosteroids

Reaction type	BR substrate	Catabolism product	Gene or enzyme	Plant	Reference
Dehydrogenation	BL	23-dehydroBL (cryptolide)	?	*Cryptomeria japonica*	Watanabe et al. (2000)
	ECS	3-dehydroECS	?	*Lycopersicon esculentum*	Hai et al. (1996)
	TE	3-dehydroTE	?	*Lycopersicon esculentum*	Abe et al. (1994)
Demethylation	BL	26-norBL	?	*Marchantia polymorpha*	Kim et al. (2000)
	CS	28-norCS	?	*Oryza sativa*	Fujioka et al. (2000)
	CS	28-norCS	?	*Arabidopsis thaliana*	Fujioka et al. (2000)
	CS	28-norCS	?	*Catharanthus roseus*	Fujioka et al. (2000)
	CS	28-norCS	?	*Lycopersicon esculentum*	Fujioka et al. (2000)
	CS	26-norCS	?	*Lycopersicon esculentum*	Kim et al. (2004)
	6dCS	6dCS-CHO	*CYP734A2*	*Oryza sativa*	Sakamoto et al. (2011)
	6dCS	6dCS-COOH	*CYP734A2*	*Oryza sativa*	Sakamoto et al. (2011)
Epimerization	EBL	3,24-diEBL	?	*Ornithopus sativus*	Kolbe et al. (1995)
	EBL	2,24-diEBL	?	*Cucumis sativus*	Nishikawa, Shida, et al. (1995)
	EBL	2,24-diEBL	?	*Triticum aestivum*	Nishikawa, Shida, et al. (1995)
	ECS	3,24-diECS	?	*Lycopersicon esculentum*	Hai et al. (1996)
	ECS	3,24-diECS	?	*Phaseolus vulgaris*	Park et al. (2000)
	ECS	3,24-diECS	?	*Ornithopus sativus*	Kolbe et al. (1995)
Esterification	EBL	3,24-diEBL-3β-laurate	?	*Ornithopus sativus*	Kolbe et al. (1995)
	EBL	3,24-diEBL-3β-palmitate	?	*Ornithopus sativus*	Kolbe et al. (1995)
	EBL	3,24-diEBL-3β-myristate	?	*Ornithopus sativus*	Kolbe et al. (1995)
	ECS	3,24-diECS-3β-laurate	?	*Ornithopus sativus*	Kolbe et al. (1995)
	ECS	3,24-diECS-3β-myristate	?	*Ornithopus sativus*	Kolbe et al. (1995)
	ECS	3,24-diECS-3β-palmitate	?	*Ornithopus sativus*	Kolbe et al. (1995)
	TE	TE-3β-myristate	?	*Lilium longiflorum*	Soeno, Asakawa, et al. (2000)
	TE	TE-3β-laurate	?	*Lilium longiflorum*	Soeno, Asakawa, et al. (2000)
Side-chain cleavage	EBL	2α,3β-dihydroxy-B-homo-7-oxa-5 α-pregnan-6,20-dione	?	*Ornithopus sativus*	Kolbe et al. (1994)
	ECS	2α,3β,6β-trihydroxy-5α-pregnan-20-one	?	*Ornithopus sativus*	Kolbe et al. (1994)
	ECS	2α,3β-dihydroxy-5α-pregnan-6,20-dione	?	*Ornithopus sativus*	Kolbe et al. (1994)
Sulfonation	EBL	EBL-22-sulfate	*BNST3*	*Brassica napus*	Rouleau et al. (1999)
	ECT	ECT-22-sulfate	*BNST3*	*Brassica napus*	Rouleau et al. (1999)
	ETE	ETE-22-sulfate	*BNST3*	*Brassica napus*	Rouleau et al. (1999)
Glycosylation	BL	23-*O*-glucosideBL	*UGT73C5*	*Arabidopsis thaliana*	Poppenberger et al. (2005)
	CS	23-*O*-glucosideCS	*UGT73C5*	*Arabidopsis thaliana*	Poppenberger et al. (2005)
	BL	23-*O*-β-D-glucopyranosyloxyBL	?	*Vigna radiata*	Suzuki et al. (1993)
	EBL	25-β-D-glucopyranosyloxy-EBL	?	*Lycopersicon esculentum*	Hai et al. (1995)
	EBL	26-β-D-glucopyranosyloxy-EBL	?	*Lycopersicon esculentum*	Hai et al. (1995)

(Continued)

TABLE 1.2 (CONTINUED)
Metabolic Reactions of Brassinosteroids

Reaction type	BR substrate	Catabolism product	Gene or enzyme	Plant	Reference
	ECS	25-β-D-glucopyranosyloxy-ECS	?	*Lycopersicon esculentum*	Hai et al. (1996)
	ECS	26-β-D-glucopyranosyloxy-ECS	?	*Lycopersicon esculentum*	Hai et al. (1996)
	TE	3-*O*-β-D-glucopyranosyl-TE	?	*Lilium longiflorum*	Soeno, Kyokawa, et al. (2000)
	ETE	3-*O*-β-D-glucopyranosyl-(1→4) -β-D-glucopyranoside-ETE	?	*Lycopersicon esculentum*	Kolbe et al. (1998)
	ETE	3-*O*-β-D-glucopyranosyl-(1→6) -β-D-galactopyranoside-ETE	?	*Lycopersicon esculentum*	Kolbe et al. (1998)
	ETE	3-*O*-β-D-glucopyranoside-ETE	?	*Lycopersicon esculentum*	Kolbe et al. (1998)
	ECS	2β-D-glucopyranosyl-3,24-diECS	?	*Lycopersicon esculentum*	Hai et al. (1996)
	ECS	3α-D-glucopyranosyl-3,24-diECS	?	*Lycopersicon esculentum*	Hai et al. (1996)
Hydroxylation	BL	26-hydroxyBL	*PAG1*	*Gossypium hirsutum*	Yang et al. (2014)
	CS	26-hydroxyCS	*PAG1*	*Gossypium hirsutum*	Yang et al. (2014)
	CS	26-hydroxyCS	*CYP734A*	*Oryza sativa*	Sakamoto et al. (2011)
	BL	26-hydroxyBL	*CYP734A7/BAS1*	*Lycopersicon esculentum*	Ohnishi, Nomura, et al. (2006)
	CS	26-hydroxyCS	*CYP734A7/BAS1*	*Lycopersicon esculentum*	Ohnishi, Nomura, et al. (2006)
	BL	26-hydroxyBL	*CYP734A1/BAS1*	*Arabidopsis thaliana*	Turk et al. (2003)
	CS	26-hydroxyCS	*CYP734A1/BAS1*	*Arabidopsis thaliana*	Turk et al. (2003)
	BL	26-hydroxyBL	*CYP72C1*	*Arabidopsis thaliana*	Nakamura et al. (2005)
	CS	26-hydroxyCS	*CYP72C1*	*Arabidopsis thaliana*	Nakamura et al. (2005)
	EBL	25-hydroxyEBL	?	*Lycopersicon esculentum*	Hai et al. (1995)
	EBL	26-hydroxyEBL	?	*Lycopersicon esculentum*	Hai et al. (1995)
	ECS	25-hydroxyECS	?	*Lycopersicon esculentum*	Hai et al. (1996)
	ECS	26-hydroxyECS	?	*Lycopersicon esculentum*	Hai et al. (1996)
	ECS	25-hydroxy-3,24-diECS	?	*Lycopersicon esculentum*	Hai et al. (1996)
	ECS	20-hydroxy-3,24-diECS	?	*Ornithopus sativus*	Kolbe et al. (1994)
	EBL	20-hydroxy-3,24-diECS	?	*Ornithopus sativus*	Kolbe et al. (1994)
	28-norCS	26-hydroxy-28-norCS	*CYP734A7/BAS1*	*Lycopersicon esculentum*	Ohnishi, Nomura, et al. (2006)

TE both to TE-3β-myristate and TE-3β-laurate occurred (Figure 1.13, Table 1.2; Soeno, Asakawa, et al. 2000), and glycosylation of TE into 3-*O*-β-D-glucopyranosyl-TE was reported (Soeno, Kyokawa, et al. 2000). Interestingly, TE can be converted to 3-dehydroTE, which is a precursor of typhasterol (TY) biosynthesis (Abe et al. 1994). Dehydrogenation of 24-epiteasterone (ETE) leads to the synthesis of 3-dehydroETE. This reaction was reported for the first time in *Lycopersicon esculentum*. ETE creates the following glucoside metabolic conjugates: 3-*O*-β-D-glucopyranosyl-(1→6)-β-D-glucopyranosyl-ETE, 3-*O*-β-D-glucopyranosyl-(1→4)-β-D-galactopyranosyl-ETE, and 3-*O*-β-D-glucopyranosyl-ETE (Figure 1.13; Kolbe et al. 1998). Interestingly, in *Brassica napus*, ETE can be inactivated via a sulfonation reaction, creating an ETE-22-sulfate (Figure 1.12; Rouleau et al. 1999).

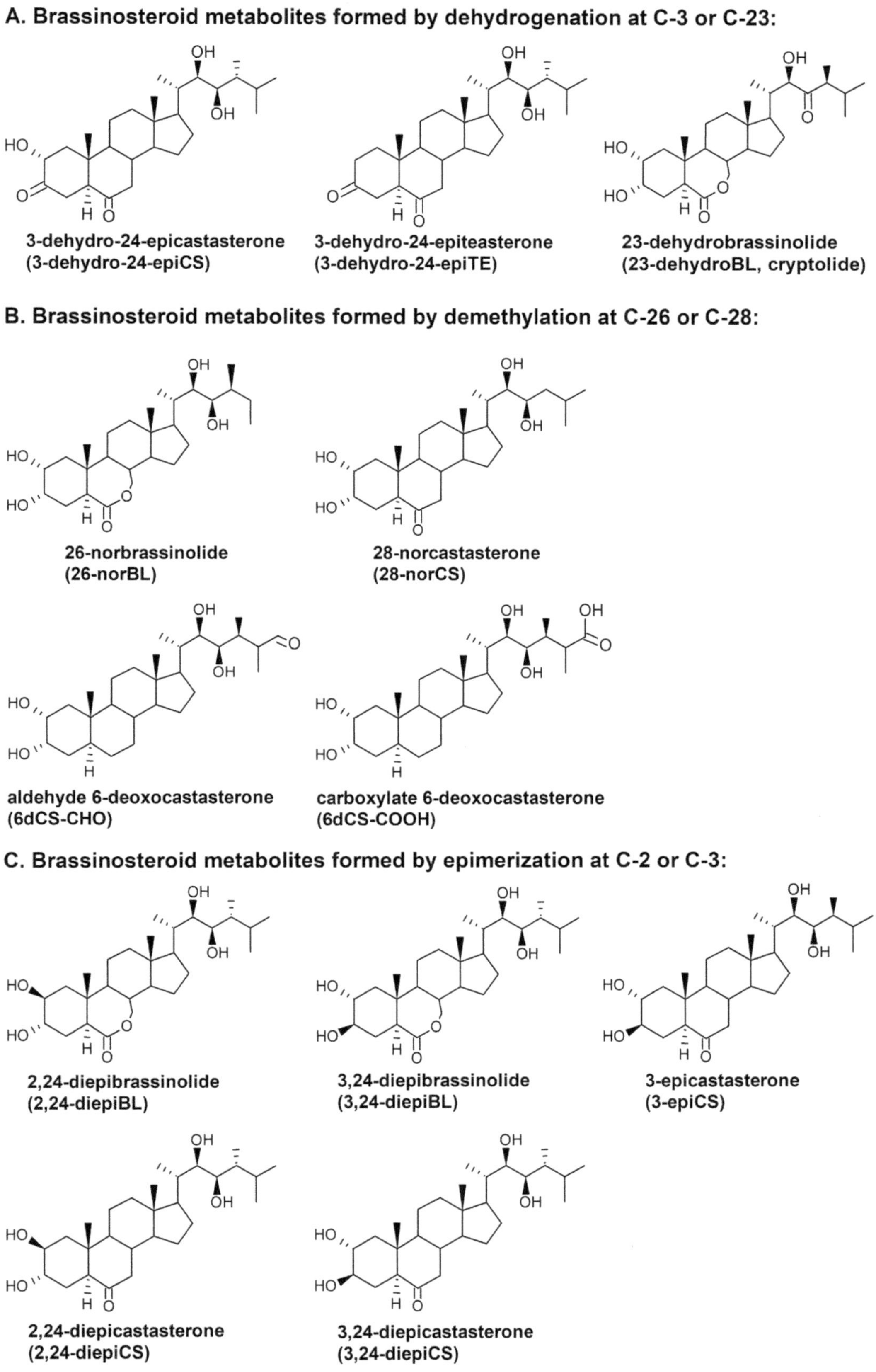

FIGURE 1.9 Chemical structures of brassinosteroid metabolites.

D. Brassinosteroid metabolites formed by esterification at C-3:

R = $C_{11}H_{23}$ (laurate)
$C_{13}H_{27}$ (myristate)
$C_{15}H_{31}$ (palmitate)

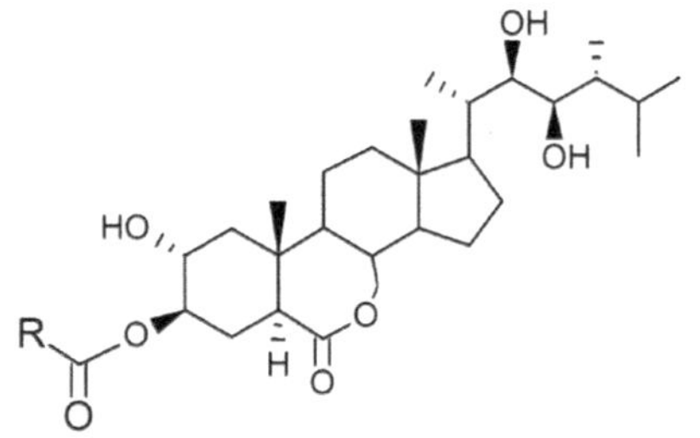

3,24-diepibrassinolide-3β-laurate (3,24-diepiBL-3β–La)

3,24-diepibrassinolide-3β-myristate (3,24-diepiBL-3β–My)

3,24-diepibrassinolide-3β-palmitate (3,24-diepiBL-3β–Pa)

R = $C_{11}H_{23}$ (laurate)
$C_{13}H_{27}$ (myristate)
$C_{15}H_{31}$ (palmitate)

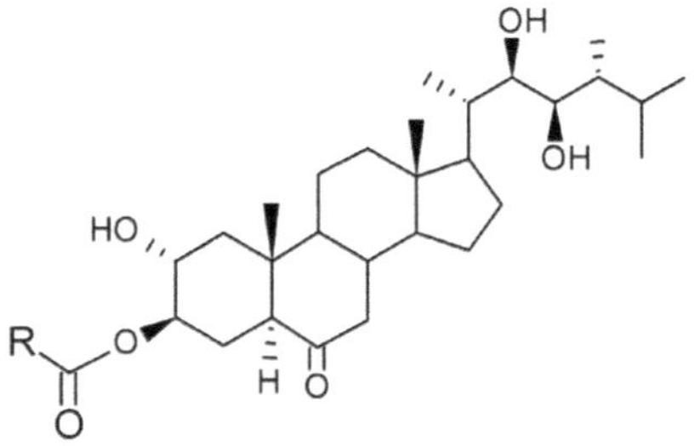

3,24-diepicastasterone-3β-laurate (3,24-diepiCS-3β–La)

3,24-diepicastasterone-3β-myristate (3,24-diepiCS-3β–My)

3,24-diepicastasterone-3β-palmitate (3,24-diepiCS-3β–Pa)

R = $C_{11}H_{23}$ (laurate)
$C_{13}H_{27}$ (myristate)

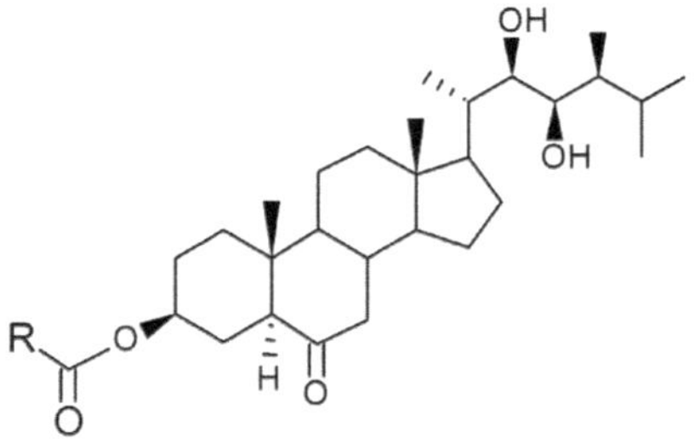

teasterone-3β-laurate (TE-3β–La)

teasterone-3β-myristate (TE-3β–My)

E. Brassinosteroid metabolites formed by glycosylation at C-2, C-3, C-23, C-25 or C-26:

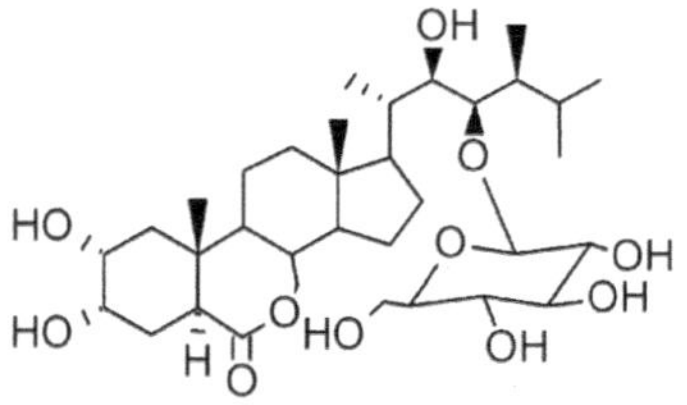

23-*O*-β-D-glucopyranosyloxy-brassinolide

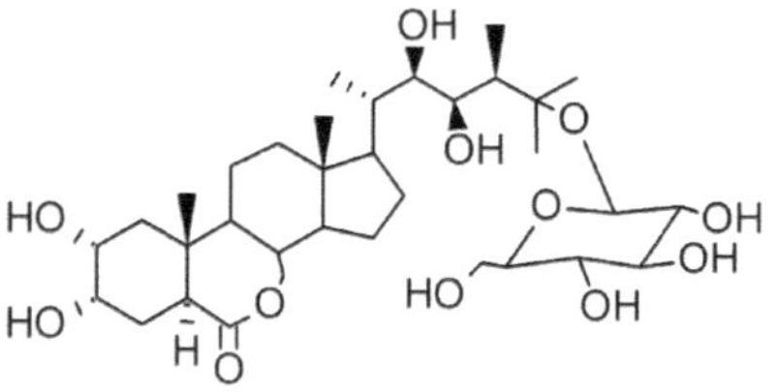

25-β-D-glucopyranosyloxy-24-epibrassinolide

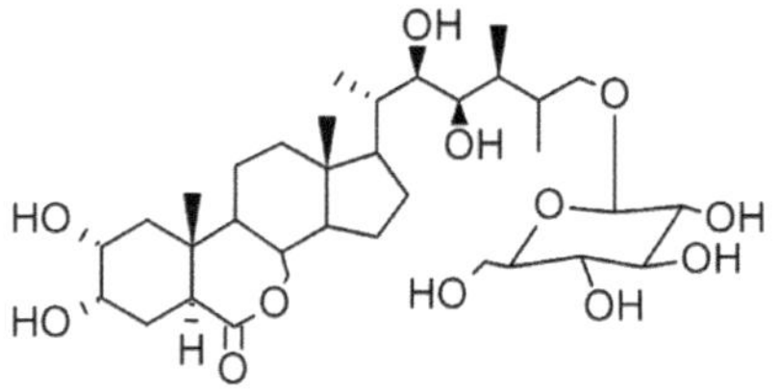

26-β-D-glucopyranosyloxy-24-epibrassinolide

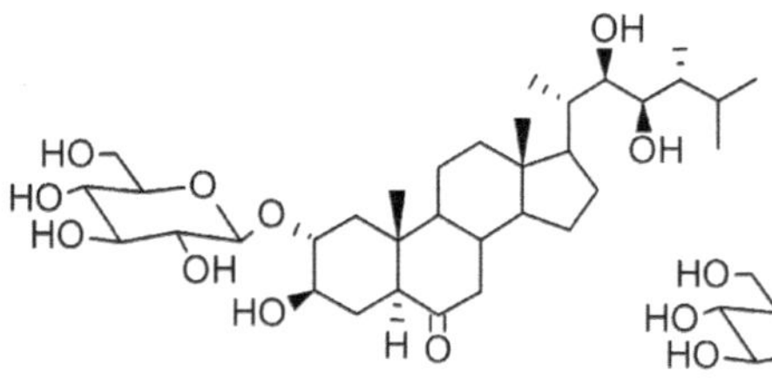

2-*O*-β-D-glucopyranosyl-3,24-diepicastasterone

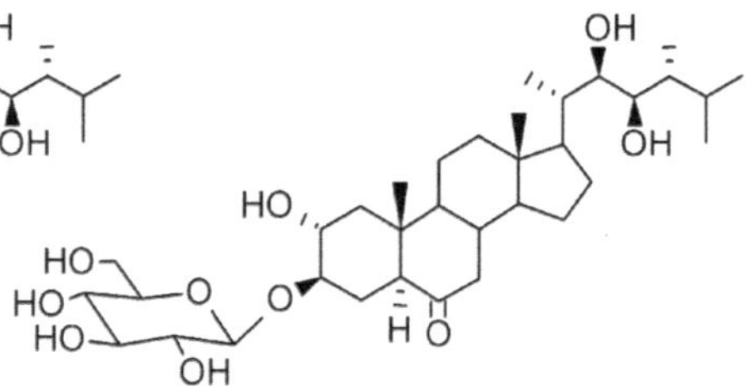

3-*O*-β-D-glucopyranosyl-3,24-diepicastasterone

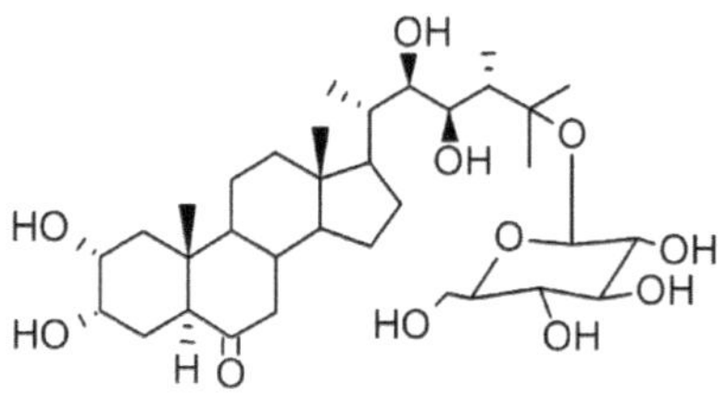

25-β-D-glucopyranosyloxy-24-epicastasterone

FIGURE 1.9 Continued.

E. *continued*

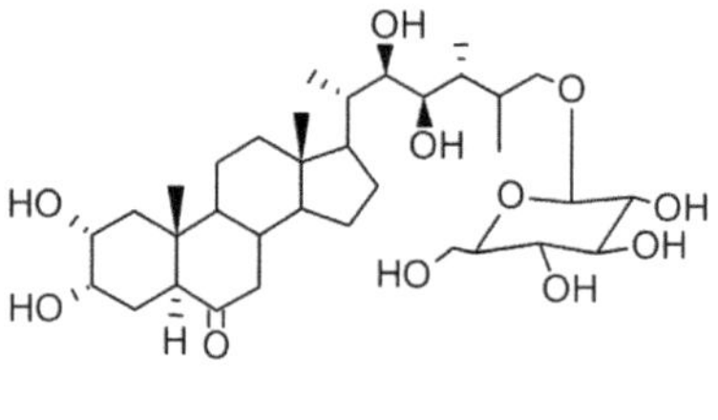

26-β-D-glucopyranosyloxy-24-epicastasterone

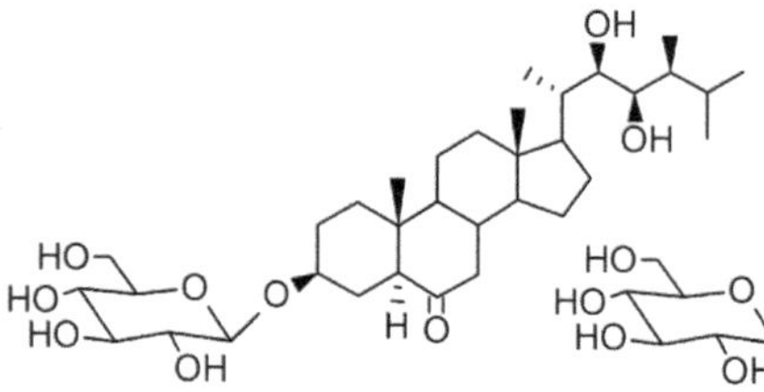

3-*O*-β-D-glucopyranosyl-teasterone

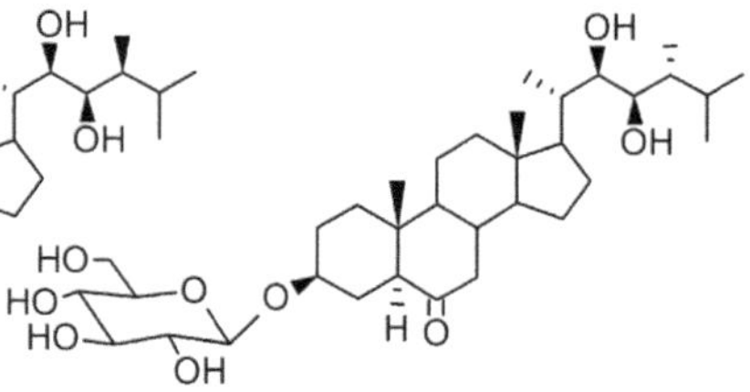

3-*O*-β-D-glucopyranosyl-24-epiteasterone

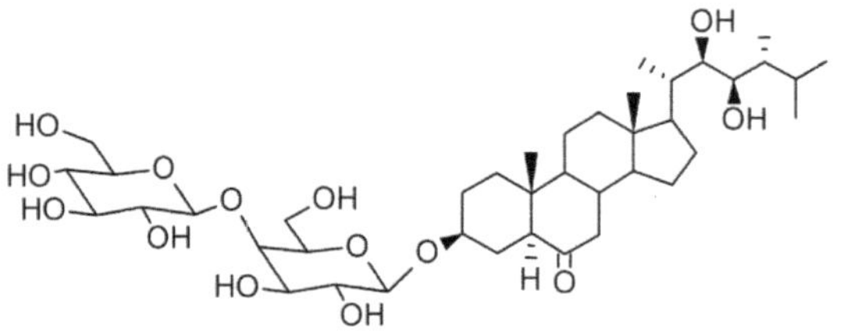

3-*O*-(β-D-glucopyranosyl-(1->4)-β-D-galactopyranosyl-24-epiteasterone

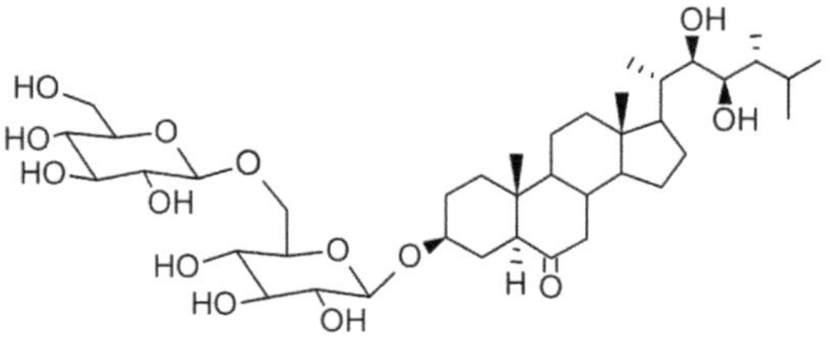

3-*O*-(β-D-glucopyranosyl-(1->6)-β-D-glucopyranosyl-24-epiteasterone

F. Brassinosteroid metabolites formed by hydroxylation at C-20, C-25 or C-26:

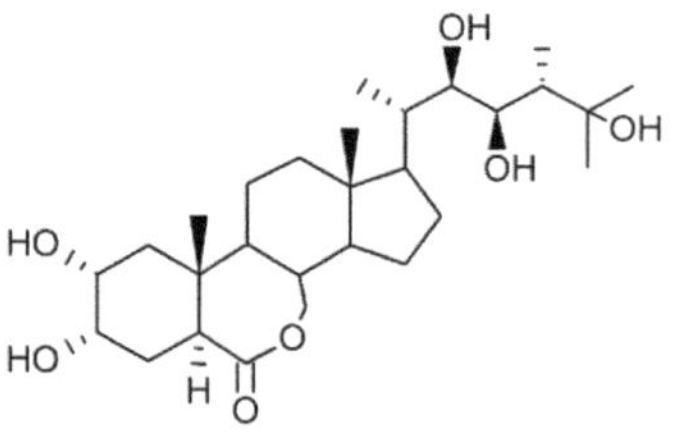

25-hydroxy-24-epibrassinolide (25-OH-24-epiBL)

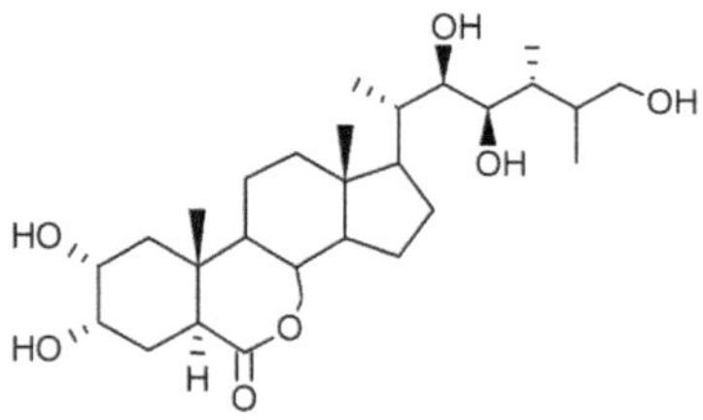

26-hydroxy-24-epibrassinolide (26-OH-24-epiBL)

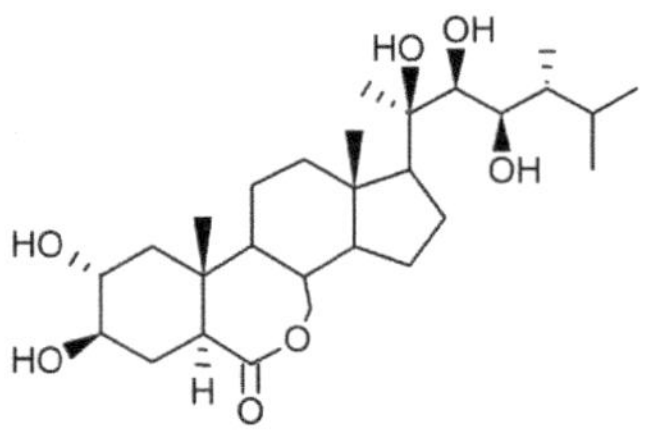

20*R*-hydroxy-3,24-diepibrassinolide (20*R*-OH-3,24-diepiBL)

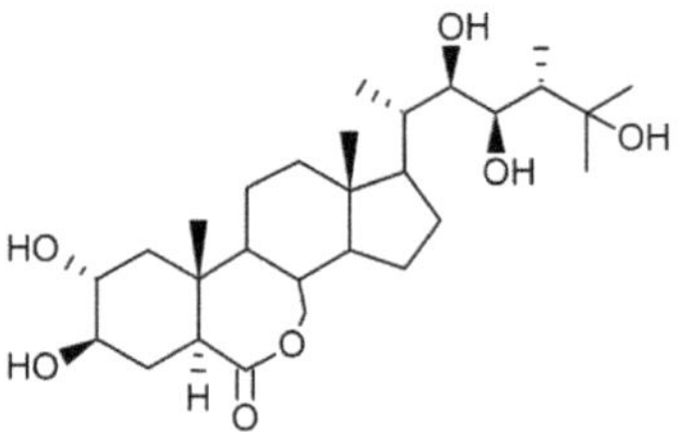

25-hydroxy-3,24-diepibrassinolide (25-OH-3,24-diepiBL)

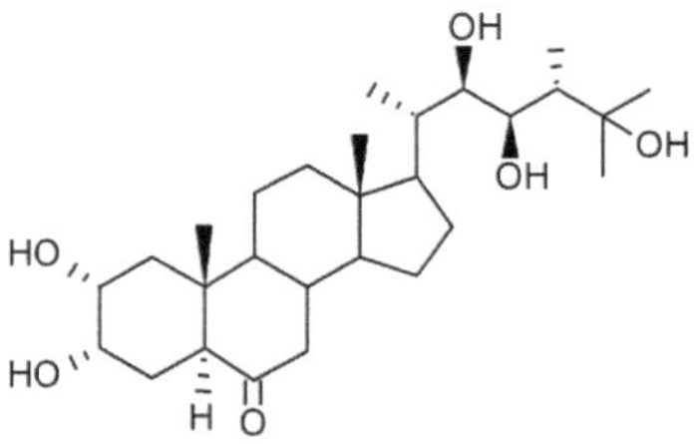

25-hydroxy-24-epicastasterone (25-OH-24-epiCS)

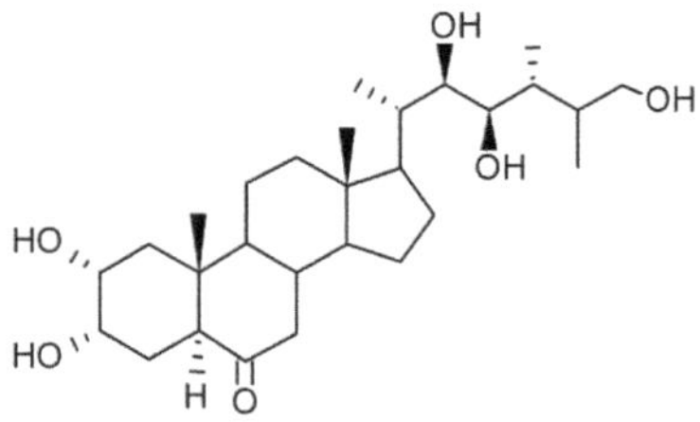

26-hydroxy-24-epicastasterone (26-OH-24-epiCS)

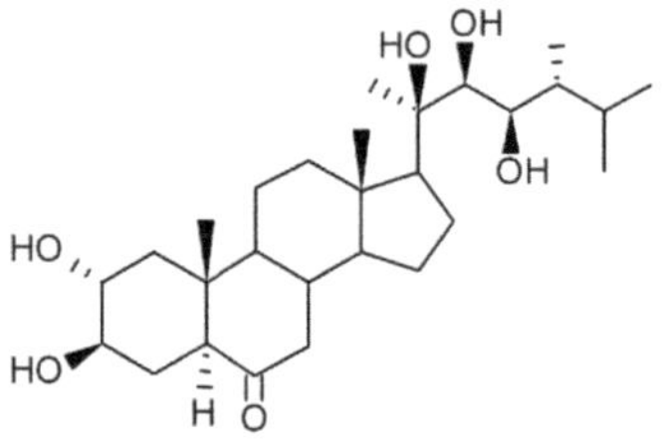

20*R*-hydroxy-3,24-diepicastasterone (20*R*-OH-3,24-diepiCS)

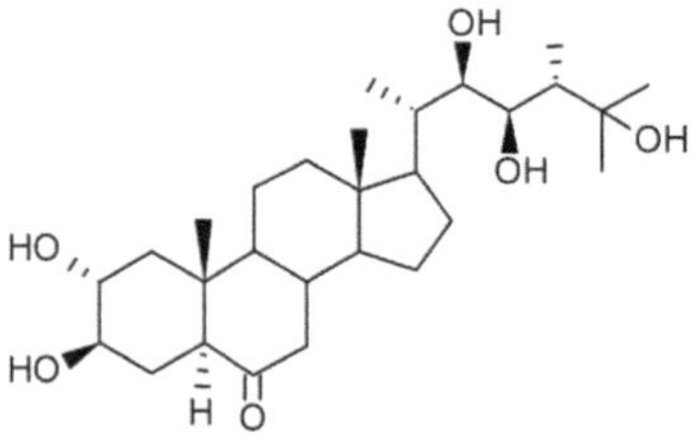

25-hydroxy-3,24-diepicastasterone (25-OH-3,24-diepiCS)

FIGURE 1.9 Continued.

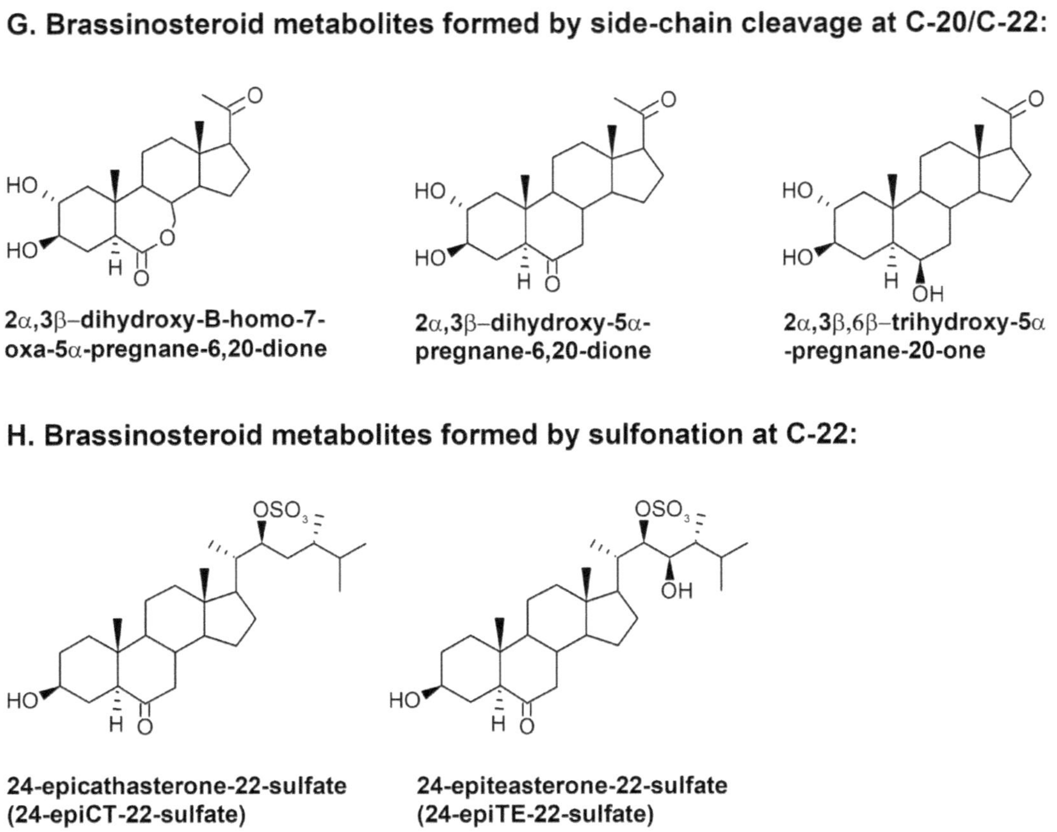

FIGURE 1.9 Continued.

1.3.4 Genetic Regulation of Brassinosteroid Catabolism, the Effect of BR Deficiency Mutants on Plants

Specific genes precisely regulate BRs action and inactivation (Table 1.2). Those encoding a cytochrome P450 monooxygenases (P450s) family play a crucial role in BR metabolism (Ohnishi, Nomura, et al. 2006, Zheng et al. 2018). The expression of CYP734As family genes regulates C-26 hydroxylation of BRs. The neighbor-joining method was used in creating a phylogenetic tree of CYP734As in various plants. The results showed the presence of CYP734A2, CYP734A4, CYP734A5, and CYP734A6 orthologs in rice; CYP734A-zm1, CYP734A-zm2, and CYP734A-zm3 in *Zea mays*; CYP734A12

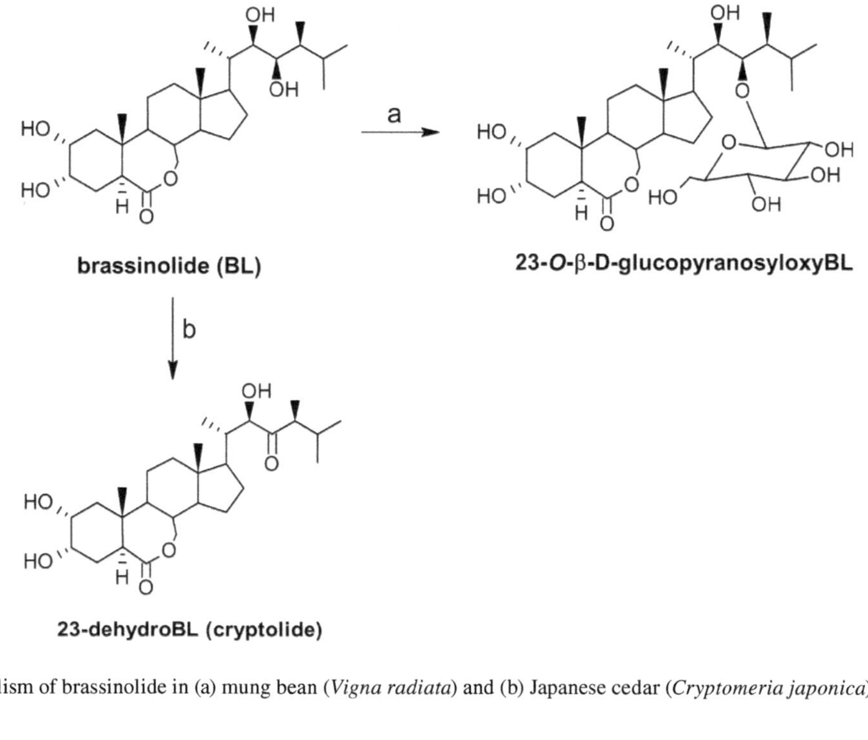

FIGURE 1.10 Metabolism of brassinolide in (a) mung bean (*Vigna radiata*) and (b) Japanese cedar (*Cryptomeria japonica*).

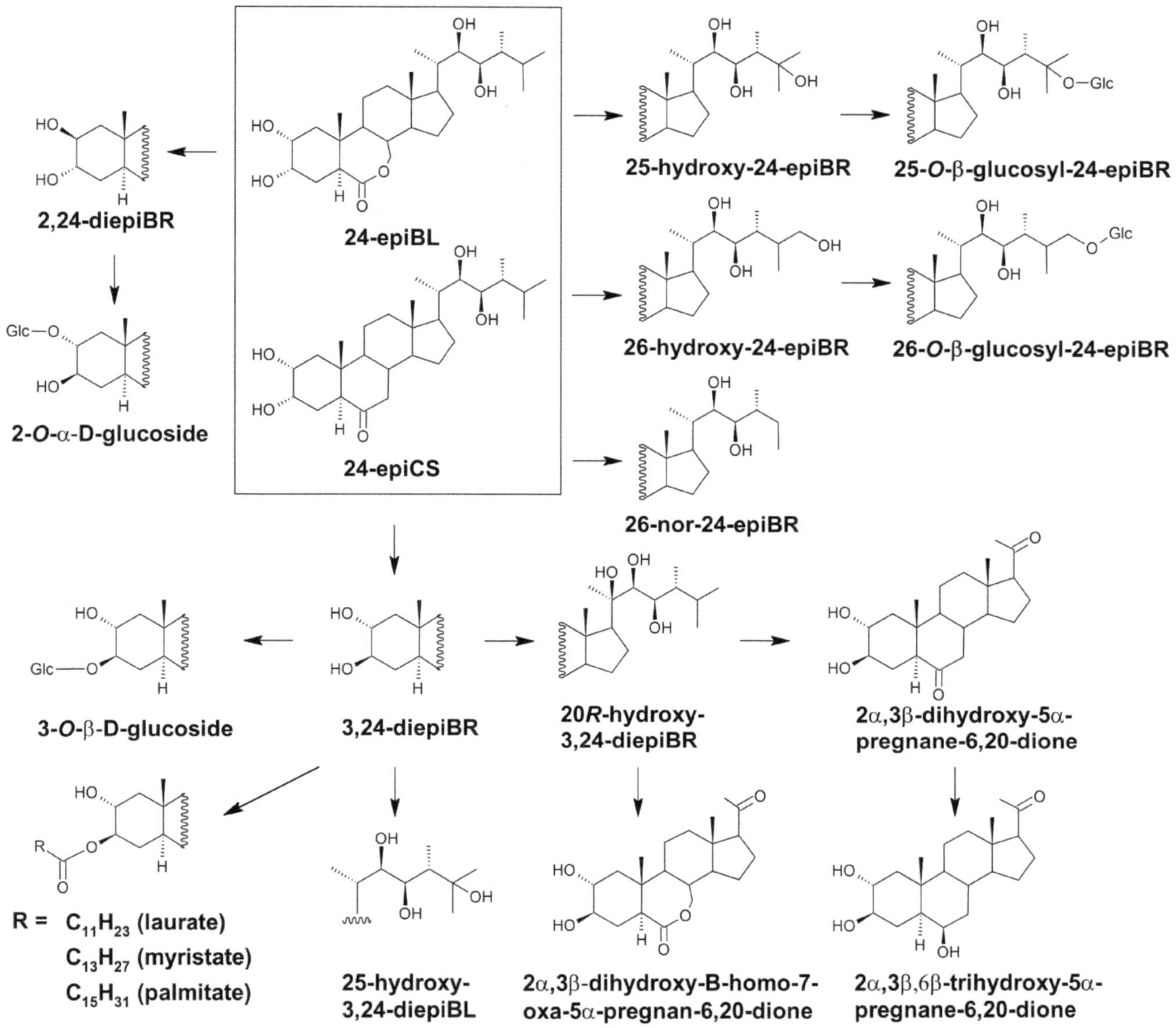

FIGURE 1.11 Metabolism of 24-epibrassinolide and 24-epicastasterone.

in *Medicago truncatula*; CYP734A11 in *Pisum sativum*; CYP734A13 and CYP734A15 in *Vitis venifera*; CYP734A-rc1 in *Ricinus communis*; CYP734A9 and CYP734A10 in *Populus trichocarpa*; CYP734A1 and CYP72C1 in *Arabidopsis thaliana*; CYP734A7 and CYP734A8 in tomato; and finally, CYP734A16 in *Picea sitchensis* (Sakamoto et al. 2011). The transgenic rice with overexpression of CYP734As genes indicated an abnormal dwarf phenotype. The plant height drastically decreased, about 90% in relation to wild-type rice, while the leaves were deformed. Moreover, the formation of flowers and seeds was not observed. These changes were caused by a deficiency of BRs. The CS presence was detected in the control group, but not in the CYP734A4 mutant. Furthermore, extremely decreased level of other BRs, derived from C-22 hydroxylation pathways of BRs biosynthesis, i.e., 6-deoxoCS, 6-deoxocathasterone (6-deoxoCT), TE, 6-deoxoTE, 3-dehydroteasterone (3DT), 3-dehydro-6-deoxoteasterone (6-deoxo-3DT), TY, and 6-deoxoTY, was reported. Similar decreases were also noted for CYP734A2 and CYP734A6 mutants. Moreover, 6-deoxoCS could be converted to 6-deoxoCS-CHO and 6-deoxoCS-COOH through the CYP734A2, which catalyzes the transformation of BRs to C-26 alcohol, aldehyde, and carboxylate derivatives (Sakamoto et al. 2011).

The expression of genes belonging to the cytochrome P450 monooxygenase family has also been studied in *Arabidopsis thaliana*. The CYP734A1/BAS1and CYP72C1 were isolated, and the phenotype of mutants (basl-D and chi2, respectively) was analyzed. Both enzymes catalyze C-26 hydroxylation of BL and CS, giving inactive metabolites. In the case of the chi2 mutant, the presence of BL and CS was not detected, while the endogenous level of BRs synthesized in early and late C-6 oxidative pathways (i.e., 6-deoxoCT, 6-deoxoTE, 6-deoxo3DT, 6-deoxoTY, 6-deoxoCS, TE, and TY) was reduced. The phenotypes of either mutant demonstrated dwarfism with hypocotyl growth inhibition (Nakamura et al. 2005, Turk et al. 2003).

In tomato, the homologs of *Arabidopsis* CYP734A1/BAS1, CYP734A7, and CYP734A8 have been identified, cloned,

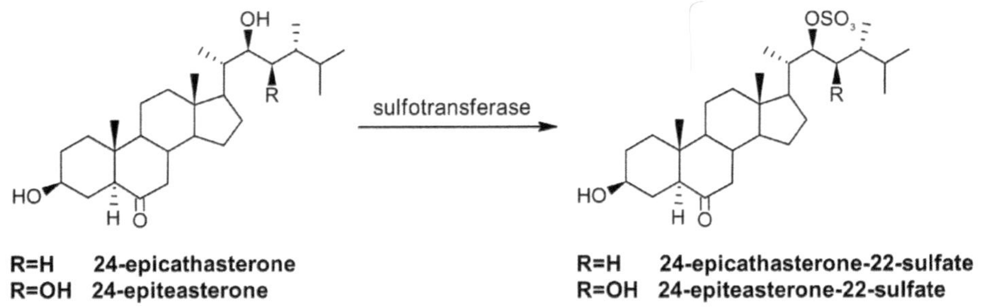

FIGURE 1.12 Sulfonation of 24-epiteasterone and 24-epicathasterone in *Brassica napus.*

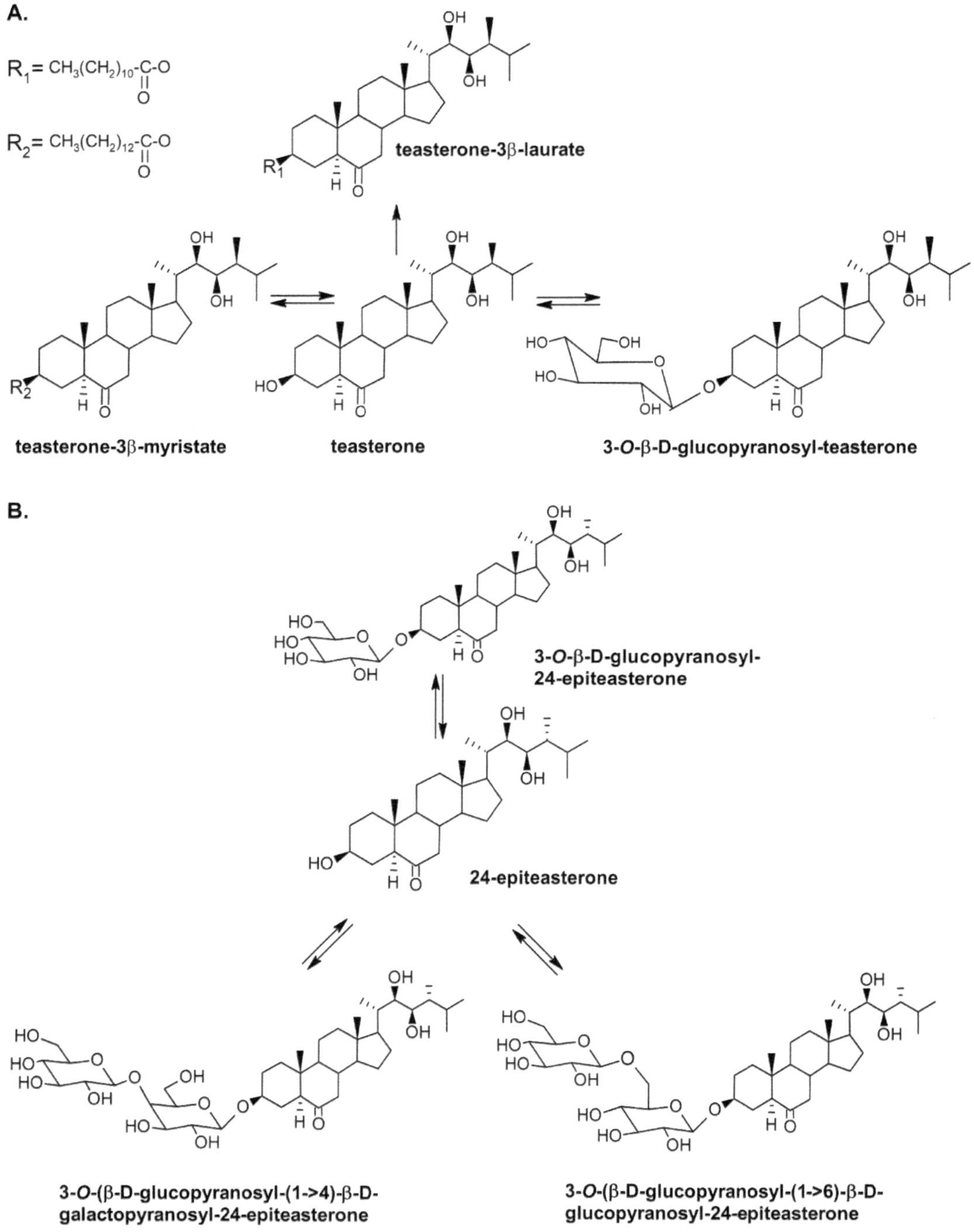

FIGURE 1.13 Metabolism of teasterone and 24-epiteasterone.

and transferred into tobacco (*Nicotiana tabacum*). The transgenic tobacco indicated a dwarf phenotype, while GC-MS analysis showed a significantly declined amount of CS and 6-deoxoCS in relation to the wild plant. Furthermore, the hydroxylation of 28-norCS and BL also occurred (Ohnishi, Nomura, et al. 2006).

The PAGODA 1 (PAG1) was identified as a gene encoding a cytochrome P450 that catabolizes active BRs in cotton. The maximum likelihood method indicated that PAG1 encodes a protein belonging to the CYP734A subfamily. The mutant of *Gossypium hirsutum* with overexpression of PAG1 showed dwarfism related to the deficiency of BRs. Expression of PAG1 caused the inhibition of stem, flower, and leaf growth and overall plant development. Moreover, the pollen vitality and fibers' length and strength were also reduced compared to wild-type cotton. The fatty acids biosynthesis was also decreased in pag1 mutants. However, the exogenous application of BL restores growth and fiber elongation of mutants. Studies confirm that the dwarf phenotype of pag1 cotton is caused by BR deficiency (Yang et al. 2014).

The sequent homologous gene of AtCYP734A1/BAS1, named as a DcBAS1, was identified in *Daucus carota*. For determining DcBAS1 expression, the carrot seedlings were treated with exogenously applied EBL. After three and five hours of treatment, DCBAS1 expression was much higher in relation to the control. *Arabidopsis thaliana* responded by dwarfism and dark green leaves with shorter petioles. Moreover, in dark conditions, the hypocotyl length was significantly reduced. However, the exposition of transgenic *Arabidopsis thaliana* on EBL overcame the inhibitory action of genes and caused increasing both leaf size and hypocotyl length. Interestingly, after 36 days of cultivation, the amount of cellulose in the mutant significantly decreased compared with the wild type of the plant. Therefore, confirming the pivotal role of BRs in plants. The family of CYP734As are multifunctional enzymes and play an important role in the metabolism and inactivation of BRs (Que et al. 2019).

Three BNST (BNS1-3) genes encoding sulfotransferases (STs) have been isolated in *Brassica napus*. STs are a family of enzymes that catalyze the *O*-sulfonation of both BRs and mammalian steroids, for example, estrogens or androgens. The enzymatic assay indicated the highest affinity to inactivate 24-ECT and 24-ETE by the BNST3. Moreover, the expression of BNST3 was also observed under high EBL concentration. C-22 sulfonation of EBL caused the inactivation of its biological activity. The addition of EBL stimulated the elongation of the internode. Thus, sulfate metabolites of BRs lose their biological activity (Rouleau et al. 1999).

The group of UGT genes encodes UGT-glycosyltransferases, which participate in the regulation of the activity of not only BRs but also mammal and insect steroid hormones. The liquid chromatography-mass spectrometry analyses confirmed that UGT73C5 catalyzes the 23-*O*-glycosylation of the BL and CS in *Arabidopsis thaliana* seedlings. Furthermore, plant mutants with overexpression of UGT73C5 were characterized by BR deficiency. The transgenic four-week-old plant was smaller and had reduced petiole and leaf length in relation to the wild type. However, their phenotype came back to normal after the application of exogenous EBL. In wild-type plants treated with BL or CS, the 23-*O*-glycosylation was indicated, whereas in mutants with silenced UGT73C5 expression, glycosylation did not occur. These results suggest that UGTs are the only enzymes catalyzing the 23-*O*-glycosylation reaction in *Arabidopsis thaliana* (Poppenberger et al. 2005).

REFERENCES

Abe et al. *Biosci. Biotechnol. Biochem.* 1994; 58:986–989.

Asami et al. *Plant Physiol.* 2000; 123:93–99.

Asami et al. *J. Biol. Chem.* 2001; 276:25687–25691.

Bajguz. *Plant Physiol. Biochem.* 2007; 45:95–107.

Choe. In *Plant Hormones: Biosynthesis, Signal Transduction, Action!*, ed. P. J. Davies. 2004; 156–178.

Choe. *Physiologia Plantarum.* 2006; 126:539–548.

Chung & Choe. *Crit. Rev. Plant Sci.* 2013; 32:396–410.

Dockter et al. *Plant Physiol.* 2014; 166:1912–1927.

Fujioka et al. *Phytochemistry.* 2000; 55:97–101.

Fujioka et al. *Plant Physiol.* 2002; 130:930–939.

Fujita et al. *Plant J.* 2006; 45:765–774.

Fujiyama et al. *Nature Plants.* 2019; 5:589–594.

Hai et al. *Phytochemistry.* 1995; 40:443–448.

Hai et al. *Phytochemistry.* 1996; 41:197–201.

Hartwig et al. *Plos One.* 2012; 7.

Hunter. *J. Biol. Chem.* 2007; 282:21573–21577.

Joo et al. *J. Exp. Bot.* 2012; 63:1823–1833.

Joo et al. *Phytochemistry.* 2015; 111:84–90.

Kanwar et al. *J. Plant Growth Regul.* 2017; 36:1002–1030.

Kim et al. *Plant Cell Physiol.* 2000; 41:1171–1174.

Kim et al. *Plant Physiol.* 2004; 135:1231–1242.

Kim et al. *Plant Physiol.* 2005; 138:2033–2047.

Kim et al. *J. Plant Biol.* 2018; 61:330–335.

Kolbe et al. *Phytochemistry.* 1994; 36:671–673.

Kolbe et al. *Phytochemistry.* 1995; 38:633–636.

Kolbe et al. *Phytochemistry.* 1998; 48:467–470.

Kwon et al. *J. Plant Biol.* 2005; 48:1–15.

Lee et al. *Bull. Korean Chem. Soc.* 2010; 31:3475–3478.

Lee et al. *Bull. Korean Chem. Soc.* 2011; 32:403–404.

Miziorko. *Arch. Biochem.* 2011; 505:131–143.

Nakamura et al. *J. Exp. Bot.* 2005; 56:833–840.

Nakano & Asami. In *Plant Chemical Biology*, ed. D. Audenaert and P. Overvoorde. 2014; 128–144.

Nelson et al. *Plant Physiol.* 2004; 135:756–772.

Nishikawa et al. *J. Plant Res.* 1995; 108:65–69.

Nishikawa et al. *J. Plant Physiol.* 1995; 147:294–300.

Oh et al. *Int. J. Mol. Sci.* 2015; 16:17273–17288.

Oh et al. *Plos One.* 2015; 10.

Ohnishi et al. *Plant Cel.* 2006; 18:3275–3288.

Ohnishi et al. *Phytochemistry.* 2006; 67:1895–1906.

Ohnishi et al. *Biosci. Biotechnol. Biochem.* 2006; 70:2071–2080.

Ohnishi. *J. Pestic. Sci.* 2018; 43:159–167.

Park et al. *Bull. Korean Chem. Soc.* 2000; 21:1274–1276.

Poppenberger et al. *Proceedings of the National Academy of Sciences of the United States of America.* 2005; 102:15253–15258.

Que et al. *J. Agric. Food Chem.* 2019; 67:13526–13533.

Roh et al. *J. Plant Biol.* 2017; 60:533–538.

Roh et al. *J. Agric. Food Chem.* 2020; 68:3912–3923.

Rouleau et al. *J. Biol. Chem.* 1999; 274:20925–20930.

Rozhon et al. *BMC Plant Biol.* 2014; 14.

Rozhon et al. *Molecules.* 2019; 24:4372.
Sakamoto et al. *Plant J.* 2011; 1 67:1–12.
Sasaki et al. *Phytochemistry.* 2013; 87:30–38.
Schaeffer et al. *Plant J.* 2001; 25:605–615.
Sekimata et al. *Tetrahedron-Asymmetry.* 2002; 13:1875–1878.
Sekimata et al. *Biosci. Biotechnol. Biochem.* 2008; 72:7–12.
Shimada et al. *Plant Physiol.* 2001; 126:770–779.
Soeno et al. *J. Pestic. Sci.* 2000; 25:117–122.
Soeno et al. *Biosci. Biotechnol. Biochem.* 2000; 64:702–709.
Son et al. *J. Korean Soc. Appl. Biol. Chem.* 2012; 55:363–366.
Sonawane et al. *Nature Plants.* 2016; 3:16205.
Suzuki et al. *Phytochemistry.* 1993; 33:1361–1367.
Symons et al. *J. Exp. Bot.* 2008; 59:17–24.
Turk et al. *Plant Physiol.* 2003; 133:1643–1653.
Verhoef et al. *J. Exp. Bot.* 2013; 64:2435–2448.
Vriet et al. *Molecular Plant.* 2013; 6:1738–1757.
Wang et al. In *Hormone Metabolism and Signaling in Plants*, ed. J. Li, C. Li and S. M. Smith. 2017; 291–326.
Watanabe et al. *J. Chem. Res.* 2000;18–19.
Yang et al. *New Phytologist.* 2014; 203:437–448.
Ye et al. *J. Integr. Plant Biol.* 2011; 53:455–468.
Zhao & Li. *J. Integr. Plant Biol.* 2012; 54:746–759.
Zheng et al. *J. Plant Physiol.* 2018; 231:68–85.
Zullo & Bajguz. In *Brassinosteroids: Plant Growth and Development*, ed. S. Hayat, M. Yusuf, R. Bhardwaj and A. Bajguz. 2019; 1–44.

2

Role of Brassinosteroids on Plant Growth and Development

Jannela Praveena, Satya Narayan Dash, Laxmipreeya Behera, and Gyana Ranjan Rout

CONTENTS

2.1 Introduction

Plant hormones are a group of naturally occurring organic substances that influence several physiological processes at a low concentration. Thus, plant hormones may play an important role in controlling the growth and development of plants (Davies, 1987). While metabolism is responsible for the power and building blocks of a plant, hormones regulate the growth of the individual plant part and integrate the growth of the whole plant. Major phytohormones such as auxins, cytokinins, gibberellins (GAs), brassinosteroids (BRs), abscisic acid (ABA), and ethylene are involved in several growth and developmental processes (Gray, 2004). BRs are a group of plant steroid hormones that regulate many processes in plant growth and development, including cell elongation, cell division, senescence, vascular differentiation, reproduction, photomorphogenesis, germination of seeds, rhizogenesis, flowering, fruit ripening, tolerance response to various biotic and abiotic stresses, and senescence (Clouse and Sasse, 1998; Divi and Krishna, 2009; Karlidag et al., 2012; Unterholzner et al., 2015; Tang et al., 2016; Wei and Li, 2016; Manoli et al., 2018). Mitchell and coworkers identified a steroidal substance in 1970 and called it "brassins." Later, it was chemically extracted by Grove et al. (1979) from bee-collected pollen of rapeseed plant (*Brassica napus* L.), which was named "brassinolide." However, they could extract only 10 mg of a solid crystalline form of brassinolide (BL) from 230 kg of pollen grains. Later, another growth substance similar to BL was discovered from the insect galls of chestnut (*Castanea crenata*) in 1982, which was named "castasterone." All these steroidal substances having growth-promoting activity were cumulatively referred to as "brassinosteroids" (BRs). BRs are a group of naturally occurring polyhydroxy steroids and are considered the sixth important group of phytohormones used for normal growth and several developmental processes in plants (Rao et al., 2002; Hayat and Ahmad, 2011; Luan et al., 2016; Khripach et al., 2000). These steroidal compounds are also involved in inducing defense mechanisms against water stress, temperature stress, oxidative stress, and salinity stresses. Coll et al. (2015) reported that BRs are natural, non-toxic, non-genotoxic, biosafe, and eco-friendly phytohormones, which can be used in agri–horti plants to improve the growth, yield, and fruit quality (Saini et al., 2015). BRs are an important class of phytohormones that can be perceived by a membrane-localized receptor-like kinase (RLK) called BRASSINOSTEROID INSENSITIVE 1 (BRI1) (Li and Chory, 1997; Hothorn et al., 2011; She et al., 2011). Upon BR binding, BRI1 forms a heterodimer with its co-receptor, BRI1-ASSOCIATED RECEPTOR KINASE 1 (BAK1), which promotes trans-phosphorylation between the cytosolic kinase domains of BRI1 and BAK1 (Li et al., 2002; Nam and Li, 2002; Wang et al., 2008; Santiago et al., 2013; Sun et al., 2013). As of now, more than 70 BRs have

DOI: 10.1201/9781003110651-2

been isolated from different plant species, out of which, only three, that is, BL, 24-epibrassinolide, and 28-homobrassinolide, are the biologically most active forms, and are extensively used in physiological studies in different plants. This chapter highlights the following: the role of BRs in plant growth and development and stress response; understanding the BR pathway; the molecular mechanism of BR signaling in various tissues; crosstalk between BRs and other phytohormones; gene involvement in the BR signaling pathway; and biosynthesis.

2.2 Structure and Occurrence of BRs and Their Regulatory Mechanisms

Naturally occurring BRs are polyhydroxylated steroidal hormones, which have a common 5α-cholestane skeleton. Based BRs have been classified as C-27, C-28, or C-29 compounds. Oxygen at C-6 and the hydroxyl group on the side chain at C-22 and C-23 positions are essential for the activity of BRs. BL is the most active BR. BRs consist of a precise class of low abundance plant steroids that bear an oxygen moiety at C-3 and additional moieties at one or more of the C-2, C-6, C-22, and C-23 carbon atoms (Bishop and Yokota, 2001; Bajguz and Hayat, 2009; Hayat and Ahmad, 2011). Several BRs have been extracted from a large number of plant species, as these steroidal chemicals are ubiquitous throughout the plant taxa. These growth-inducing steroidal hormones were isolated from both reproductive parts (pollen grains, anthers) and vegetative parts (stem, leaf, root, flower, and seed). In addition, BRs have also been extracted from the crown galls of the chestnut plant (*C. crenata*). BRs have been intensively studied and established a complex BR signal transduction pathway that plays an important role in plant growth and development. The signal transduction pathway shows that BR is perceived by BRASSINOSTERIOD INSENSITIVE 1 (BRI1) receptor kinase at the cell surface and activates BRASSINAZOLE RESISTANT 1 (BZR1) and BRI1-EMS SUPPRESSOR 1 (BES1) transcription factors to induce stress tolerance. Exogenously applied BR binds to BRI1, inducing an association with BRI1-ASSOCIATED RECEPTOR KINASE 1 (BAK1) and disassociation of BRI1 KINASE INHIBITOR 1 (BKI1). Sequential transphosphorylation between BRI1 and BAK1 is necessary to activate BRI1 and to phosphorylate BR SIGNALING KINASE 1 (BSK1) and enhance BRI1 SUPPRESSOR 1 (BSU1) activity. The activated BSU1 inhibits BRASSINOSTEROID INSENSITIVE 2 (BIN2) through dephosphorylation of the phospho-tyrosine residue of BIN2, which allows the accumulation of unphosphorylated BZR1 and BZR2/BES1 transcription factors (Figure 2.1).

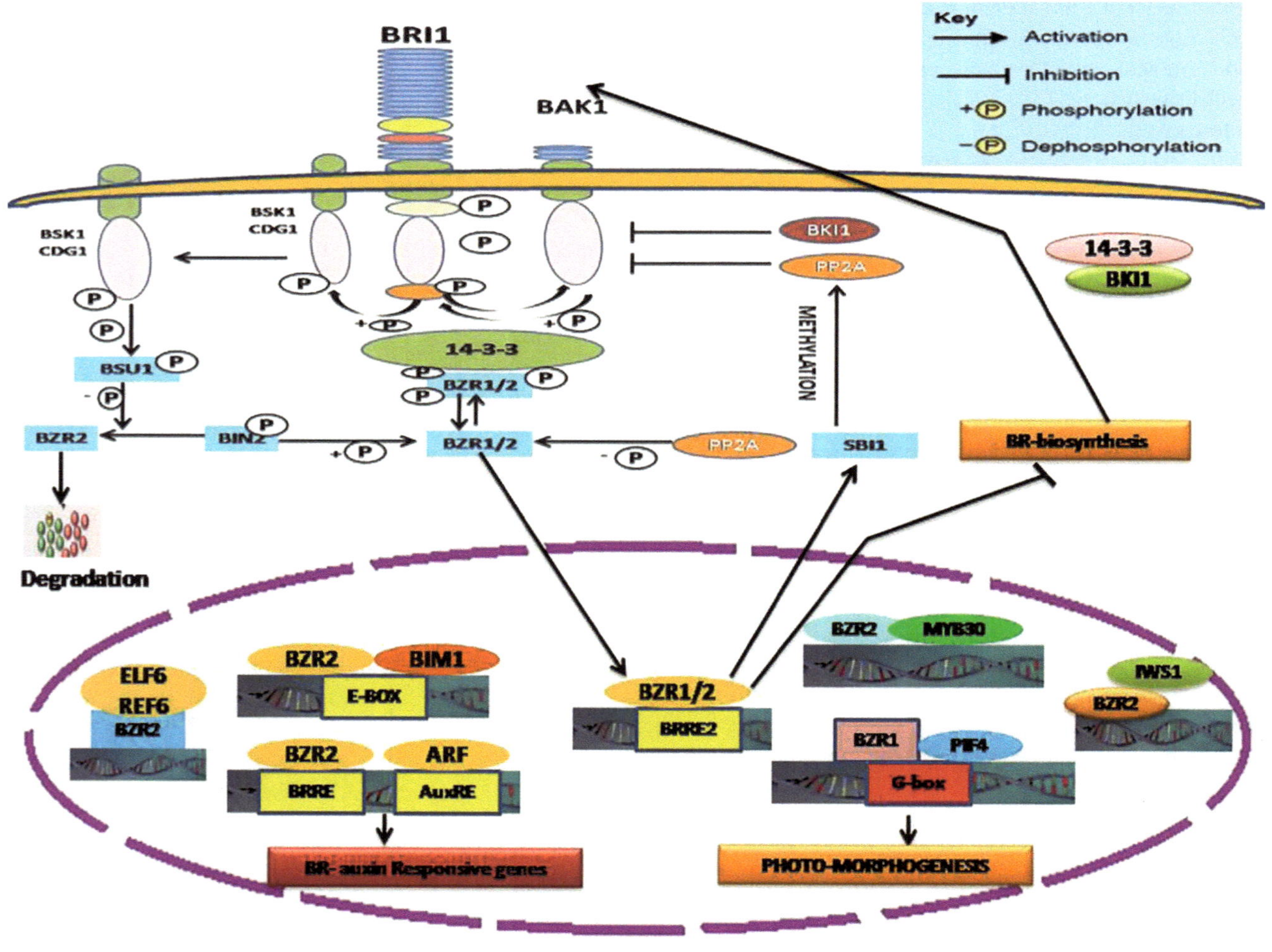

FIGURE 2.1 The BR signal transduction pathway regulates various aspects of plant development and physiology. The BR signal is transduced by a receptor kinase-mediated signal transduction pathway, which ultimately results in the altered expression of numerous genes.

The dephosphorylated BZR1 and BES1 enter the nucleus and regulate BR-targeted genes to enhance plant stress tolerance (Vardhini et al., 2015; Li et al., 2009; Takeuchi et al., 1996; Vardhini et al., 2015), regulating the accumulation of endogenous hormones (Wei et al., 2015; Wu et al., 2017) and upregulating a number of genes (Li et al., 2016).

2.3 Brassinosteroids and Different Plant Stress Responses

Both abiotic and biotic stresses have been recognized as the prime potential threats to the normal growth of plants and agricultural productivity. Stresses including water, salinity, alkalinity, lodging, metals, UV radiation, ozone, and temperature alter the plant growth by affecting the growth and metabolism. BRs have an active involvement in photosynthesis, antioxidant metabolism, osmolyte accumulation, nitrogen metabolism, and plant–water relations under normal and stressful conditions (Ali et al., 2007, 2008a, b; Hayat et al., 2012; Krumova et al., 2013; Fariduddin et al., 2014). Genetic and molecular studies have witnessed their active role in various developments including vegetative and reproductive differentiation and cellulose biosynthesis (Yang et al., 2011; Clouse, 2015). BRs are known to exist in a wide range of organisms including lower and higher plants. These steroidal plant hormones mediate growth elevation via the activation of different mechanisms (Bajguz and Hayat, 2009). BRs have been shown to be beneficial for growth regulation by imparting the induction of genes related to stress and other defense mechanisms for better growth adaptability (Zhao et al., 2016). BRs have been reported to have a positive role in modulating plants from oxidative damage of reactive oxygen species (ROS), as well detrimental effects on pigments and photosynthesis, amelioration of various components of the antioxidant defense system, osmoprotectant regulation and the aptitude to assist plants for the synthesis of protective substances, expression of genes involved in defense responses, and the biosynthesis of other plant growth regulators (PGRs). It has been reported that BRs impart growth stimulation and stress mitigation in a concentration-dependent manner. It has been reported that the exogenous application of BRs to stressed plants either through spraying, priming, or feeding along with a nutrient solution showed stress tolerance mechanisms (Anuradha and Rao, 2007; Hayat et al., 2007, 2010; Ali et al., 2008a, b; Fariduddin et al., 2009).

2.3.1 Drought Stress

One of the main challenges for sustainable agricultural production is the frequent occurrence of drought stress, particularly in arid and semi-arid regions of the globe. Often characterized by reduced relative water content (RWC), drought stress causes a considerable reduction in crop growth and yield (Jatav et al., 2014; Ahanger and Agarwal, 2017a,b). Drought stress-induced growth reduction generally occurs due to reduced photosynthetic rate, alteration in nitrogen and antioxidant metabolism, accumulation of secondary metabolites, and mineral nutrition (Jatav *et al.*, 2014). However, drought-induced adverse effects on plant growth and metabolism can be mitigated by exogenous application of a variety of growth-regulating substances, including BRs. Upreti and Murti (2004) reported that the application of either EBL or HBL to water-stressed plants of *Phaseolus vulgaris* resulted in improved growth and yield by enhancing root nodulation and nitrogenase activity, thereby providing strength to the plants to withstand drought. Fariduddin et al. (2009) reported that the application of 0.01 μM HBL to drought-stressed plants of *Brassica juncea* at two developmental stages enhanced the photosynthetic efficiency by improving RWC, stomatal conductance, and water use efficiency by increasing the accumulation of proline. It has been widely reported that exogenous BR application upregulates the activity of antioxidant enzymes and the levels of non-enzymatic antioxidants for mediating the efficient removal of ROS and thereby providing protection to membrane lipids for maintaining membrane integrity and functioning (Li et al., 2012a, b; Shahana et al., 2015; Ahmad et al., 2017a). Behnamnia (2015) reported that the application of EBL reduced drought-induced oxidative damage in tomato plants by downregulating lipoxygenase activity and upregulating the antioxidant defense system by enhancing the expression of antioxidant isozymes. Sahni et al. (2016) reported overexpressing the BR biosynthetic gene, DWF4, in transgenic *Brassica napus* to enhance drought stress tolerance. It has also been observed that the application of EBL augments drought tolerance by upregulating the transcription factors regulating the expression of the drought-responsive element, DRE, as observed in *Arabidopsis thaliana* and *Brassica napus* (Kagale et al., 2007).

2.3.2 Salt Stress

Salt stress is another limiting factor to reduce plant growth and yield performance. Due to high concentrations of salt, many agricultural lands are unproductive wastelands. Salinity reduces plant productivity by adversely affecting growth through the induction of osmotic and ionic imbalances (Ahanger and Agarwal, 2017; Ahmad et al., 2017b, 2018b; Kaur et al., 2018). Salinity-induced deleterious effects include ionic toxicity, osmotic stress, reduced nitrogen metabolism, elevated production of ROS leading to oxidative damage, hampered photosynthetic functioning and photosynthate translocation, impeded uptake, and translocation of mineral nutrients. The application of BRs reduces salinity stress as reported in several crops like *Oryza sativa* (Anuradha and Rao, 2003), *Cicer arietinum* (Ali *et al.*, 2007), *Brassica juncea* (Ali *et al.*, 2008), *Vigna sinensis* (El-Mashad and Mohamed, 2012), and *Mentha piperita* (Coban and Baydar, 2016). The most studied metabolic mechanism, with respect to the effect of BRs, is oxidative defense. Shahbaz and Ashraf (2007) reported that the foliar application of EBL in wheat cultivars counteracted the adverse effects of salinity by enhancing the activity of peroxidase and catalase with contrasting degrees of salt tolerance. Similarly, in salt-stressed *Cucumis sativus*, supplementation of EBL caused a marked enhancement in growth by upregulating the activity of superoxide dismutase (SOD), peroxidase, and catalase (CAT) concomitant with reduced electrolyte leakage and malondialdehyde (MDA) content as reported by Song et al. (2006). BRs are also known to regulate genes involved in some key metabolic processes in plants exposed to saline stress. Sharma et al. (2013) observed that the application of EBL enhanced the expression of stress-responsive genes and the *OsBRI1* conferred stress tolerance in *Oryza sativa*. In *Arabidopsis thaliana*, application of EBL enhanced the expression of phytohormone marker genes and it rescued the ethylene-insensitive *ein2* mutant and the ABA-deficient *aba-1* mutants of *Arabidopsis* from salt stress, suggesting that BRs share transcriptional targets with other hormones (Divi et al., 2010).

2.3.3 Temperature Stress

Climate change is widely witnessed in the current era, resulting in considerable fluctuation in mean temperatures. Both high and low (cold and chilling) temperatures are potential environmental factors affecting both physiological and biochemical aspects of metabolism, and such changes are regulated at the molecular as well as genetic levels (Kazemi-Shahandashti et al., 2014; Siboza et al., 2014; Hatfield and Prueger, 2015; Calzadilla et al., 2016). However, BR application has been reported to benefit crop plants by enhancing the efficiency of growth mediating key metabolic pathways. Exogenous application of BR has been widely reported to mitigate the adverse effects of high temperature regimes to a considerable extent in different crops by regulating a variety of metabolic processes (Ogweno et al., 2008; Hayat et al., 2010; Mazorra et al., 2011; Sahni et al., 2016; Yadava et al., 2016) as well as low temperature (Janeczko et al., 2007; Liu et al., 2009; Aghdam et al., 2012; Aghdam and Mohammadkhani, 2014; Calzadilla et al., 2016). Both high and low temperatures can effectively retard plant growth but the effects of low and high temperatures differ considerably in plants. For example, low temperature, particularly below freezing, can freeze cells, thereby causing a marked disruption in the uptake and translocation of both water and nutrients. However, the ameliorative effect of BRs has been assessed in different crops exposed to low temperature stress. Xi et al. (2013) reported that the application of BR in grapevines reduces the cold-induced ion leakage by maintaining membrane integrity through improvement in antioxidant and osmoregulatory components. In *Cucumis sativus*, the foliar spray of HBL mediated growth enhancement under chilling stress by improving the activities of antioxidant enzymes, thereby providing protection to the photosynthetic system from the ROS-induced oxidative damage (Fariduddin et al., 2011). Exogenous application of BR activated photosynthetic enzymes and antioxidant enzymes, leading to an improvement in the rate of photosynthesis through the alleviation of chilling induced pho-oxidative damage (Jiang et al., 2013). Heat shock proteins have been widely studied in plants because of their potential role in high-temperature tolerance. However, it is evident that BRs can promote the expression of heat shock proteins in heat-stressed plants. For example, exogenously supplied EBL improved the growth of *Brassica napus* and tomato seedlings by enhancing the expression of heat shock proteins (hsp100, hsp90, hsp70) and the low molecular weight hsps under thermal stress (Dhaubhadel et al., 1999).

2.3.4 Nutrient Stress

The optimal availability of mineral nutrients is believed to regulate growth by mediating changes at physiological, biochemical, and molecular levels, and the deficiency of essential inorganic nutrients affects growth adversely ((Ahanger et al., 2015, 2016, 2017a). Although phytohormones, including BRs, cannot replace nutrients in regulating the physiological and biochemical aspects of plant metabolism, they can compensate the need for nutrients to some extent. Janeczko et al. (2010) reported that the foliar application of EBL caused a significant enhancement in the uptake of potassium and calcium, thereby enabling *Triticum aestivum* cultivars S-24 and MH-97 to counteract salt stress by maintaining the K/Na ratio. In addition, exogenous foliar application of EBL increased uptake of potassium, magnesium, and calcium, and reduced sodium and iron in wheat (Janeczko et al., 2010). The foliar application of EBL (0.01 M) in *Cucumis sativus* cv. "Jinyou No. 4" mitigated the uptake of mineral nutrients like potassium, phosphorous, and manganese. EBL-treated plants maintained a higher K^+/Na^+ ratio and improved activity of Ca^{2+} ATPase, which mediated extrusion of excess Ca^{2+} from the cells to prevent toxicity (Yuan et al., 2015a, b). Song et al. (2016) demonstrated that the application of EBL to *Arachis hypogeal* mitigated Fe-deficiency-induced oxidative stress by upregulating the activity of nitrate reductase and the antioxidant system, and the accumulation of osmolytes, thereby reducing the production of ROS including superoxide and H_2O_2. Zhao et al. (2016) demonstrated that treatment of BR reduced ammonium toxicity by downregulating the expression of *ammonium transporter-1* (AMT1) expression in the roots of *Arabidopsis*, which was confirmed by its reversion using the BR receptor BRI1 mutant bri1-5. The expression of AMT1 transporters (AMT1;1, AMT1;2, AMT1;3) is directly regulated by the BR signaling transcription factor, BES1, and NH_4^+-mediated repression of AMT1 transporters was observed to suppress in a gain-of-function ammonium-sensitive BES1 mutant (besl-D). They also concluded that BR-induced regulation of nitrogen uptake and assimilation occurs via the BR signaling pathway.

2.3.5 Heavy Metal Stress

Brassinosteroids have the ability to regulate the uptake of ions into plant cells and can be used to reduce the accumulation of heavy metals and radioactive elements in plants. Moreover, BRs also minimize the toxic effects and symptoms generated by an excess of heavy metals (Bajguz and Hayat, 2009). Ameliorative properties of 28-HBL were deliberated in *Raphanus sativus* L. under chromium toxicity mediated by regulating antioxidant enzymes and helping in mitigating the toxic effects of chromium (Sharma and Bhardwaj, 2007; Sharma et al., 2011). They also affirmed that 28-HBL treatments in radish seedlings lowered Cr uptake. Hayat et al. (2007) studied the role of HBL in cadmium stress in *Brassica juncea* and reported that it helps the biosynthesis of chlorophyll, alters water balance, and decreases activities of various enzymes to check oxidative stress. BRs enhanced proline accumulation and CAT, APX, GPX, and SOD activities, whereas reduced POX and ascorbic acid oxidase activities. Moreover, lipid peroxidation induced by cadmium was reduced with the supplementation of BRs (Anuradha and Rao, 2007). Aluminum toxicity is the major growth-limiting factor for crop cultivation on acidic soils. Seedlings of mung bean were subjected to aluminum (Al) stress and sprayed with EBL or HBL (Ali et al., 2008a). HBL- or EBL-treated plants showed higher activity of SOD, CAT, and POX, and proline content under Al stress. The increase in Al resistance conferred by BRs was reflected in the improvement of plant growth, photosynthesis, and related processes in the presence of Al. EBL significantly increases the fresh mass of shoots and roots, and chlorophyll content in mung bean under Al stress (Ali et al., 2008a). Although nickel is an essential element, its high concentration is toxic and inhibits photosynthesis, respiration, the activities of enzymes, and protein content. Spraying with HBL partially neutralizes the toxic effect of nickel in *Brassica juncea* as reported by Sharma et al. (2008). The growth of seedlings was inhibited by Ni and this reduction was restored by HBL treatment. The protein content and activities of CAT, GR, APX, SOD, and GPX were also increased by HBL treatment.

2.3.6 Biotic Stress

In field conditions, plants face different kinds of stress conditions, both biotic (damage by any living organism: bacteria, viruses, fungi, parasites, beneficial and harmful insects) and abiotic (environmental factors such as low and high temperature stress, salinity, etc.). Like animals, plants also have an immune system, which provides resistance to external stressors (biotic and abiotic stress). To effectively combat invasion by infectious pathogens and herbivorous pests, plants make use of pre-existing physical and chemical barriers, as well as inducible defense mechanisms, which become activated upon attack; plant defenses function as a unit to reduce the harmful effects of biotic stresses. The induced defense system of plants against biotic stresses is similar to the defense against abiotic stress. For both types of stress, the induced defense system is regulated by complex interconnected signal transduction pathways in which plant hormones (ABA, ETH, JA, SA, and BR) play a fundamental role (Wu et al., 2017; Hu et al., 2017). Application of BRs at low concentrations significantly improves growth and yield and increases resistance to viral and fungal pathogens in cucumber, tobacco, and tomato (Bajguz and Hayat, 2009). However, the levels of protection and effectiveness depend on the application of BRs. Lu et al. (2017) reported that the flavonoid compounds anthocyanin and catechin, and transcript levels of induced *MYB* genes (*MYB30*), were increased in rust infections in apple. The *MYB30* genes directly regulated *BES1* in *Arabidopsis. BES1* is a key gene of the BR signal transduction pathway, and *AtMYB30* mutants and *BES1* interact with each other and promote BR-targeted genes both in vitro and in vivo (Kim and Wang, 2010). Furthermore, plant hormones ABA, ethylene, jasmonic acid (JA), and salicylic acid (SA) were at the highest level in rust-infected apple plants (Lu et al., 2017). Li et al. (2016) reported that the contents of SA, JA, and ethylene were significantly enhanced with the application of BR in pepper plant exposed to chilling stress. It was suggested that BR functions via synergistic crosstalk with SA, JA, and ethylene signaling pathways to respond to chilling stress. The study also indicated that BRs play an essential role in biotic stress tolerance by activating enzymes, biotic resistance genes, antioxidants, hormones, transcriptional factors, and signaling pathways to reduce biotic stress damage.

2.4 Brassinosteroid: Phytohormones Crosstalk

BRs perform diverse functions due to their interplay with other phytohormones. In response to environmental cues, BRs interact with different phytohormones such as abscisic acid (ABA), auxin, cytokinin (CK), ethylene, GAs, JA, polyamines (PA), and SA, and regulate myriad aspects of plant growth and developmental processes in plants (Choudhary et al., 2012; Gruszka, 2013). Crosstalk between BR and auxin regulates myriad aspects of plant growth and development (Hao et al., 2013; Saini et al., 2013; Liu et al., 2014; Chaiwanon and Wang, 2015). Interactive effects of BRs and auxin involved physiological processes such as hypocotyl elongation or root development. The role of auxin and BR interaction in regulating stress responses has remained elusive (Kissoudis et al., 2014). In recent times, it has been exhibited that the BR signaling part BIN2 can specifically collaborate with an auxin signaling segment ARF2, an individual from the auxin response factor group of transcriptional controllers (Vert et al., 2008). Furthermore, in yucca mutants, a 40% higher transcript level in BR upregulation of genes has been reported (Nemhauser et al., 2004) and revealed the BR and auxin crosstalk point. Despite the fact that the relationship between BR and auxin has been archived in plant development and formative procedures, examinations are essential to comprehend the component of auxin and BR crosstalk associated with tolerance. BRs also interact with gibberellic acid to coordinate different physiological processes as reported by various authors (Sun et al., 2010; Li et al., 2012a). However, evidence indicates BR–GA antagonistic interaction in defense-related processes against root oomycete *Pythium graminicola*. It has been demonstrated that in several GA-deficient and/or -insensitive mutants disease development was more severe. It implies a positive role of GA in providing resistance against *P. graminicola*. Further, it has been reported that susceptibility similar to those observed in BR-treated plants was detected when endogenous GA level was disrupted using the GA biosynthesis inhibitor, uniconazole (De Vleesschauwer et al., 2012). However, the application of BR and uniconazole did not cause any additive effect. But treatment of uniconazole along with brassinazole, a BR inhibitor, negated the resistance-inducing effect of brassinazole. It indicates that BR has an effective immune response led by GA. Further, it has been demonstrated that the abundance of GA repressors, DELLA and SLR1, is positively regulated by BR. This phenomenon leads to the BR-mediated suppression of GA biosynthetic genes such as *GA20ox* and *GA3ox3* inducing *GA2ox* expression, which is involved in the suppression of GA signaling and its deactivation (De Vleesschauwer et al., 2012). Recently, crosstalk between BR and GA has been established in regulating plant cell elongation in rice (Tong et al., 2014). It has been suggested that BR promotes GA accumulation by inducing the expression of *D18/GA3ox-2*, one of the GA biosynthetic genes. However, the application of an exogenous high concentration of BR leads to the activation of *GA2ox-3*, a GA inactivation gene, resulting in inhibition of cell elongation. Moreover, GA inhibits BR signaling as well as its biosynthesis in a feedback inhibiting loop but facilitates cell elongation through activating the primary BR signaling pathway upon applying an exogenous high GA concentration, indicating BR-GA crosstalk in regulating cell elongation (Tong et al., 2014). The interaction between BR, IAA, and GA on cotton fiber development has been studied in *Gossypium hirsutum* (Hu et al., 2011). A class of DELLA proteins *GhGAI1* was downregulated by BR and auxin treatment during cotton fiber initiation and elongation, suggesting its importance in cotton fiber improvement through genetic modulation of the phytohormone strategy. Allen and Ptashnyk (2017) reported new mathematical models for the BR signaling pathway and for the crosstalk between the BR and GA signaling pathways. They suggested that the interaction between transcription factors BZR and DELLA exerts more influence on the dynamics of the signaling pathways than BZR-mediated GA biosynthesis (Figure 2.2). There is an interaction between transcription factors that may constitute the principal mechanism of the crosstalk between BR and GA signaling pathways. In general, perturbations in the GA signaling pathway have larger effects on the dynamics of components of the BR signaling pathway than perturbations in the BR signaling pathway on the dynamics of the GA pathway. The perturbation in the crosstalk mechanism also has a larger effect on the dynamics of the BR pathway than the GA pathway. Large changes in the dynamics of GA signaling processes can only be observed in the case where there are disturbances in both the BR signaling pathway and the crosstalk mechanism.

CK and BR indirectly crosstalk through the modulation of auxin transport at the molecular level in regulating lateral root development. BR enhances the expression of auxin efflux carriers such as *PIN* genes, which probably facilitate the maintenance of local auxin maxima required for root primordial development (Bao *et al.*, 2004). Conversely, cytokinin inhibits the establishment of lateral root primordia and disturbs auxin accumulation by downregulating the expression of *PIN* genes, indicating an indirect interaction between BR and CK (Benjamins and Scheres, 2008). Research studies indicate that in *Chlorella vulgaris*, CK stimulates the accumulation of endogenous BR, suggesting the synergistic interaction between BR and CK (Bajguz and Piotrowska-Niczyporuk, 2014). Upon exogenous treatment of 10 nM trans-zeatin (*t*Z) to the *C. vulgaris* culture, there was a considerable increase in the level of all endogenous BR by 27–46%. Moreover, the co-application of both BL and *t*Z leads to the highest stimulation in the number of *C. vulgaris* cells and endogenous accumulation of proteins, chlorophylls, and mono-saccharides, whereas the lowest was observed with treatment with 28-homocastasterone and 1,3-diphenylurea (DPU) indicating a BR–CK crosstalk point ((Bajguz and Piotrowska-Niczyporuk, 2014).

Brassinosteroid and ethylene crosstalk regulates different aspects of plant growth and developmental processes. BR has been identified as a negative regulator of shoot gravitropism, whereas ethylene has been shown to promote gravitropic reorientation in light-grown seedlings (Vandenbussche et al., 2013).

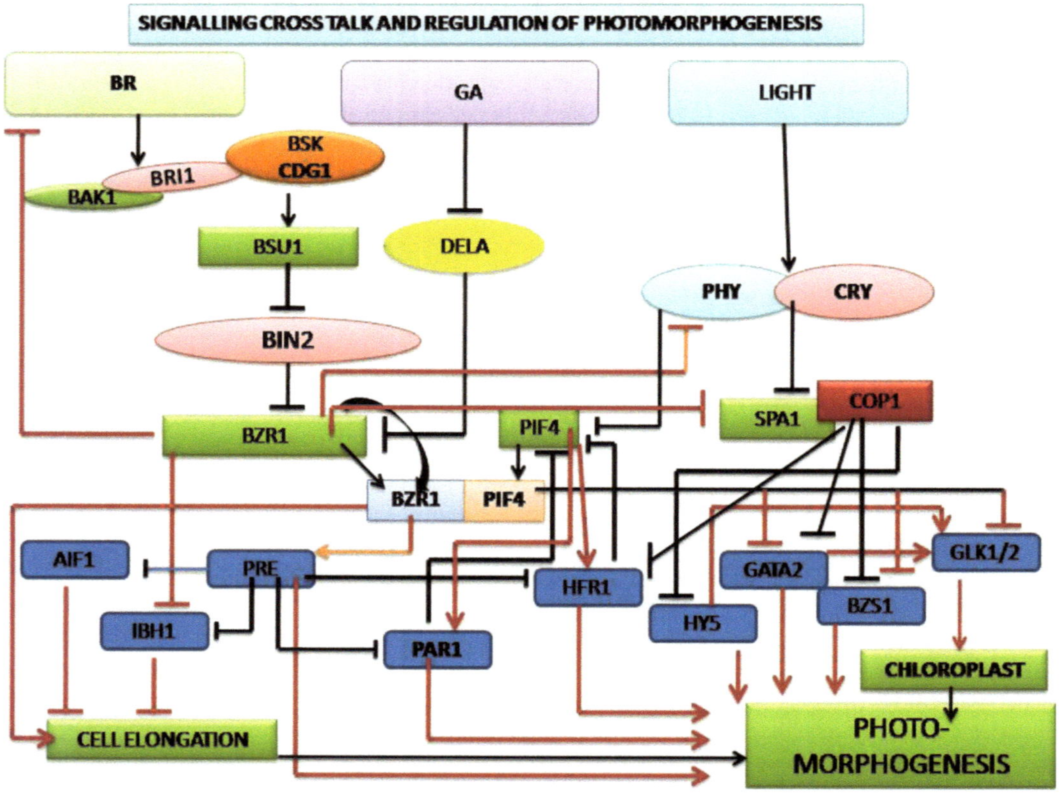

FIGURE 2.2 Signaling crosstalk and regulation of photomorphogenesis. Suppression of photomorphogenesis in the dark requires BR and two other hormones: GA and auxin. A large body of evidence shows that BR antagonizes light signals, has similar physiological effects to GA, and interacts synergistically with auxin during cell elongation and gene expression.

It has been suggested that BR and ethylene interact indirectly in regulating shoot gravitropic responses by involving auxin signaling genes (Guo et al., 2008). BR activates *AUX/IAA* (negative regulator of auxin signaling) and inactivates *ARF7* and *ARF19* (positive regulator of auxin signaling), thus inhibiting shoot gravitropic responses. Alternatively, ethylene downregulates *AUX/IAA* and enhances the *ARF7* and *ARF19* genes to positively regulate shoot gravitropic responses (Figure 2.2; Vandenbussche et al., 2013). Therefore, ethylene and BR have been found to have opposite effects on the upward growth of etiolated shoots. Furthermore, ethylene–BR antagonism has also been observed in the case of roots. Ethylene reduces root gravitropic responses (Buer et al., 2006), while BR enhances root gravitropic bending, probably by modulating auxin transport (Kim et al., 2007; Vandenbussche et al., 2013). In BR-insensitive mutants, *bri1-301* and *bak1* delayed root growth, and a reduced response to the gravitropic stimulus was revealed (Kim et al., 2007). However, in ethylene insensitive mutants, *ein2-5* and *etr1-3* reduced inhibition toward root gravitropic responses (Buer et al., 2006), indicating antagonistic interaction between BR and ethylene in regulating gravitropic responses in plants. It has been observed that the exogenous application of BR enhanced ethylene biosynthesis in *Arabidopsis* seedlings (Hansen et al., 2009). BR upregulates the expression of *1-aminocyclopropane-1-carboxylate synthase* (*ACS*), the key gene required for ethylene production (Muday et al., 2012). Further, BR acts post-transcriptionally and also increases the stability of ACS proteins such as ACS5, ACS6, and ASC9 by preventing its ubiquitination mediated by 26S proteasome. Therefore, in response to various endogenous and exogenous signals, *ACS* is regulated by BR to continuously adjust ethylene biosynthesis in various tissues (Hansen et al., 2009). The synergistic interaction between ethylene and BR in regulating hyponastic growth has also been demonstrated (Polko et al., 2012). Ethylene is a key regulator of hyponastic growth, which is employed by plants to cope with biotic and abiotic stresses. *ROT3/CYP90C1* encodes an enzyme that mediates the C-23 hydroxylation of BR. A mutation in *ROT3* reduces hyponastic growth leading to the impairment of local cell expansion and inhibition of BR biosynthesis, indicating that hyponastic growth induced by ethylene is mainly regulated by BR (Polko et al., 2012).

It is well documented that abscisic acid (ABA) is required to inhibit seed germination and is also mandatory to establish seed dormancy during embryo maturation. On contrary, BR promotes seed germination indicating the antagonistic interaction between both these hormones (Steber and McCourt, 2001; Finkelstein et al., 2008). Genetic, physiological, and biochemical studies have revealed that BR and ABA can co-regulate the expression of hundreds of genes (Nemhauser et al., 2006; Zhang et al., 2009a). BR and ABA signaling mutants have been analyzed to investigate how ABA inhibits BR signaling (Zhang et al., 2009a). It has been observed that in BR biosynthetic and signaling mutants, such as *det2-1* and *bri1*, respectively, the effect of ABA on BR signaling does not rely upon BR perception, but depends on BIN2, a negative regulator of BR signaling (Zhang et al., 2009a). However, on analyzing ABA signaling mutants, it has been demonstrated that the regulatory effect of ABA on BR signaling largely depends on ABI2 and slightly on ABI1, a PP2C family serine/threonine phosphatases (Zhang et al., 2009a). This study indicates that ABA and BR crosstalk through BR signaling components such as BIN2 and ABA signaling components such as ABI1 and ABI2. Furthermore, BR and ABA have been suggested to play antagonistic roles in regulating seed germination and post-germinative growth processes (Hu and Yu, 2014). ABA inhibits while BR enhances seed germination and post-germinative growth processes. It has been observed that BIN2 positively regulates ABA responses by physically interacting with ABI5. Hence, the study confirms that BIN2 stabilizes ABI5 by phosphorylating it, thus mediating ABA responses during seed germination. However, BR application inhibits the regulation of ABI5 by BIN2 to antagonize ABA-mediated inhibition (Hu and Yu, 2014). Recently, mutant studies indicate a synergistic correlation between BR and ABA in inducing responses such as H_2O_2 production, *respiratory burst oxidase homolog1* (*RBOH1*) gene expression, NADPH oxidase activity, and mediating heat and oxidative stress tolerance (Zhou et al., 2014). In ABA biosynthetic mutant, *not*, BR induces a transient increase in these responses, however, in BR biosynthetic mutant *d^im*, ABA induced a strong and prolonged increase in these responses. These results indicate that ABA biosynthesis plays a key role in sustaining stress tolerance in BR-induced pathways in plants (Zhou et al., 2014). Preliminary studies indicate that BR and PA (polyamine) crosstalk is involved in enhancing the ability of stress tolerance potential of plants (Liu and Moriguchi, 2007). Furthermore, BR treatment maintains the optimum amount of spermidine concentration required for normal plant growth and specifically increases the production of putrescine required for stress tolerance but decreases the concentration of cadaverine which generates oxidative burst to counteract heavy metal stress (Takahashi and Kakehi, 2010). Furthermore, the co-application of Cu and BR also decreases cadaverine content, enhancing superoxide dismutase activity required for stress tolerance (Kuznetsov et al., 2009). It indicates the key role of BR–PA interaction in providing abiotic stress tolerance.

2.5 Brassinosteroids: SA and JA Crosstalk

The potential crosstalk between BR and SA is mediated via *non-expressor of pathogenesis-related genes 1* (*NPR1*) and *WRKY70*, encoding a transcription factor working downstream of NPR1 (Divi et al., 2010). SA mediated gene NPR1 is an essential module of 24-epibrassinolide mediated increase in thermo- and salinity tolerance in *Arabidopsis thaliana*. The existence of the crosstalk between BR and SA plays a pivotal role in the response of plants to biotic as well as abiotic stresses. It has been demonstrated that in tobacco as well as in rice that BR acts as an inducer of a broad range of disease resistance. It has been observed that BR enhances resistance to the fungal pathogen *Magnaporthe grisea* and the bacterial pathogen *Xanthomonas oryzae* in rice. Moreover, it has been suggested that in tobacco, enhancement in BR-mediated resistance does not require SA, which has further been consolidated by measuring SA accumulation and its analysis using *NahG* transgenic tobacco (Nakashita et al., 2003). It indicates that

BR and SA act independently in providing resistance against pathogens. Although crosstalk may exist between BR and SA signaling pathways for inducing resistance, it is distinct and acts independently (Nakashita et al., 2003). To enhance abiotic tolerance, the synergistic connection of BR and JA assumes key parts in plant growth. It has been shown that BR improves the JA level in rice under stress (Kitonaga et al., 2006), which unequivocally advances the declaration of thionin qualities encoding antimicrobial peptides, demonstrating a potential crosstalk point with these two phytohormones. Strikingly, the hindrance of JA incited accumulation of anthocyanins by brassinazole in *Arabidopsis* has been additionally detailed by BRs motioning on the JA pathway (Peng et al., 2001). The transcript levels of JA biosynthesis quality and the JA-initiated signaling gene were down-controlled when the BR focus was low. It has been reported that BR biosynthesis is controlled by the improved JA-antecedent, 12-oxo-phytodienoic destructive, and subsequently joins BR and JA pathway initiation (Nahar et al., 2013).

2.6 Physiological Roles of BRs in Plant Growth

BRs play important roles in regulating plant growth and development at very low concentrations, ranging from nanomolar to micromolar (Clouse and Sasse, 1998). They affect a multitude of physiological and metabolic processes, including the coordination of morphogenic events throughout plant ontogeny, from seed germination and seedling elongation to maturity and seed development. Specific physiological processes affected by BRs include cell elongation, division and differentiation, enhancement of crop yield, reproductive biology (flowering), senescence, induction of ethylene biosynthesis, root growth and development, pollen tube growth, proton pump activation, photosynthesis activation, and the antioxidant system (Cao et al., 2005; Houimli et al., 2008; Shahbaz et al., 2008). Brassinosteroids are known to promote the elongation of shoot tissues in a number of plants at very low concentrations. Wang and co-workers (1993) demonstrated that BRs could stimulate hypocotyl elongation by increasing wall relaxation without a concomitant change in wall mechanical properties in pak choi (*Brassica chinensis* L.). Plant hormones are thought to regulate the biosynthesis and activity of cell wall modifying enzymes and other proteins such as xyloglucan endotransglucosylase/hydrolase (XTHs), cellulose synthase, expansins, sucrose synthase, and glucanases, thereby regulating cell elongation. It has been reported that BRs are involved in the regulation of genes encoding XTHs in *Arabidopsis*, tomato (*Solanum lycopersicum* L.), soybean, and rice. In addition, physiological measurements revealed that BRs could stimulate wall loosening in epicotyls of soybean and hypocotyls of *Brassica chinensis* and *Cucurbita maxima* (Bishop and Koncz, 2002; Clouse and Sasse, 1998; Sakurai, 1999). The dwarf nature of BR-deficient mutants and the ability to return to normal phenotype with the application of BRs show the key role of BRs in plant growth and development. Some studies have demonstrated that BRs may also promote cell elongation by regulating the transport of water via aquaporins as well as regulating the activity of a vacuolar H^+-ATPase subunit (Friedrichsen and Chory, 2001; Morillon et al., 2001). It has been suggested that BRs play a key role in *Arabidopsis* cell division in mutant *det2* (*de-etiolated2*) suspension cultures, where it was shown that epibrassinolide (24-EBR) caused an increase in transcript levels of the gene encoding cyclin-D3, a regulatory protein of the cell cycle. Cyclin-D3 is also regulated by cytokinins, and it may be significant that 24EBR can efficiently substitute for zeatin (a naturally occurring cytokinin) in the growth of *Arabidopsis* callus and suspension cultures (Hu et al., 2000).

BRs have to control the two cellular processes, that is, cell division and cell expansion. BRs induce elongation of hypocotyls, epicotyls, and peduncles of dicots, as well as coleoptiles and mesocotyls of monocots (Cheon et al., 2010; Gonzalez-Garcia et al., 2011; Zhiponova et al., 2013). BR-induced cell expansion is accompanied by proton extrusion and the hyperpolarization of the cell membrane and stimulates and accelerates the growth of cell the cycle (Bajguz and Czerpak, 1996). BRs are involved in the process of cell enlargement through their effects on gene expression and enzyme activity (Mussig and Altmann, 1999). BRs increase cell division by increasing the transcript levels of genes encoding cyclin D3, a regulatory protein of the cell cycle. Cyclin-D3 is also regulated by cytokinins, and it may be significant that BRs can efficiently substitute for cytokinin in the growth of *Arabidopsis* callus and suspension cultures (Riou-Khamlichi et al., 1999; Hu et al., 2000). The role of cyclins and CDK genes has also been investigated in the early fruit development of tomato (Joubes et al., 1999, 2000).

2.7 Impact of BRs on Photosynthesis

Photosynthesis is the basis of plant growth and development, where plants convert light energy to chemical energy that is then used in various developmental activities. Chlorophyll is one of the basic units of photosynthesis and appropriate levels of chlorophyll are required for proper photosynthesis. Chlorophyll is an important component frequently used as an indicator of chloroplast development and photosynthetic activity (Shu et *al*., 2016). Chlorophyll is highly sensitive to external stimuli (stress) that decrease total chlorophyll a, b, and carotenoid contents in leaves (Rehman et al., 2016). Exogenous application of BRs significantly increased the chlorophyll content and enhanced photosynthetic characteristics of plants under low temperature and low light stresses (Yang et al., 2016; Deslauries et al., 2010). Moreover, BRs can regulate the combination of the chlorophyll molecule (by regulating chlorophyllase activity) with membrane protein and maintain stability of the thylakoid membranes. BRs regulate and maintain photosynthetic activity under water stress conditions in tomato seedlings. BRs alleviate the adverse effect of different stress conditions and regulate the defense system by regulating transcription levels of defense genes in cucumber (Li et al., 2013). Moreover, BRs regulate the Rubisco activate (*RCA*) gene, which plays a key role in photosynthesis under drought and heat stress in wheat. In addition, BR significantly increases the activities of antioxidant enzymes and the process of photosynthesis (Zhao et al., 2017), also BR-treated seedlings increased in O_2 assimilation and the quantum yield of photosystem II

(PSII) and ribulose-1, 5-bisphosphate carboxylase/oxygenase (Rubisco) activities and expression of Rubisco large subunit (*rbcL*) and rubisco small subunit (*rbcS*) genes to increase photosynthetic capacity (Xia et al., 2009). Thus, BRs promote the accumulation of chlorophylls and photosynthetic capacity by regulating a variety of enzymes including chlorophyllase and rubisco; transcript levels of encoded genes involved chlorophyll and photosynthesis under stress. Similarly, the total chlorophyll content or its fractions increased in the leaves of *Vigna radiata* (Bhatia and Kaur, 1997), *Brassica juncea* (Hayat et al., 2001a), and *Cucumis sativus* (Yu et al., 2004) by direct application of 28-homobrassinolide and 24-epibrassinolide respectively to their foliage. The treatment of EBL to the seeds of *Vigna radiata* (Yusuf et al., 2012) or foliar treatment of HBL to *Brassica juncea* (Hayat et al., 2000, 2001b) significantly increased the activity of carbonic anhydrase. Application of EBL also reduced the toxic effect of salinity and restored the activity of carbonic anhydrase in *Cucumis sativus* (Fariduddin et al., 2013). Under various abiotic stress conditions, BRs restored the activity of the nitrate reductase enzyme. They play a pivotal role in the supply of nitrogen and the growth and productivity of plants. Stress conditions like salinity inhibit nitrate transport due to the interference with nitrate uptake and xylem loading, which is finally attributed to reduced NR activity in leaves (Anuradha and Rao, 2003). After the application of BRs, the level of NR is increased in wheat (Hayat et al., 2001b), *Lens culinaris* (Hayat and Ahmad, 2003), pea (Shahid et al., 2011), *Vigna radiata* (Yusuf et al., 2012), and *Cucumis sativus* (Fariduddin et al., 2014b). Liu et al. (2014) reported that BR increased essential inorganic ions, decreased toxic ions, and promoted ion homeostasis, especially in leaves, roots, and epicotyls of canola under salt stress. Shu et al. (2016) reported that the application of 24-epibrassinolide enhanced the activity of nitrate reductase, nitrite reductase, glutamine synthetase, glutamate synthase, and glutamate dehydrogenase enzymes, and induced photosynthetic characteristics of tomato seedlings under stress condition. Steber (2001) noted that BR application increased H^+- ATPase and $Ca2^+$- ATPase activities in root and leaf, which are responsible for establishing an electrochemical potential gradient to maintain ion balance in plants to alleviate the stress effect. Foliar application and seed treatments of BR significantly enhanced the growth of tomato, as well as the number of fruits and the weight per plant (Zaharah et al., 2012; Thussagunpanit et al., 2015). BR played a positive role in fruit ripening and fruit growth of mango and in the quality of pitaya (Zaharah et al., 2012; Li et al., 2013). Brassinolide has been used to improve the yield of lettuce, radish, bush bean, and pepper (Meudt et al., 1983, 1984). Foliar application of dilute aqueous solution of BL also improved the yield of wheat and mustard (Braun and Wild, 1984), rice, corn, and tobacco (Yokota and Takahashi, 1986). BRs were also found to increase the growth and yield of sugar beet (Schilling et al., 1991), legumes (Kamuro and Takatsuto, 1991), and rape seed (Hayat et al., 2000, 2001b, 2003b). Application of 28-homobrassinolide and 24-epibrassinolide significantly increased the yield of potato, mustard, rice and cotton, *Lens culinaris*, and *Vigna radiata*, and that of corn, tobacco, watermelon, cucumber, and grape, respectively (Ikekawa and Zhao, 1991; Ramraj et al., 1997; Hayat and Ahmad, 2003a, b; Fariduddin et al., 2003).

2.8 Role of BRs in Ion Homeostasis

Ionic concentration in cell compartments and tissues is kept within the limits that are ideal for cell performance. Homeostasis signifies the ability of living organisms to regulate the concentration of mineral ions within a defined space despite their fluctuating concentrations in their surroundings. Ion homeostasis provides optimum conditions for enzyme activity, maintains the turgor pressure of cells, and plays a vital role in cell signaling (Ashraf, 1994, 2004; Munns et al., 2006). Salinity stress adversely affects ion homeostasis in cells. BRs directly or indirectly help plants to maintain ion homeostasis. BRs have been found to have a positive effect on the activity of high-affinity K^+ transporters and are associated with the reduction in Na^+ and enhancement in K^+ concentration. Brassinosteroids have also been found to improve the Ca^{2+}/Na^+ and K^+/Na^+ ratios of wheat cultivars by enhancing Ca^{2+} and K^+ uptake, and thus enhancing salt tolerance (Qasim et al., 2006). Treatment with EBRs increased the expression of various hormone marker genes in both wild-type and mutant seedlings. BRs exert antistress effects independently and through interactions with other hormones as well as sharing transcriptional targets with other hormones.

2.9 BR Molecular Mechanism and Mode of Action

BRs play an important role in the regulation of growth and differentiation in plants. The identification of BR-responsive genes proved to be highly useful for the exploration of the molecular mode of action of BRs as described by several authors (Müssig and Altmann, 2003; Nemhauser and Chory, 2004; Vert et al., 2005). Many genes involved in BR signal perception and downstream signaling have been identified. Some of the researchers reported that the BRI1 gene interacts with the LRR receptor kinase BAK1 (Li, 2003; Vert et al., 2005; Wang and He, 2004). BAK1 and its homologs may be co-receptors of BRI1 or represent direct BRI1 targets that initiate signaling events (Vert et al., 2005). BR binding to BRI1 inactivates the BIN2 kinase via unknown intermediate steps. The BES1 transcription factor acts downstream from BIN2 (Li and Deng, 2005). BIN2 phosphorylates BES1, thus targeting the protein for ubiquitination and subsequent proteasome-dependent degradation. The BZR1 protein is closely related to BES1 and represses BR biosynthetic genes. The abundance of the BRZ1 protein is also affected by BIN2 kinase. Furthermore, BES1 homologs (BEH1-4, BES1/BZR1 homolog 1-4) function redundantly in BR signaling (Yin et al., 2005). BR biosynthetic pathways consist of two major parts: sterol biosynthesis and a BR-specific pathway. Analysis of BR metabolic gene expressions indicates that BR homeostasis is maintained through feedback expressions of multiple genes, each of which is involved not only in BR-specific biosynthesis and inactivation, but also in sterol biosynthesis (Tanaka et al., 2005). Nine BR *dwarf* loci in *Arabidopsis*, *dwf1* to *dwf8*, and *dwf12*, have been reported. The *dwf1*, *dwf5*, and *dwf7* mutants are defective in sterol biosynthesis, and a second group including *dwf3*, *dwf4*, *dwf6*, and *dwf8* belong to the BR-specific pathway. Only *dwf2* and *dwf12* mutants are insensitive to bioactive BRs. The *dwf1* was the first BR-related gene identified (Choe et al., 1998). The three alleles *dwarf1* (*dwf1*), *dim*, and *cbb1* were defective in the synthesis of campesterol from 24-methylenecholesterol (Takahashi et al., 1995; Klahre et al., 1998). The *dwf3* mutants have only been rescued by 23-alpha-hydroxylated BRs. These mutants have also been found to be allelic to *cpd* (Choe et al., 1998). The *dwf4* mutant was also shown to be defective in the BR biosynthetic pathway, more specifically in a steroid 22α-hydroxylation (CYP90B1), presenting 513 amino acids and 43% identity and 66% similarity with the *cpd* gene (Choe et al., 1998), which catalyzed a key regulatory step in BRs biosynthesis (Choe et al., 2001). Choe and co-workers (2001) have shown that transgenic *Arabidopsis* plants overexpressing *dwf4* (*aod4*) presented a dramatic increase in hypocotyls length in both light and dark grown as compared with wild type; *dwarf5* (*dwf5*) has been shown to be defective in the reduction of 5-dehydroepisterol to 24-methylenecholesterol (Bishop and Yokota, 2001). Clouse and colleagues (1996) identified the first BL insensitive mutant, the BL insensitive1 (*bri1*) gene. BRI1 has been cloned and shown to encode a receptor kinase with an extracellular domain (Li and Chory, 1997), which appears to contain 24 rather than 25 leucine rich repeats (LRRs), with LRR21 (formerly LRR22) being an unusual methionine-rich repeat (Vert et al., 2005). Friedrichsen and colleagues (2000) demonstrated that a BRI1::GFP (GFP, green fluorescent protein) fusion protein was located at the plasma membrane, which, along with the protein acting Ser/Thr phosphorylation, suggested that BRs were perceived at the cell surface. A dramatic increase in BL binding activity in the membrane fractions of the BRI1::GFP transgenic plants was also found, which was due to an increase in binding sites with similar binding affinities (Wang et al., 2001). Since all of the over 20 BR-insensitive mutants reported to date, such as *cbb2*, 18 *bin*, and three alleles of *dwf2*, were all allelic to *bri1* (Clouse and Sasse, 1998; Li and Chory, 1999), one might conclude that BRI1 was the only unique and specific component of the BRs signal transduction pathway; Li and co-workers (2002) identified in *Arabidopsis* a dominant genetic suppressor of *bri1*, *bak1-1d* (bri1-associated receptor kinase1-1 dominant). This gene encodes an LRR-RLK serine/threonine protein kinase, which interacts with *bri1*. *bri1* and *bak1* can phosphorylate each other, being the autophosphorylation activity of *bak1* enhanced by *bri1*. Expression of a *bak1* dominant-negative mutant allele results in a severe dwarf phenotype, mimicking the phenotype of null *bri1* alleles. The BRI1-BAK1 receptor complex is now thought to initiate BR signaling (Russinova et al., 2004). *bak1* acts as a co-receptor and/or downstream target of *bri1* (Vert et al., 2005). He et al. (2005) noted that *bzr1-1d* suppresses BR-deficient and BR-insensitive (*bri1*) phenotypes. BZR1 is a transcriptional repressor that binds directly to the promoters of feedback-regulated BR biosynthetic genes (He et al., 2005). The BZR1 protein accumulates in the nucleus of elongating cells of dark-grown hypocotyls and has been shown to be a positive regulator of the BR signaling pathway (Wang et al., 2002). Thus, BZR1 coordinates BR homeostasis and signaling by playing dual roles in regulating BR biosynthesis and downstream BR responses. The BZR1-BES1 family of proteins directly binds to and regulates BR-responsive genes, which establish a link between hormonal signal transmission in the cytoplasm and transcriptional status change in the nucleus (Li and Deng, 2005). The *exordium* (*exo*) protein has been identified as a regulator of BR-responsive genes in *A. thaliana* (Coll-Gracia et al., 2004). The *exo* gene was characterized as a BR-upregulated gene; *exo* overexpression resulted in increased transcript levels of the BR-up-regulated *kcs1*, *exp5*, *delta-tip*, and *agp4* genes, thought to be involved in the mediation of BR-promoted growth. It has also been noted that *exo* overexpressing lines showed enhanced vegetative growth, resembling the features of BR-treated plants. Song et al. (2019) reported that GmBZL3 acts as a major BR signaling regulator through crosstalk with multiple pathways in *Glycine max*. Transcription factors like BZR1 and BES1/BZR2 are well defined as downstream regulators of the BR signaling pathway in *Arabidopsis* and rice. Soybean contains four BZR1-like proteins (GmBZLs), and it was reported that GmBZL2 plays a conserved role in BR signaling regulation. GmBZL3 might play conserved roles during soybean development. The overexpression of GmBZL3P219L in the *Arabidopsis* BR-insensitive mutant bri1–5 partially rescued the phenotypic defects including BR insensitivity, which provides further evidence that GmBZL3 functions are conserved between soybean and the homologous *Arabidopsis* genes. In addition, the identification of the GmBZL3 target genes through ChIP-seq technology revealed that BR has broad roles in soybean and regulates multiple pathways, including other hormone signaling and disease-related and immune response pathways. Moreover, BR-regulated GmBZL3 target genes are a major transcription factor responsible for BR-regulated gene expression and soybean growth. A comparison of GmBZL3 and AtBZR1/BES1 targets demonstrated that GmBZL3 might play conserved as well as specific roles in the soybean BR signaling network. Jia et al. (2020) reported the molecular mechanism and mode of action of brassinosteroids in the reproductive process of *Arabidopsis thaliana*. They suggested that outer integument growth and embryo sac development were impaired in the ovules of the *Arabidopsis thaliana* BR receptor null mutant bri1-116. On the basis of gene expression and RNA-seq analyses, it was shown that the expression of INNER NO OUTER (INO), an essential regulator of outer integument growth, was significantly reduced in the bri1-116 mutant. INO expression was increased as a result of overexpression of transcriptional activity of BZR1 in the mutant that alleviated the outer integument growth defect in bri1-116 ovules, suggesting that BRs regulate outer integument growth partially via BZR1-mediated transcriptional regulation of INO. INO expression in *bzr-h*, a null mutant for all BZR1 family genes, was barely detectable; and the outer integument of *bzr-h* ovules had much more severe growth defects than those of the bri1-116 mutant. They noted that BRs have a new role in regulating ovule development and suggest that BZR1 family transcription factors might regulate outer integument growth through both BRI1-dependent and BRI1-independent pathways.

2.10 Conclusion

BRs can act efficiently in plants as immune modulators when applied at the appropriate concentration at the definite stage of plant development. BRs are implicated in plant responses to abiotic and biotic environmental stresses. BRs regulate the stress response as a result of a complex sequence of biochemical reactions such as the activation or suppression of key enzymatic reactions, induction of protein synthesis, and the production of various chemical defense compounds. BRs open up new approaches for plant resistance against hazardous environmental conditions. It is concluded that BRs are effectively involved in diverse metabolic processes including photosynthesis, antioxidant metabolism, osmotic regulation, nitrogen metabolism, and plant–water relations under normal and stressful conditions. The exogenous application of BRs has considerable influence on cell division, cell elongation, reproductive development, vascular differentiation, stress tolerance and pathogen resistance, fruit set, yield, quality, and so on. Based on the evidence of various research publications, it has been concluded that BRs and their analogs are the sixth group of phytohormones and help in the proliferation of root systems with the uptake of water and nutrients that can sustain crop yield. Recent progress in the chemical synthesis of brassinosteroids and their analogs has led us to economically feasible approaches for large-scale applications to give higher crop yields.

REFERENCES

Aghdam et al. *Sci Hort.* 2012; 144:116–120.
Ahanger and Agarwal. *Protoplasma.* 2017a; 254:1471–1486.
Ahanger and Agarwal. *Plant Physiol Biochem.* 2017b; 115:449–460.
Ahmad et al. *J Plant Inter.* 2017a; 12:429–437.
Ahmad et al. *J Plant Growth Regul.* 2017b; 37:309–322.
Ali et al. *Environ Exp Bot.* 2008; 62:153–159.
Allen and Ptashnyk. *J Theoretical Biol.* 2017.432:109–131.
Anuradha and Rao. *Plant Growth Regul.* 2003; 40:29–32.
Anuradha and Rao. *Plant Soil Environ.* 2007; 53:465–472.
Ashraf. *Crit Rev Plant Sci.* 1994; 13:17–42.
Ashraf. *Flora.* 2004; 199:361–376.
Bajguz and Czerpak. *J Plant Growth Regul.* 1996; 15:153–156.
Bajguz and Hayat. *Plant Physiol Biochem.* 2009; 47:1–8.
Bajguz and Piotrowska-Niczyporuk. *Plant Physiol Biochem.* 2014; 80:176–183.
Bao et al. *Plant Physiol.* 2004; 134:1624–1631.
Behnamnia. *Int J Agri Crop Sci.* 2015; 8:455–462.
Benjamins and Scheres. *Annu Rev Plant Biol.* 2008; 59:443–465.
Bishop and Koncz. *Plant Cell.* 2002; 14:97–110.
Bishop and Yokota. *Plant Cell Physiol.* 2001; 42:114–120.
Braun and Wild. *J Plant Physiol.* 1984; 116:189–196.
Buer et al. *Plant Physiol.* 2006; 140:1384–1396.
Calzadilla et al. *Front Plant Sci.* 2016; 7:374.
Cao et al. *Physiol Plant.* 2005; 123:57–66.
Chaiwanon and Wang. *Curr Biol.* 2015; 25:1031–1042.
Cheon et al. *BMC Plant Biol.* 2010; 10:270.
Choe et al. *Plant J.* 2001; 26 573–582.
Choudhary and Yu. *Trends Plant Sci.* 2012; 17:594–605.
Clouse. *J Plant Growth Regul.* 2015; 34:828–844.
Clouse and Sasse. *Annu Rev Plant Physiol Plant Mol Biol.* 1998; 49:427–451.
Clouse et al. *Plant Physiol.* 1996; 111:671–678.
Coban and Baydar. *Ind Crops Prod.* 2016; 86:251–258.
Coll et al. *Biologia.* 2015; 70:726–732.
Coll-Garcia et al. *FEBS-Letters.* 2004; 563:82–86.
Davies. Martinus Nijhoff Publishers, Dordrecht. 1987.
De Vleesschauwer et al. *Plant Physiol.* 2012; 158:1833.
Dhaubhadel et al. *Plant Mol Biol.* 1999; 40:333–342.
Divi and Krishna. *Nat Biotechnol.* 2009; 26:131–135.
Divi et al. *BMC Plant Biol.* 2010; 10:151.
El-Mashad and Mohamed. *Protoplasma.* 2012; 249:625–635.
Fariduddin et al. *Acta Bot Croatia.* 2003; 65:19–23.
Fariduddin et al. *Acta Physiol Plant.* 2009; 31:889–897.
Fariduddin et al. *Environ Monitoring Assessment.* 2013; 185:7845–7856.
Fariduddin et al. *Photosynthetica.* 2014; 52:464–474.
Finkelstein et al. *Annu Rev Plant Biol.* 2008; 59:387–415.
Friedrichsen. *Plant Physiol.* 2000; 123:1247–1256.
Friedrichsen and Chory. *Bioassays.* 2001; 23:1028–1036.
Gonzalez-Garcia et al. *Development.* 2011; 138:849–859.
Gray. *PLoS Biol.* 2004; 2: E311.
Grove et al. *Nature.* 1979; 281:216–217.
Gruszka. *Int J Mol Sci.* 2013; 14 8740–8774.
Guo et al. *Plant Mol Biol.* 2008; 66:345–360.
Hansen et al. *Plant J.* 2009; 57:606–614.
Hao. *Plant Cell Rep.* 2013; 32:1017–1030.
Hatfield and Prueger. *Weather Climate Ext.* 2015; 10:4–10.
Hayat and Ahmad. *Acta Agron Hung.* 2003; 51:381–387.
Hayat and Ahmad. Springer, Dordrecht. 2011.
Hayat et al. *Photosynthetica.* 2000; 38:469–471.
Hayat et al. *Acta Physiol Plant.* 2001; 23:27–30.
Hayat et al. *Photosynthetica.* 2001; 38:469–471.
Hayat et al. *Environ Exp Bot.* 2007; 0:33–41.
Hayat et al. *Environ Exp Bot.* 2010; 69(2):105–112.
Hayat et al. *Saudi J Biol Sci.* 2012; 19:325–335.
Hothorn et al. *Nature.* 2011; 474:467–471.
Houimli et al. *Eur Asia J BioSci.* 2008; 2:83–90.
Hu and Yu. *Plant Cell.* 2014; 26:4394–4408.
Hu et al. *Plant J.* 2000; 24:693–701.
Hu et al. *Agric Sci China.* 2011; 10:1168–1176.
Hu et al. *Front Plant Sci.* 2017; 8:1039.
Janeczko et al. *Biol Plant.* 2007; 51:355–358.
Janeczko et al. *J Agron Crop Sci.* 2010; 196:311–321.
Jatav et al. *J Indian Bot Soc.* 2014; 93:177–189.
Jiang et al. *Physiol Plant.* 2013; 48:133–145.
Jia et al. *J Integr Plant Biol.* 2020; 62 (8):1093–1111.
Joubes et al. *Plant Physiol.* 1999; 121:857–869.
Kagale et al. *Planta.* 2007; 225:353–364.
Karlidag et al. *Sci Hortic.* 2012; 130:133–140.
Kaur et al. *Sci. Rep.* 2018; 8:8735.
Kazemi-Shahandashti et al. *J Plant Physiol.* 2014; 171:1106–116.
Khripach et al. *Annals Botany.* 2000; 86:441–447.
Kim and Wang. *Annu Rev Plant Biol.* 2010; 61:681–704.
Kim et al. *Plant Cell Environ.* 2007; 30:679–689.
Kissoudis et al. *Front Plant Sci.* 2014; 5:207.
Kitanaga et al. *Biosci Biotechnol Biochem.* 2006; 70:2410–9.
Klahre et al. *Plant Cell.* 1998; 10:1677–1690.
Krumova et al. *J Photochem Photobiol.* 2013; B126:97–104.
Kuznetsov et al. *J Plant Physiol.* 2009; 166:40–51.
Li. *Curr Opin Plant Biol.* 2003; 6:494–499.
Li and Chory. *Cell.* 1997; 90:929–938.
Li and Chory. *J Exp Bot.* 1999; 50:275–282.
Li and Deng. *Trends Plant Sci.* 2005; 10:266–268.
Li and Nam. *Science.* 2002; 295:1299–1301.

Li et al. *Plant J.* 2009; 58:275–86.
Li et al. *Sci Signal.* 2012a; 5(244):72–82.
Li et al. *Bio Plant.* 2012b; 56:192–196.
Li et al. *Nanfang Nongye Xuebao.* 2013; 44:1150–1153.
Li et al. *Front Plant Sci.* 2016; 7:1–16.
Liu and Moriguchi. *Plant Cell Rep.* 2007; 26:125–131.
Liu et al. *Plant Growth Regul.* 2009; 59:207–214.
Liu et al. *Plant Biol.* 2014; 16:440–50.
Luan et al. *South Africa J Enol Vitic.* 2016; 34:196–203.
Manoli et al. *Plant Growth Regul.* 2018; 84:423–436.
Mazorra et al. *Plant Physiol Biochem.* 2011; 49:1420–1428.
Meudt et al. *Proc Plant Growth Reg Soc Am.* 1983; 10:312–318.
Meudt et al. Canadian Patent No. 1173659. *Assigned to USA.* 1984; 11.
Morillon et al. *Planta.* 2001; 212:199–204.
Muday et al. *Trends Plant Sci.* 2012; 17:181–195.
Munns et al. *J Exp Bot.* 2006; 57:1025–1043.
Mussig and Altmann. *Plant Physiol Biochem.* 1999; 37:363–372.
Mussig and Altmann. *J Plant Growth Regul.* 2003; 22:313–324.
Nahar et al. *Molecular Plant Microbe Interaction.* 2013; 26:106–15.
Nakashita et al. *Plant J.* 2003; 33:887–898.
Nam and Li. *Cell.* 2002; 110:203–212.
Nemhauser and Chory. *J Exp Bot.* 2004; 55:265–270.
Nemhauser et al. *Cell.* 2006; 126:467–475.
Ogweno et al. *J Plant Growth Regul.* 2008; 27:49–57.
Peng et al. *J. Integr Plant Biol.* 2001; 53:632–40.
Polko et al. *J. Exp. Bot.* 2012; 64:613–624.
Qasim et al. *Pak J Bot.* 2006; 38:1621–1632.
Ramraj et al. *J Agri Sci.* 1997; 128:405–413.
Rao et al. *Curr Sci.* 2002; 82:1239–1245.
Riou-Khamlichi et al. *Science.* 1999; 283:1541–1544.
Russinova et al. *Plant Cell.* 2004. 16:3216–3229.
Sahni et al. *Scientific Reports.* 2016; 6:28298.
Saini et al. *Front Plant Sci.* 2015; 6:1–17.
Saini et al. *Plant Cell Rep.* 2013; 32:741–757.
Sakurai, ed. Springler- Verlag, Tokyo. 1999; 91–111.
Santiago et al. *Science.* 2013; 341:889–892.
Schilling et al. *Bioactivity and Applications.* 1991; 208–219.
Shahbaz and Ashraf. *Pak J Bot.* 2007; 39:513–522.
Shahbaz et al. *Plant Growth Regul.* 2008; 55:51–64.
Shahid et al. *Aust J Crop Sci.* 2011; 5:500–510.
She et al. *Nature.* 2011; 474:472–476.
Shu et al. *Plant Physiol Biochem.* 2016; 107:344–53.
Siboza et al. *J Plant Physiol.* 2014; 171:1722–1731.
Song et al. *Biol Plant.* 2016; 60:329–342.
Song et al. *BMC Plant Biol.* 2019; 19:86–101.
Sun et al. *Dev Cell.* 2010; 19:765–777.
Sun et al. *Cell Res.* 2013; 23:1326–1329.
Takahashi and Kakehi. *Ann Bot.* 2010; 105:1–6.
Takahashi et al. *Genes Develop.* 1995; 9:97–107.
Takeuchi et al. *Proc Jpn Soc Chem Regul Plants.* 1996; 31:100–1.
Tang et al. *BMC Biol.* 2016; 14:113–118.
Thussagunpanit et al. *Photosynthetica.* 2015; 53(2):312–20.
Tong et al. *Plant Cell.* 2014; 26:4376–4393.
Unterholzner et al. *Plant Cell.* 2015; 27:2261–2272.
Upreti and Murti. *Biol Plant.* 2004; 48:407–411.
Vandenbussche et al. *Am J Bot.* 2013; 100:215–225.
Vert et al. *Annu Rev Cell Dev Biol.* 2005; 21:177–201.
Vert et al. *Proc Natl Acad Sci.* 2008; 105:9829–9834.
Wang and He. *Trends Plant Sci.* 2004; 9:91–96.
Wang et al. *Plant Physiol.* 1993; 101:965–968.
Wang et al. *Nature.* 2001; 410:380–383.
Wang et al. *Dev Cell.* 2002; 2:505–513.
Wang et al. *Dev Cell.* 2008; 15:220–235
Wei and Li. *Mol Plant.* 2016; 9:86–100.
Wei et al. *Front Plant Sci.* 2015; 6:982.
Wu et al. *Front Plant Sci.* 2017; 8:1017.
Xi et al. *Plant Growth Regul.* 2013; 71:57–65.
Yadava et al. *Nat Sci.* 2016; 8:171–179.
Yang et al. *Mol Plant.* 2011; 4:588–600.
Yokota and Takahashi. In: M. Bopp (eds.), *Plant Growth Substances.* 1986; 129–138.
Yuan et al. *J Plant Physiol.* 2015a; 188:29–36.
Yuan et al. *Biol Plant.* 2015b; 59:99–105.
Yusuf et al. *Plant Physiol Biochem.* 2012; 57:143–153.
Zaharah et al. *J Plant Growth Regul.* 2012; 31:363–72.
Zhang et al. *Proc Natl Acad Sci.* 2009; 106:4543–4548.
Zhao et al. *Biol Plant.* 2016; 60:563–571.
Zhiponova et al. *New Phytol.* 2013; 197:490–502.
Zhou et al. *J Exp Bot.* 2014; 65:4371–4383.

3

Brassinosteroids: Crucial Regulators of Growth under Stress

Neha Dogra, Gurvarinder Kaur, Isha Madaan, Sarvajeet Singh Gill, and Geetika Sirhindi

CONTENTS

3.1 Introduction: Discovery and Physiological Roles

Plants are known to synthesize numerous biologically important compounds that enable them to flourish throughout their life span. These phytohormones regulate a diverse array of physiological responses and also help the plant to develop strategies against unfavorable environmental conditions. Yet, among the known distinct phytohormonal classes, no steroidal compounds were reported in plants until the mid-twentieth century, even though a number were found in animals. However, 1970 proved to be the golden era in the history of the plant community as the foundations for the search for plant steroids were laid. Mitchell et al. (1970) reported the presence of new compounds in the pollens of rape seeds (*Brassica napus* L.) that demonstrated growth-promoting activities even at low concentrations. They named this new class of compounds "Brassins." This discovery of the novel plant hormone triggered research in the exploration of its biological activities and the milestone of the discovery of the first plant steroidal was achieved nine years later. Grove et al. (1979) reported the role of newly discovered compounds "Brassinolides" in bee collected pollens of *Brassica napus* L. and further unveiled the structure of these compounds, which was a steroidal lactone (22*R*, 23*R*, 24*S*)-2α, 3α, 22, 23-tetrahydroxy-24-methyl-B-homo-7-oxa-5α-cholesta n-6-one) with the help of stereoscopic analysis and X-ray diffraction. The author also performed bean second intermodal bioassay in order to test the biological significance of the compound. Crystalline forms of brassinolides (10 ng per plant) led to a 200% increase in the inter-node elongation as compared with untreated. The research in the field of plant steroidal has been quick and promising. Within three years of the discovery of brassinolides, the biological precursor of brassinolides was discovered from the insect galls of *Castanea crenata* and was named "castasterone" (Yokota et al., 1982). Today, numerous plant steroids have been identified that are produced naturally

DOI: 10.1201/9781003110651-3

and these constitute the newer class of plant steroidal hormones named "brassinosteroids."

Brassinosteroids (BRs) are the class of polyhydroxy steroidal lactones that regulate multidimensional roles in plants from germination to senescence not only in normal circumstances but also under adverse conditions of biotic and abiotic stress (Bajguz and Hayat, 2009). The discovery of a young sixth novel class of hormones led to excitement in plant researchers whose properties and biological activities can be compared with that of animal steroids like ecdysteroids (Bajguz and Tretyn, 2003). It was later discovered that BRs have plasma membrane-based receptor kinases whereas animal steroids have nuclear receptors where they bind and are directly involved in transcriptional level modulation in order to produce a physiological response (Caño-Delgado et al., 2004). Out of the numerous known BRs, only three BRs (brassinolide, 28-homobrassinolide, and 24-epibrassinolide) are imperative to play a role in regulating major biological activities (Figure 3.1; Vardhini and Anjum, 2015) and it is the steroidal lactone that is responsible for the biological activity of BRs (Rao et al., 2002). BRs are 5α-cholestane derivatives along with a carbon side chain. They display large structural differences in their side chain from which numerous BRs have been identified and are still under research. BRs can be classified on the basis of the number of carbon atoms as C_{27}, C_{28}, and C_{29} BR compounds; the most studied belong to the C_{27} family (Kanwar et al., 2017). BRs exist as both free and conjugated forms. Conjugated forms of BRs with fatty acids and sugars can be found (Hayat et al., 2003).

BRs are widespread in the plant kingdom and are profusely distributed among angiosperms, gymnosperms, and lower plants like pteridophytes, bryophytes, and algae. They are found at a higher concentration in the juvenile parts of the plants as compared with mature tissues. They can evidently be found in flowers, pollens, seeds, stems, leaves, and roots. It was reported that the content of BRs was much higher in pollens (1–100 ng g^{-1}fresh weight) compared with a lower concentration (0.01–0.1 ng g^{-1}fresh weight) in shoots and leaves (Bajguz and Tretyn, 2003). High concentrations like 6.4 mg of BRs (Typhasterol) per kilogram of pollens have also been reported in *Cupressus arizonica* (Fujioka, 1999).

BRs initiate a cascade of bio-signaling pathways even at low concentrations in order to produce a response to a stimulus. The light to BR signaling pathway will be shown later in the chapter. It helps in controlling various biochemical, physiological, and molecular processes. The mechanisms underlying this cascade from cell surface reception to the regulation of transcription factors by BRs have been elucidated in numerous studies. The earliest experimental studies conducted in understanding the role of BRs have thrown light on the cell elongating function of BRs, however, in-depth research and further exploration have led to the unraveling of many more functions of BRs that make them a distinct class of hormones. Numerous reports regarding BR-insensitive or BR-deficient mutants suggest that BRs regulate the growth of the plant endogenously as well as show positive results when applied exogenously even in low concentrations. Vriet et al. (2012) reported the functions of plant steroids in various developmental as well as physiological processes, such as germination, photomorphogenesis

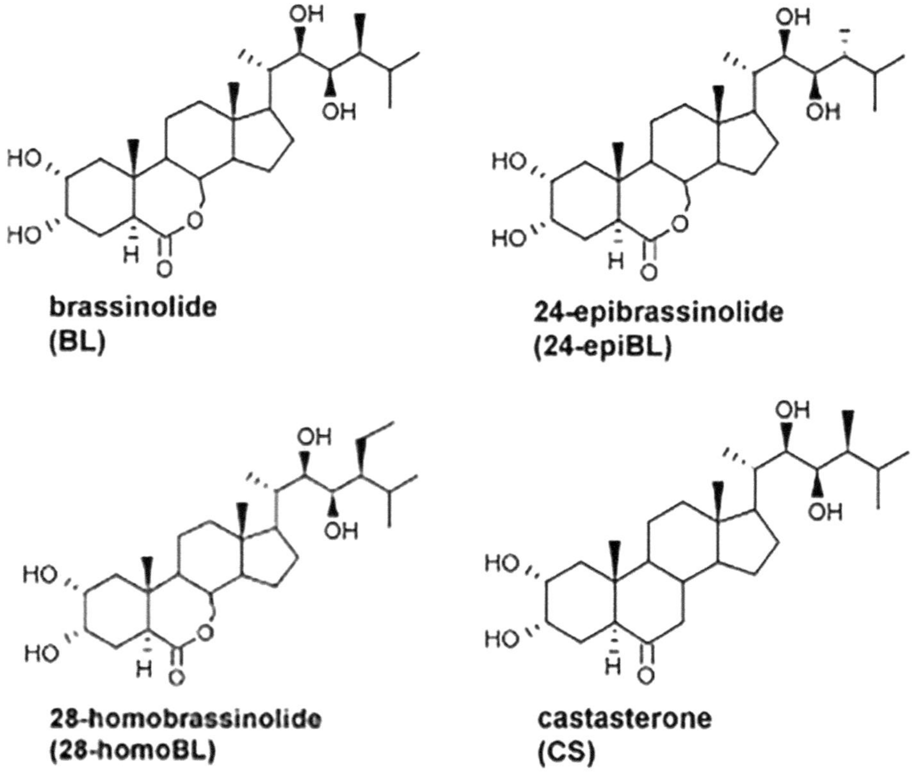

FIGURE 3.1 The biologically important ring structure of brassinolide, 24-epibrassinolide, 28-homobrassinolide, and the BR biosynthetic precursor, castasterone (Bajguz, 2011).

of seedlings, stem and root elongation, growth and differentiation, male fertility, epinasty, and flower and fruit induction and their development. The endogenous level of BRs or their exogeneous application have proved to be fruitful during the germination of seeds just like the gibberellic acid in tobacco seeds (Leubner-Metzger, 2003). BRs also promote cell division and cell elongation in plants and both these processes are important in the overall growth of the plant (Gudesblat and Russinova, 2011). BR signaling plays a significant role in regulating various developmental processes like seed development (Jiang et al., 2013), pollen development (Ye et al., 2010), and time of flowering (Domagalska et al., 2010). BRs have an important role to play in agricultural crops, such as regulating the height of the plant and the leaf angle as well as the inflorescence; these three processes are potential determinants of the yield in plants (Yang et al., 2018; Sakamoto et al., 2006 and Yamamuro et al., 2000). Another important process BRs are involved in is the accumulation of nutrients in plants. For instance, in the case of *Vitis vinifera*, BRs induce the accumulation of total soluble sugars in berries where genes encoding mono- and disaccharide transporters are upregulated. BRs also aid in positively regulating the activities of various enzymes like suc synthesases and invertases for sugar accumulation (Xu et al., 2015). BRs also stimulate the production of wood as reported in *Populus trichocarpa* (Jin et al., 2017) as well as fibers in *Gossypium hirsutum* (Yang et al., 2014). In addition, plants face numerous challenges in their natural environment due to fluctuating environmental conditions, and in these stress conditions, BRs have proven themselves as potent amelioraters. The various conditions of stress along with their adverse effects and the role of BRs in stress tolerance are explained further in the text.

3.2 Insights into BR Signaling

BRs are perceived by BR-INSENSITIVE 1 (BRI1) and also by its homologs BRI1-LIKE 1 (BRL1) and BRI1-LIKE 3 (BRL3), a family of plasma membrane localized leucine-rich repeat receptor kinases (Clouse et al., 1996, Caño-Delgado et al., 2004, Li and Chory, 1997) along with co-receptor BRI1-ASSOCIATED KINASE 1 (BAK1) and SOMATIC EMBRYOGENESIS RECEPTOR KINASES (SERKs; Nam and Li, 2002, Li et al., 2002, Gou et al., 2012). When levels of BR are low, BR signaling is restrained by several mechanisms (Figure 3.2, left side). Initially, BR KINASE INHIBITOR 1 (BKI1) associates with BRI1 and then prevents BRI1–BAK1 interactions (Wang and Chory, 2006). Second, a GSK3-like kinase, BR-INSENSITIVE 2 (BIN2), phosphorylates a collection of substrates (Youn and Kim, 2015) counting BES1/BZR1 family transcription factors and functions as master regulators of the BR pathway. The phosphorylation of BES1 and BZR1 leads to inactivation through mechanisms, incorporating cytoplasmic retention via an interface with 14-3-3 proteins (Gampala et al., 2007; Ryu et al., 2007), reduction in binding of DNA (Vert and Chory, 2006) and also protein degradation (Yin et al., 2002, Wang et al., 2011). When there is a presence of BR, it binds to BRI1 and BAK1 to initiate a

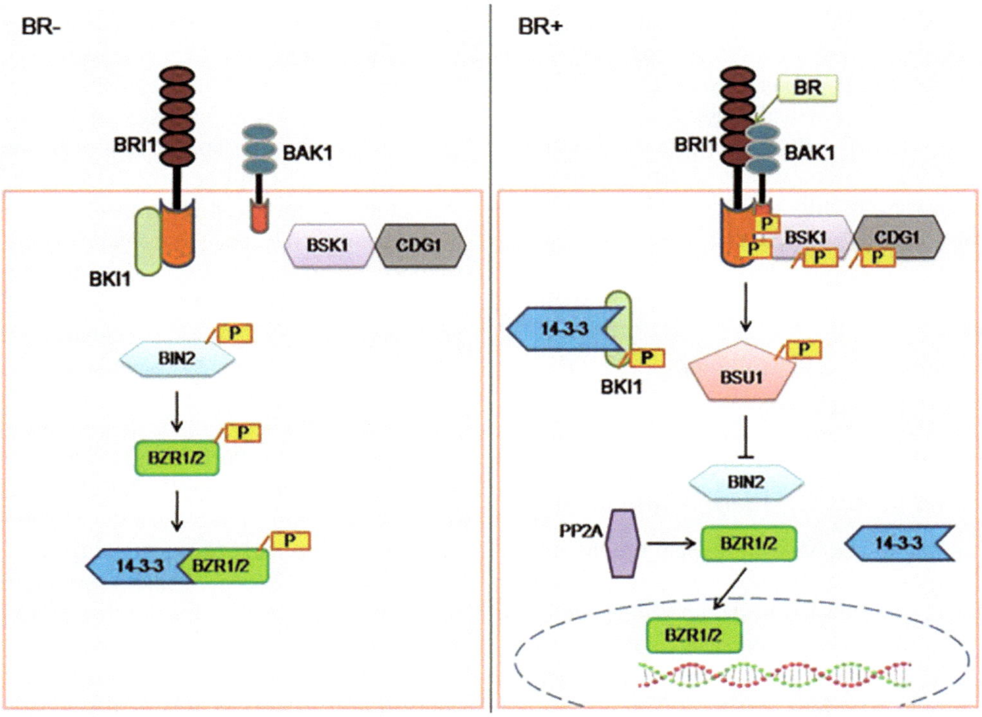

FIGURE 3.2 Brassinosteroid signal transduction pathway in the absence (a) and presence (b) of BR. BR binding induces BRI1 heterodimerization with co-receptor BAK1, resulting in the activation of BRI1, phosphorylation of BKI1 by BRI1, and then dissociation of BKI1 from BRI1. These proceedings lead to the activation of BRI1, which phosphorylates BSK1 and CDG1. Activated BSK1 and CDG1 promote activation of phosphatase BSU1 and thus dephosphorylating and inactivating BIN2. Inactivated BIN2 allows the presence of the dephosphorylated form of BZR1/2, which can move into the nucleus and regulate the transcription of many target genes. Whereas phosphorylated BZR1/2 are retained in the cytoplasm by the 14-3-3 protein.

series of signaling events that ultimately activate BES1/BZR1 family transcription factors (Figure 3.2, right side). Binding of BR to the BRI1–BAK1 complex causes BRI1 to quickly phosphorylate BKI1 (Wang and Chory, 2006), which leads to the dissociation of BKI1 from BRI1 and also the sequestration of BKI1 by 14-3-3 proteins (Wang et al., 2011, Jaillais et al., 2011). BRI1 and BAK1 sequentially are phosphorylated and activate one another (Wang et al., 2005, Wang et al., 2008, Oh et al., 2009). The activated BRI1 phosphorylates receptor-like cytoplasmic kinases BR SIGNALING KINASES (BSKs) and CONSTITUTIVE GROWTH (CDG1) further activate phosphatase BRI1-SUPPRESSOR 1 (BSU1; Kim et al., 2009, Kim et al., 2011, Tang et al., 2011, Sreeramulu et al., 2013). BSU1 is projected to dephosphorylate BIN2 and that inactivates BIN2 kinase activity (Kim et al., 2009). Multiple additional mechanisms that also regulate BIN2 have been reported, which enclose targeted protein degradation in the subsistence of BRs by F-box E3 ubiquitin ligase KINK SUPPRESSED IN BZR1-1D (KIB1; Peng et al., 2008, Zhu et al., 2017)—the action of PROTEIN PHOSPHATASE 2A (PP2A) direct to the dephosphorylation of BES1/BZR1 family transcription factors (Tang et al., 2011). The dephosphorylated BES1/BZR1 translocates from the cytoplasm to the nucleus where they perform functions along with a suite of transcription factors and cofactors to regulate the expression of thousands of BR-regulated genes (He et al., 2002, Yin et al., 2002, Wang et al., 2002, Zhao et al., 2002, Yin et al., 2005, Yu et al., 2011). These transcription factors are themselves synchronized by BES1 and/or BZR1 and also interact with them to carry out BR-regulated gene expression (Guo et al., 2013, Li, 2010). PHYTOCHROME-INTERACTING FACTORS (PIFs) are light-regulated transcription factors and they function as key regulators of growth and responses to the environment (de Lucas and Prat, 2014). PIF4 physically interacts with BES1 and BZR1 and shares over 2000 common target genes determined by genome-wide ChIP analysis (Oh et al., 2012). Similar to PIFs, auxin-responsive transcription factors ARF6 and ARF8 also interact with BZR1 (Oh et al., 2009).

3.3 Stress Tolerance: BR Services to the Plant Community

3.3.1 Abiotic Stress

3.3.1.1 Appraisal of BRs for Thermo Tolerance

The structure, as well as composition of plant life throughout the planet, has changed since the last ice age and a parallel magnitude of change is expected in the upcoming century if emissions persist at high rates (Nolan et al., 2018). The temperature across the globe is continuously changing for the worse for biotic beings, viz., flora and fauna. These rapid temperature changes are a consequence of the continuous interference of human beings during rapid industrialization and urbanization (Vardhini, 2019). Ever increasing and decreasing ranges of temperature in the environment are leading to a malfunctioning of various physiological and biochemical processes in plants in terms of high temperature or heat stress as well as low-temperature stress (chilling and freezing stress). Therefore, considering how plants can better withstand changing environments represents a crucial challenge. Plants have evolved several pleiotropic and intricate regulatory functions to defend against biotic and abiotic stress situations (Rehman et al., 2016, Kagale et al., 2010). During stress, various phytohormones cooperate and play a fundamental role in signal transduction pathways and arouse the defense mechanisms of plants (Br and Sm, 2009). BRs are a novel group of plant growth regulators (PGRs) with significant growth-promoting activity. Beyond their roles in growth, BRs also control the response to stress conditions (Nolan et al., 2017a). Between BRs and stress responses, there is the presence of a difficult relation, but the application of BRs promotes tolerance to numerous stress conditions (Bajguz and Hayat, 2009; Kagale et al., 2007; Yuan et al., 2010; Anjum et al., 2011; Divi et al., 2016). Studies of both BR-deficient and -insensitive mutants showed that destruction of the BR pathway is recurrently associated with amplified survival in the face of different stress conditions (Feng et al., 2015; Northey et al., 2016; Nolan et al., 2017b).

3.3.1.2 Low Temperature/Chilling Stress

Seasonal and diurnal temperature changes are prominent environmental factors that influence plant distribution and can powerfully limit crop productivity. While chilling sensitive plants from tropical and subtropical regions go through damage even above the freezing temperature (Eremina et al., 2016). Chilling stress is the chief abiotic stress that is responsible for restrictive plant development worldwide, strongly damaging plant physiological processes (Xi et al., 2013). Low-temperature stress leads to muddled photosynthetic processes, detained plant growth, negative effects on chlorophyll contents, and flower bud abortion that results in major yield and economic losses. Chilling stress also has a direct impact on disrupting the thylakoid electron transport process, the carbon reduction cycle, and the stomatal management of the CO_2 supply, with an amplified accretion of sugars, peroxidation of lipids, and water balance (Allen and Ort, 2001). Exogenous application of BR significantly enhanced resistance against low-temperature stress by modulating the morphological, physiological, and biochemical characteristics of cucumber (Wei et al., 2015), tomato (Shu et al., 2016), and pepper (Jie et al., 2015, Yuan et al., 2016). BRs treatments improved seedling tolerance (Wang and Zeng, 1993) and ameliorated root length, root biomass, total biomass, and the height of rice plant under low-temperature stress (Kim and Sa, 1989, Hirari et al., 1991). In another study, Krishna (2003) observed the same results in maize. It was postulated that the application of BRs promoted growth recovery of seedlings following chilling treatment (0–3°C). Low-temperature stress ameliorated the proline, glycine betaine, soluble protein, and soluble sugar contents of plants (Ashraf et al., 2007, Burbulis et al., 2011). It was found that BRs enhanced proline content and thus improved the chilling resistance and cell membrane stability of plants (Liu et al., 2009, Liu et al., 2011, Hu et al., 2007, Fariduddin et al., 2011). Stress triggered the production of antioxidant enzymes to prevent chilling injury (Oidaira et al., 2008). As a result of BR treatment, improvement in the activities of antioxidative enzymes might occur with the increase in de novo synthesis or activation of enzymes by being mediated through the transcription or translation of precise genes to gain tolerance (Bajguz, 2000).

The authors also showed that BRs assuage the negative effects of low temperature as well as chilling stress by maintaining photosynthetic activities and carbohydrate metabolism, escalating chlorophyll contents, reducing toxic ion contents, inducing changes in defense enzymes, activating gene expression, increasing the content of endogenous plant hormones, and activating signal transduction pathways (Li et al., 2015, Eremina et al., 2016). Crosstalk between BRs and the alternative oxidase (AOX) pathway plays a crucial role in improving plant tolerance to low temperature (Deng et al., 2015). Evidence strongly supports the contribution of AOX in reducing the generation of several reactive oxygen species (ROS) under stress conditions (Wang et al., 2012, Panda et al., 2013). BR-enhanced AOX synthesis has been verified to limit ROS synthesis to protect photo systems under low-temperature stress (Deng et al., 2015). It was suggested that BR and AOX coordinate in balancing the chloroplast-to-mitochondria electron transfer, which inhibits superfluous ROS accumulation. Application of 24-EpiBL was observed to amplify cell membrane permeability in oilseed rape grown at 20°C, while exposure to low temperature (2°C) significantly decreased the permeability of the plasma membrane (Janeczko et al., 2007). BRs mediated the encouragement of cold stress tolerance through the accumulation of active and unphosphorylated forms of both BZR1 and BES1 and thus promoting expression of C-REPEAT/DEHYDRATION-RESPONSIVE ELEMENT BINDING FACTOR1 (CBF1) and CBF2 that positively regulated low-temperature stress responses (Li et al., 2017). CESTA is a positive controller of BR signaling (Poppenberger et al., 2011), which promotes responses against cold stress. It is dephosphorylated and then stimulated in response to BRs, which escorts CESTA-mediated activation of cold responsive genes through both CBF-dependent and independent pathways. This promoted basal and acquired freezing tolerance (Eremina et al., 2016). BIN2 also plays a significant role in BR-mediated regulation of cold-stress responses by the phosphorylating inducer of CBF EXPRESSION1 during extended exposure to cold stress and promoting its deprivation to attenuate CBF induction (Ye et al., 2019). In addition, authors reported that BRs amplified endogenous jasmonic acid and salicylic acid and also improved the ethylene biosynthesis pathway, signifying that BRs function via the synergistic crosstalk with jasmonic acid, salicylic acid, and the ethylene signaling pathway to act in response to cold stress (Wei et al., 2015, Li et al., 2013). BRs kindled the expression of cellular redox homeostasis-related genes (*PER72*, *peroxidase, 72*; *GSTX1*, *probable glutathione S-transferase*; and *CAT2*, *catalase isozyme 2*) to reduce the harmful effects of chilling stress (Li et al., 2016). This suggests that BRs play a major role against chilling/cold stress by activating the responsive genes (in *Brassica* and *Arabidopsis*), signal transduction pathways such as BR and ethylene signaling pathways, the transcriptional levels of several stress responsive genes (*SOD, POD, CAT, GR*), and defense systems (Li et al., 2016, Shu et al., 2016).

3.3.1.3 High Temperature or Heat Stress

In general, a transitory increase in temperature (10–15°C above natural) causes heat stress (Bajguz, 2008). High-temperature condition effects are seen at the biochemical and molecular levels in plants (especially in the leaves). Due to heat stress, a decrease in the duration of developmental phases leads to fewer and smaller organs, diminishing light perception over a shortened life cycle, and losing product (Stone, 2001, Rane and Chauhan, 2002, Hussain and Mudasser, 2006). Heat shock often induces overproduction of ROS (Janeczko et al., 2011), which can cause membrane lipid peroxidation, nucleic acid damage, and protein denaturation (Yin et al., 2008, Bartwal et al., 2012). Numerous studies have verified that ROS scavenging mechanisms play an imperative role in protecting plants from high-temperature stress (Hu et al., 2010, Asthir et al., 2012). BRs were found to play a positive role in extenuating high temperature or heat stress in plants (Confraria et al., 2007, Zhou et al., 2014). They were reported to control growth and this is a wide spectrum of physiological responses to several abiotic stress conditions (Figure 3.3; Wu et al., 2017). BRs revealed a significant response to heat stress in *Ficus concinna* (Jin et al., 2015), banana (Nasser et al., 2004), *Brassica*, and *Arabidopsis* (Kagale et al., 2009) by maintaining the physiological and antioxidant defense system. BR appliance reduces ROS content and augments antioxidant enzyme activities to afford thermotolerance to rising temperatures (Wu et al., 2014). Seedlings of *Ficus concinna were* treated with BR and then exposed to high-temperature conditions (28°C, 35°C, 40°C) for 48 hours, which produced a significant increase in oxidized glutathione (GSSG), reduced glutathione (GSH), oxidized ascorbate (DHA), and ascorbate (AsA) contents, as well as enhancement in antioxidant enzyme activity (SOD, POD, CAT, GR, APX; Jin et al., 2015). Cukor et al. (2018) observed that the application of BRs positively regulated seed germination of Scots pine cultivated under heat stress conditions. Hayat et al. reported that treatment of 28-homoBL to *Vigna radiata* mitigated heat stress by improved leaf water potential (ψ), membrane stability index (MSI), activities of antioxidative enzymes, and proline levels. Furthermore, Mazorra et al. observed that the application of 24-EpiBL induced tolerance to heat shock (HS) in tomato. HomoBL was found to mitigate negativity of high temperature in the expansion of apical meristems of banana shoots cultured *in vitro* conditions. Application of 24-EpiBL mitigated heat stress and enhanced physiological functions of barley (Janeczko et al.). Dhaubhadel et al. reported that 24-EpiBL resulted in improved basic thermotolerance of tomato seedlings due to the protection of translational machinery as well as heat shock protein synthesis by BR application. BRs mitigated heat-induced inhibition of photosynthetic capability by enhanced carboxylation efficiency as well as antioxidative enzyme system in *Lycopersicon esculentum.* Recently, BRs enhanced the lipid productivity and stress tolerance of *Chlorella* cells subjected to heat stress (Liu et al.). Under high-temperature stress, both BES1 and BZR1 hoard and function with PIF4 to endorse thermogenic growth (Ibañez et al., 2018; Martínez et al., 2018). The enhanced levels of BES1 and BZR1 promote expression of PIF4 and this augments PIF4 levels, permitting the de-repression of BRs

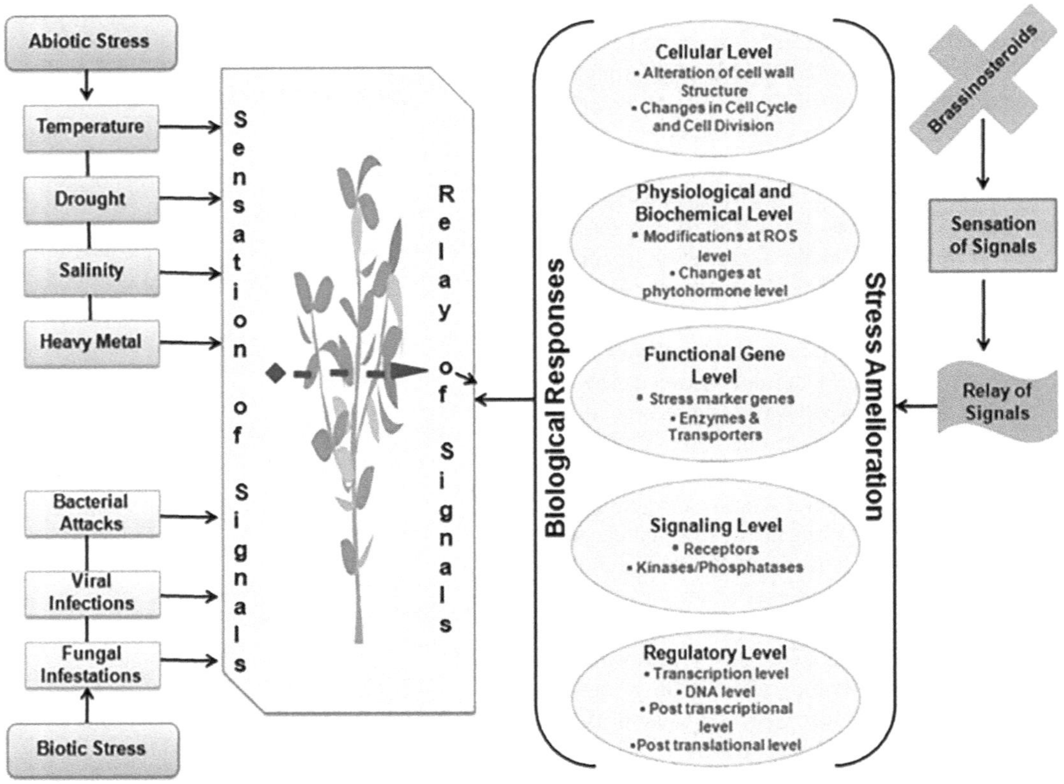

FIGURE 3.3 Brassinosteroid-mediated cascade of remedial strategies for abiotic stress responses at different levels in order to protect plants from harsh environmental conditions.

biosynthesis by switching the BES1 from repressive homodimer to PIF4-BES1 heterodimer that activates the transcription (Martínez et al., 2018). In an apparently opposite manner, increased temperatures decrease BRI1 levels, which tempers BRs signaling and enhances root growth (Martins et al., 2017). BRI1 endures ubiquitination, endocytosis, and deprivation (Martins et al., 2015; Zhou et al., 2018), which are requisites for heat-induced reduction in BRI1 accretion (Martins et al., 2017). PUB12 and PUB13 ubiquitinate BRI1 after BR perception (Zxsazhou et al., 2018) and E3 ubiquitin ligase, which is liable for this ubiquitination during heat stress and remains to be identified.

3.3.2 Potential of BRs for Drought Stress Tolerance

Unprecedented global climate change has led to conditions of water insufficiency, that is, drought. Water-deficient conditions not only disturb cellular homeostasis but also cause an abrupt burst of ROS that damages the protein as well as the genetic machinery of the plants. When exposed to conditions of drought, plants respond by accumulating osmoprotectants that are compatible solutes and aid in maintaining the cellular homeostatic balance (Rontein et al., 2002). BRs are known to induce stress tolerance in plants without compromising overall growth. They are recognized to regulate the level of these osmolytes. They also enhance ADS activity and increase the water content in leaves when under drought conditions. This ADS system scavenges the ROS generated and keeps their levels under tight control. BRs also increase the contents of NO (nitric oxide) in mesophyll cells during conditions of drought that crosstalk with the ABA synthesis pathway that help in drought tolerance (Zhang et al., 2011). Increased growth of roots was observed in BR-treated plants under drought conditions as compared with the control and the same has been reported in different studies. Both drought tolerant as well as resistant varieties of wheat were treated with BRs to check tolerance against drought. Both the varieties responded positively against drought, however, drought-tolerant varieties treated with BRs behaved better in terms of growth (Krishna, 2003). It was extrapolated from studies in cucumber that BR application protected the plant from desiccation and also increased levels of compatible solutes like amino acids in leaves (Pustovoitova et al., 2001). In the case of *Brassica napus*, with seedlings raised in 1 μM EBL and then grown under water-deficient conditions, it was observed that EBL supported the growth of the seedlings under drought (Kagale et al., 2007). Similar results were observed in the case of *Robinia pseudoacacia* L. where the root dipping treatment of EBL enabled it to withstand drought stress (Li et al., 2007). Another study in rice cultivars reported the positive effects of BRs (both HBL and EBL) under drought conditions (Farooq et al., 2009). In this study, BR treatment of 0.01 μm was given as seed soaking and foliar spray. The BR treatment ameliorated the toxic effects of drought, enhanced the biomass of the plants, and increased the efficiency of CO_2 assimilation with the accumulation of osmolytes, thus improving water use efficiency and membrane properties of the plants. Increased CO_2 assimilation with BR treatment was also observed in soybean (Zhang et al., 2008), cauliflower (Hniličká et al., 2010), and tomato (Yuan et al., 2010). The study also reported better performance of foliar spray over that of seed treatment of BRs. Similar effects of BRs in minimizing drought toxicity were observed in *Solanum lycopersicum* and osmolyte production maintained cell homeostasis (Behnamnia et al., 2009). The role of stress-responsive transcription factors like WRKY has been acknowledged and it has also been established that they are positively regulated along with the BR-signaling cascade in order to inculcate drought stress tolerance in *Arabidopsis* (Chen et al., 2017). Lee et al. (2018) reported that BR-deficient mutants showed susceptibility toward osmotic stress. The study also reported the regulatory potential of BRs in penalizing growth under drought conditions, proving a connection between BR-signaling and stress-responsive genes. Mahesh et al. (2013) further reported the diminished effect of water stress on seed germination and seedling growth. Tolerance was associated with increased contents of nucleic acids and proteins and further enhances membrane integrity under desiccation. Another report extrapolated that treatment with 24-EBL in *Capsicum annuum* harmonized growth and protected the plant from photoinhibition under drought; the excitation energy was dissipated from the light harvesting complex in photosystem II in the form of heat (Hu et al., 2013). Reports of *Salvia miltiorrhiza*, which is a Chinese medicinal plant, suggest that BR treatment under drought minimized the contents of MDA and increased proline along with activities of ADS (Zhu et al., 2014). Many more such reports have been added to the literature that state that BR treatments in the form of seed soaking or foliar spray have been useful in mitigating the effects of drought and this further fine-tunes the overall growth of economically important crops.

3.3.3 Alleviation of Plant Salinity Stress by BRs

Being sessile in nature, plants are exposed to diverse biotic and abiotic stresses throughout their life cycle. Among the various stresses experienced by plants, soil salinity is one of the major constraints to the survival and productivity of foods as well as cash crops. Salinity stress affects a vast array of plant physiological and metabolic responses such as seed germination, ion homeostasis, water relations, photosynthesis, transpiration, lipid metabolism, protein synthesis, and overall crop yield. Accelerating modernization and urbanization involving the use of inorganic fertilizers, gypsum, irrigation with brackish waters, and poor drainage has posed a serious threat to agricultural soils, largely contributing to soil salinization. Plants in order to survive under such salt stress conditions develop various inherent mechanisms such as accumulation of compatible solutes, sequestration of sodium (Na^+) ions in vacuoles, and Na^+ ion exclusion from leaf blades (R Munns and Tester, 2008). However, the problem of salinity still confers deleterious effects on the productivity of economically as well nutritionally beneficial crops. Therefore, supplementation of a number of phytohormones is employed in order to develop salt tolerance in various plants (Figure 3.4). Among the various phytohormones, BRs are known to regulate a number of salinity stress tolerance mechanisms and ensure the survival of the crop under harsh conditions. The effects of certain active BRs on crop plants are discussed in future sections.

Inhibition of seed germination and seedling growth of salt-stressed *Oryza sativa* seedlings was reciprocated by exogenous application of millimolar concentrations of 28-HBL and 24-EBL independently. Both BRs enhanced the synthesis of nucleic acids and soluble proteins in rice seedlings under salt stress, developing tolerance in them (Anuradha and Rao, 2001). The three most common and active BRs, including BL, 28-HBL, and 24-EBL, when employed on groundnut seedlings under conditions of NaCl stress, reversed the inhibitory effects of stress by elevating seedling growth and the overall biomass of seeds (Rao, 1997). Foliar spray with 28-HBL to salt-stressed wheat plants displayed positive modulation in the growth and productivity of the crop; 28-HBL improved

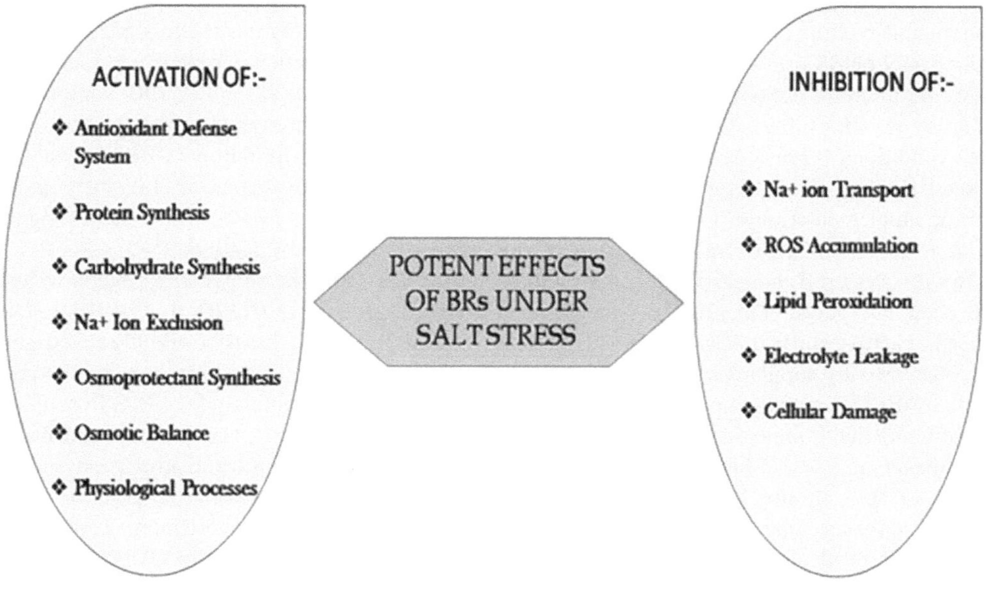

FIGURE 3.4 Under salt-stressed conditions, abrupt ROS levels lead to cellular damage. BRs are involved to maintain ion homeostasis and ROS scavenging mechanisms in order to protect plants.

the growth parameters, photosynthetic efficiency, content of photosynthetic pigments, and overall yield even under stress conditions (Eleiwa et al., 2011). Foliar application of 24-EBL to wheat plants significantly enhanced the total biomass of the plant and hence proved ameliorative under the conditions of salt stress (Shahbaz and Ashraf, 2007).

Salt stress also has detrimental effects on cell divisions, particularly on root and shoots meristems. Seed priming with 24-EBL in barley seeds caused a reduction in chromosomal abnormalities during mitotic cell divisions of root meristems, thereby alleviating salt stress (Tabur and Demir, 2009). Seeds of different varieties of *Medicago sativa* when primed with BL under conditions of high soil salinity displayed improvement in germination index, morphological characters (root and shoot length), physiological parameters (fresh and dry weight), and activities of antioxidant enzymes, proving BL potential in conferring salt tolerance in the plant (Zhang et al., 2007). Exogenous supplementation of mung bean plants with low concentrations of BL enhanced the overall growth in terms of seedling length, chlorophyll content, photosynthetic efficiency, and nitrate reductase activity under the diminishing growth environment of salinity (Lalotra et al., 2017); 24-EBL application improved the growth status of oilseed rape plants by enhancing total biomass, protein synthesis, osmolyte content, and maintaining osmotic balance and ion homeostasis even under harsh conditions of NaCl stress (Efimova et al., 2014).

Different modes of application of BRs confer different effects on the growth of plants under salt stress conditions. Research studies revealed that independent root application and leaf-root application of BRs are superior to only leaf application in enhancing chlorophyll content, biomass, and root activity in salt-stressed cotton plants (Shu et al., 2014). Foliar spray of low concentrations of 24-EBL was found to have ameliorative effects on salt stress by amplifying growth as well as the mineral content of different wheat cultivars; 24-EBL improved the $Ca2^+/Na^+$ and K^+/Na^+ ion status of wheat plants when treated even under saline conditions (Grown, 2006). Deterioration in total carbohydrate, sugar, and protein content under saline growth conditions was counteracted by foliar spray of BRs in wheat plants. In addition, uptake of macro and micronutrients enhanced to significant levels in wheat straw as well grains on application of 24-EBL under salt stress conditions (Eleiwa et al., 2011). Soil salinity lowers the seed yield of oilseed crops, causing commercial loss. However, supplementation of low concentrations of 28-HBL at foliar levels enhanced overall crop productivity and seed yield in salt-stressed *B. juncea* plants, proving to be beneficial to the economy (Hayat et al., 2007).

Salinity-induced oxidative stress in terms of enhanced ROS production was overcome by supplementation of 24-EBL in rice plants, where 24-EBL stimulated upregulation of various enzymatic and non-enzymatic antioxidants involved in ROS scavenging (Sharma et al., 2013). Similarly, tomato plants were acclimatized to soil salinity by supplementing them with 24-EBL, which enhanced the levels of enzymatic antioxidants and hence improved plant growth under conditions of salt stress (Mazzora et al., 2002). Saline growth conditions enhance lipid peroxidation and hence cause cellular damage to plants, leading to electrolyte leakage, ion imbalance, and the production of harmful ROS. Such deleterious growth effects were significantly reciprocated by exogenous application of 28-HBL in a dose-dependent manner in maize plants where 28-HBL stimulated antioxidant enzymes and other enzymes involved in photosynthesis and metabolic processes (Arora et al., 2008). Both 24-EBL and 28-HBL enhanced the level of osmoprotectants, total phenols and flavonoids acting as non-enzymatic antioxidants, and the expression of enzymatic oxidants in response to salinity stress in maize seedlings (Rattan et al., 2020); 28-HBL is involved in the tight regulation of ROS accumulation by elevating the gene expression of antioxidant enzymes in *B. juncea* plants grown under growth deteriorating saline conditions (Kaur et al., 2018).

BRs are found to show synergistic effects with other phytohormones, certain secondary metabolites, and biotic agents in alleviating salt stress in plants. Co-application of BRs and kinetin (KN), a synthetic phytohormone, mitigated salt stress in tomato plants through stimulation of the antioxidant defense system and osmolyte and secondary metabolite accumulation (Ahanger et al., 2020). Also, co-supplementation of 24-EBL and putrescine (Put), a polyamine (PA), counteract the diminishing effect of saline stress by reducing lipid peroxidation and MDA content and upregulating the mechanisms involved in the regulation of protein synthesis and antioxidant enzyme activities (Shummu et al., 2012). Interactive effects of 24-EBL and arbuscular mycorrhizal (AM) fungi, *Glomus mosseae*, have also been reported to mitigate salt stress in wheat plants, where enhancement in overall plant growth was observed in terms of root, shoot length, leaf area, and total plant biomass (Tofighi et al., 2017). Co-application of 24-EBL and sodium nitroprusside (SNP) reversed the hampered growth of *B. juncea* plants by positively modulating photosynthetic and chlorophyll fluorescence parameters and nitrogen metabolism, and the downregulation of mechanisms causing electrolyte leakage and lipid peroxidation (Gupta et al., 2017).

Molecular evidence in support of the imperative role of BRs in conferring salt tolerance to plants was supported by studies on the *Arabidopsis* mutant with mutation in the BSK5 gene that encodes a brassinosteroid-signaling kinase protein, where substantial reduction in growth was found under salinity stress conditions (Li et al., 2012). BRs confer salinity stress tolerance in plants through the accumulation of certain heat shock proteins (HSPs). Inhibitors of BR biosynthesis downregulated the expression of numerous HSPs playing an imperative role in salt tolerance in *Zea mays,* hence signifying the potential of BRs in stimulating HSP synthesis (Derevyanchuk et al., 2016). Studies on gene expression of abiotic stress genes (*WAK*, *HvPIP1.1*, *HvPIP1.2*, *HvPIP1.3*, *HvPIP1.5*, *CYCD3*, *DREB2*) and the brassinosteroid-related gene (*DWARF4*) in barley under conditions of salt stress showed another remarkable point in the potent role of BRs in conferring salt tolerance to plants. Abiotic stress-related and BR-related genes were upregulated in significant proportions in 28-HBL treatment owing to their imperative role in mitigating salt stress conditions (Marakli and Gozukirmizi., 2018).

Hence, BRs increase the tolerance of plants to salt stress through up/downregulation of various plant physiological and metabolic processes. Being eco-friendly in nature, BRs can be applied in agricultural use after unveiling the various mechanisms regulated by BRs to confer salt resistance in plants. This would be an effective approach to ensure food and environmental safety. The regulation of a number of genes by BRs remains unexplored. A number of research studies are being carried out to explore the role of the BR signaling pathway and its crosstalk with other hormones at cell, tissue, and organ levels for its utilization in the agriculture sector.

3.3.4 BRs as Potent Ameliorates of Heavy Metal Stress

BRs belonging to the class of polyhydroxy steroids are well known for their crucial roles in the regulation of plant growth and development under normal as well as stressed conditions. BRs display protective behavior under various kinds of abiotic stresses in plants. One of the major and dominant plant stressors being added to the environment these days is heavy metals. Plants are exposed to various forms of heavy metals present in soil, water, and air. Heavy metals impose various detrimental effects on the normal physiology of plants, either directly by impairing the activity of many essential enzymes catalyzing primary photosynthetic reactions and other developmental processes or indirectly through the elevated accumulation of ROS. Various BRs such as 28-HBL and 24-EBL aid in the amelioration of heavy metal stress encountered by plants during their growth through stimulating a number of stress-tolerant mechanisms (Figure 3.5).

The influence of BRs in mitigating heavy metal stress largely depends on the stage of plant development at which they are being applied. Most BRs are known to reciprocate the toxic effects of various heavy metals by stimulating either or all of the following tolerant mechanisms:

- Reduction in uptake and accumulation of heavy metals by plant roots.
- Inhibition of apoplastic and symplastic transport of heavy metals from the roots to the aerial parts of plants.
- Activation of transcription factors involved in the sequestration of heavy metals into vacuoles.
- Enhancement of activities of various enzymes involved in primary metabolic processes, viz. photosynthesis, respiration in plants.
- Stimulation of phytochelatin and metallothionein synthesis.
- Amplification of the antioxidant defense system to counter oxidative stress (ROS formation) induced by heavy metals.

Nevertheless, various signaling mechanisms induced by BRs in order to inculcate heavy metal tolerance remain unexplored, yet the protective potential of BRs against heavy metals cannot be overlooked. Various research studies reveal the important role of BRs in responding to the deterioration of plant growth due to the incorporation of different kinds of heavy metals in their tissues. BRs and their ameliorative potential against different heavy metal toxicity are discussed in the following sections.

3.3.4.1 BRs and Aluminum Toxicity

The role of 28-HBL and 24-EBL in the amelioration of aluminum (Al) stress has been explored through the enhancement of the antioxidant defense system, that is, catalase, peroxidase,

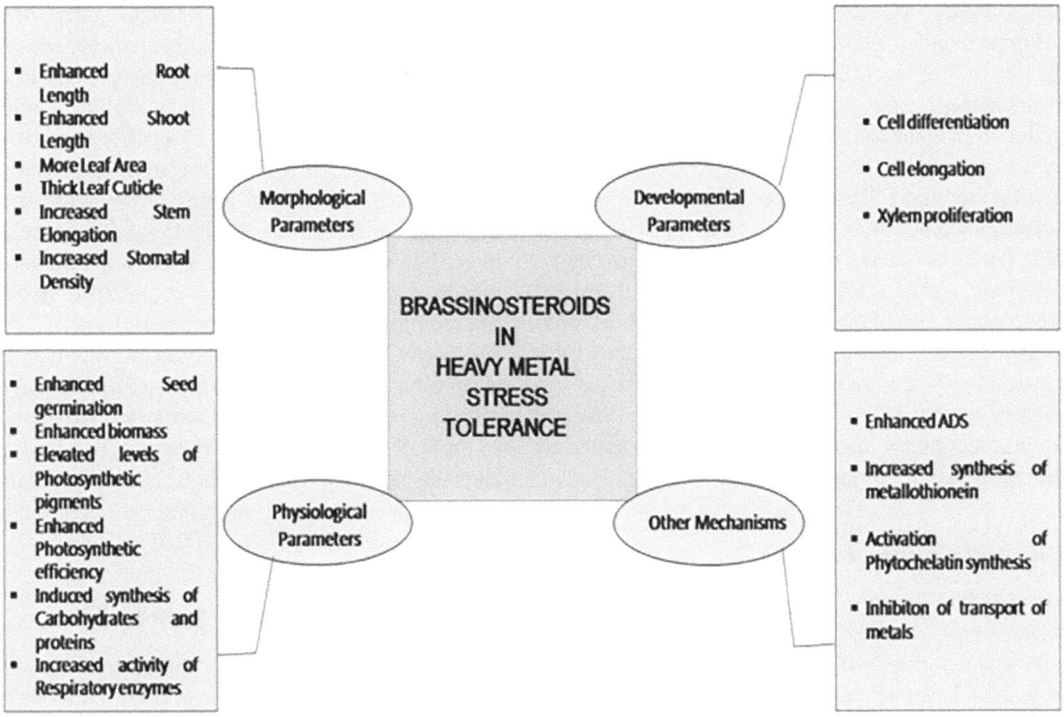

FIGURE 3.5 Heavy metals present in the soil act as major pollutants for plants and cause deleterious effects on growth. However, BRs act as the perfect ameliorators and allow plants to enjoy their genetic potential.

and superoxide dismutase activity in *Vigna radiata* and *Cajanus cajan* plants (Ali et al., 2008). BRs induced plant growth improvement, enhanced photosynthetic efficiency, and the antioxidant enzyme defense system under Al stress, which substantiate their effective potential against heavy metal stress. It has also been revealed that crosstalk of BRs with calcium (Ca) signal transduction pathways results in the stimulation of the MAPKs/CDPKs cascade of phosphorylation in order to activate various transcription factors involved in cellular protection under Al stress (Ashraf et al., 2007). The inhibitory effect of Al on plant growth in *C. cajan* was counteracted by exogenous application of 24-EBL, while application resulted in a remarkable increase in chlorophyll, nucleic acid, and soluble protein content in Al-stressed *C. cajan* plants (Sri et al., 2016).

3.3.4.2 BRs and Cadmium Toxicity

BRs are known to actively stimulate the antioxidant defense system by inducing NADPH-based H_2O_2 production and photosynthesis in cucumber and tomato plants under cadmium (Cd) stress (Ahammed et al., 2013). Cd-stressed plants sprayed with 28-HBL displayed a remarkable rise in level of chlorophyll pigments, net photosynthetic rate, antioxidative enzyme activities, and nitrate reductase activity in *Brassica juncea* plants (Hayat et al., 2007). Radish seeds supplemented with 24-EBL and 28-HBL independently under Cd stress showed elevated levels of seed germination, antioxidant enzyme activities, fresh weight, and decreased lipid peroxidation in their plants (Anuradha et al., 2007). Foliar application of 24-EBL and 28-HBL to tomato plants neutralized Cd toxicity by amplifying growth, quality, and yield of fruits (Hayat et al., 2012), and also by enhancing the photosynthetic machinery and antioxidant defense system (Hasan et al., 2011). Application of 24-EBL to *Vigna unguiculate* plants under Cd stress caused a significant rise in leaf, root, and total dry matter as well as chlorophyll fluorescence and gas exchange parameters along with a reduction in oxidant compounds leading to cell damage. The detrimental consequences caused due to Cd toxicity on gas exchange factors and PSII activities (chlorophyll fluorescence) are diminished by foliar application of EBR (Santos et al., 2018). BRs have also been reported to upregulate the activities of H^+-ATPase and NADPH oxidase enzymes to impart Cd tolerance to plants (Jakubowska and Janicka, 2017). Foliar spray of 24-EBL aided in neutralizing the photosynthetic inhibition and oxidative stress caused by Cd-phenanthrene co-contamination in tomato plants by enhancing the antioxidant defense system, secondary metabolism, and the xenobiotic detoxification system (Ahammed et al., 2013).

3.3.4.3 BRs and Copper Toxicity

Seed priming with BRs in *B. juncea* exposed to copper (Cu) stress impeded Cu uptake and accumulation as well as improved shoot generation and biomass production (Sharma and Bhardwaj, 2007). Studies revealed the effective role of co-application of 24-EBL and polyamines to *Raphanus sativus* plants in overcoming Cu toxicity by reducing Cu uptake and assimilation and modulating gene expression of metabolism of indole-3-acetic acid (IAA) and abscisic acid (ABA; Choudhary et al., 2012); 24-EBL has also been reported to downregulate lipid peroxidation and enhance dry weight, chlorophyll content, proline content, and relative water content in *Helianthus annus* plants under Cu stress conditions (Filova et al., 2013).

3.3.4.4 BRs and Lead Toxicity

Research studies by Khripach et al. (1999) showed a 50% reduction in absorption of lead (Pb) after the exogenous application of 24-EBL in *Beta vulgaris*. Even low concentrations of 28-HBL were found to mitigate Pb toxicity in fenugreek plants by enhancing the biomass, photosynthetic pigments, and rate of photosynthesis (Swamy et al., 2014). The negative effect of Pb on germination and seedling growth in *B. juncea* was reciprocated by 24-EBL in a dose-dependent manner through regulation of the antioxidant defense system (Soares et al., 2020). BL application to *Chlorella vulgaris* cultures restored their growth arrested due to Pb toxicity. Moreover, BLs also augmented the chlorophyll, monosaccharides, protein, and phytochelatin levels in these cultures, thereby counteracting the deleterious effects of Pb (Bajguz, 2011). Supplementation of 24-EBL substantially ameliorated the deteriorating effect of Pb toxicity in the aromatic plant geranium by enhancing its chlorophyll levels (Rao and Raghu, 2016).

3.3.4.5 BRs and Chromium Toxicity

The oxidative stress in chromium (Cr)-stressed *R. sativus* seedlings is counteracted with 24-EBL by improving growth in terms of root and shoot length and fresh weight. The synergistic effect of 24-EBL and polyamines was reported in these seedlings with respect to the mitigation of Cr toxicity by attenuating ROS production and enhancing the levels of photosynthetic pigments, total soluble sugars, and ROS scavenging systems (Choudhary et al., 2012). Similarly, alleviation of Cr toxicity by both 24-EBL and 28-HBL was reported by Sharma et al. (2018) in radish plants through regulation of the expressions of key antioxidant genes. Pre-sowing seed treatments of 24-EBL aided in the diminution of Cr stress by enhancing overall growth, protein content, and antioxidant enzyme activities in *B. juncea* plants (Arora et al., 2010). In addition to this, foliar application of low concentrations of 24-EBL greatly assuaged Cr-induced deterioration of growth and photosynthesis of tobacco seedlings. It was also found to have a prominent effect on reducing Cr ion uptake, oxidative stress, and damage to the cell structure (Bukhari et al., 2016); 24-EBL remodulates the physiological, metabolic, and antioxidant defense system in a positive direction in order to reciprocate Cr stress in tomato seedlings (Jan et al., 2020).

3.3.4.6 BRs and Nickel Toxicity

Foliar application of minute concentrations of 24-EBL and 28-HBL helps in the detoxification of Ni in *B. juncea* plants. Nickel (Ni) generally leads to electrolyte leakage, lipid peroxidation, and reduced relative water content and photosynthetic pigments. The inhibitory effect of Ni stress was overcome by

24-EBL by enhancing antioxidant enzyme activities and other metabolic pathways (Ali et al., 2008, Kumari and Malik., 2011). Reduction in uptake and accumulation of Ni was found to be stimulated by 24-EBL along with counteracting the Ni toxic effects by enhancing the antioxidant defense system and quantum yield of PS II photochemistry in *Eucalyptus urophylla* plants (Ribeiro et al., 2020).

3.3.4.7 BRs and Zinc Toxicity

Application of zinc oxide nanoparticles (ZnO-NPs) leads to oxidative stress and decreased growth of tomato seedlings. BRs efficiently stimulated antioxidant enzyme activities, redox homeostasis, and related gene expression in order to manage ZnO-NP-induced oxidative stress in tomato seedlings (Li et al., 2016). BRs proved to be beneficial in ensuring food safety by enhancing metal stress tolerance in various edible crops; 24-EBL convalesces root development, photosynthetic efficiency, and activities of various antioxidant enzymes in Zn-stressed soybean plants, thereby overcoming the inhibitory effect of Zn on growth and development (Santos et al., 2020). Both 24-EBL and 28-HBL when sprayed over the leaves of radish plants exposed to Zn toxicity alleviated toxic effects by recovering plant growth in terms of enhanced root and shoot length and biomass. Also, improvement in the activities of photosynthetic enzymes and antioxidant enzymes was reported with the application of BLs (Ramakrishna and Rao, 2015).

Hence, BRs being the youngest phytohormones are now well established for their prominent roles in controlling a plant's fundamental growth and development under normal as well as stressed conditions. BRs have been explored to regulate a wide range of processes, including source/sink relationship, seed germination, photosynthesis, senescence, photomorphogenesis, flowering, and responses to abiotic and biotic stresses. Thus, BRs can be considered as potent ameliorators of heavy metal stress in plants.

3.4 Deciphering the Role of BRs against Different Biotic Attacks

Plants in their natural environment have to face a large number of natural invaders. They have various inbuilt mechanisms that they activate during such biotic attacks. These mechanisms include various physical barriers, chemical protectors, and activation of molecular levels of defense like pathogen-related genes. Plants can act both locally, that is, at the site of a wound or infection as well as distantly. This process of acquiring resistance in the tissues away from the site of infection is called systemic acquired resistance (SAR). BRs protect plants from the outburst of ROS by elevating the efficiency of ADS, however, this may not be the same in case of biotic stress where ADS may be suppressed in order to protect plants from pathogen infection. In such cases, ROS act as signal transmitters that carry the signal of infection to distant parts of local infected tissue in order to enable the plants to acquire systemic resistance (Irfan et al., 2010). Neill et al. (2002) reported a gradual increase in H_2O_2 which is thought to be an essential modulator of SAR. The role of various phytohormones in inducing resistance has been acknowledged.

BRs have proved themselves as promising candidates in the field of agriculture not only by enhancing the yield but also by regulating various physiological activities during stressful conditions. The stature of BRs as stress managers is not confined to abiotic conditions but also includes protection against unpredictable biotic attacks (Figure 3.6). These biotic attacks may include pathogenic infestations or herbivore injuries. Any plant may be infected by several bacterial, viral, and fungal pathogens that may cause significant loss in the total yield and leads to drastic economic losses worldwide (Culbreath et al., 2003; Rodoni, 2009). However, BRs have proved their potential in this regard too. BRs have been involved in various signal transduction pathways and activate various responses against pathogens (Dong, 1998). Application of BRs enhanced

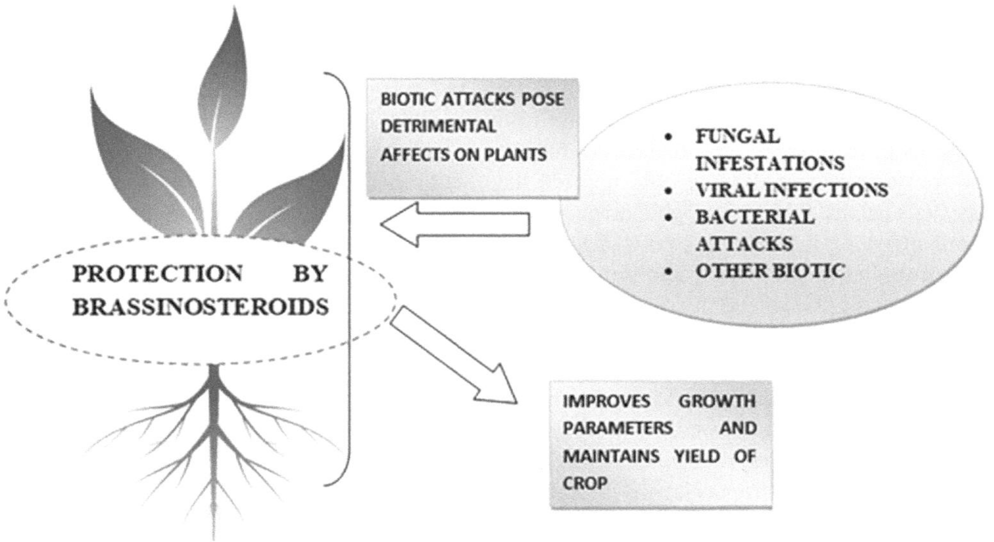

FIGURE 3.6 Various biotic attacks pose harmful effects on plants and deteriorate their yield. However, BRs mitigate the harmful effects of these biotic attacks and allow plants to flourish.

the plant's tolerance against pathogenic attacks. Crosstalk of BRs with other hormones like ethylene and abscisic acid allows it to confer protection against harmful pathogens and enable plant survival. BR also affects the levels of salicylic acid that is involved in pathogen-related (PR) protein synthesis. BAK1 acts as a co-receptor for the perception of pathogen-associated molecular patterns (PAMP; Kemmerling and Nurnberger, 2008). Furthermore, BRs enhance the levels of antioxidants during such conditions, specifically the peroxidases that directly participate in polyphenol metabolism. Most studies on the role of BRs against biotic stress involve fungal infestations, while studies related to other biotic attacks are in their infancy.

3.4.1 Fungal Infestations

The efficiency of BRs in inducing fungal pathogenic tolerance is reported to be higher than many commercial fungicides. The effect of a single dosage of 28-HBL (20 mg ha^{-1}) was as good as a commercial fungicide, Arcerid (composed of ridomil and polycarbacine) at 2 kg ha^{-1}. In addition, it was also found that the yield of plants treated with 28-HBL was higher compared with those treated with Arcerid as well as untreated plants (Khripach et al., 2000). Zhu et al. (2010) studied the protective action of BRs against blue mold rot in the harvested jujube fruit. It was found that BRs were effective in nullifying the effect of *Penicillium expansum* by increasing the activity of antioxidants. However, it was further reported that BRs were able to maintain fruit quality by slowing down ethylene production and hence delaying senescence; BRs did not show in vitro anti-fungal action. Similar results were obtained in the case of cucumber. Churikova and Vladimirova (1997) reported that the application of EBL twice in cucumber reduced the effect of various fungal pathogens. The first application was done at seed soaking (0.1 mg l^{-1}) and the second as a foliar spray (25 mg ha^{-1}) during the flowering stage. The resistance against fungal pathogens was reported along with the increase in activity of various antioxidants like peroxidases. Transcriptome analysis revealed that BR-related genes were upregulated during grey mold infection caused by *Botrytis cinerea* in rose petals (*Rosa hybrida*), indicating the regulatory potential of BRs in the activation of pathogen-related genes (Liu et al., 2018). In order to confirm these results, BR was exogenously applied in rose petals before inoculating *B. cinerea*, and the results further consolidated the transcriptome analysis as BRs uplifted the protection of the petals against this necrotrophic fungal pathogen. *Phytophthora infestans* infects and causes major crop losses in potato. But studies reveal that the application of certain doses of BRs (10–20 mg/ha) potentially reduces the levels of infection of the pathogen; however, the results varied with the time and method of BR treatment. It was only at post-harvest stage that BRs protected the potato tubers from late blight; BRs suppressed the immunity and allowed mycelial growth (Korableva et al., 2002). This was due to different stimulating points in the plant and pathogen (Khripach et al., 1996). Also, it was observed in the case of tomato, when the 14-day long treatment was given to the plants, that EBL protected from attacks of *Verticillium dahlia* and also reduced the severity of disease symptoms, but when this treatment of EBL was given to the plants 24 hours before inoculation, no effect was seen (Krishna, 2003). Hence, the time, stage, and method of application are detrimental to BR pathogen-related resistance. In the case of barley, when epibrassinolide was exogenously applied at a dosage of 5 mg ha^{-1} during the tillering phase, it resisted the attack of *Helminthosporium teres* Sacc. In addition, it also increased the grain yield (Pshenichnaya et al., 1997). The protective action of brassinolide (BL) was reported in tobacco plant against the fungal pathogen *Oidium* sp. An enhanced resistance against *Magnaporthe grisea* was found in rice plants with BL application. Ali et al. (2013) reported that the extent of the damage caused by *Fusarium culmorum* in wheat was reduced by 86% on BR application. Furthermore, a reduction of 33% was reported in barley caused due to *Fusarium* head blight. Recently, in 2018, a study on strawberry reported that BR-induced tolerance against *Colletotrichum acutatum*, which causes anthracnose disease (Furio et al., 2018). The authors used 24-EBL and brassinosteroid spirostanic analog DI-31, and the latter produced more positive effects than the former at low concentrations (0.1 mg l^{-1}).

3.4.2 Viral Infections

Plants may face major threats from viral pathogens that can cause major or minor losses. However, BRs can be acknowledged for their antiviral activities. The application of BRs in plants can be adopted as an eco-friendly approach, enabling plants to tackle unfavorable environmental cues. BRs have also been reported to have potential in inculcating resistance against viral infections (Rodkin et al., 1997). Zhang et al. (2015) studied the positive role of BRs in inculcating tolerance against cucumber mosaic virus (CMV) in *Arabidopsis* with the help of BR signaling mutant *bri1-5* and *bes1-D*. It was reported that BRs not only provided tolerance against CMV but also protected the plant against photo-oxidative damage by increasing the efficiency of the antioxidant defense system. BES1 and BZR1 directly target various PR genes like *WRKY30* and regulate their expression (Sun et al., 2010; Yu et al., 2011). It is actually the complex signaling cascade of BRs that activates the target genes to confer stress tolerance in plants (Zhang et al., 2015). Xia et al. (2009) studied the positive role of BR in helping *Cucumis sativus* to survive attacks of CMV and also protected the plant against oxidative burst. It was further observed that when the EBL was applied to seeds of *Lychnis viscaria,* which shows a lower number of pathogen-related (PR) proteins (chitinase, B-1, 3 glucanase, and peroxidase), it induced resistance against certain virus pathogens like tobacco mosaic virus (TMV) and various fungal pathogens (Roth et al., 2000). Similar results were obtained in the case of tobacco plant where BL treatment protected from TMV attack. But this study unveiled the aspect that BR does not interact directly with either SA endogenous levels or PR genes and BR does not induce resistance through SAR (Krishna, 2003). Khripach et al. (2000) documented the antiviral role of 24-EBL and 28-HBL in the case of potato cuttings and it was found that BRs inculcated resistance against several viruses in potato. Deng and others (2016) reported that treatment of BRs in the form of foliar application (0.1 μM) confers tolerance against TMV in *Nicotiana benthamiana.* Rice black-streaked dwarf virus (RBSDV) belongs to the family of *Reoviridae* and infects rice, maize, barley, and wheat. It causes severe disease in the case of maize and rice. It was reported that combination treatment of methyl jasmonate (MeJA) and 24-EBL reduced the severity of the virus infection, however, treatment of only BR augmented plant susceptibility toward the virus, indicating the importance of hormonal crosstalk during stress conditions (He et al., 2017). Besides plant viral diseases, BRs can also act against Junin IV 4454 strain, poliovirus type I, tacaribe TRLV 11573 strain, and herpes simplex virus (type I and II) (reviewed by Bhardwaj et al., 2011).

3.4.3 Bacterial Attacks

BRs are known to protect rice plants from several bacterial diseases (Nakashita et al., 2003). In the case of *Nicotiana tabacum*, it was found that BL treatment was able to protect it from the attacks of *Pseudomonas syringae* pv. *tabaci* (Pst), a bacterial pathogen. Further, protection against the attacks of *Xanthomonas oryzae* pv. *Oryzae* was also reported in rice plants (Friebe, 2006). Canales et al. (2011) reported BR to lower the amounts of *Candidatus Liberibacter asiaticus*, which is the causal organism of citrus greening or HLB disease, along with inducing resistance against causal bacterium. BR also upregulates the expression of defense-related genes like 3-glucanase, glutathione peroxidase, chitinase, β-1 allene oxidase synthase, phenylalanine ammonia lyase, and hydroperoxide lyase (Canales et al., 2011). Besides such studies, there is a need to explore the role of BRs as anti-bacterial agents that may be able to fill large lacunae in the reservoir of knowledge related to the anti-bacterial property of BRs.

3.4.4 Other Biotic Attacks

Insect pests are also considered major attacks on the plant community. In the case of *Spodoptera littoralis* (action leaf worm), which causes major losses in cotton and other vegetable and ornamental crops, it was observed that application of BRs could increase the mortality rate of the larvae (Smagghe et al., 2002). Ohri et al. (2002) studied the effect of different concentrations of brassinolides against root knot nematodes (*Meloidogyne incognita*) in tomato seedlings. It was observed that at higher concentrations of BL, a lower number of egg masses and fewer galls developed. Nahar et al. (2013) reported the resisting effect of BRs toward *Meloidogyne graminicola* infecting the roots of rice. The immune-modulating property of BRs has enabled them to provide tolerance against many pathogens in plants. It is an undeniable fact that BRs can potentially replace traditional and harmful chemicals used for pathogen attacks.

3.5 Concluding Remarks

From this study, it can be concluded that research on the chemistry, physiology, and molecular biology of BRs affords a convincing body of support that these plant steroids are essential regulators of plant growth and development. The application of BRs assists in maintaining favorable cellular conditions in stress-subjected plants by regulating ion metabolism, amplifying the antioxidant defense system, and enhancing osmoprotectant accumulation while helping to overcome the deleterious effects of stress.

Plants need assured environmental cues for their normal growth and development, while extreme weather dealings, as well as environmental pollution, can negatively affect crop production. Cellular homeostasis, detoxification, and revival of growth are three major kinds of responses driven by plants to overcome stress events. BRs play a fundamental role in mediating these responses by regulating specific sets of genes. BRs have been revealed to regulate the transcription of such genes that encode defensive proteins vital for stress tolerance. Although BR effects on plants are less prominent under control (normal) conditions, their beneficial effects are well recognized in stressful conditions. BR-enhanced stress tolerance is closely associated with BR-induced improvements in photoprotection, CO_2 assimilation, antioxidant potential (enzymatic and non-enzymatic), defense response, redox homeostasis, ROS scavenging, secondary metabolism, detoxification potential, and autophagy. As multiple stressors frequently occur under natural conditions, BRs have significant implications on crop production in the face of a changing climate.

BR research is accelerating rapidly, and with the propagation of cloned genes and advances in micro-chemical techniques, the range of experimental approaches to understand BR action continues to inflate. Much remains to be done, however. Continual research in the field of plant steroids has made BR signaling the paramount studied pathway. Various advances like genetic studies, such as mutant screens, have established BRs as key components of growth and many of the aspects and substantial components of BR signaling, as well as biosynthesis, have been revealed (Clouse, 2015). All events from PM-localized RKs to the regulation of DNA based gene expression, which is under the tight control of BES1 and BZR1, have been elucidated via biochemical, structural, genomic, and molecular studies (Kim and Wang, 2010; Clouse, 2011; Dejonghe et al., 2014; Nolan et al., 2017). All these approaches have also unveiled the fact that BRs are involved in crosstalk mechanisms with several other hormones and various stress responses and do not operate in seclusion (Nolan et al., 2017). Moreover, BR signaling varies among different cells and tissues, which can be manipulated to improve plant growth and stress responses (Fàbregas et al., 2018). Although a lot of research studies have been conducted on plants' own steroidal hormones, and this research has proved to be fruitful in displaying the regulatory role of BRs, many questions and aspects, such as how BRs mechanistically control large sets of genes, are unaddressed. Extensive research on how a number of genes at the same time are activated or repressed in order to sustain growth in unfavorable environmental cues is needed in order to uncover more about BRs. Various technological advances like proteomic approaches (Song et al., 2018) and single cell genomics (Shahan, 2019) can prove to be fruitful in designing various computational models that can better elucidate and uncover new aspects. Many characteristics of BR biosynthesis and potential modes of transport still need to be studied, although the bulk of the information has been added to the literature (Vukašinović and Russinova, 2018). Future studies should aim to obtain evidence on how BRs are transported out of cells and also on identifying BR transporters. Such approaches will deepen our understanding of BR signaling and the regulatory potential of environments such as shade, extreme temperatures, and drought. Furthermore, this will open up many more fields of research and the major focus of researchers shall remain on the manipulation of BR signaling or the biosynthetic pathway in order to provide plants with greater opportunities to flourish in their natural environment, which can be invaded by environmental constraints.

REFERENCES

Ahammed et al. *Journal of Experimental Botany* 2013; 64(1):199–213.

Ahanger et al. *Plant Physiology and Biochemistry* 2020; 147:31–42.

Ali et al. *Environmental and Experimental Botany* 2008; 62(2):153–159.

Ali et al. *Chemosphere* 2008; 72(9):1387–1392.

Ali et al. *Phytopathology* 2013; 103(12):1260–1267.

Allen et al. *Trends in Plant Science* 2001; 6(1):36–42.

Anjum et al. *Journal of Agronomy and Crop Science* 2011; 197(3):177–185.

Anuradha et al. *Plant Growth Regulation* 2001; 33(2):151–153.

Anuradha et al. *Plant, Soil and Environment* 2007; 53:465.

Arora et al. *Acta Physiologiae Plantarum* 2008; 30(6):833.

Arora et al. *Brazilian Journal of Plant Physiology* 2010; 22(3):159–165.

Ashraf et al. *Environmental and Experimental Botany* 2007; 59(2):206–216.

Asthir et al. *Biologia Plantarum* 2012; 56(4):757–761.

Bajguz et al. *Plant Physiology and Biochemistry* 2000; 38(3):209–215.

Bajguz et al. Kluwer Academic Publishers, Dordrecht, 2003; 1–44.

Bajguz et al. *Plant Physiology and Biochemistry* 2009; 47(1):1–8.

Bajguz et al. *Journal of Plant Physiology* 2009; 166(8):882–886.

Bajguz et al. Springer, 2011; 1–27.

Bajguz et al. *Archives of Environmental Contamination and Toxicology* 2011; 60(3):406–416.

Bartwal et al. *Journal of Plant Growth Regulator* 2012; 32:216–232.

Behnamnia et al. *Turkish Journal of Botany* 2009; 33:417–428.

Bhardwaj et al. In *Brassinosteroids: A Class of Plant Hormone* 2011; 439–458.

Br et al. *Plant Molecular Biology* 2009; 69(4):451–462.

Bukhari et al. *Environmental Science and Pollution Research International* 2016; 23(18):18229–18238.

Burbulis et al. *Journal of Food, Agriculture and Environment* 2011; 9:371–374.

Canales et al. *PLOS ONE* 2011; 11:e0146223.

Caño-Delgado et al. *Development* 2004; 131(21):5341–5351.

Chen, J et al. *The Plant Cell* 2017; 29(6):1425–1439.

Choudhary et al. *PLOS ONE* 2012; 7(3):33210.

Choudhary et al. *Journal of Experimental Botany* 2012; 63(15):5659–5675.

Churikova et al. *Brassinosteroids- A New Class of Plant Hormones*, Academic Press, California, 1997.

Clouse et al. *Plant Physiology* 1996; 111(3):671–678.

Clouse et al. *Plant Cell* 2011; 23(4):1219–1230.

Clouse et al. *Journal of Plant Growth Regulation* 2015; 34(4):828–844.

Confraria et al. *Molecular and Integrative Physiology* 2007; 146:279–279.

Cukor et al. *Baltic Forestry* 2018; 24:60–67.

Culbreath et al. *Annual Review of Phytopathology* 2003; 41:53–75.

de Lucas et al. *New Phytologist* 2014; 202(4):1126–1141.

Dejonghe et al. *Plant Biology* 2014; 22:48–55.

Deng et al. *Journal of Experimental Botany* 2015; 66(20):6219–6232.

Deng et al. *Scientific Reports* 2016; 6:20579.

Derevyanchuk et al. *Plant Growth Regulation* 2016; 78(3):297–305.

Dhaubhadel et al. *Plant Molecular Biology* 1999; 40(2):333–342.

Dhaubhadel et al. *Plant Journal: For Cell and Molecular Biology* 2002; 29(6):681–691.

Divi et al. *Plant Biotechnology Journal* 2016; 14(1):419–432.

Domagalska et al. *PLOS ONE* 2010; 5:14012.

Dong et al. *Current Opinion in Plant Biology* 1998; 1(4):316–323.

dos et al. *Journal of Soil Science and Plant Nutrition* 2020; 20(1):105–124.

Efimova et al. *Russian Journal of Plant Physiology* 2014; 61(6):733–743.

Eleiwa et al. *Australian Journal of Basic and Applied Sciences* 2011; 5:58–65.

Eremina et al. *Proceedings of the National Academy of Sciences of the United States of America* 2016; 113:982–991.

Fàbregas et al. *Nature Communications* 2018; 9(1):4680.

Fariduddin et al. *Photosynthetica* 2011; 49(1):55–64.

Farooq et al. *Journal of Agronomy and Crop Science* 2009; 195(4):262–269.

Feng et al. *Plant Science* 2015; 234:163–173.

Filová et al. *Acta Universitatis Agriculturae et Silviculturae Mendelianae Brunensis* 2013; 61(3):623–629.

Fujioka et al. *Plant Hormones* 1999; 21–45.

Furio et al. *European Journal of Plant Pathology* 2019; 153(1):167–181.

Gampala et al. *Developmental Cell* 2007; 13(2):177–189.

Gou et al. *PLOS Genetics* 2012; 8(1).

Griffin et al. *Functional Ecology* 2018; 32(7):1746–1756.

Grove et al. *Nature* 1979; 281(5728):216–217.

Grown et al. *Pakistan Journal of Botany* 2006; 38:1621–1632.

Gudesblat et al. *Current Opinion in Plant Biology* 2011; 14(5):530–537.

Guo et al. *Current Opinion in Plant Biology* 2013; 16(5):545–553.

Gupta et al. *Plant and Soil* 2017; 411(1–2):483–498.

Hasan et al. *Chemosphere* 2011; 84(10):1446–1451.

Hayat et al. In *Brassinosteroids*, Springer, Dordrecht 2003; 231–246.

Hayat et al. *Indian Journal of Plant Physiology* 2007; 12:207.

Hayat et al. *Environmental and Experimental Botany* 2007; 60(1):33–41.

Hayat et al. *Environmental and Experimental Botany* 2010; 69(2):105–112.

Hayat et al. *Saudi Journal of Biological Sciences* 2012; 19(3):325–335.

He et al. *Proceedings of the National Academy of Sciences of the United States of America* 2002; 99(15):10185–10190.

He et al. *New Phytologist* 2017; 214(1):388–399.

Hniličkа et al. *Scientia Agriculturae Bohemica* 2010; 41:15–20.

Honjo et al. *Japanese Journal of Crop Science* 1991; 60(1):29–35.

Hu et al. *Photosynthetica* 2010; 48(4):537–544.
Hu et al. *Biologia Plantarum* 2010; 54(4):761–765.
Hu et al. *Scientia Horticulturae* 2013; 150:232–237.
Hussain et al. *Agricultural Systems* 2007; 94(2):494–501.
Irfan et al. *Protoplasma* 2010; 241(1–4):3–17.
Jaillais et al. *Genes and Development* 2011; 25(3):232–237.
Jakubowska et al. *Plant Science* 2017; 264:37–47.
Jan et al. *Journal of Plant Growth Regulation* 2020; 1–18.
Janeczko et al. *Biologia Plantarum* 2007; 51:355–358.
Janeczko et al. *Acta Physiologia Plantarum* 2011; 33(4):1249–1259.
Jiang et al. *Plant Physiology* 2013; 162(4):1965–1977.
Jie et al. *Journal of Nuclear Agriculture and Biology* 2015; 29:1001–1008.
Jin et al. *AoB Plants* 2015; 7.
Jin et al. *Plant Biotechnology Journal* 2017; 15(10):1309–1321.
Kagale et al. *Planta* 2007; 225(2):353–364.
Kagale et al. *Plant Physiology* 2010; 152(3):1109–1134.
Kanwar et al. *Journal of Plant Growth Regulation* 2017; 36(4):1002–1030.
Kaur et al. *Scientific Reports* 2018; 8(1):1–13.
Kemmerling et al. *Plant Signaling and Behavior* 2008; 3(2):116–118.
Khripach et al. Academic Press, San Diego. 1999.
Khripach et al. *Annals of Botany* 2000; 86(3):441–447.
Kim et al. *Res. Rep. Rural Develop. Admin. Rice* 1989; 31:49–53.
Kim et al. *Nature Cell Biology* 2009; 11(10):1254–1233.
Kim et al. *Molecular Cell* 2011; 43(4):561–571.
Korableva et al. *Biologia Plantarum* 2002; 45(1):39–40.
Krishna et al. *Journal of Plant Growth Regulation* 2003; 22(4):289–297.
Kumari et al. *IUP Journal of Life Sciences* 2011; 5.
Lalotra et al. *Annual Research and Review in Biology* 2017:1–8.
Lee et al. *Plant Physiology and Biochemistry* 2018; 127:553–560.
Leubner-Metzger et al. In *Brassinosteroids*, Springer, Dordrecht, 2003; 119–128.
Li et al. *Cell* 1997; 90(5):929–938.
Li et al. *Cell* 2002; 110(2):213–222.
Li et al. *New Forests* 2008; 35(3):255–266.
Li et al. *Current Opinion in Plant Biology* 2010; 13(5):540–547.
Li et al. *Biochemical and Biophysical Research Communications* 2012; 426(4):522–527.
Li et al. *PLOS ONE* 2013; 8(6):e66582.
Li et al. *Acta Physiologia Plantarum* 2015; 37:22.
Li et al. *Frontiers in Plant Science* 2016; 7:1–16.
Li et al. *Frontiers in Plant Science* 2016; 7:615.
Liu et al. *Journal of Applied Phycology* 2018; 30(1):253–260.
Liu et al. *BMC Genetics* 2018; 19(1):1–10.
Liu et al. *Plant Growth Regulation* 2009; 59(3):207–214.
Liu et al. *Journal of Plant Physiology* 2011; 168(9):853–862.
Mahesh et al. *American Journal of Plant Sciences* 2013.
Marakli et al. *Plant Biosystems - An International Journal Dealing with all Aspects of Plant Biology* 2018; 152(3):324–332.
Mazorra et al. *Biologia Plantarum* 2002; 45(4):593–596.
Mazorra et al. *Plant Physiology and Biochemistry* 2011; 49(12):1420–1428.
Mitchell et al. *Nature* 1970; 225(5237):1065.
Munns et al. *Annual Review of Plant Biology* 2008; 59:651–681.
Nahar et al. *Molecular Plant–Microbe Interactions* 2013; 26(1):106–115.
Nakashita et al. *The Plant Journal* 2003; 33(5):887–898.
Nam et al. *Cell* 2002; 110(2):203–212.
Nasser et al. *International Journal of Agriculture and Biology* 2004; 6:771–775.
Neill et al. *Plant Physiology* 2002; 128(1):13–16.
Nolan et al. *Biochemical Journal* 2017; 474(16):2641–2661.
Northey et al. *Nature Plants* 2016; 2:1–7.
Ogweno et al. *Journal of Plant Growth Regulation* 2008; 27(1):49–57.
Oh et al. *Proceedings of the National Academy of Sciences of the United States of America* 2009; 106(2):658–663.
Oh et al. *Nature Cell Biology* 2012; 14(8):802–809.
Ohri et al. *Indian Journal of Nematology* 2002; 32:175–178.
Oidaira et al. *Journal of Plant Physiology* 2008; 156:811–813.
Panda et al. *Molecular Biotechnology* 2013; 54(2):551–563.
Peng et al. *Molecular Plant* 2008; 1(2):338–346.
Pirzadah et al. In Plant Signaling: Understanding the Molecular Crosstalk 2014; 193–226.
Pshenichnaya et al. In *Problems of Exp. Botany.* Belorussian Science, Minsk, 1997; 210–217.
Pustovoitova et al. In *Doklady Biochemistry and Biophysics* 2001; 376:36–38. Kluwer Academic Publishers-Plenum Publishers.
Ramakrishna et al. *Protoplasma* 2015; 252(2):665–677.
Rane et al. *Indian Journal of Agricultural Sciences* 2002; 72:581–585.
Rao et al. *Indian Journal of Plant Physiology* 1997; 2:156–157.
Rao et al. *Current Science* 2002:1239–1245.
Rao et al. *Journal of Medicinal Plants* 2016; 4:06–08.
Rattan et al. *Journal of Plant Growth Regulation* 2020; 1–11.
Rehman et al. *Crop and Pasture Science* 2016; 67(7):712–718.
Ribeiro et al. *Annals of Forest Science* 2020; 77:1–19.
Rodkin et al. In *Plant Growth and Development Regulators.* Moscow. 1997; 317–318.
Rodoni et al. *Virus Research* 2009; 141(2):150–157.
Rontein et al. *Metabolic Engineering* 2002; 4(1):49–56.
Roth et al. *Zeitschrift für Naturforschung Section C* 2000; 55(7–8):552–559.
Ryu et al. *Plant Cell* 2007; 19(9):2749–2762.
Sakamoto et al. *Nature Biotechnology* 2006; 24(1):105–109.
Santos et al. *Photosynthetica* 2018; 56(2):591–605.
Shahan et al. *Plant Cell* 2019; 31(5):933–934.
Shahbaz et al. *Pakistan Journal of Botany* 2007; 39:513.
Sharma et al. *Plant Physiology and Biochemistry* 2013; 69:17–26.
Sharma et al. *Brazilian Archives of Biology and Technology* 2018; 61.
Sharma et al. *Acta Physiologiae Plantarum* 2007; 29(3):259–263.
Shu et al. *Agricultural Science and Technology* 2014; 15:1433.
Shu et al. *Plant Physiology and Biochemistry* 2016; 107:344–353.
Shummu et al. *American Journal of Plant Sciences* 2012.
Smagghe et al. *Insect Biochemistry and Molecular Biology* 2002; 32(2):199–204.
Soares et al. *Ecotoxicology and Environmental Safety* 2020; 193:110296.
Song et al. *Plant Journal: For Cell and Molecular Biology* 2013; 74(6):905–919.
Sri et al. *American Journal of Plant Sciences* 2016; 7:1618–1628.
Stone et al. Food Products Press, Binghamton, NY, 2001; 302.
Sun et al. *Developmental Cell* 2010; 19(5):765–777.
Swamy et al. *International Journal of Multidisciplinary and Current Research* 2014; 2:317–321.
Tabur et al. *Plant Growth Regulation* 2009; 58(1):119–123.
Tang et al. *Science* 2008; 321(5888):557–560.

Tang et al. *Nature Cell Biology* 2011; 13(2):124–131.
Tofighi et al. *Journal of Plant Nutrition* 2017; 40(8):1091–1098.
Vardhini et al. *Plant Stress* 2010; 4:56–61.
Vardhini et al. *Frontiers in Environmental Science* 2015; 2:67.
Vardhini et al. *Int. J. Earth Sci. Geol.* 2019; 1(2):59–65.
Vert et al. *Nature* 2006; 441(7089):96–100.
Vriet et al. *The Plant Cell* 2012; 24(3):842–857.
Vukas_inovic´ et al. *Trends in Plant Science* 2018; 23:285–292.
Wang et al. *Journal of Plant Physiology and Molecular Biology* 1993; 19:38–42.
Wang et al. *Developmental Cell* 2002; 2(4):505–513.
Wang et al. *Plant Cell* 2005; 17(6):1685–1703.
Wang et al. *Science* 2006; 313(5790):1118–1122.
Wang et al. *Developmental Cell* 2008; 15(2):220–235.
Wang et al. *Developmental Cell* 2011; 21(5):825–834.
Wang et al. *Planta* 2012; 235(1):53–67.
Wei et al. *Frontiers in Plant Science* 2015; 6:982.
Wu et al. *Acta Physiologia Plantarum* 2014; 36(2):251–261.
Wu et al. *Frontiers in Plant Science* 2017; 8:1017.
Xi et al. *Plant Growth Regulation* 2013; 71(1):57–65.
Xia et al. *Plant Physiology* 2009; 150(2):801–814.
Xu et al. *Plant Physiology and Biochemistry* 2015; 94:197–208.
Yamamuro et al. *The Plant Cell* 2000; 12(9):1591–1605.
Yang et al. *New Phytologist* 2014; 203(2):437–448.
Yang et al. *The Plant Cell* 2018; 30(1):48–66.
Ye et al. *Proceedings of the National Academy of Sciences* 2010; 107(13):6100–6105.
Yin et al. *Cell* 2002; 109(2):181–191.
Yin et al. *Cell* 2005; 120(2):249–259.
Yin et al. *Plant Growth Regulation* 2008; 54:45–54.
Yokota et al. *Tetrahedron Letters* 1982; 23:1275–1278.
Youn et al. *Molecular Plant* 2015; 8(4):552–565.
Yu et al. *Plant Journal: For Cell and Molecular Biology* 2011; 65(4):634–646.
Yuan et al. *Scientia Horticulturae* 2010; 126(2):103–108.
Yuan et al. *Horticultural Plant Journal* 2016; 2(2):105–113.
Zhang et al. *Australian Journal of Agricultural Research* 2007; 58(8):811–815.
Zhang et al. *Plant Growth Regulation* 2008; 56(3):257–264.
Zhang et al. *Plant and Cell Physiology* 2011; 52(1):181–192.
Zhang et al. *Planta* 2015; 241(4):875–885.
Zhao et al. *Plant Physiology* 2002; 130(3):1221–1229.
Zhou et al. *Journal of Experimental Botany* 2014; 65(15):4371–4383.
Zhu et al. *Postharvest Biology and Technology* 2010; 56(1):50–55.
Zhu et al. *Journal of Research in Agriculture and Animal Science* 2014; 2:1–6.
Zhu et al. *Molecular Cell* 2017; 66(5):648–657.

4

Role of Brassinosteroids During Abiotic Stress Tolerance in Plants

Navdeep Kaur, Shivani Saini, and Pratap Kumar Pati

CONTENTS

4.1 Introduction

The production of food crops needs to be increased by 70% to sustain the growing food demand of an expected increase of about 2.3 billion people by the end of the year 2050 (Tilman et al., 2011; Wani et al., 2016). However, a sizable portion of food crops is lost annually due to the sensitivity of plants towards abiotic stresses. Different abiotic factors such as increased salinity, drought, extreme temperatures, and so on impose a serious threat to the optimum growth and development of plants (Sharma et al., 2019; Vakilian, 2020). These stresses are estimated to cause an almost $14–19 million loss annually to the global economy (Martinez et al., 2018). Moreover, as the world has reached a consensus that the global climate is changing, the problem of abiotic stresses is expected to be further escalated (Raza et al., 2019). It is predicted that in the coming three to five decades, the earth's temperature will change by almost 2–3°C, which will have a huge impact on crop productivity (Hatfield and Pruegar, 2015; Parry et al., 2007). The most significant direct effect of this global change on plants will be the imposition of heat stress (Bita and Gerates, 2013). In addition, it will influence the availability of water to plants as there will be an increased occurrence of drought stress in different regions of the world (Ferguson et al., 2019; Raza et al., 2019). Moreover, rising sea levels due to global warming will lead to increased contamination of inland water with saline water (Jeppesen et al., 2015; Kaur et al., 2019a). Thus, the exposure of plants to abiotic stresses in the current scenario of rapid climate change has become a matter of concern for plant biologists all over the world.

Researchers have explored different strategies to decipher a practical solution for the problem of abiotic stresses in plants (Marothia et al., 2020; Nguyen et al., 2018). Classically, plant biologists focused on developing abiotic stress-resilient crops using breeding. In the past few decades, interest has mainly shifted towards genetic engineering. Recently, the use of genome editing has been in the limelight for modulating abiotic stress endurance in plants (Zafar et al., 2020). However, there are still many limitations to the successful use of these approaches in the present scenario of the dual challenge of abiotic stresses and climate change. Breeding is laborious and time demanding, whereas genetically engineered crops lack public acceptance due to possible health risks associated with them (Ashkani et al., 2015; Maghari and Ardekani, 2011; Zambounis et al., 2020). Genome editing has been recognized as a powerful approach for improving various traits including abiotic stress tolerance in plants, however, again, there are several technical and practical concerns about its successful use in agriculture (Jansing et al., 2019). In this context, an eco-friendly approach to solve the problem of abiotic stresses in plants is needed.

Phytohormones are chemical compounds that act as signaling molecules and regulate almost all the processes associated with plant growth and development (Khan et al., 2020). Initially, phytohormones were classified into five main groups: auxins, gibberellins, cytokinins, abscisic acid, and ethylene. Later, brassinosteroids, salicylic acid, polyamines, jasmonates, and strigolactones were added to this list (Khan et al., 2020; Santner and Estelle, 2009). Along with their well-established role in the regulation of a plant's growth processes, phytohormones also activate various vital pathways that mediate stress acclimatization. Different phytohormones either act alone or in crosstalk with each other for regulating a plethora of events related to plant's responses to stress (Wani et al., 2016). The exogenous application of phytohormones has been successfully used to ameliorate the negative effects of abiotic

DOI: 10.1201/9781003110651-4

stresses in plants (Nawaz et al., 2017). Brassinosteroids (BRs), which are recognized as the sixth class of phytohormones, play a significant role in the modulation of different traits associated with growth and development in plants (Alhaithloul et al., 2020). They have been well documented in the literature for their potential as protective agents against a range of abiotic stresses (Sharma et al., 2017). In the past few years, considerable progress has been made toward cognizance of the molecular events associated with BR-mediated stress adaptation in plants. Thus, the present chapter focuses on the potential of BRs in the amelioration of the negative effects of different abiotic stresses in plants.

4.2 Brassinosteroids: An Important Phytohormone

BRs are vital plant growth regulators that play a significant role in the modulation of different morphological, physiological, anatomical, and biochemical processes of plants (Hussain et al., 2020; Kaur et al., 2019b). They are reported to be omnipresent in all plants and regulate almost all major processes involved in the life cycle, ranging from seed germination to senescence (Gudesblat et al., 2011; Nolan et al., 2020; Saini et al., 2015). They were discovered by Mitchell et al. in 1970 from *Brassica* pollens. They identified specific plant components from *Brassica* pollens that showed significant growth-enhancing effects (Mitchell et al., 1970). Preliminary histological studies conducted by Mitchell and his coworkers showed that biological reactions induced by the extracted components were different from that of gibberellins. Thus, they identified the isolated ingredients as the sixth class of plant hormones and named them "brassins" (Mitchell et al., 1970). Later, a research group from the US Department of Agriculture (USDA) purified the brassins from *Brassica* pollens collected from bees. They used 500 pounds of pollens to purify 4 mg of brassins followed by a prediction of its crystal structure that led to the identification of brassinolide (BL) as the active constituent (Grove et al., 1979). After these initial studies, almost 70 different kinds of brassinolide analogs have been identified from various tissues in a range of plant species (Kutschera et al., 2012). Based on their chemical structure, BRs have been classified into three major classes: C_{27}, C_{28}, and C_{29} (Bajguz and Hayat, 2009; Vardhini and Anjum, 2015). They are found to be widely distributed among various plant tissues. The highest content of BRs has been observed in immature seeds, pollens, flowers, and roots, whereas the lowest levels are reported in leaves and shoots (Takatsuto, 1994). Grafting experiments conducted using BR-deficient mutants have shown that endogenous BRs function in an autocrine or paracrine manner and do not move from one tissue to another (Bishop et al., 2001; Symons et al., 2008). It has been speculated that BRs could be synthesized *in situ* in various plant organs as BR biosynthetic genes have been identified from a range of plant tissues. For inducing long-distance actions, BRs may crosstalk with other plant growth regulators, including gibberellins and auxins (Lacombe et al., 2016).

BRs are recognized at the cellular membranes by the Brassinosteroid insensitive 1 (BRI1) receptor, which has shown a high degree of conservation among various plant species (Clouse, 2015). In *Arabidopsis*, three homologs of BRI1 have been identified, viz. BRL1, BRL2, and BRL3 (Cano-Delgado et al., 2004). Once activated, BRI1 interacts with Brassinosteroid insensitive 1-associated receptor kinase 1 (BAK1) and thus initiates a cascade of events for activating BR-regulated events in the plant cell (Belkhadir and Jaillais, 2015). BRs play a critical role in regulating almost all the vital processes involved in the growth and development of plants. They promote cell elongation and expansion in association with auxins. BRs are also reported to participate in the regulation of cell division and the regeneration of cell walls. Moreover, they are involved in the promotion of vascular differentiation, pollen elongation during the formation of the pollen tube, and seed germination. Furthermore, they play an important role in plant reproduction, photosynthesis, senescence, epinasty, and so on (Hussain et al., 2020; Nolan et al., 2020; Saini et al., 2015). Apart from these well-established roles, BRs counteract the negative effects of various abiotic stresses in plants (Sharma et al., 2017).

4.3 Role of Brassinosteroids in Abiotic Stress Tolerance

BRs have gained the attention of researchers for their potential to confer abiotic stress endurance in plants (Sharma et al., 2017). Their exogenous use has been well-documented to counteract the negative impact of various abiotic stresses in plants (Nawaz et al., 2017). It has been realized that BR-mediated modulation to abiotic stress endurance in plants is dependent on a range of factors, viz. plant species, type and duration of stress, growth stage and conditions of growth of the plant, dose of BRs, and so on (Ahammed et al., 2020; J Ahammed et al., 2015). Moreover, different mutant studies have confirmed their involvement in the regulation of abiotic stress responses in plants and have also given insights into the crosstalk of BR signaling and abiotic stress-responsive pathways (Ahammed et al., 2020). In the following sections, the potential of BRs in the amelioration of different abiotic stresses will be discussed in detail.

4.3.1 Salinity

Salt stress negatively affects plant growth and development, thus significantly contributing to worldwide crop loss (Arif et al., 2020; Kaur et al., 2017; Kaur et al., 2020). Several studies have justified the role of BRs in coping with the harmful effects of salt stress through the regulation of multiple processes in plants (Vázquez et al., 2019). BRs protect plants against the negative impact of salt stress predominantly by enhancing the activities of different antioxidant enzymes to protect the plants against salinity-induced oxidative stress (Arora et al., 2008; Dong et al., 2017; El-Khallal et al., 2009). In maize plants, BR application enhanced the levels of antioxidant enzymes such as superoxide dismutase (SOD), glutathione reductase (GR), catalase (CAT), glutathione peroxidase (GPX), and ascorbate peroxidase (APX) (El-Khallal et al., 2009). Furthermore, the application of BR also increased the content of plant hormones

(IAA, GA, Zeatin) and promoted the accumulation of several defense-related proteins implicated in providing salt stress resistance (El-Khallal et al., 2009). In another study, seedlings of *Zea mays* L. (var. Partap-1) subjected to salt stress exhibited an enhancement in the activities of antioxidative enzymes such as SOD, guaiacol peroxidase, CAT, and APX with 28-homobrassinolide (HBL) treatment (Arora et al., 2008). In *Cucumis sativus* seedlings, exogenous BR application significantly enhanced the activities of SOD, POD, and CAT, and subsequently decreased the salt stress index, mortality rate, malondialdehyde (MDA) content, and electrolyte leakage (Qingmao et al., 2006). In *Cucumis sativus* L. seedlings, exogenous BR application resulted in a lowering of the salt injured index under the effect of salt stress (Qingmao et al., 2006). Furthermore, BR significantly enhanced the levels of free-proline, soluble sugar, and the activity of antioxidative enzymes such as SOD, POD, and CAT in the leaves of *Cucumis sativus* L. (Qingmao et al., 2006). BR treatment increased the activities of leaf SOD, POD, and CAT, while decreasing the leaf superoxide anion production rate, hydrogen peroxide level, MDA content, and cell membrane permeability in cucumber seedlings (Xiao-min and Yang, 2013). On BR application, seedling growth was enhanced as leaves showed a higher photosynthetic rate, leading to effective mitigation of the damage caused by salinity stress (Xiao-min and Yang, 2013). BR application also significantly improved the germination percentage, germination index, seedling growth, and vigor index of the seeds of *Medicago sativa* under salt stress (Zhang et al., 2007). Moreover, BR treatment significantly enhanced the activities of antioxidant enzymes such as POD, SOD, and CAT, while reducing the MDA accumulation in *M. sativa* plants exposed to salinity (Zhang et al., 2007). Exogenously applied 24-epibrassinolide (EBL) to *Oryza sativa* var Pusa Basmati-1 resulted in better growth, proline concentration, levels of protein content, and antioxidant enzyme activity while decreasing the MDA content (Sharma et al., 2013). Further, 24-epibrassinolide (EBL) upregulated the expression of different oxidative stress marker genes (Sharma et al., 2013). In addition, the application of EBL and salt together resulted in the elevated expression level of *OsBRI1,* a BR signaling gene (Sharma et al., 2013). In another study, the positive effect of EBL on wheat growth and development has been highlighted under NaCl toxicity (Dong et al., 2017). EBL mediated the augmentation in the content of free proline and soluble proteins. It also enhanced antioxidant enzyme activities, chlorophyll content, root activity, and H^+-ATPase activity. Furthermore, exogenous EBL application enhanced the K, Ca, Mg, Fe, and Zn uptake to promote plant growth under salinity stress conditions (Dong et al., 2017). In *Malus hupehensis* (a widely grown apple rootstock), it was observed that exogenous brassinolide (BL) application maintained photosynthetic capacity, improved the activities of SOD and CAT, and promoted the accumulation of proline, soluble sugar, leading to the alleviation of salinity stress (Su et al., 2020). In addition, BL application also increased K^+ content in shoots and roots by regulating the expression levels of $Na^+(K^+)/H^+$ antiporter genes (*MhNHXs*) through *MhBZRs*. It was also indicated that *MhBZR1* and *MhBZR2* can directly bind to the promoter of *MhNHX4-1* and inhibit its expression (Su et al., 2020). Furthermore, studies suggest crosstalk between BR and ethylene in H_2O_2 production that positively contributes to alleviating oxidative damage caused by salinity stress in *Solanum Lycopersicum* L (tomato; Zhu et al., 2016). It was conjectured that ethylene signaling molecules, such as ethylene insensitive 2 (EIN2), ethylene insensitive 3-like (EILs), and ethylene response factors (ERFs), also participated in BR-mediated salt stress tolerance (Zhu et al., 2016). Mutant analysis has revealed that oxidative damage caused by salinity stress is higher in *SlEIN2*, *SlEILs*, and *SlERFs* silenced plants than control plants under BL treatment (Zhu et al., 2016). BR-induced salt stress tolerance has also been correlated with the EIL-dependent signaling pathway through the action of mitogen-activated protein kinases (MAPK), which promotes ethylene accumulation, triggering salt stress tolerance. Thus, it was speculated that BR-mediated salt stress tolerance is regulated by some important components of ethylene signaling molecules in tomato (Zhu et al., 2016).

The critical function of BR signaling in regulating salt stress tolerance is primarily regulated by the BR receptor, BRI1 (Cui et al., 2012). Mutation of the ubiquitin-conjugating enzyme (UBC32) {a functional endoplasmic reticulum [ER]-associated protein degradation [ERAD] component} resulted in the accumulation of *bri1-5* and *bri1-9*, the mutant forms of BRI1. These mutant forms subsequently activated BR signal transduction for promoting BR-mediated salt stress tolerance in plants (Cui et al., 2012). Thus, this study indicated the critical role of UBC32 in the regulation of the BR signaling pathway during salt stress. Moreover, high salinity also causes growth retardation by suppressing the brassinazole resistant 1 (BZR1) transcription factor and subsequent BR signaling-related functions (Geng et al., 2013). In *bzr1-D* mutants, growth was observed to be strongly repressed for several hours, indicating the critical role of BR signaling in salt stress amelioration (Geng et al., 2013). In *Arabidopsis*, the role of endogenous BR in salt stress adaptation was analyzed using BR mutants such as *det2-1* and *bin2-1* (Zeng et al., 2010). These mutants were found to be hypersensitive to salinity stress compared with Columbia wild type and showed significantly reduced seed germination and seedling growth. These growth-related defects were found to be partially rescued by the exogenous application of BRs (Zeng et al., 2010). Thus, the study demonstrated that modulation of endogenous BR levels and BR signaling are positively involved in the salt stress tolerance mechanism in *Arabidopsis*.

4.3.2 Heavy Metal Stress

BRs play diverse functions in regulating a plant's responses to heavy metal stress (Anwar et al., 2018; Rajewska et al., 2016). BRs alter the accumulation of different heavy metals such as cadmium (Cd), lead (Pb), copper (Cu), and zinc in plants including tomato, barley, and radish (Anuradha and Rao, 2007a,b; Hasan et al., 2011; Hayat et al., 2010; Ramakrishna and Rao, 2013, 2015). They also mitigate the damaging effects of heavy metals by augmenting photosynthetic rate, activating antioxidant defense mechanisms, and upregulating stress-responsive genes and transcription factors in various plants (Anwar et al., 2018). In tomato, foliar EBL application alleviated the negative impact of Cd stress by enhancing enzymatic activities and the expression levels of various antioxidant enzyme encoding genes and mitigated photosynthetic inhibition (Ahammed et al., 2013). In addition, detoxification of Cd was also enhanced with EBL treatment, suggesting its critical role in reducing Cd residues and alleviating heavy metal stress (Ahammed et al., 2013). In rape cotyledons, BR mitigated the negative impact of cadmium on photosynthetic pathways by enhancing the activities of O_2 evolving complexes and preventing the damage caused to photochemical reaction centers (Janeczko et al., 2005). The study displayed the critical role of BRs in improving photosynthetic rate by maintaining the efficiency of the photosynthetic electron transport system and protecting the activities of various enzymes involved in photosynthesis-related processes in rape cotyledons (Janeczko et al., 2005). Furthermore, it has also been demonstrated that pre-soaking of seeds in 28-HBL later increases the content of chlorophyll (Chl), different proteins, and proline, decreases malondialdehyde (MDA) level, reduces metal uptake, and improves growth in *Raphanus sativus* under Cr stress (Sharma et al., 2011).

In plants, heavy metal stress promotes ROS production that imposes oxidative stress (Ali et al., 2019; Kohli et al., 2017). Plants generally overcome these oxidative damages by the activation of the antioxidant defense system. BRs have been recognized as of particular importance in the regulation of antioxidant defense mechanisms in plants (Sharma et al., 2011, 2018; Sytar et al., 2019). BRs promoted the expression of *Cu/Zn SOD*, *MnSOD*, *CAT1*, *CAT2*, *CAT3*, and *FeSOD* under Cd stress, while augmenting transcript levels of *SOD* and *CAT* (except *CAT2*) under Cr stress in radish seedlings (Sharma et al., 2018b). Moreover, the application of Cr stress in combination with 28-HBL significantly induced the activities of all the antioxidant enzymes except guaiacol peroxidase in *Raphanus sativus* (Sharma et al., 2011). Although the role of exogenous application of BRs in the alleviation of HM stress has been well documented in the literature, the effect of heavy metal stress on BR biosynthesis, signaling, and the transport mechanism is still lacking. Therefore, deeper studies are required to investigate the role of heavy metals in the modulation of BR homeostasis in the future.

4.3.3 Drought

Drought stress is one of the major environmental constraints that severely affects the productivity of several agricultural crops (Anwar et al., 2018). In recent years, numerous studies have explored the critical role of BR in mitigating the effects of drought stress on plants (Ahammed et al., 2020; Anwar et al., 2018; Planas-Riverola et al., 2019). In tomato, EBL treatment alleviated drought stress by increasing the net photosynthetic rate, relative water content, and the activities of antioxidant enzymes such as CAT, APOX, and SOD (Yuan et al., 2010). On the contrary, a decline in the levels of H_2O_2, MDA, stomatal conductance, and intercellular CO_2 concentration was found upon its application in drought-exposed tomato plants. Interestingly, considerable enhancement in the content of ABA was observed in the two tomato genotypes Mill. cv. Ailsa Craig and ABA-deficient mutant notabilis upon EBL application under drought stress (Yuan et al., 2010). This investigation suggested the critical role of EBL-mediated enhancement of endogenous ABA content and the elevated activities of antioxidant enzymes in ameliorating the damaging effects of drought stress. Moreover, BR priming enhanced the yield of peanut plants and decreased the inhibitions mediated by drought stress (Huang et al., 2020). Further, it enhanced the levels of defense-related plant hormones such as ABA and salicylic acid that participate in the process of drought stress adaptation (Huang et al., 2020).

In recent years, it has been revealed that BR signaling plays a pivotal role in providing drought stress resistance through regulating the transcript levels of various drought-responsive genes in plants. Overexpression of brassinosteroid LRR receptor kinase (*BRL3)*, a member of the BR receptor family, can prevent the arrest of growth during drought stress, and hence may be used to engineer drought-tolerant crops (Fàbregas et al., 2018). Further, increased *BRL3* expression triggers the accumulation of osmoprotectants such as proline and sugars under drought stress (Fàbregas et al., 2018). In *Brachypodium distachyon*, the loss-of-function of *BRASSINOSTEROID INSENSITIVE 1*-RNA interference (*BdBRI1-RNAi*) mutants exhibited reduced plant height, shortened internodes, and narrow and short leaf under drought stress (Feng et al., 2015). BR signaling genes such as *BdBES1*, *BdBZR1*, and *brassinolide enhanced 2* (*BdBLE2)* displayed reduced expression, on the contrary, the transcript levels of BR biosynthesis genes *BdD2*, *BdCPD*, and *BdDWF4* were highly elevated in *BdBRI1*-RNAi mutants (Feng et al., 2015). Interestingly, enhanced drought stress resistance and significantly upregulated expression of drought-responsive genes such as *BdP5CS* and *BdCOR47/BdRD17* were also observed in these mutants, suggesting the key role of BR signaling in drought stress tolerance (Feng et al., 2015). BRI1-EMS suppressor (BES)/brassinazole-resistant (BZR) family transcription factors are implicated in the regulation of a variety of physiological processes. In the recent past, the critical role of BES/BZR transcription factors in mitigating the damaging effects of drought stress has been investigated in wheat (Cui et al., 2019). In *Triticum aestivum*, overexpression of *brassinazole-resistant 2* (*TaBZR2*) leads to drought resistance, on the contrary, RNAi mediated down-regulation of *TaBZR2* promoted drought sensitivity (Cui et al., 2019). It was suggested that *TaBZR2* interacts with the promoter of *glutathione s-transferase-1 (TaGST1)* to activate it, which participates in the scavenging of superoxide anions accumulated as a result of drought stress (Cui et al., 2019). This study provided novel insights into the key role of *TaBZR2* in ameliorating the effects of drought stress in wheat by eliminating superoxide anions to promote drought stress tolerance (Cui et al., 2019). Hence, BRs play a significant role in ameliorating the damaging effects of drought stress; however, several molecular mechanisms regulated by BR in mitigating drought stress still remain to be elucidated.

4.3.4 Temperature

Temperature is one of the most crucial factors in regulating plant growth and developmental processes. Thus, a small fluctuation in the optimum temperature necessary for plant growth can seriously hamper the overall growth of plants (Bita and Gerats, 2013). BRs enhance the survival of plants growing under extreme temperatures by regulating multidirectional activities (Vardhini et al., 2019). They primarily modulate the structural/functional properties of cellular membranes for enhancing stress adaptation in plants exposed to low/high temperatures (Sadura and Janeczko, 2018). One of the pronounced effects of temperature extremities, particularly low temperature, is the generation of free radicals that result in lipid peroxidation of the cellular membranes (Aghdam et al., 2012; Cui et al., 2016). BRs significantly reduce the levels of membrane lipid peroxidation in different plants, including cucumber, eggplant, and pepper (Hu et al., 2010, Jiang et al., 2013, Wang et al., 2012, Wu et al., 2014). Further, *Chorispora bungeana* cultured in the presence of 24-epibrassinolide showed reduced MDA levels under low temperature conditions (Liu et al., 2009; 2011). Moreover, in studies conducted using BR mutant plants, it was observed that BR insensitive *bri1-116(-/-)* mutants showed significantly enhanced levels of MDA as compared with wild-type plants growing under cold stress. Alternatively, plants overexpressing *Dwarf*, a BR biosynthetic gene, were found to show decreased oxidative stress under cold stimuli (Qu et al., 2011; Xia et al., 2018). Similarly, plants exposed to heat stress also showed decreased levels of MDA after BR treatment (Sadura and Janeczko, 2018). Besides regulating the levels of MDA, BRs reduces the membrane damage caused by electron leakage in plants exposed to low temperature (Agdham et al., 2012; Janeczko, 2007). BRs also modulate the levels of unsaturated fatty acids in plants exposed to low temperature, which improves membrane fluidity, thus enhancing stress acclimation in plants (Li et al., 2012). Moreover, BRs promote the expression of TIL, ASR, TSD, REM, and ATP-synthase b subunit encoding genes that play a critical role in the formation as well as repair of membranes in plants exposed to heat/cold stress (Sadura and Janeczko, 2018).

Photosynthesis is another major process regulated by BRs in plants growing under heat/cold stress (Sadura and Janeczko, 2018). High/low temperature leads to a reduction in levels of photosynthetic pigments. BR application restores the photosynthetic pigment content in plants experiencing temperature variations. They modulate either the biosynthesis or the activities of various enzymes involved in the degradation of chlorophyll (Sadura and Janeczko, 2018). BRs also positively regulate the biosynthesis of chlorophyll. Studies conducted on barley BR-deficient mutants suggested that BRs regulate the process of chlorophyll biosynthesis in crosstalk with cytokinins (Sadura and Janeczko, 2018). They also enhance the net photosynthetic rate in plants by modulating the activities of photosynthetic enzymatic systems under low/high temperature stress. It positively stimulates the activities of Rubisco, RCA, and enzymes involved in the Calvin cycle for the regeneration of Rubisco (Jiang et al., 2013; Xia et al., 2009b). Apart from the regulation of pigment biosynthesis/degradation phenomenon, BRs participate in the regulation of gaseous exchange parameters in plants exposed to extreme temperatures. In cucumber, the application of BRs was observed to counteract the negative effects of cold stress on stomatal conductance, transpiration rate, and CO_2 concentration (Farridudin et al., 2011; Hu et al., 2010). During the process of cold resistance, the accumulation of sugars is an important protective phenomenon that results in lowering the freezing point of cellular solutions, thus protecting them from frost injury (Sadura and Janeczko, 2018). BRs modulate the levels of sucrose and other protective sugars in an attempt to provide low temperature stress tolerance in plants (Pociecha et al., 2017; Singh et al., 2012). Along with sugars, BRs also modulate the levels of proline in plants that play an important role in antioxidant defense and osmoprotection during temperature stress (Sadura and Janeczko, 2018).

At the molecular level, BRs regulate the transcript levels of cold-responsive (COR) proteins and heat shock proteins (HSP) encoding genes that are involved in cold and heat stress adaptation, respectively (Ritonga et al., 2020; Yadav et al., 2020). BRs regulate the expression of COR genes through constitutive activation of *C-repeat/dehydration responsive element binding factor* (*CBF*) transcription factors via the BR-controlled basic helix–loop–helix transcription factor (CES) (Eremina et al., 2016; Sadura and Janeczko, 2018). During exposure to heat stress, BRs result in the accumulation of HSPs that serve chaperonic activity in plants, thus protecting cellular proteins against heat stress (Dhaubhadel et al., 1999, 2002; Sadura and Janeczko, 2018; Samakovli et al., 2014).

4.3.5 Pesticides

BRs are well-known plant hormones that are extensively utilized for the management of pesticide stress in plants (Sharma et al., 2018). Plants are often exposed to several pests, resulting in the reduction of crop production. Various synthetic pesticides are utilized to prevent crop loss mediated by pest attacks. However, the continuous application of pesticides mediates resistance in pests and also promotes the disruption of various metabolic processes, thus negatively impacting plant yield and productivity. These pesticides may also persist in various plant parts, causing a threat to sustainable agriculture (Sharma et al., 2018a). To combat the negative impact of pesticide stress, BRs play a critical role in maintaining the homeostasis of plants by regulating physiological, biochemical, and molecular processes in plants (Hou et al., 2018; Shahzad et al., 2018, Xia et al., 2009a; Wang et al., 2017). In rice, HBL maintained the homeostasis of the plant and promoted growth under the effect of pesticide stress by enhancing levels of proline, protein, and chlorophyll while reducing MDA content (Sharma et al., 2015). Exogenous application of HBL also promoted the activity of antioxidant enzymes to reduce $O^{\cdot -}_2$ and H_2O_2 accumulation to alleviate the negative impact of pesticides (Sharma et al., 2015). In another study, it has been shown that EBL treatment counteracts the negative effect of pesticide application on the quantum yield of PSII and net photosynthetic rate in cucumber (Xia et al., 2009a). EBL application promotes the metabolism of pesticide chlorpyrifos by enhancing the activity and expression of enzymes such as glutathione reductase (GR), glutathione *S*-transferase (GST), P450 monooxygenase (P450), peroxidase (POD), and MRP-type transporter (MRP; Xia et al., 2009a). In grapevine, treatment with EBL stimulated the accumulation of soluble protein and free proline, while reducing MDA content under the effect of chlorothalonil (CHT; Wang et al., 2017). Furthermore, upon EBL application, the transcript levels of key genes such as

GST, *GR*, *P450*, and *MRP* that participate in CHT degradation were also found to be elevated (Wang et al., 2017). In another study, foliar as well as root application of BRs in tomato plants exposed to pesticides improved the photosynthetic machinery and enhanced the activities of GPX, CAT, and GR (Ahammed et al., 2012). In addition, BRs also reduced glutathione (GSH), oxidized glutathione (GSSG), and the content of MDA, indicating the critical role of BR in accelerating detoxification activity against the pesticides for plant protection (Ahammed et al., 2012). In *Brassica juncea* L., EBL application raised the GSH content and activities of GR, POD, and GST under the effect of the commonly used pesticide imidacloprid (IMI) to mitigate its damaging impact (Sharma et al., 2016). Recently, the significant role of EBL in metabolizing pesticides by inducing modest oxidative burst has been highlighted in tomato (Hou et al., 2018). In tomato plants, EBL application significantly enhanced the expression of *RESPIRATORY BURST OXIDASE HOMOLOG1* (*RBOH1*) and H_2O_2 accumulation under the effect of CHT (Hou et al., 2018). Further, upon silencing *RBOH1*, the expression of *GRXS16* was downregulated and the metabolism of CHT was reduced. It has been suggested that *GRXS16* increases glutathione content, the activity of GST, and transcripts of *GST1*, resulting in the enhanced metabolism of CHT in tomato (Hou et al., 2018). Hence, it is clearly evident that BR protects crops from the toxic effects of pesticides by metabolizing and detoxifying them and thereby protecting the environment and biodiversity.

4.4 Conclusion

BRs are hugely important phytohormones that in very low concentrations can protect plants against different abiotic stresses. However, most of the progress in deciphering the role of BRs in the modulation of abiotic stress responses has been made through their exogenous application. Thus, in the future, emphasis should be given to gaining deeper insights into the possible influence of abiotic stress stimuli on BR biosynthesis/degradation and signaling. Moreover, future studies should be focused on the dissection of molecular crosstalk between BR signaling and other phytohormones during abiotic stress adaptation in plants. This will lead to an advancement of knowledge on the role of BRs in the amelioration of adverse effects of abiotic stresses in plants.

REFERENCES

Aghdam et al. *Scientia Horticulturae* 144 (2012): 116–120.

Ahammed et al. *Chemosphere* 86, no. 5 (2012): 546–555.

Ahammed et al. *Journal of Experimental Botany* 64 (2013): 199–213.

Ahammed et al. *Journal of Plant Growth Regulation* (2020): 1–14.

Ahanger et al. *Journal of Plant Growth Regulation* 37, no. 4 (2018): 1007–1024.

Ali et al. *Production, Metabolism, Signaling and Defense Mechanisms* (2019): 353–370.

Alhaithloul, Haifa Abdulaziz, Abdelghafar Abu-Elsaoud, and Soliman. *Abiotic Stress in Plants*. 2020.

Anuradha and Seeta Ram Rao. *Indian Journal of Plant Physiology* 12 (2007a): 396–400.

Anuradha et al. *Plant Soil and Environment* 53 (2007b): 465.

Anwar et al. *Biological Research* 51 (2018): 46.

Arif et al. *Plant Physiology and Biochemistry* 156 (2020): 64–77.

Arora et al. *Acta Physiologiae Plantarum* 30 (2008): 833.

Ashkani et al. *Frontiers in Plant Science* 6 (2015): 886.

Bajguz et al. *Plant Physiology and Biochemistry* 47 (2009): 1–8.

Belkhadir et al. *New Phytologist* 206 (2015): 522–540.

Bishop et al. *Plant and Cell Physiology* 42 (2001): 114–120.

Bita et al. *Frontiers in Plant Science* 4 (2013): 273.

Caño-Delgado et al. *Development* 131 (2004): 5341–5351.

Clouse. *Journal of Plant Growth Regulation* 34 (2015): 828–844.

Cui et al. *The Plant Cell* 24 (2012): 233–244.

Cui et al. *Functional & Integrative Genomics* 16 (2016): 29–35.

Cui et al. *Plant Physiology* 180 (2019): 605–620.

Dhaubhadel et al. *Plant Molecular Biology* 40 (1999): 333–342.

Dhaubhadel et al. *The Plant Journal* 29 (2002): 681–691.

Dong et al. *Journal of Soil Science and Plant Nutrition* 17 (2017): 554–569.

El-Khallal et al. *Research Journal of Agriculture and Biological Sciences* 5 (2009): 391–402.

Eremina et al. *Proceedings of the National Academy of Sciences* 113 (2016): E5982–E5991.

Fàbregas et al. *Nature Communications* 9 (2018): 1–13.

Fariduddin et al. *Photosynthetica* 49, no. 1 (2011): 55–64.

Feng et al. *Plant Science* 234 (2015): 163–173.

Ferguson. *Emerging Topics in Life Sciences* 3 (2019): 165–181.

Geng et al. *The Plant Cell* 25, no. 6 (2013): 2132–2154.

Grove et al. *Nature* 281, no. 5728 (1979): 216–217.

Gudesblat et al. *Current Opinion in Plant Biology* 14 (2011): 530–537.

Hasan et al. *Chemosphere* 84 (2011): 1446–1451.

Hatfield et al. *Weather and Climate Extremes* 10 (2015): 4–10.

Hayat et al. *Protoplasma* 239 (2010): 3–14.

Hou et al. *Environmental Pollution* 240 (2018): 227–234.

Hu et al. *Photosynthetica* 48 (2010): 537–544.

Huang et al. *Genes* 11 (2020): 919.

Hussain et al. *Plant Growth Regulation* (2020): 1–16.

Jahammed et al. *Current Protein and Peptide Science* 16 (2015): 462–473.

Janeczko et al. *Photosynthetica* 43 (2005): 293–298.

Janeczko et al. *Biologia Plantarum* 51 (2007): 355–358.

Jansing et al. *International Journal of Molecular Sciences* 20 (2019): 2888.

Jeppesen et al. *Hydrobiologia* 750 (2015): 201–227.

Jiang et al. *Physiologia Plantarum* 148 (2013): 133–145.

Kaur et al. *Frontiers in Environmental Science* 5 (2017): 42.

Kaur et al. In *Advances in Rice Research for Abiotic Stress Tolerance*, pp. 113–132. Woodhead Publishing, 2019a.

Kaur et al. *Plant Growth and Development*, pp. 407–423. Springer, Singapore, 2019b.

Kaur et al. *Scientific Reports* 10 (2020): 1–11.

Khan et al. *Plant Growth Regulation* (2020): 1–15.

Kohli et al. In *Reactive Oxygen Species and Antioxidant Systems in Plants: Role and Regulation under Abiotic Stress*, pp. 185–214. Springer, Singapore, 2017.

Kutschera et al. *Journal of Experimental Botany* 63 (2012): 3511–3522.

Lacombe et al. *Current Opinion in Plant Biology* 34 (2016): 1–8.

Li et al. *Amino Acids* 43 (2012): 2469–2480.

Liu et al. *Plant Growth Regulation* 59 (2009): 207–214.
Liu et al. *Journal of Plant Physiology* 168 (2011): 853–862.
Maghari et al. *Avicenna Journal of Medical Biotechnology* 3 (2011): 109.
Marothia et al. Current Knowledge and Future Prospects. *Abiotic Stress in Plants* (2020).
Martinez et al. *Molecules* 23 (2018): 535.
Mitchell et al. *Nature* 225 (1970): 1065–1066.
Nawaz et al. *Environmental Science and Pollution Research* 24 (2017): 15959–15975.
Nguyen et al. *Physiologia Plantarum* 164 (2018): 452–466.
Nolan et al. *The Plant Cell* 32 (2020): 295–318.
Parry et al. Cambridge University Press, Cambridge, UK (2007).
Planas-Riverola et al. *Development* 146, no. 5 (2019).
Pociecha et al. *Journal of Plant Growth Regulation* 36, no. 3 (2017): 618–628.
Qingmao et al. *Scientia Agricultura Sinica* (2006).
Qu et al. *Cryobiology* 63, no. 2 (2011): 111–117.
Rajewska et al. *Frontiers in Plant Science* 7 (2016): 629.
Ramakrishna et al. *Environmental and Experimental Botany* 96 (2013): 52–58.
Ramakrishna et al. *Protoplasma* 252 (2015): 665–677.
Raza et al. *Plants* 8 (2019): 34.
Ritonga et al. *Plants* 9 (2020): 560.
Sadura et al. *Biologia Plantarum* 62 (2018): 601–616.
Saini et al. *Frontiers in Plant Science* 6 (2015): 950.
Samakovli et al. *New Phytologist* 203 (2014): 743–757.
Santner, Aaron, and Mark Estelle. *Nature* 459, no. 7250 (2009): 1071–1078.
Shahzad et al. *Ecotoxicology and Environmental Safety* 147 (2018): 935–944.
Sharma et al. *Ecotoxicology* 20, no. 4 (2011): 862–874.
Sharma et al. *Plant Physiology and Biochemistry* 69 (2013): 17–26.
Sharma et al. *Journal of Plant Growth Regulation* 34 (2015): 509–518.
Sharma et al. *Environmental Science and Pollution Research* 23 (2016): 14518–14525.
Sharma et al. *Frontiers in Plant Science* 8 (2017): 2151.
Sharma et al. *Brazilian Archives of Biology and Technology* 61 (2018b).
Sharma et al. *Cogent Food & Agriculture* 4 (2018a): 1436212.
Sharma et al. *Journal of Plant Growth Regulation* (2019): 1–23.
Singh et al. *Physiology and Molecular Biology of Plants* 18 (2012): 229–236.
Su et al. *Frontiers in Plant Science* 11 (2020): 38.
Symons et al. *Journal of Experimental Botany* 59 (2008): 17–24.
Sytar et al. *Journal of Plant Growth Regulation* 38 (2019): 739–752.
Takatsuto et al. *Journal of Chromatography A* 658 (1994): 3–15.
Tilman et al. *Proceedings of the National Academy of Sciences* 108, no. 50 (2011): 20260–20264.
Vakilian et al. *Scientific Reports* 10 (2020): 1–10.
Vardhini et al. *Frontiers in Environmental Science* 2 (2015): 67.
Vardhini et al. *International Journal of Earth Sciences Geology* 1 (2019): 59–65.
Vázquez et al. *Australian Journal of Crop Science* 13 (2019): 115.
Wang et al. *Scientia Horticulturae* 144 (2012): 195–200.
Wang et al. *Plant Growth Regulation* 81 (2017): 455–466.
Wani et al. *The Crop Journal* 4 (2016): 162–176.
Wu et al. *Biologia Plantarum* 58 (2014): 185–188.
Xia et al. *Journal of Agricultural and Food Chemistry* 57 (2009a): 8406–8413.
Xia et al. *Planta* 230 (2009b): 1185.
Xia et al. *Plant, Cell & Environment* 41 (2018): 1052–1064.
Xiao-min et al. *Yingyong Shengtai Xuebao* 24 (2013).
Yuan et al. *Scientia Horticulturae* 126 (2010): 103–108.
Zafar et al. *Journal of Experimental Botany* 71(2020): 470–479.
Zambounis et al. In *Genomic Designing of Climate-Smart Fruit Crops*, pp. 385–404. Springer, Cham, 2020.
Zeng et al. *Journal of Plant Growth Regulation* 29, no. 1 (2010): 44–52.
Zhang et al. *Australian Journal of Agricultural Research* 58 (2007): 811–815.
Zhu et al. *Scientific Reports* 6 (2016): 35392.

5

Crosstalk of Reactive Oxygen Species and Brassinosteroids in Plant Abiotic Stress Mitigation

Navdeep Kaur, Shivani Saini, Deeksha Marothia, and Pratap Kumar Pati

CONTENTS

Abbreviations: BR: brassinosteroid, ROS: reactive oxygen species, EBL: 24-epibrassinolide, BL: brassinolide, BRI1: brassinosteroid insensitive 1, BAK1: brassinosteroid Insensitive 1-associated receptor kinase 1, BRL3: BRI1 LIKE3, BES: BRI1-EMS suppressor, BZR: brassinazole-resistant, SOD: superoxide dismutase, CAT: catalase, GR: glutathione reductase, APX: ascorbate peroxidase, GPX: glutathione peroxidase, DHAR: dehydroascorbate reductase, MDHAR: monodehydroascorbate reductase, MDA: malondialdehyde, Rboh: respiratory burst oxidase homologs, MAPK: mitogen-activated protein kinase.

5.1 Introduction

Being sessile, plants frequently encounter different abiotic stresses such as drought, cold, heat, salinity, and heavy metal toxicity that negatively impact their growth and survival (Ahammed et al., 2020; Saddhe et al., 2020). Such environmental constraints significantly reduce crop productivity worldwide mainly by affecting the stress tolerance properties of plants (Yadav et al., 2020). In response to these environmental cues, plants trigger the expression of various genes and transcription factors (TFs) that participate in the activation of different biochemical, physiological, and molecular defense pathways that enhance the survival of plants (Choudhary et al., 2011; Golldack et al., 2011; Legris et al., 2017; Planas-Riverola et al., 2019; Saini et al., 2018). ROS such as superoxide radical (O^{2-}), hydroxyl radical (OH^-), and hydrogen peroxide (H_2O_2) are the reactive products of oxygen that are naturally produced in minute amounts in plants during photorespiration, photosynthesis, and mitochondrial respiration (Mittler et al., 2017; Saini et al., 2018). In recent years, much emphasis has been laid on understanding the critical role of ROS molecules in the regulation of various signaling pathways associated with plant growth, development, and stress tolerance (Foyer and Noctor, 2009; Mittler, 2017; Noctor et al., 2017; Singh et al., 2020). Different studies have uncovered the key role of ROS in mediating responses to abiotic stresses, which is tightly controlled through the regulation of their production and breakdown process (Huang et al., 2019; Kaur et al., 2016a). In plants, apoplastic NADPH oxidases (respiratory burst oxidase homologs, Rbohs) are the key enzymes that participate in the production of ROS (Kaur et al., 2016b; Saini et al., 2018). On the contrary, ROS scavenging enzymes such as superoxide dismutase (SOD), catalases (CATs), ascorbate peroxidases (APXs), glutathione S-transferases (GSTs), and glutathione peroxidases (GPXs) detoxify ROS, thus, tightly regulating ROS levels (Ahmad et al., 2010; Bela et al., 2015; Dietz, 2011; Dixon and Edwards, 2010). The tight regulation of ROS production and scavenging process maintains the homeostasis and threshold content of ROS required for regulating plant physiological, biochemical, and molecular processes involved in responses against various environmental stresses (Kaur et al., 2016a; Nath et al., 2016; Saini et al., 2018; Saxena et al.,2016).

ROS also crosstalks with plant hormones such as brassinosteroids (BRs), auxin, salicylic acid (SA), jasmonic acid (JA), ethylene (ET), cytokinin (CK), abscisic acid (ABA), and gibberellic acid (GA) in regulating stress responses (Huang et al., 2019; Tripathi et al., 2020; Xia et al., 2009). Among these phytohormones, brassinosteroids (BRs) are a group of polyhydroxy steroidal plant hormones that were first extracted from the pollens of *Brassica napus* and influence diverse physiological processes in plants including abiotic stress tolerance (Grove et al., 1979). The present chapter focuses on understanding the implication of BRs in the modulation of the ROS

DOI: 10.1201/9781003110651-5

cascade to ameliorate the detrimental effects of environmental challenges for promoting crop production.

5.2 Brassinosteroids

BRs are a unique group of polyhydroxy steroidal phytohormones that play diverse physiological functions in plants (Ahammed et al., 2020; Kaur et al., 2019; Sharma et al., 2017). They regulate cell elongation, cell division and differentiation, rhizogenesis, senescence, leaf expansion, photomorphogenesis, flower development, male sterility, stomatal development, vascular differentiation, and seed germination (Bajguz and Piotrowska-Niczyporuk et al.; 2014; Rao et al., 2002; Saini et al., 2015). Furthermore, BR triggers several stress adaptive mechanisms in plants, and hence, offers tolerance against a wide spectrum of stresses reflecting its dynamic roles (Saini et al., 2015; Xia et al., 2009). Various studies have revealed that BR biosynthesis occurs in the endoplasmic reticulum through precursor campesterol, which synthesizes the most active form of BR, brassinolide (Bajguz et al., 2020). BR signaling commences with the BR perception by the BRASSINOSTEROID INSENSITIVE1 (BRI1) membrane receptor and corresponding homologs BRI1-LIKE1(BRL1) and BRI1 LIKE3 (BRL3) (Planas-Riverola et al., 2019; Saini et al., 2015). Upon binding of BR to BRI1 and co-receptors, BR-regulated genes involved in the growth and defense of plants are activated by BRI1-EMS-SUPPRESSOR1 (BES1), BRASSINAZOLE-RESISTANT1 (BZR1), and the subsequent transcription factors (Nolan et al., 2020; Planas-Riverola et al., 2019; Tong and Chu, 2018). 24-epibrassinolide and 28-homobrassinolide are the most bioactive forms of BR that play a critical role in the modulation of abiotic stress responses in plants (Kagale et al., 2007; Rajewska et al., 2016; Xia et al., 2018). Further, BR participates in maintaining cellular homeostasis, regulates growth, and promotes the detoxification potential of plants to overcome abiotic stresses (Ahammed et al., 2020; Anwar et al., 2018).

BRs have also been shown to mitigate the damaging effects of abiotic stresses by activating various molecular pathways involved in the process of stress adaptation (Ahammed et al., 2020; Anwar et al., 2018). Among these, the ROS network is recognized as one of the major pathways that play a significant role in the BR-mediated regulation of abiotic stress responses. BRs have been shown to actively participate in the modulation of ROS generation as well as ROS scavenging systems for the amelioration of the toxic effects of different abiotic stresses in plants (Vardhini and Anjum, 2015).

5.3 BR-Mediated Regulation of the ROS Generating System

BRs are known to stimulate ROS production upon stress exposure and this BR-induced ROS generation is of significant importance for the subsequent stress acclimation in plants (Ahammed et al., 2020; Sun et al., 2019). The prime sites where ROS is generated inside the plant cell are mitochondria, cytosol, plasma membrane, chloroplast, peroxisome, and the apoplastic space. In addition to this, ROS is also generated in some less commonly known sites like the nucleus and endoplasmic reticulum (Dumanovic et al., 2020, Shapiguzov et al., 2012). Moreover, the generation of ROS occurs in the apoplastic space and plasma membrane, where an NADPH oxidase complex is primarily involved in ROS production (Mhamdi and Van Breusegem, 2018). Suzuki and Mittler (2006) interpreted that ROS, such as O_2^{-}, are produced by NADPH oxidases during stress conditions and trigger downstream stress-responsive pathways. Besides this, ROS leads to oxidative bursts in the cell, which activate various defense-related mechanisms. These include MAPK signaling, heat shock proteins, antioxidant enzymes production, and synthesis of pathogenesis-related proteins required for combating both abiotic and biotic stresses (Apel and Hirt, 2004; Bhatt et al., 2020; Gechev et al., 2006).

Exogenous application of BRs resulted in increased activity of NADPH oxidase leading to BR-induced H_2O_2 accumulation in plants exposed to different abiotic stresses (Ahammed et al., 2020). H_2O_2 is one of the most stable ROS that behaves both as an oxidant as well as a reductant (Smirnoff and Arnaud, 2018). However, due to its strong oxidant property, it may cause localized oxidative damage that negatively affects the vital metabolic functions of plants Conversely, in order to mediate ROS signaling, H_2O_2 plays several roles by acting as a signal, a mediator, and an effector molecule (Smirnoff and Arnaud, 2018). For instance, in tomato, a transient H_2O_2 production in the apoplast via NADPH oxidase, which is encoded by the *RBOH1* gene, has been linked to BR-induced heat stress tolerance (Zhou et al., 2014). However, when *RBOH1*, *MPK2*, or *MPK1/2* genes were silenced, the accumulation of H_2O_2 was significantly suppressed and the BR-induced tolerance in response to heat stress was compromised (Nie et al., 2013). Moreover, the silencing of *MPK1* did not show the same effect, thereby suggesting that MPK2 is of more significance than MPK1 in BR-mediated generation of H_2O_2 during heat tolerance (Ahammed et al., 2020). Further, RBOH1, 2-cysteine peroxiredoxin (2-Cys Prx) and GLUTAREDOXIN (GRX) have been suggested to participate in a signaling cascade involved in BR-induced cold tolerance in tomato (Xia et al., 2018). Furthermore, in response to chilling temperatures, plants activate the BZR1, which up-regulates the transcript levels of *RBOH1*, thus resulting in H_2O_2 production (Fang et al., 2019). However, repression of *RBOH1* or mutation in *BZR1* eliminated the BR-induced photoprotection, leading to enhanced photoinhibition during cold stress (Ahammed et al., 2020; Fang et al., 2019). Additionally, EBR, which is a bioactive epibrassinolide, restored the NADPH oxidase activity in brassinazole (a BR biosynthesis inhibitor) treated cucumber plants (Xia et al., 2009). Thus, these studies clearly indicate that BRs have certain regulatory effects on the ROS generating system that largely contribute to the mitigation of deleterious effects of abiotic stresses in plants.

5.4 BR-Mediated Regulation of the ROS Scavenging System

The ROS scavenging machinery plays a critical role in maintaining the optimum levels of ROS in plants exposed to abiotic stress conditions (Ahmad et al., 2010; Foyer and Noctor, 2016;

Raja et al., 2017). There are two arms of the ROS antioxidant system in plants including the enzymatic and non-enzymatic (Das and Roychoudhury, 2014). BRs modulate the activities and levels of both the types of antioxidant processes for maintaining harmony between ROS generation and detoxification in plants exposed to different abiotic stress conditions (Anwar et al., 2018; Yusuf et al., 2019). In the following sections, the role of BRs in the regulation of different types of antioxidants during abiotic stress conditions in plants is discussed in detail.

5.4.1 Enzymatic Antioxidants

Enzymatic antioxidants are the direct molecular players that primarily protect plants against ROS-induced oxidative damage (Ahmad et al., 2010; Soares et al., 2019). These antioxidants include SOD, APX, CAT, GST, glutathione reductase (GR), dehydroascorbate reductase (DHAR), monodehydroascorbate reductase (MDHAR), and so on (Gill et al., 2010; Hasanuzzaman et al., 2012; Hasanuzzaman et al., 2020; Soares et al., 2019). These enzymes are present in almost all the cellular compartments and generally, in an organelle, more than one antioxidant enzyme is present to efficiently scavenge ROS molecules (Ahmad et al., 2010). SOD is the foremost important enzyme of the enzymatic antioxidant machinery that catalyzes the dismutation of O^{2-} into H_2O_2 (Saibi et al., 2018). Based upon the co-factor required for their activity and protein fold, there are three isoforms of SOD found in plants: Cu/Zn-SOD localized in chloroplast and cytosol; Mn-SOD in mitochondria; and Fe-SOD in the chloroplast (Ahmad et al., 2010; Saibi et al., 2018). H_2O_2 generated as a result of SOD action on O^{2-} is an important ROS molecule in terms of plant growth, development, and stress adaptation (Smirnoff et al., 2019). At optimum levels, it is responsible for evoking a series of abiotic stress-responsive pathways in plants, and once its beneficial job is done, H_2O_2 needs to be immediately scavenged from the cells to prevent it from imposing oxidative stress in plants (Kaur et al., 2016a; Saini et al., 2018). CAT is one of the most critical enzymes that detoxify H_2O_2. It converts H_2O_2 into O_2 and H_2O, which are stable beneficial molecules for the cell (Ahmad et al., 2010). CAT has been classified into three major classes in plants: class 1 which detoxifies H_2O_2 generated during the process of photorespiration; class 2 is reported to be present in the vascular tissues and is implicated to be involved in lignification, however, their exact role is still not known; and classes 3 are predominantly present in young plants and seeds and scavenge H_2O_2 generated due to fatty acid degradation in glyoxisomes during the glyoxylate cycle (Ahmad et al., 2010; Willekens et al., 1994). Apart from CAT, the ascorbate-glutathione cycle is the major mechanism for the detoxification of H_2O_2 from plants (Bartoli et al., 2017). APX, GR, DHAR, and MDHAR enzymes are the key players that mediate efficient scavenging of H_2O_2 in the ascorbate-glutathione cycle (Ahmad et al., 2010; Bartoli et al., 2017). APX converts H_2O_2 into H_2O using ascorbate as an electron donor (Ahmad et al., 2010). The resulting oxidized ascorbate known as monodehydroascorbate is again recycled to ascorbate with the advent of DHAR, MDHAR, and GR enzymes (Bartoli et al., 2017). Besides these enzymes, GPX enzymes also play a significant role in ROS detoxification in plant cells (Ahmad et al., 2010). These enzymes convert different hydroperoxides including H_2O_2 present in the cell into alcohols. In addition, GPXs have been implicated to play a critical role in the detoxification of products formed during ROS-induced lipid peroxidation (Ahmad et al., 2010). There are three different types of GPX enzymes that have been identified in plants. These comprise of GPx (selenium-dependent), PHGPX (non-selenium-dependent phospholipids hydroperoxidases GPX), and GST-GPx (glutathione transferases that show GPx activity) (Ahmad et al., 2010).

Exogenous as well as pre-sowing treatment with different BRs viz. 28-homobrassionolide (HBL) and 24-epibrassinolide (EBL) in different concentrations modulate the enzymatic activities of various antioxidant enzymes including SOD, CAT, GPX, APX, DHAR, MDHAR, GR, and peroxidase (POD) in plants exposed to salinity, drought, pesticides, heavy metals stress, extreme temperatures, and so on (Barros et al., 2020; Chen et al., 2019; Jiang et al., 2013; Kaur et al., 2018; Sharma et al., 2013; 2015; Vardhini et al., 2015; Wang et al., 2020). Moreover, at the genetic level, BRs have been implicated to regulate the transcript levels of different antioxidant enzymes encoding genes (Sharma et al., 2013; 2015). In response to salinity and pesticide stress, BR application enhanced the levels of a range of antioxidant enzymes accompanied by significant up-regulation in the expression of genes encoding for different isoforms of SOD viz. *Fe-SOD*, *Mn-SOD*, and *Cu/Zn-SOD*, along with *CAT* and *APX* (Sharma et al., 2015). In *Brassica juncea* plants, seed priming with 28-homobrassinolide attenuated the detrimental effects of combined salinity and temperature stress by significantly enhancing the enzymatic activities of SOD, APOX, CAT, GR, DHAR, and MDHAR, along with significant up-regulation of transcript levels of genes encoding for these enzymes (Kaur et al., 2018). In addition to this, exogenous application of EBR of about 0.1 μM improved CO_2 assimilation and alleviated the photoinhibition of PSII under chilling temperatures by regulating the activities of key enzymes that were critical in the ascorbate-glutathione (AsA-GSH) cycle and redox homeostasis (Jiang et al., 2013). In *Vitis vinifera* (grapes) seedlings, EBR application regulated the constituents of the AsA-GSH cycle, resulting in elevated chilling tolerance (Chen et al., 2019). Similarly, BR treatment remarkably improved the antioxidant defense machinery of plants exposed to heat, heavy metals, and drought stress (Jiang et al., 2013; Barros et al., 2020). The BR-mediated stimulation of enzymatic antioxidants primarily protects the plants by decreasing the levels of lipid peroxidation through the efficient detoxification of ROS (Vardhini et al., 2015). Lipid peroxidation is one of the direct and foremost effects of ROS in plants under stress conditions. It disturbs the lipid bilayer structure of the plasma membrane thus amending their structural-functional properties. (Ahammed et al., 2020; Sharma et al., 2013). BR application has been reported to significantly lower the levels of MDA, which is considered a biomarker for measuring the extent of lipid peroxidation in plants (Ahammed et al., 2020; Sharma et al., 2013; Vardhini et al., 2015)

The BR-mediated regulation of the ROS enzymatic antioxidant genetic network might operate through a complex pathway activated through BZR1/BES1 transcription factor (Wang et al., 2020). In tomato, down-regulation of the *SLB3* gene (that belongs to the BES1 TF family member) using virus-induced gene silencing (VIGS) resulted in a significant decrease in SOD activity and further resulted in an enhancement in the levels of H_2O_2, O^{2-}, and MDA under drought stress (Wang et al., 2020). Further, the transcript levels of BR pathway genes including *SlDWARF, BIN2*-related genes, and *SlCPD* were decreased in *SLB3* downregulated plants under water-deficient conditions thus suggesting the critical role of BR signaling in the regulation of ROS network under stress conditions (Wang et al., 2020). Moreover, $Ca2^+$ and calcium/calmodulin-dependent protein kinase (CCaMK) play a critical role in the regulation of BR-mediated antioxidant defense (Liu et al., 2020; Yan et al., 2015). In maize, overexpression, as well as silencing of the ZmCCaMK gene, demonstrated that Ca^{2+} accumulation and ZmCCaMK activation are prerequisites for BR-induced modulation of antioxidant defense machinery (Yan et al., 2015). A recent study indicated that two amino acid residues including Thr 420 and Ser 454 are critical for phosphorylation of ZmCCaMK during BR-regulated induction of antioxidant enzymes (Liu et al., 2020).

5.4.2 Non-Enzymatic Antioxidants

Non-enzymatic antioxidants play an important role in regulating the levels of ROS in plants (Soares et al., 2019; Yusuf et al., 2019). The major non-enzymatic antioxidants functional during the process of abiotic stress acclimatization are proline, ascorbic acid, glutathione, α-tocopherols, carotenoids, and different phenolic compounds (Ahmad et al., 2010). Among these, proline is one of the most critical compounds that play a significant role in ROS detoxification and osmotic adjustment during stress conditions (Furlan et al., 2020). In several reports, a positive co-relation between BR application and enhanced levels of proline has been well established. BRs increase the net levels of proline in plants that aids up in the process of stress acclimation (Rattan et al., 2020; Sharma et al., 2013; 2015; Soliman et al., 2020). Recent studies highlight that BRs modulate the metabolic dynamics of proline in plants. In *Elymus nutans* plants, 1μM EBR treatment preceding cold stress exposure enhanced the activities of key enzymes involved in proline biosynthesis including Δ1-pyrroline-5-carboxylate reductase (P5CR) and Δ1-pyrroline-5-carboxylate synthetase (P5CS), whereas decreased the activities of those involved in proline catabolism viz. Δ1-pyrroline-5-carboxylate dehydrogenase (P5CDH) and proline dehydrogenase (PDH). Moreover, the transcript levels of genes encoding for the above-mentioned enzymes were found to correlate with their respective enzyme activities (Fu et al., 2019).

Ascorbic acid and glutathione are the antioxidant compounds that play a central role in the ascorbate-glutathione cycle involved in ROS detoxification (Bartoli et al., 2017). Ascorbate is present in different forms in plants including reduced ascorbate, also known as ascorbic acid, and two oxidized forms that comprise mono- and dehydroascorbate. The key factor that regulates the antioxidant potential of ascorbate is the ratio of reduced to the oxidized form of ascorbate (Conklin et al., 2000; Cruz-Rus et al., 2012; Zechmann, 2011). It may directly detoxify ROS in plants or may exert its effect through the well-known glutathione-ascorbate cycle. Besides, ascorbate is involved in redox signaling in plants (Zechman, 2011). Glutathione is another major antioxidant that is found in two distinct forms in plants, including reduced glutathione (GSH) and the oxidized disulfide (GSSG) (Shu et al., 2011). A high proportion of GSH/GSSG ratio is critical for maintaining the antioxidative potential of glutathione (Noctor et al., 1998; 2012; Mhamdi et al., 2010). BRs modulate the levels of ascorbate and glutathione to accelerate the ascorbate-glutathione cycle during stress conditions for efficient scavenging of ROS molecules (Ahanger et al., 2020; Hasan et al., 2020; Morales et al., 2014). Recently, researchers have reported that in pepper plants exposed to cadmium stress, BR treatment improves the stress acclimatization by inducing the production of nitric oxide that in turn up-regulates the ascorbate-glutathione cycle (Kaya et al., 2020). Moreover, studies indicate that BRs regulate the biosynthesis of ascorbate and glutathione in plants (Morales et al., 2014; Zhou et al., 2015). Tocopherols belong to lipid-soluble compounds found only in photosynthetic organisms that possess significant antioxidative properties (Waskiewicz et al., 2014). BRs have been confirmed to stimulate the production of tocopherols in plants exposed to different stresses. In soybean, seed priming with BRs enhanced the tocopherol content to 42.25% as compared with control plants (Soliman et al., 2020). In *bes1* loss-of-function mutant Arabidopsis plants, the exogenous application of BRs increased the content of tocopherols, providing clear evidence of the role of BRs in the modulation of these lipophilic compounds (Setsungnern et al., 2020). Carotenoids are the isoprenoid-derived compounds that are critical for plant metabolism (Waskiewicz et al., 2014). In tomato, exogenous application of BRs increased the overall content of carotenoids. Moreover, transgenic tomato plants overexpressing Arabidopsis BZR1-1D TF showed significantly enhanced levels of carotenoids (Liu et al., 2014). Thus, it is clearly evident that BRs regulate the levels of non-enzymatic antioxidants during abiotic stress tolerance in plants.

5.5 Conclusion

The ROS pathway is evoked in plants in response to a wide spectrum of abiotic stresses. BRs directly or indirectly participate in the modulation of the ROS network during the process of stress acclimation. However, an in-depth understanding of the molecular cascade involved in the BR-mediated regulation of the ROS stress adaptive mechanism is still lacking. Thus, future research efforts are critical for generating a new wealth of knowledge that will provide valuable inputs on the molecular dynamics of crosstalk between BRs and ROS.

Acknowledgments

Financial assistance from the Rashtriya Uchchattar Shiksha Abhiyan (RUSA-II) Program, Ministry of Human Resource Development (MHRD), Government of India, New Delhi, is duly acknowledged

REFERENCES

Ahammed et al. *Journal of Plant Growth Regulation* (2020): 1–14.

Ahanger et al. *Plant Physiology and Biochemistry* 147 (2020): 31–42.

Ahmad et al. *Critical Reviews in Biotechnology* 30, no. 3 (2010): 161–175.

Anwar et al. *International Journal of Molecular Sciences* 19, no. 9 (2018): 2497.

Anwar et al. *Biological Research* 51 (2018).

Apel and Hirt. *Annual Review of Plant Biology* 55 (2004): 373–399.

Bajguz and Piotrowska-Niczyporuk. *Plant Physiology and Biochemistry* 80 (2014): 176–183.

Bajguz et al. *Frontiers in Plant Science* 11 (2020): 1034.

Barros Junior et al. *Physiologia Plantarum* (2020).

Bartoli et al. In *Ascorbic Acid in Plant Growth, Development and Stress Tolerance*, pp. 177–200. Springer, Cham, 2017.

Bela et al. *Journal of Plant Physiology* 176 (2015): 192–201.

Bhatt et al. *Protective Chemical Agents in the Amelioration of Plant Abiotic Stress: Biochemical and Molecular Perspectives* (2020): 254.

Chen et al. *Scientia Horticulturae* 244 (2019): 379–387.

Choudhary et al. *Chemosphere* 84 (2011): 592–600.

Conklin et al. *Genetics* 154, no. 2 (2000): 847–856.

Cruz-Rus et al. *Biotechnology Journal* 7, no. 9 (2012): 1110–1121.

Das et al. *Frontiers in Environmental Science* 2 (2014): 53.

Dietz. *Antioxidants & Redox Signaling* 15, no. 4 (2011): 1129–1159.

Dixon et al. *The Arabidopsis Book/American Society of Plant Biologists* 8 (2010).

Dumanović et al. *Frontiers in Plant Science* 11 (2020).

Fang et al. *Plant Physiology* 180, no. 4 (2019): 2061–2076.

Foyer et al. *Antioxidants & Redox Signaling* 11, no. 4 (2009): 861–905.

Foyer et al. *Plant, Cell & Environment* 39, no. 5 (2016): 951–964.

Fu et al. *Environmental and Experimental Botany* 167 (2019): 103831.

Furlan et al. *Plant Physiology and Biochemistry* 151 (2020): 566–578.

Gechev et al. *Bioessays* 28, no. 11 (2006): 1091–1101.

Gill et al. *Plant Physiology and Biochemistry* 48, no. 12 (2010): 909–930.

Golldack et al. *Plant Cell Reports* 30, no. 8 (2011): 1383–1391.

Grove et al. *Nature* 281, no. 5728 (1979): 216–217.

Hasan et al. *Journal of Plant Interactions* 15, no. 1 (2020): 371–385.

Hasanuzzaman et al. In *Crop Stress and Its Management: Perspectives and Strategies*, pp. 261–315. Springer, Dordrecht, 2012.

Hasanuzzaman et al. *Antioxidants* 9, no. 8 (2020): 681.

Huang et al. *Frontiers in Plant Science* 10 (2019): 800.

Jiang et al. *Physiologia Plantarum* 148, no. 1 (2013): 133–145.

Kagale et al. *Planta* 225, no. 2 (2007): 353–364.

Kaur, Navdeep, and Pratap Kumar Pati. "Harnessing the potential of brassinosteroids in abiotic stress tolerance in plants." In *Brassinosteroids: plant growth and development*, pp. 407–423. Springer, Singapore, 2019.

Kaur et al. *Scientific Reports* 8, no. 1 (2018): 1–13.

Kaur et al. *BMC Plant Biology* 16, no. 1 (2016a): 1–13.

Kaur et al. *Plant Signaling & Behavior* 11, no. 12 (2016b): e1247136.

Kaya et al. *Ecotoxicology and Environmental Safety* 196 (2020): 110483.

Legris et al. *The Plant Journal* 90, no. 4 (2017): 683–697.

Liu et al. *Plant Biotechnology Journal* 12, no. 1 (2014): 105–115.

Liu et al. *Biochemical and Biophysical Research Communications* (2020).

Mhamdi et al. *Plant Physiology* 153, no. 3 (2010): 1144–1160.

Mhamdi et al. *Development* 145, no. 15 (2018).

Mittler. *Trends in Plant Science* 22 (2017): 11–19.

Morales et al. *Plant physiology and Biochemistry* 74 (2014): 315–322.

Nath et al. *Frontiers in Plant Science* 7 (2016): 1574.

Nie et al. *Plant, Cell & Environment* 36, no. 4 (2013): 789–803.

Noctor et al. *Journal of Experimental Botany* 49 (1998): 623–647.

Noctor et al. *Plant, Cell & Environment* 35 (2012): 454–484.

Noctor et al. In *Seminars in Cell & Developmental Biology*, vol. 80, pp. 3–12. Academic Press, 2018.

Nolan et al. *The Plant Cell* 32 (2020): 295–318.

Planas-Riverola et al. *Development* 146 (2019).

Raja et al. *Environmental and Experimental Botany* 137 (2017): 142–157.

Rajewska et al. *Frontiers in Plant Science* 7 (2016): 629.

Rao et al. *Current Science* (2002): 1239–1245.

Rattan et al. *Journal of Plant Growth Regulation* (2020): 1–11.

Saddhe et al. *Physiologia Plantarum* (2020).

Saibi et al. Superoxide Dismutase: Structure, Synthesis and Applications; Magliozzi, S., Ed. (2018): 101–142.

Saini et al. *Frontiers in Plant Science* 6 (2015): 950.

Saini et al. *Analytical Biochemistry* 550 (2018): 99–108.

Saxena et al. *Frontiers in Plant Science* 7 (2016): 570.

Setsungnern et al. *Plant Science* 296 (2020): 110470.

Shapiguzov et al. *Frontiers in Plant Science* 3 (2012): 292.

Sharma et al. *Plant Physiology and Biochemistry* 69 (2013): 17–26.

Sharma et al. *Frontiers in Plant Science* 8 (2017): 2151.

Sharma et al. *Journal of Plant Growth Regulation* 34 (2015): 509–518.

Shu et al. *Plant Physiology and Biochemistry* 49, no. 10 (2011): 1228–1237.

Singh, Baldev, Navdeep Kaur, Paramdeep Kumar, Vipin Hallan, and Pratap Kumar Pati. "Reactive oxygen species generating and scavenging systems play critical role in conferring leaf spot disease resistance in *Withania somnifera* (L.) Dunal." *Industrial Crops and Products* 157 (2020): 112889.

Smirnoff, Nicholas, and Dominique Arnaud. "Hydrogen peroxide metabolism and functions in plants." *New Phytologist* 221, no. 3 (2019): 1197–1214.

Smirnoff et al. *New Phytologist* 221, no. 3 (2019): 1197–1214.

Soares et al. *Environmental and Experimental Botany* 161 (2019): 4–25.

Soliman et al. *Physiology and Molecular Biology of Plants* 26, no. 3 (2020): 501–511.

Sun et al. *Plant Signaling & Behavior* 14 (2019): 1657343.

Suzuki et al. *Physiologia Plantarum* 126 (2006): 45–51.

Tong et al. *Trends in Plant Science* 23, no. 11 (2018): 1016–1028.

Tripathi et al. *Journal of Hazardous Materials* (2020): 124820.

Vardhini et al. *Frontiers in Environmental Science* 2 (2015): 67.

Wang et al. *Journal of Integrative Agriculture* 19, no. 11 (2020): 2699–2708.
Waśkiewicz et al. In *Oxidative Damage to Plants*, pp. 201–234. Academic Press, 2014.
Willekens et al. *Febs Letters* 352 (1994): 79–83.
Xia et al. *Plant Physiology* 150, no. 2 (2009): 801–814.
Xia et al. *Plant, Cell & Environment* 41 (2018): 1052–1064.
Yadav et al. *Sustainable Crop Production* (2020).
Yan et al. *Plant and Cell Physiology* 56, no. 5 (2015): 883–896.
Yusuf et al. In *Brassinosteroids: Plant Growth and Development*, pp. 323–348. Springer, Singapore, 2019.
Zechmann et al. *Plant Signaling & Behavior* 6 (2011): 360–363.
Zhou et al. *Journal of Experimental Botany* 65, no. 15 (2014): 4371–4383.
Zhou et al. *Scientific Reports* 5, no. 1 (2015): 1–7.

6

Brassinosteroid Signaling in Adaptative Responses to Abiotic Stress

R. D. Myrene and V. R. Devaraj

CONTENTS

6.1 Introduction

The earliest report on brassinosteroids can be traced to the 1930s when the first documented evidence of pollen extracts promoting plant growth was reported (Mandava, 1988). A decade later, Mitchell and Whitehead (1941) reported that the application of hexane extracts of maize pollen to the first internode of young bean seedlings produced elongation of the treated internode. The 1960s saw research focused on the discovery of new plant hormones by the US Department of Agriculture (USDA; Maugh II, 1981)—pollen extracts of about 60 plant species were tested by employing the bean second internode bioassay for stimulating curvature (Mandava and Mitchell, 1971). A new group of lipidic plant hormones called *brassins* rich in glucosyl esters of fatty acids was reported (Mitchell et al., 1970; Mandava et al., 1973). Treatment with the brassin fraction increased plant growth, crop yield, and seed viability (Meudt et al., 1984). Mandava et al. (1978) attempted to isolate the brassins active compounds using 250 kg of bee collected rape pollen extracted with isopropanol and partitioned between carbon tetrachloride, methanol, and water. Final purification, yielding 10 mg of crystalline brassinolide, was accomplished by column chromatography and high-performance liquid chromatography (HPLC) (Figure 6.1). Spectroscopic methods, including X-ray analysis, were used to elucidate the structure as (22R, 23R,24S)-2a,3a,22,23-tetrahydroxy-24-methyl-B-homo-7-oxa-5a-cholestan-6-one. Castasterone, the putative precursor of brassinolide, was isolated by Japanese scientists a few years later (Yokota et al., 1982a). The first chemical syntheses of brassinolide and brassinosteroids were attempted by Fung and Siddall (1980). At least 50 natural brassinosteroids were isolated from various organs of plants by gas or liquid chromatography combined with mass spectrometry (Adam and Marquardt, 1986; Singh and Bhardwaj, 1986; Abreu, 1991; Takatsuto, 1994; Fujioka, 1999).

Based on various alkyl-substitution patterns of side chains, brassinosteroids are classified as C_{27}, C_{28}, or C_{29}; 2-OH groups at ring A and a 6-ketone or 7-oxa-6-ketone system at ring B are found to be necessary for activity (Taiz and Zeiger, 2010). The chemical structure of brassinolide, the most active BR, is illustrated in Figure 6.1. Other BRs are derived from BL through changes in the area (a) and (b) enclosed by dashed lines (Figure 6.1) (Fujioka and Yokota, 2003). Conjugation of BR with sugars and fatty acids results in inactivation (Fujioka and Yokota, 2003).

6.2 BR Signaling in Plants

The BR signal transduction pathway (Figure 6.2) plays a vital role in plant growth and development via the modulation of gene expression, cell division and elongation, vascular differentiation, and reproductive development (Bajguz, 2007). BR perceived by the plasma membrane-associated receptor kinase, BRASSINOSTERIOD INSENSITIVE 1 (BRI1) activates two major transcription factors of the BR

DOI: 10.1201/9781003110651-6

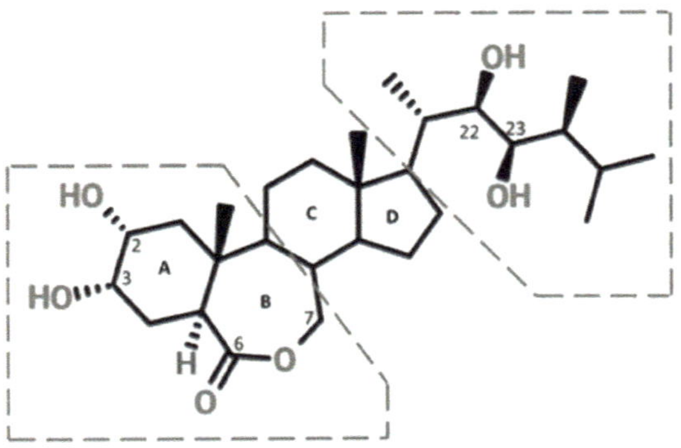

FIGURE 6.1 The chemical structure of brassinolide (BL). The steroid rings are labeled A–D. The areas within the dashed lines can be substituted with different groups (Tang et al., 2016).

signaling pathway namely, BRASSINAZOLE RESISTANT 1 (BZR1) and BRI1-EMS SUPPRESSOR 1 (BES1). BR binding at the island domain of its receptor BRI1 fixes the ectodomain with respect to the leucine-rich repeat (LRR) (Hothorn et al., 2011), creating a docking platform for its coreceptor BRI1-ASSOCIATED RECEPTOR KINASE 1 (BAK1). Activation of the receptor and coreceptor stimulates the phosphorylation of BRI1 KINASE INHIBITOR 1 (BKI1; Wang and Chory, 2006), causing its dissociation from BRI1. Activated BRI1 phosphorylates CONSTITUTIVE DIFFERENTIAL GROWTH 1 (CDG1) and BR-SIGNALING KINASE 1 (BSK1). CDG1, in turn, phosphorylates BRI1 SUPPRESSOR 1 (BSU1), causing dephosphorylation of BRASSINOSTEROID INSENSITIVE2 (*BIN2*). Dephosphorylated *BIN2* is subsequently restrained by KINK SUPPRESSED IN BZR1-1D (KIB1), preventing the association of *BIN2* with BZR1/BES1 and facilitating its ubiquitination and degradation (Zhu et al., 2017). A consequence of this inhibition is the accumulation of unphosphorylated BZR1 and BZR1/BES1 transcription factors and the regulation of BR-targeted genes responsible for enhancing plant stress tolerance (Li et al., 2009) via different physiological processes, such as protein metabolism, cellular transport and signaling, cell wall biosynthesis, chromatin and cytoskeleton components, environmental responses, hormone responses, and immunity (Sun et al., 2010; Hohmann et al., 2018). Furthermore, PHOSPHATASE 2A (PP2A) enhances BR signaling by dephosphorylating BZR1 and BES1, whereas SUPPRESSOR OF BRI1 (SBI1) deactivates BRI1 through the methylation of PP2A. The methylated PP2A relocates to the plasma membrane facilitating its association with the BR-activated BRI1, leading to BRI1 dephosphorylation and degradation, and ultimately termination of BR signaling (Yu et al., 2011). Hence, PP2A and SBI1 provide a negative feedback mechanism that triggers BRI1 turnover after activation of the BR signaling pathway (Wu et al., 2011). Studies have shown that BRs can induce or repress the expression of ~5,000 to 8,000 genes (Nolan et al., 2017).

In the absence of BR, the receptor kinase BRI1 does not heterodimerize with its coreceptor BAK1. Consequently, *BIN2* is free to constitutively phosphorylate numerous substrates including BZR1 and BES1 (Youn and Kim, 2015), inducing their interactions with 14-3-3 proteins that, in turn, promote the cytoplasmic retention of BZR1/BES1 (Ryu et al., 2007), suppressing their DNA-binding activity (Vert and Chory, 2006), and stimulating their degradation (Kim et al., 2019). Additionally, BZR1/BES1 are maintained in the inactive state through binding with BRZ-SENSITIVE-SHORT HYPOCOTYL1 (BSS1; Shimada et al., 2015).

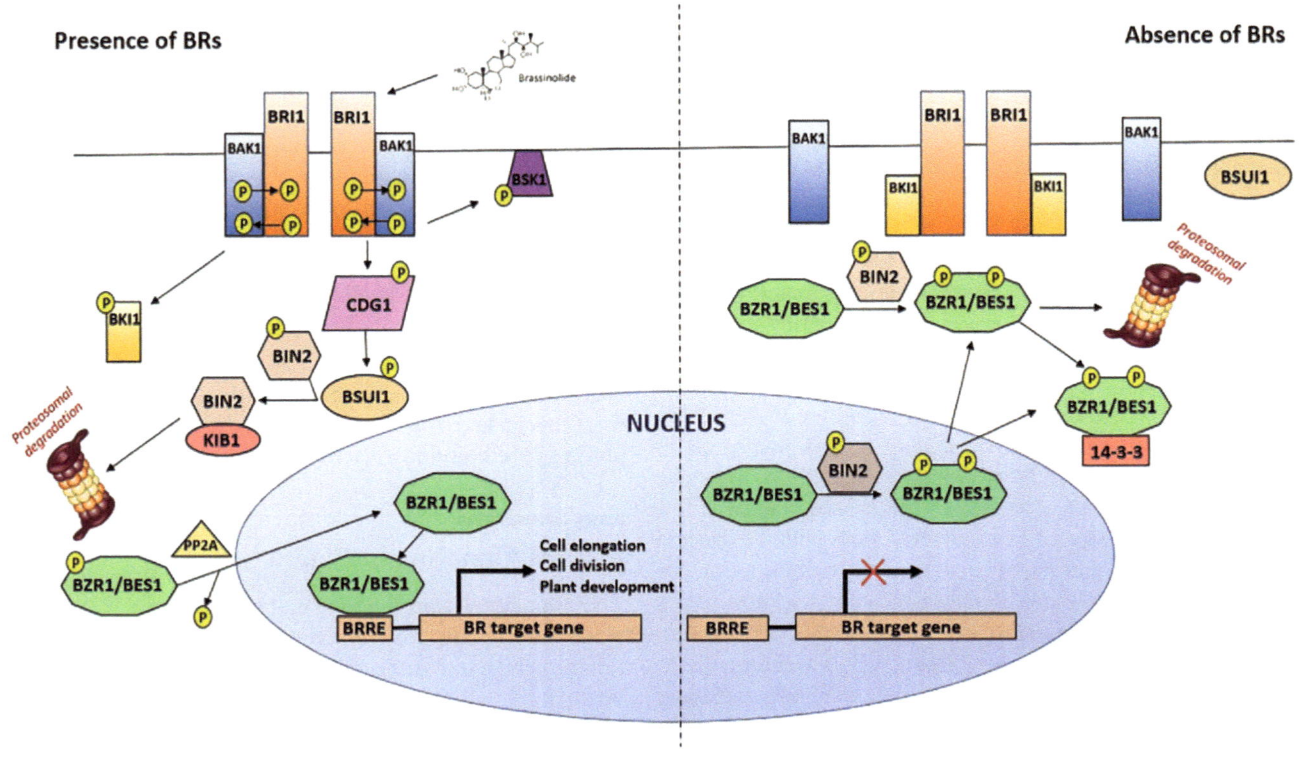

FIGURE 6.2 Model for the brassinosteroid (BR) signaling pathway in the presence and absence of BR.

6.3 Role of BR-Mediated Stress Responses at Different Levels of Organization

In order for a plant to respond appropriately to environmental changes, it must first be able to sense the changing environment. This involves the use of multiple sensors for signal perception, resulting in the activation of signal transduction pathways that culminate in the activation or repression of various stress-responsive genes. Thus, generating the initial response needed for adaptation to abiotic stress. For this to occur, BR signaling must be integrated with various abiotic stress responses taking place at different levels (Figure 6.3).

6.3.1 Role of BRs at the Cellular Level

6.3.1.1 Cell Cycle and Cell Division

The manifestation of the adverse effects of stress, that is, inhibition of cell proliferation and cell expansion, is governed by BR levels. BRs are known to modulate meristematic proliferation, crucial for optimal cell expansion (Chaiwanon and Wang, 2015). The first demonstration of this relationship was seen in an *Arabidopsis* BR-deficient mutant, constitutive photomorphogenesis and dwarfism (cpd)—exhibiting a reduction in the size of the leaf blade due to a prolonged cell division phase and delayed differentiation (Noguchi et al., 1999). Abiotic stress results in reduced expression of cyclin-dependent kinase (CDK) genes and increased expression of inhibitors of mitotic activity, the cyclin-dependent kinase inhibitor (ICK1, EL2; Rodríguez et al., 2013). BRs induce the expression of cell cycle-related genes (CYCA, CYCB, CYCD3;1, CYCD3;2) encoding cyclin-dependent kinases (CDKs) responsible for cell proliferation and differentiation (Fu et al., 2008). BES1 regulates the expression of U-type cyclin CYC U4;1 and GSK-kinase, thereby reducing the proliferation of the abaxial sclerenchyma cell number and controlling leaf erectness (Sun et al., 2015). A consequence of this is enhanced photosynthetic efficiency, reduced transpiration rates, and better yields (Lang et al., 2004). BRs regulate the expression of the transcription factors MULTIPASS (OsMPS) (Schmidt et al., 2013) and BRASSINOSTEROIDS AT VASCULAR AND ORGANIZING CENTER (BRAVO) cells (Chaiwanon and Wang, 2015). BES1 interacting with BRAVO counteracts BR-mediated cell division in a quiescent center cell (QC) of the primary root of *Arabidopsis* (Vilarrasa-Blasi et al., 2014), providing stress acclimatization in plants. The short roots of bes1-D (gain-of-function) mutants with impaired BR biosynthesis were rescued by the addition of low concentrations of BR (Chaiwanon and Wang, 2015). Similarly, BRs were found to promote root length in wild-type roots subjected to abiotic stress and treated with low concentrations of BRs (Chaiwanon and Wang, 2015). Mathematical and computational modeling suggests that BRI1 facilitates size-based cell elongation termination (Pavelescu et al., 2018).

Early studies have indicated the role of BR in the development of vascular tissues, particularly tracheary element differentiation (Yamamoto et al., 2001). Cpd mutants in *Arabidopsis* have abnormal xylem development (Choe et al., 1999). BR suppresses radial vascular cell division (Kang et al., 2017) and triple mutants (brl1 brl3 bak1-3) hypersensitive to BR treatment exhibit greater stele narrowing than that of wild type (Fàbregas et al., 2013). Vascular patterning supported by mathematical modeling studies has attributed the synergistic role of BR and auxin in the development of a periodic pattern of vascular bundles in *Arabidopsis* shoots (Ibanes et al., 2009). Yet, further research is needed to establish the relationship between BR receptors and downstream transcriptional factors in overall organ growth.

6.3.1.2 Cell Wall and Cell Membrane Modification

Abiotic stress renders critical challenges to the functional preservation of plasma membrane and endomembranes of the plant cell. Perception of abiotic stress also results in the modification of the structure of the plant cell wall (Tenhaken, 2015). The roles of BR signaling contributing to the remodeling of the cell wall are summarized in Figure 6.4. BRs induce the expression of cell wall extension and loosening enzymes such as xyloglucan endotransglucosylase/hydrolase (XTHs), pectin lyase-like (PLLs), expansins (EXPs; Uozu et al., 2000), and pectin methylesterases (PMEs; Yang et al., 2013). XTHs catalyze endocleavage of

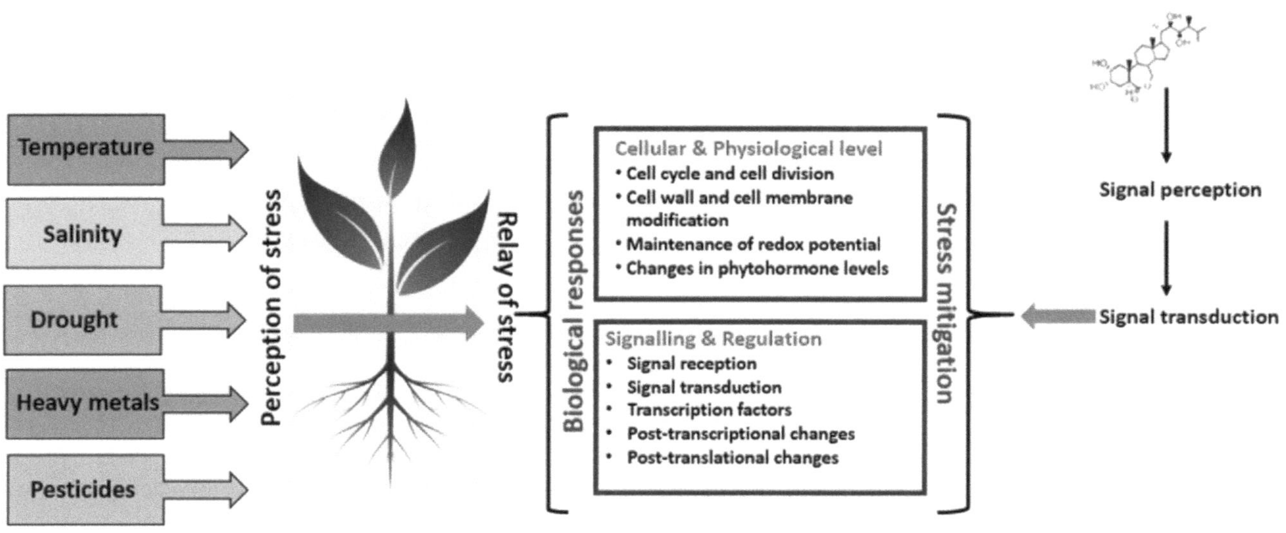

FIGURE 6.3 Brassinosteroid mediated mitigation strategies in response to abiotic stress. Adapted from Sharma et al. (2017).

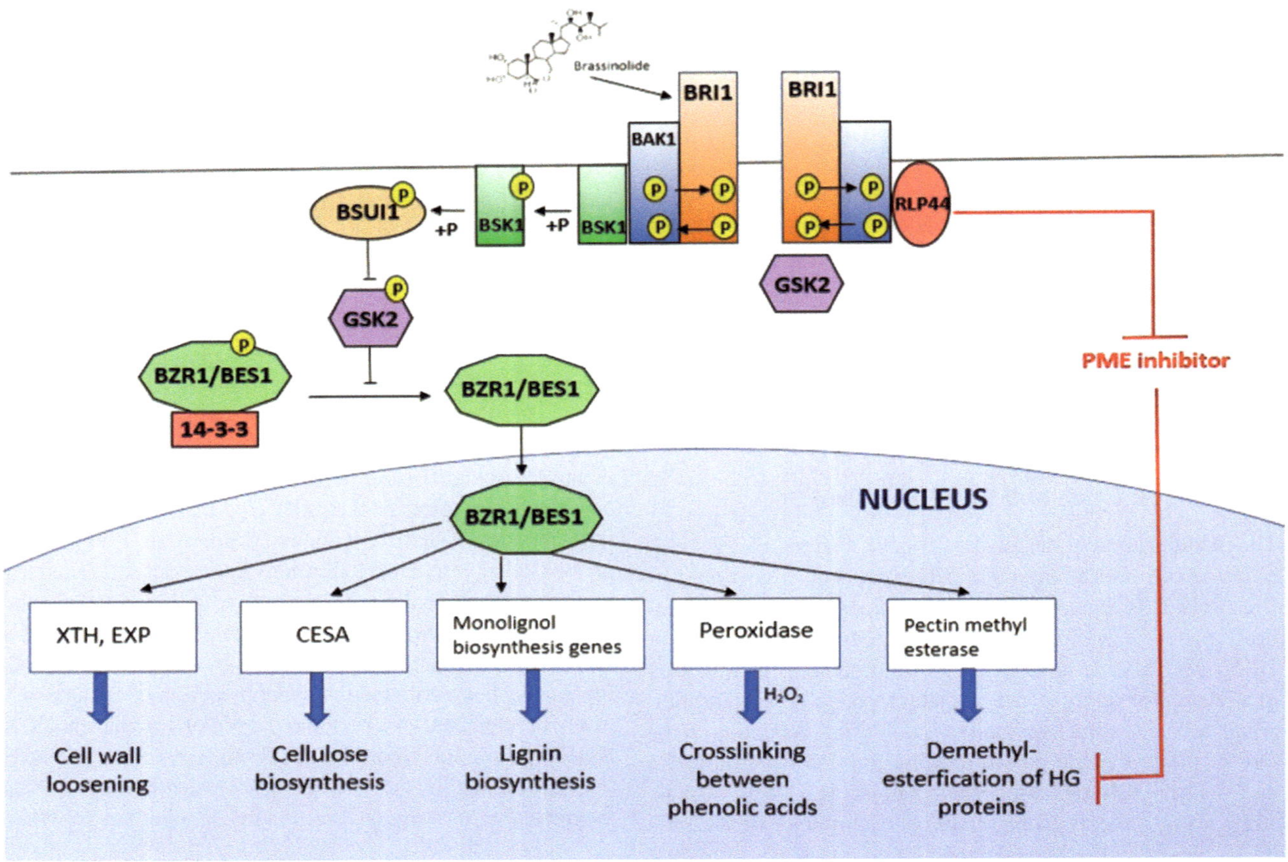

FIGURE 6.4 Cell wall remodeling mediated by BR. Adapted from Vriet et al. (2013).

xyloglucan polymers and transfer the newly generated ends to other xyloglucan chains (Eklof and Brumer, 2010), while expansins loosen linkages between cellulose microfibrils (Yennawar et al., 2006). Deposition of cellulose microfibrils is responsive to changing environmental conditions (Wang et al., 2016). The CELLULASE SYNTHASE A (CESA) gene superfamily, required for cellulose synthesis, is a downstream target of the transcription factor BES1 of the BR signaling pathway (Xie et al., 2011). CESA1 kinase activity is increased by the degradation of the inhibitor protein *BIN2* (Sánchez-Rodríguez et al., 2017). The loss of cellulose caused by abiotic stresses may be compensated by increased deposition and accumulation or better orientation of cellulose microfibril deposition (Bashline et al., 2014) via treatment with exogenous BRs or overexpression of BR receptor genes (Zhang et al., 2014).

Pectic polysaccharides binding to the cellulose and hemicellulose network maintain extensibility and inhibit collapse of the cellulose matrix (Voxeur and Hofte, 2016). Among various pectic polysaccharides, the degree of methyl esterification in homogalacturonan determines the stiffness of the pectic matrix. This is balanced by the activity between pectin methylesterase enzymes (PMEs) and PME inhibitors (PMEIs). PME demethylates HG chains, leading to a decrease in stiffness of the wall and consequent cell growth (Hofte, 2015). The BR-receptor kinase BAK1 interacts with RECEPTOR-LIKE PROTEIN (RLP44), a cell wall surveillance protein, to repress the activity of PME inhibitors and therefore reduce the stiffness of the pectic matrix, promoting cell wall loosening under stress conditions (Wolf et al., 2014).

Lignin biosynthesis is often induced as a defensive response to biotic and abiotic stresses. BR treatment induces the accumulation of lignin-rich S units in switchgrass suspension cells (Shen et al., 2013). In *Arabidopsis*, the BR-activated transcription factor BES1 regulates lignin biosynthesis by promoting the expression of VASCULAR-RELATED NAC-DOMAIN6 and 7 (VND6 and VND7), responsible for the transition of xylem cells to form tracheary elements (Li et al., 2016). Exogenous applications of BR were significantly found to increase antioxidant enzyme activities (Yan et al., 2015; Sharma et al., 2016), leading to both lignin polymerization and cross-linking of components in the cell wall (Tenhaken, 2015). Thus, BRs not only enhance the antioxidative mechanism but also alleviate oxidative damage by cross-linking phenolic compounds in the cell wall (O'Brien et al., 2012).

6.3.2 Role of BR at Physiological and Biochemical Level

6.3.2.1 Maintenance of Redox Potential

Reactive oxygen species (ROS) in plant stress signaling pathways possess bimodal activity, causing cell death at high ROS levels and playing a regulatory role at low ROS levels. Exogenous application of BR induces H_2O_2, superoxide radical (O_2^-) and hydroxyl radical (•OH) production, activating the mitogen-activated protein kinase (MAPK) cascade and positively amplifying the stress signal (Yang et al., 2019). The excess H_2O_2 serves to upregulate antioxidant enzymes, transcription factors, heat shock proteins,

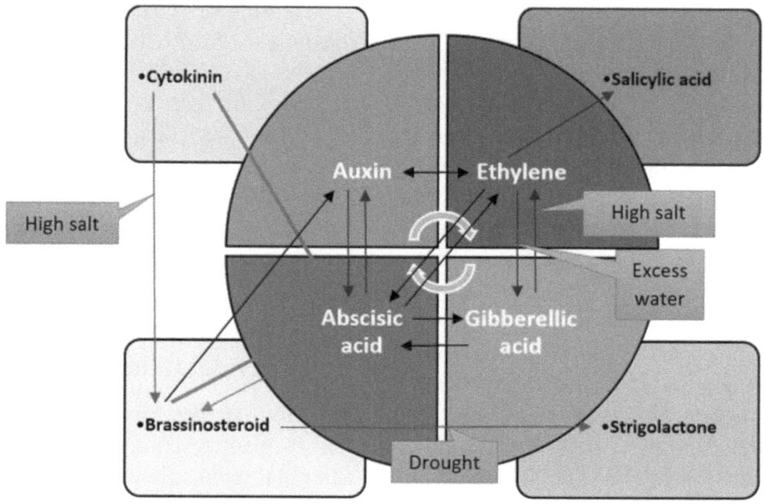

FIGURE 6.5 A generalized model of hormone interaction under biotic and abiotic stress. Adapted from Ahmed et al. (2018).

and various other proteins responsible for scavenging ROS. BRs act locally but induce plant systemic stress tolerance via increased H_2O_2 (Xia et al., 2011). This relationship was reported in the tomato plant, where apoplastic H_2O_2 was found to increase in the presence of exogenous BR. Furthermore, MPK2 rather than MPK1 was found to be linked to BR-induced H_2O_2 accumulation (Nie et al., 2013). Silencing respiratory burst oxidase homologs (RBOH) in *Nicotiana benthamiana* made the plant more susceptible to abiotic stress (Deng et al., 2015). Additionally, a spike in H_2O_2 through activation of NADPH oxidase enhances ABA biosynthesis, and the resultant increase in H_2O_2 provides sustained stress tolerance. It has been reported that BR mutants have lower GSH/GSSG and AsA/DHA ratios (Zhou et al., 2015). Cucumber seedlings treated with 24-epibrassinolide (EBR) under autotoxicity conditions exhibited higher activities of SOD, CAT, POD, and APX, and enhanced AsA and GSH levels (Yang et al., 2019). Removal of peroxidation products and the reduction in MDA level with exogenous BR treatment are also indicative of a decrease in cell membrane injury (Yang et al., 2019).

Nitric oxide (NO), the downstream signaling molecule of H_2O_2 in BR signaling, is important for vital processes such as hormone responses, programmed cell death, ion homeostasis, disease resistance, and response to abiotic stress (Zhao et al., 2004). NO production reduced by silencing of nitrate reductase (NR) and nitric oxide synthase (NOS) genes prevents the BR-induced alternative respiratory pathway and blockage of the above-mentioned responses. NO-induced activation of ABA and NIA1 protein biosynthesis trigger stomatal closure in response to abiotic stress (Shi et al., 2015).

6.3.2.2 Interplay of Brassinosteroids and Other Phytohormones

The pleiotropic and intricate regulatory mechanisms of phytohormones have been highlighted in plants subjected to abiotic stress (Br and Sm, 2009). Their role is central to several adaptation processes such as growth, allocation of nutrients, and source-sink interactions (Tiwari et al., 2017). The crosstalk seen between various hormonal signaling pathways and stress signaling pathways generates a customized stress response best suited for the given plant species under a particular environmental condition. Exogenous application of BR (Wani et al., 2016), as well as genetic deactivation of negative regulators of the BR signaling pathway (Koh et al., 2007), have shown positive effects on stress tolerance. BR and ABA assume opposing multi-faceted roles: BR triggers seed germination and post-germinative development while ABA hinders it (Hu and Yu, 2014). Chemical inhibition of ABA biosynthesis results in a reduction in BR-induced stress tolerance (Zhang et al., 2011). The antagonistic relationship between the two phytohormones is seen at the targeted gene *BIN2* of the BR signaling pathway, enabling fine-tuning of resource utilization between stress response and growth (Chung et al., 2014). However, a synergistic relationship occurs between ABA and BR with respect to the production of H_2O_2, respiratory burst oxidase homolog 1 (RBOH1), and NADPH oxidase activity (Zhou et al., 2014) (Figure 6.5).

An interaction between BR and auxin possesses differential spatiotemporal responses concerning tissue and organ specificity—BR-induced genes and auxin-related genes are present largely in epidermal cells of the basal meristem (Singh and Savaldi-Goldstein, 2015), while BR downregulated genes occur in the stele of the apical meristem zone (Vragovi´c et al., 2015). In *Arabidopsis*, the level of BR needed for development is governed by BRAVIS RADIX (BRX), which, in turn, is triggered by an increase in auxin levels (Mouchel et al., 2006). Further substantiating the role of auxin in BR synthesis, CPD and DWF4 (dwarf4) genes are controlled by BRX (Tanaka et al., 2005). Moreover, a collaboration of *BIN2* and auxin reaction factors (ARF2), a group of transcriptional controllers, has been reported (Vert et al., 2008). Optimal root growth is regulated by the spatiotemporal balance between stem cell division and differentiation maintained by an antagonistic relationship between the two phytohormones (Chaiwanon and Wang, 2015).

Crosstalk was envisaged between BR and cytokinin (CK) when researchers discovered an enhanced expression of isopentenyl transferase, a key enzyme in CK synthesis. The gene for this enzyme provides drought tolerance when under a drought-inducible promoter and a subsequent increase in proteins responsible for BR biosynthesis (Ahmad et al., 2018). BR has also been reported to react with gibberellic acid (GA). In rice, exogenous GA application induced the gene OsGSR1 while exogenous BR treatment repressed it (Wang et al., 2009). BZR1, the transcription factor activated upon BR signaling, is known to interact with REPRESSOR OF GA1-3 (RGA), belonging to the DELLA protein family and responsible for GA signaling. An antagonistic relationship is observed when overexpression of DELLA proteins reduces BZR1 activity (Li et al., 2012). GAI protein, a negative regulator of the GA signaling pathway, binds to BZR1 causing the deactivation of its transcriptional regulatory activity (Gallego- Bartolomé et al., 2012). Negative crosstalk was seen in SA- and GA-mediated immunity of rice against the oomycete *Pythium graminicola* antagonized by BR, subsequently preventing DELLA degradation (De Vleesschauwer et al., 2012).

Ethylene biosynthesis is regulated by BR via ACC synthase and ACC oxidase activities (Hansen et al., 2009). Soil flooding causes hyponastic growth mediated by ethylene, which, in turn, regulates ROTUNDIFOLIA3/CYP90C1 expression, a protein involved in the hydroxylation of several BRs at the C23 position (Polko et al., 2013). Ethylene also regulates the submergence response in rice via the submergence tolerance gene, SUB1A, which encodes an ERF protein, a transcription factor that differentially regulates BR biosynthetic genes. Pre-treatment with EBR was found to increase tolerance to submergence (Schmitz et al., 2013). BRs mitigate oxidative damage caused by abiotic stress by inducing ethylene biosynthesis (Wei et al., 2015). Microarray studies have demonstrated overlapping complex regulation of BR genes with other phytohormones like ethylene, ABA, auxin, and jasmonic acid (JA; Divi et al., 2016).

Genes responsible for JA biosynthesis and JA-mediated signaling are downregulated when BR levels are low. Exogenous foliar application of JA showed counter communication between BR and JA in the rice roots as observed by downregulation of BR biosynthesis and signaling genes, OsDWF4 and OsBRI1 (Nahar et al., 2013). Similarly, salicylic acid (SA) and BR together provide salt and thermotolerance through crosstalk and interaction with other stress hormones (Divi et al., 2010; Ahmad et al., 2018).

Strigolactones (ST) are second messengers implicated in the dynamic feedback loop of the auxin signaling pathway. ST controls auxin levels by regulating its synthesis and/or its polar transport. An increase in auxin transport inhibits root growth in ST mutants (Zhang et al., 2013). ABA-deficient tomato mutants demonstrated significantly lower levels of ST (López Ráez et al., 2010). ST also influences ROS levels in many hormone signaling pathways; however, evidence is needed to ascertain this crosstalk (Gómez-Cadenas et al., 2014).

6.3.3 The Role of BR Signaling and Regulation in Adaptations to Abiotic Stress

Numerous studies have suggested that BRs play an important role in plant responses to various stresses such as drought, abnormal temperatures, high osmotic pressure, and attack by pathogens (Krishna, 2003). Plants tolerate biotic and abiotic stress by switching between growth activation and repression under unfavorable conditions (Bechtold and Field, 2018). The elucidation of extensive crosstalk between BRs and stress-responsive hormones such as abscisic acid, ethylene, and JA also indicates the importance of BRs in the intricate balance between normal growth and resistance against environmental assaults (Zhang et al., 2009). At the forefront of the interactions controlling the response to environmental stresses is the abscisic acid (ABA) pathway (Zhu et al., 2017). Proposed mechanisms of BR signaling as an adaptation to stress include: (i) regulation of stress-responsive transcription machineries (Ye et al., 2017); (ii) activation of antioxidant systems (Zou et al., 2018); and (iii) production of osmoprotectants (Fàbregas et al., 2018).

6.3.3.1 *Heat Stress*

Plant adaptation to heat stress is as shown in Figure 6.6A. Regulation of thermomorphogenesis upon temperature

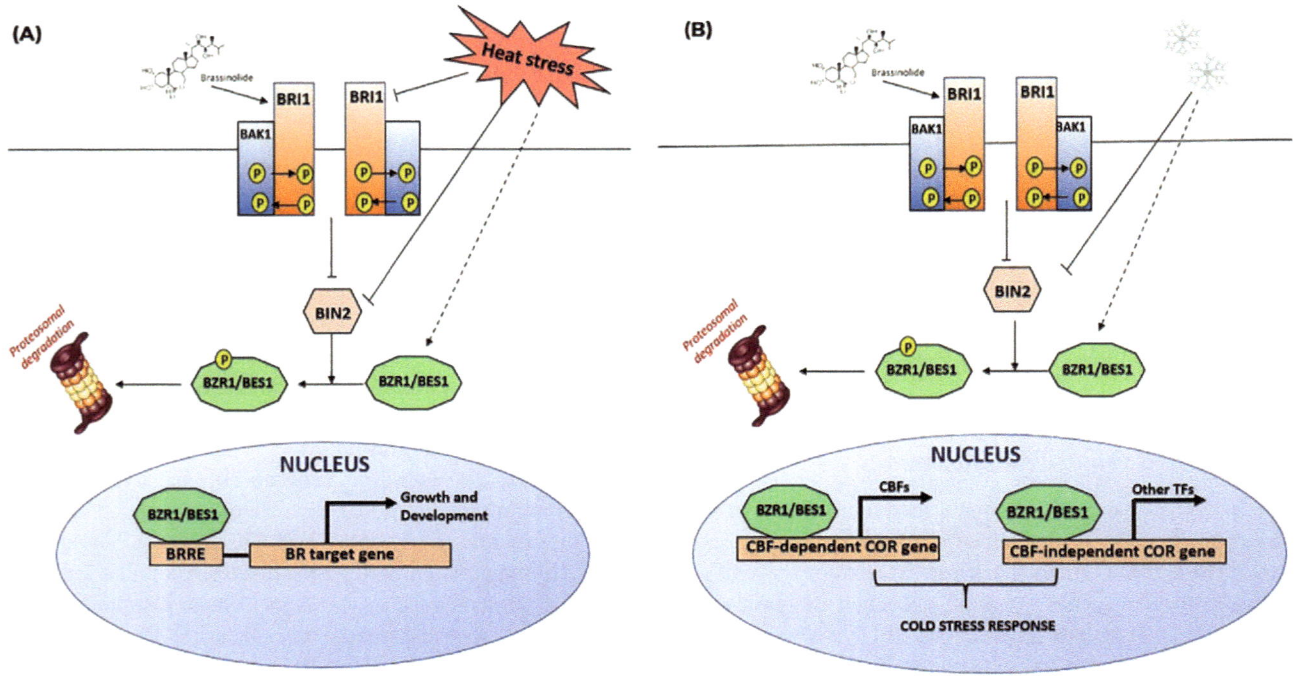

FIGURE 6.6 BR signaling controls the switch between development and abiotic stress responses. Schematic representation of BR response under high temperature (A) and cold (B) stress.

elevation is by the association of BZR1 with BR-responsive genes either directly or via binding to the promoter of PHYTOCHROME INTERACTING FACTOR 4 (PIF4; Ibañez et al., 2018). Also, increased levels of PIF4 shifts the equilibrium toward the accumulation of the nuclear protein complex BES1-PIF heterodimers instead of BES1 homodimers (Martinez et al., 2018). The resultant decline in the homodimeric BES1 causes de-repression of BR biosynthesis and feedback inhibition of BR signaling output. High temperatures also reduce BRI1 levels by ubiquitination, endocytosis, and degradation (Zhou et al., 2018), thereby affecting primary root elongation (Martins et al., 2017). Following BR perception, PUB12 and PUB13 ubiquitinate BRI1 (Zhou et al., 2018), however, the E3 ubiquitin ligase responsible for ubiquitination under heat stress is still not yet identified.

6.3.3.2 Cold Stress

In cold stress adaptation (Figure 6.6B), three C-REPEAT/DEHYDRATION-RESPONSIVE ELEMENT BINDING FACTORS (CBF) play pivotal roles by controlling the transcription of CBF-dependent and CBF-independent core cold-responsive (COR) genes (Eremina et al., 2016). These are activated by INDUCER OF CBF EXPRESSION 1 (ICE1), CALMODULIN-BINDING TRANSCRIPTION ACTIVATOR 3 (CAMTA3), or BZR1 (Li et al., 2017), whereas MYB15 and ETHYLENE-INSENSITIVE 3 (EIN3) negatively regulate CBF expression (Shi et al., 2012). Under cold stress, activated OPEN STOMATA 1 (OST1) phosphorylates ICE1, thereby preventing degradation mediated by the E3 ligase HIGH EXPRESSION OF OSMOTICALLY RESPONSIVE GENE 1 (HOS1). Dephosphorylated BZR1/BES1 regulates COR genes by binding to CBF1 and CBF2, thereby affecting the transcription of their downstream targets (Li et al., 2017; Figure 6.2B). Another BR-regulated transcription factor, BR-ENHANCED EXPRESSION 1 (BEE1), promotes cold acclimation. Overexpression of *BIN2* is linked to hypersensitivity to freezing stress, whereas gain-of-function bzr1-1D and bes1-D mutants have enhanced tolerance for freezing stress (Li et al., 2017). Even though CBF-dependent cold signaling pathways have been studied extensively, the mechanism of fine-tuning cold signaling remains unclear. For this, plasma membrane-localized COLD-RESPONSIVE PROTEIN KINASE 1 (CRPK1) phosphorylates 14-3-3 protein, which, in turn, translocates from the cytoplasm into the nucleus to promote the degradation of CBFs via the 26S proteasome pathway (Li et al., 2017).

6.3.3.3 Drought Stress

There are numerous reports on the role of BR and related compounds in drought tolerance (Planas-Riverola et al., 2019). It is known to function via H_2O_2- and NO-mediated machinery, wherein ABA induction by BR under oxidative stress takes place (Zhang et al., 2011). NADPH oxidase triggers H_2O_2 secretion, which, in turn, triggers ABA biosynthesis, acting as a positive-feedback mechanism for prolonged tolerance to oxidative stress (Zhou et al., 2014). H_2O_2-mediated modification enhances the transcriptional activity of BZR1 while thioredoxin TRXh5 catalyzes its reductive degradation (Tian et al., 2018) (Figure 6.6C). Overexpression of the BR receptor gene BRL3 driven by a constitutive promoter was found to increase survival during drought. The BRL3 protein accumulates primarily in root vascular tissue, stimulating the accumulation of osmolytes and other stress-responsive gene products (Fàbregas et al., 2018). BES1 inhibits drought response in *Arabidopsis*, while BES1 and BR trigger the expression of a subset of drought-responsive genes (Ye et al., 2017). A BES1 and BZR1 homolog in *Triticum aestivum* called TaBZR2 promotes superoxide scavenging by glutathione S-transferase induction (Cui et al., 2019). Studies in *Solanum lycopersicum* indicate that overexpression of the BR receptor gene SlBRI1 reduced drought tolerance (Nie et al., 2019). Thus, BR is operational at various levels: modulation of ABA, stress-responsive gene expression, and production of H_2O_2, antioxidants, and osmolytes (Planas-Riverola et al., 2019).

6.3.3.4 Salt Stress

Tolerance to salinity is via the regulation of ethylene (ET) biosynthesis and signaling (Figure 6.7B). BR pre-treatment enhances the activity of 1-aminocyclopropane-1-carboxylate synthase (ACS) needed for ethylene synthesis (Zhu et al., 2016). *BIN2* causes the phosphorylation of BZR1, which then undergoes proteasomal degradation (Figure 6.7B). ET activity ultimately results in the activation of the transcription factor ERF1 that binds to the DRE-box gene and initiates tolerance to salinity. Blocking ethylene synthesis has demonstrated an inhibition of the antioxidant enzyme activity, which subsequently reduces the ability of the plant to tolerate salt (Zhu et al., 2016). In *Solanum lycopersicum*, overexpressing the genes *DWARF* and *BRI1* responsible for BR biosynthesis and signaling, respectively, was found to elevate ethylene production and post-harvest ripening (Li et al., 2016; Nie et al., 2017). Conversely, silencing of the *BRI1* gene prevented changes in ethylene accumulation and ACS activity during pre-treatment with BR (Zhu et al., 2016; Lv et al., 2018). High salinity suppresses the nuclear accumulation of BZR1 in roots, subsequently reducing BR signaling functions (Geng et al., 2013). Application of EBR to mutants deficient and insensitive to ET increased their survival rates and reverted from hypersensitivity to salt to that of WT plants response (Divi et al., 2010). Treatment of lettuce with the brassinosteroid analog, DI-31, exhibited an increase in ET production and avoidance of premature death when subjected to salt stress (Peres et al., 2019).

6.3.4 Brassinosteroid Homeostasis and Its Regulation

Activation of stress signaling pathways by BRs is directly or indirectly regulated via the negative regulator *BIN2* and transcription factors BZR1/BES1, acting as broad range regulators for various developmental processes (Wang et al., 2013). Many recent discoveries have proposed an integration hub enabling the coordinated regulation of physiological processes when challenged by environmental stress conditions (Li et al., 2018). For example, BZR1 undergoes phosphorylation by the MAPK complex through a highly complex mechanism due to the presence of 25 putative *BIN2*-mediated phosphorylation sites and 11 phosphorylation sites for kinases (Wang et al., 2013). In addition, phosphorylation of BZR1 may also be regulated by other proteins such as cyclophilin protein CYP20-2 (Zhang et al.,

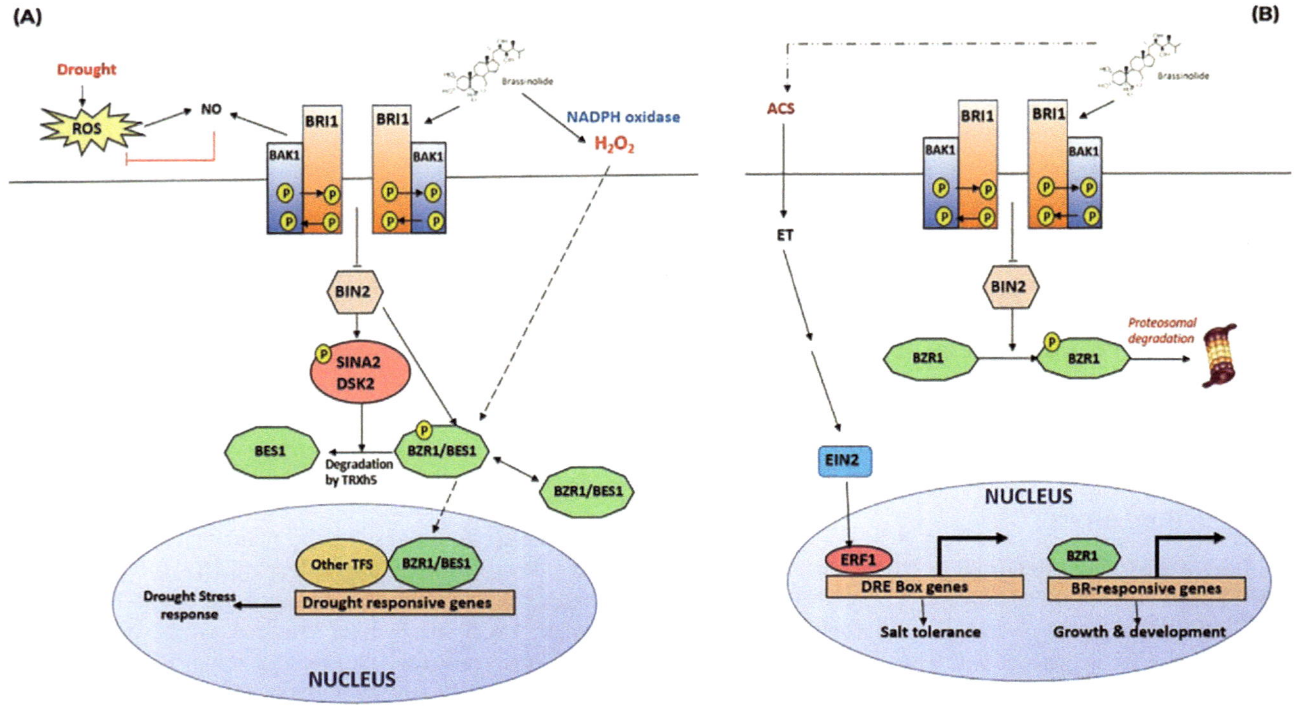

FIGURE 6.7 BR signaling controls the switch between development and abiotic stress responses. Schematic representation of BR response under drought (A) and salinity (B) stress.

2013). BZR-SENSITIVE-SHORT HYPOCOTYL1, also known as BLADE ON PETIOLE1 (BSS1/BOP1) has been recently identified as a negative regulator of BR signaling by the formation of a cytosolic complex with BZR1 and BES1, thereby inhibiting its transport to the nucleus. Exogenous BR treatment dissolves the protein complex and decreases the expression of the BSS1/BOP1 gene (Shimada et al., 2015).

H_2O_2 positively regulates the function of BZR1. BR activation and BRI1-mediated signaling increase the accumulation of H_2O_2, leading to oxidation of BZR1, and triggering cell division and elongation. Oxidation also promotes BZR1 interaction with oxygen and light signaling pathways. It can be concluded that both BR and H_2O_2 control the activity of BZR1 via nuclear localization/DNA binding (Tian et al., 2018).

Target of rapamycin (TOR) kinase, known to integrate nutrient and energy signaling, ensures sugar-induced plant growth in the dark. Inhibition of TOR activity causes decreased expression of BR-responsive genes and repression of plant growth. It can be assumed that BZR1 accumulation and BR signaling is promoted by TOR mediation. However, BZR1 protein degradation is induced by starvation. This fine-tuning is required for maintaining a balance between plant growth and the availability of carbon (Zhang et al., 2016).

The BR signal output is also maintained by phosphorylation status, stability of transcription factors, and cytoplasm-nucleus shuttling. The ubiquitin receptor protein dominant suppressor of KAR2 (DSK2), constitutively photomorphogenic1 (COP1), Skp-Cullin-F-box (SCF) E3 ubiquitin ligases—more axillary growth locus2 (MAX2) and SINA of *Arabidopsis thaliana* (SINAT)—are directly involved in the degradation of BZR1/BES1 (Nolan et al., 2017; Wang et al., 2013; Kim et al., 2014; Yang et al., 2017). Autophagy-mediated degradation during drought and starvation stress is triggered by the interaction of BZR1/BES1 with DSK2 and autophagy-related 8 (ATG8) protein, required for autophagosome assembly and cargo recruitment (Nolan et al., 2017). Inhibition of plant shoot branching occurs by the interaction of MAX2 with BZR1 and BES1. Light-regulated stability of BZR1 is mediated by COP1 and SINAT. COP1 captures and degrades the inactive/phosphorylated form of BZR1, consequently increasing the BZR1 active form. The presence of the active form stimulates plant growth (Kim et al., 2014). On the other hand, ubiquitination and degradation of dephosphorylated forms of BZR1 and BES1 are triggered by SINAT proteins, which are stimulated by light. This mechanism of light/dark conditions plays an important role in regulating BZR1/BES1 protein stability and plant morphogenesis in the presence of hormones and environmental cues (Yang et al., 2017).

The BR-dependent gene expression including BR biosynthesis, BR perception, and BR signaling is regulated by a complicated network of interactions coordinated by BZR1 and BES1 (Gruszka et al., 2013). Basic helix–loop–helix (bHLH) transcription factors regulate cell proliferation and differentiation. ATBS1, an atypical bHLH, stimulates BR response by sequestering transcription factors that would otherwise negatively affect BR signaling (Wang et al., 2009). The PRE1 protein and increased lamina inclination1 (ILI1) inactivate bHLH transcription factors by heterodimerization and sequestering the negative regulators of plant growth: ILI1 binding bHLH1 (IBH1), LONG HYPOCOTYL IN FAR-RED1 (HFR1), and PHYTOCHROME RAPIDLY REGULATED1 (PAR1; Zhang et al., 2009). Suppression of PAR1 and HFR1 activity occurs through BR-mediated activation of PREs while IBH1 gene transcription is repressed in a BZR1-dependent manner. Auxin and gibberellins also stimulate PRE1 gene expression;

an interesting finding is that the PRE1 gene is found largely in young and growing tissues, while the IBH1 gene is mainly expressed in mature organs.

Cell elongation is regulated by antagonization between RE1 and IBH1. PRE1 suppresses the activity of IBH1, enabling BRs to exert a double inhibitory effect on IBH1 at both the transcript and protein levels. The regulation of plant development occurs due to the integration of multiple signaling pathways in which the PRE1/ILI1 group plays an important role (Zhang et al., 2009). The tri-antagonistic system (ATBS1/PRE-AIFs/IBH1-HBI1/ACE1) interacts with BR, gibberellin, and light signaling to regulate plant growth and development (Kim et al., 2017). This system can be either induced or depressed through a complex feedback mechanism via BZR1/BES1. *BIN2* kinase phosphorylates atypical bHLH encoded by the AIF2 gene, suppressing plant growth (Kim et al., 2017). Dephosphorylation of AIF2, directing it to proteasomal degradation, is triggered by BR and ABA and other phytohormones.

Epigenetic modifications by phytohormones alter gene expressions. BR-regulated gene activity is also mediated by histone modifications like methylation and demethylation. One-third of BR-regulated genes are positively regulated by EARLY FLOWERING6 (ELF6) and RELATIVE OF EARLY FLOWERING6 (REF6; Yu et al., 2008). REF6, a H3K27 demethylase mediated by the polycomb repression complex2 (PRC2) family demethylates the double- and triple-methylated H3K27 residue (Lu et al., 2011). BES1 along with the Interacting-With-Spt61 (IWS1) factor responds to BR and recruits REF6 to remove the histone H3K27 di- and tri-methylation repression mark, thus permitting transcription and RNA export. H3K36 di- and trimethylation occurs by BES1 and SET DOMAIN GROUP8 (SDG8) histone methyltransferase (Wang et al., 2014), however, the precise mechanism of H3K36 gene regulation is still not understood. BZR1 interacts with histone deacetylase HDA19 and TOPLESS (TPL) to form a trimeric complex that regulates cell elongation (Oh et al., 2014).

6.4 Conclusion

The process of conferring abiotic stress tolerance is a complex one as diverse mechanisms need to be explored. The phytohormone brassinosteroid regulates several processes during the plant's life cycle. The BR signalosome forms a hub in this complicated network of multi-signaling crosstalk linking phytohormones with environmental cues. As a result, coordinated regulation of physiological processes enabling rapid adaptation to abiotic conditions is possible. Even though substantial progress has been made in the field of BR signal perception, the interlinking of these signals to plant tolerance is yet to be thoroughly explored. A holistic understanding of complicated regulatory mechanisms necessitates further research. Rapid developments in the fields of genomics and proteomics can be employed to identify key genes and proteins involved in BR signaling for the amelioration of stress effects. Nevertheless, recent findings and a wealth of data suggest that BR signaling is an ideal target to increase crop yield under prevailing environmental stress.

REFERENCES

Abreu. *Química Nova*, 1991; 14:44–48.

Adam and Marquardt. *Phytochemistry* 1986; 25:1787–1799.

Ahmad et al. *J. Appl. Biol. Biotechnol.* 2018; 6(1):56–62.

Bajguz. *Plant Physiol. Biochem.* 2007; 45, 95–107.

Bashline et al. 2014; 7:586–600.

Bechtold and Field. *J. Exp. Bot.* 2018; 69:2753–2758.

Br and Sm. *Plant Mol. Biol.* 2009; 69:451–62.

Chaiwanon and Wang. *Curr. Biol.* 2015; 25:1031–1042.

Choe et al. *Plant Cell* 1999; 11:207–221.

Chung et al. *Mol. Cells* 2014; 37:795–803.

Cui et al. *Plant Cell* 2012; 24:233–244.

De Vleesschauwer et al. *Plant Physiol.* 2012; 158:1833–1846.

Deng et al. *J. Exp. Bot.* 2015; 66:6219–6232.

Divi et al. *BMC Plant Biol.* 2010; 10:151.

Divi et al. *Plant Biotechnol J.* 2016; 14:419–432.

Eklof and Brumer. *Plant Physiol.* 2010; 153:456–466.

Eremina et al. *Proc. Natl. Acad. USA* 2016; 113:5982–91.

Fàbregas et al. *Plant Cell* 2013; 25:3377–3388.

Fàbregas et al. *Nat. Commun.* 2018; 9:4680.

Fu et al. *J. Exp. Bot.* 2008; 59:2299–2308.

Fujioka. In: Sakurai A, Yokota T, Clouse SD (eds), *Brassinosteroids: Steroidal Plant Hormones*, pp. 21–45. Springer, Tokyo.

Fujioka and Yokota. *Annu. Rev. Plant Biol.* 2003; 54:137–64.

Fung and Siddall. *J. Am. Chem. Soc.* 1980; 102:6580–6581.

Gallego-Bartolomé et al. *Proc. Natl. Acad. Sci. USA.* 2012; 109:13446–13451.

Geng et al. *Plant Cell* 2013; 25:2132–2154.

Gómez-Cadenas. In: *Phytohormones: A Window to Metabolism, Signaling 289 and Biotechnological Applications.* Springer, New York. 2014.

Gruszka. *Int. J. Mol. Sci.* 2013; 14:8740–8774.

Hansen et al. *Plant J.* 2009; 57:606–614.

Hofte. *Plant Cell Physiol.* 2015; 56:224–231.

Hohmann et al. *Nat. Plants* 2018; 4:345–351.

Hothorn et al. *Nature* 2011; 474:467–471.

Hu and Yu. *Plant Cell* 2014; 26:4394–408.

Ibanes et al. *Proc. Natl. Acad. Sci. USA* 2009; 106:13630–13635.

Ibañez et al. *Curr. Biol.* 2018; 28:303–310.

Kang et al. *Development* 2017; 144:272–280.

Kim et al. *Plant J.* 2014; 77:737–747.

Kim et al. *Plant Cell Physiol.* 2017; 58:227–239.

Kim et al. *Plant Cell* 2019; 31:791–808.

Koh et al. *Plant Mol. Biol.* 2007; 65:453–466.

Krishna. *J. Plant Growth Regul.* 2003; 22:289–297.

Lang et al. *Acta Agron. Sin.* 2004; 30:883–887.

Li et al. *Plant J.* 2009; 58:275–86.

Li et al. *Sci. Signal* 2012; 5:72.

Li et al. *Front. Plant Sci.* 2016; 7:1–16.

Li et al. *Mol. Plant* 2016; 9:113–125.

Li et al. *Mol. Plant* 2017; 10:545–559.

Li et al. *Biochim. Biophys. Acta* 2018; 1861:561–571.

López-Ráez et al. *New Phytol.* 2010; 187:343–354.

Lu et al. *Nat. Genet.* 2011; 43:715–719.

Lv et al. *PLoS Genet.* 2018; 14:e1007144.

Mandava. *Annu. Rev. Plant Physiol. Plant Mol. Biol.* 1988; 39:23–52.

Mandava and Mitchell. *Indian Agri.* 1971; 15:19–31.

Mandava et al. *Ind. Eng. Chem. Prod. Res. Dev.* 1973; 12:138–139.

Mandava et al. *Ind. Eng. Chem. Prod. Res. Dev.* 1978; 17:351–354.
Martinez et al. *EMBO J.* 2018; 37:e99552.
Martins et al. *Nat. Commun.* 2017; 8:309.
Maugh. *Science.* 1981; 212:33–34.
Meudt et al. *Canadian Patent No. 117659.* 1984.
Mitchell et al. *Nature.* 1970; 225:1065–1066.
Mitchell and Whitehead. *Bot Gaz.* 1941; 102:770–790.
Mouchel et al. *Nature* 2006; 443:458.
Nahar et al. *Mol. Plant Microbe Interact.* 2013; 26:106–15.
Nie et al. *Plant Cell Environ.* 2013; 36:789–803.
Nie et al. *Front. Plant Sci.* 2017; 8:1386.
Nie et al. *Plant Physiol. Biochem.* 2019; 138:36–47.
Noguchi et al. *Plant Physiol.* 1999; 121:743–752.
Nolan et al. *Dev. Cell* 2017; 41:33–46.
O'Brien et al. *Planta* 2012; 236:765–779.
Oh et al. *Nat. Commun.* 2014; 5:4140.
Pavelescu et al. *Mol. Syst. Biol.* 2018; 14: e7687.
Peres et al. *Review Int. J. Mol. Sci.* 2019; 20(2):331.
Planas-Riverola et al. *Development* 2019; 146:151894.
Polko et al. *J. Exp. Bot.* 2013; 64:613–624.
Rodríguez et al. *Biotechnol. Apl.* 2013; 30:178–181.
Ryu et al. *Plant Cell* 2007; 19:2749–2762.
Sánchez-Rodríguez et al. *Proc. Natl. Acad. Sci. USA.* 2017; 114:3533–3538.
Schmidt et al. *Plant J.* 2013; 76:258–273.
Schmitz et al. *New Phytol.* 2013; 198:1060–1070.
Sharma et al. *Environ. Exp. Bot.* 2016; 122:1–9.
Sharma et al. *Front. Plant Sci.* 2017; 8:24–36.
Shen et al. *Plant Cell* 2013; 25:4342–4361.
Shi et al. *Plant Cell* 2012; 24:2578–2595.
Shi et al. *Plant J.* 2015; 82:280–301.
Shimada et al. *Plant Cell* 2015; 27:375–390.
Singh and Bhardwaj. *Indian J. Chem. B.* 1986; 25:989–998.
Singh, and Savaldi-Goldstein. *J. Exp. Bot.* 2015; 66:1123–1132.
Sun et al. *Dev. Cell* 2010; 19, 765–777.
Sun et al. *Dev Cell.* 2015; 34:220–228.
Taiz and Zeiger. *Plant Physiol.* 5th ed. Sunderland: Sinauer Associates; 2010.
Takatsuto. *J. Chromatogr. A.* 1994; 658:3–15.
Tanaka et al. *Plant Physiol.* 2005; 138:1117–25.
Tang et al. *BMC Biol.* 2016; 14:113.
Tenhaken. *Front. Plant Sci.* 2015; 5:771.
Tian et al. *Nat. Commun.* 2018; 9:1063.
Tiwari et al. *Curr. Genomics* 2017; 18:469–482.
Uozu et al. *Plant Physiol.* 2000; 122:53–859.
Vert and Chory. *Nature* 2006; 441:96–100.
Vert et al. *Proc. Natl. Acad. Sci.* 2008; 105:9829–34.
Vilarrasa-Blasi et al. *Dev. Cell* 2014; 30:36–47.
Voxeur and Hofte. *Glycobiology* 2016; 26:950–960.
Vragovi´c et al. *Proc. Natl. Acad. Sci. USA* 2015; 112:923–928.
Vriet et al. *Mol. Plant.* 2013; 6:1738–1757.
Wang and Chory. *Science* 2006; 313, 1118–1122.
Wang et al. *Bangladesh J. Bot.* 2016; 45:143–50.
Wang et al. *Plant Cell* 2009; 21:3781–3791.
Wang et al. *Plant J.* 2009; 57:498–510.
Wang et al. *Dev. Cell* 2013; 27:681–688.
Wang et al. *Mol. Cell. Proteomics* 2013; 12:3653–3665.
Wang et al. *Mol. Plant* 2014; 7:1303–1315.
Wani et al. *Crop J.* 2016; 4:162–176.
Wei et al. *Front. Plant Sci.* 2015; 6:982.
Wolf et al. *Proc. Natl. Acad. Sci. USA* 2014; 111:15261–15266.
Wu et al. *Sci. Signal.* 2011; 4, 1–25.
Xia et al. *New Phytol.* 2011; 191:706–720.
Xie et al. *J. Exp. Bot.* 2011; 62:4495–4506.
Yamamoto et al. *Plant Physiol.* 2001; 125:556–563.
Yan et al. *Plant Cell Physiol.* 2015; 56:883–896.
Yang et al. *Physiol. Plant.* 2013; 148:502–511.
Yang et al. *Dev. Cell* 2017; 41:47–58.
Yang et al. *Agronomy* 2019; 9(5):265.
Ye et al. *Plant Cell* 2019; 31(11):2682–2696.
Ye et al. *Nat. Commun.* 2017; 8:14573.
Yennawar et al. *Proc. Natl. Acad. Sci. USA* 2006; 103:14664–14671.
Yokota et al. *Tetrahedron Lett.* 1982; 23:1275–1278.
Youn and Kim. *Mol. Plant* 2015; 8:552–565.
Yu et al. *Proc. Natl. Acad. Sci. USA* 2008; 105:7618–7623.
Yu et al. *Plant J.* 2011; 65:634–646.
Zhang et al. *Plant Cell* 2009; 21:3767–3780.
Zhang et al. *Plant Signal. Behav.* 2009; 4:1117–1120.
Zhang et al. *Plant Cell Physiol.* 2011; 52:181–192.
Zhang et al. *Plant Cell* 2013; 25:2504–2521.
Zhang et al. *Plant Cell Rep.* 2014; 33:683–696.
Zhang et al. *Curr. Biol.* 2016; 26:1–7.
Zhao et al. *Plant Physiol.* 2004; 134:849–857.
Zhou et al. *J. Exp. Bot.* 2014; 65:4371–83.
Zhou et al. *Sci. Rep.* 2015; 5:9018.
Zhou et al. *Proc. Natl. Acad. Sci. USA* 2018; 115:1906–1915.
Zhu. *Mol. Cell* 2017; 66:648–657.
Zhu et al. *Sci. Rep.* 2016; 6:35392.
Zou et al. *Physiol. Plant* 2018; 163, 196–210.

7

Protective Role of Brassinosteroids in Plants During Abiotic Stress

Mona Gergis Dawood

CONTENTS

Abbreviations

BRs: Brassinosteroids
DET2: Deetiolated2
CPD: Constitutive photomorphogenesis and dwarfism
BL: Brassinolide
28-HomoBL: 28-homobrassinolide
24-EpiBL: 24-epibrassinolide
BRI1: Brassinosteroid insensitive 1
BAK1: BRI1 associated receptor kinase 1
BRZ: Brassinazole
BIN2: Brassinosteroid insensitive 2
BZR1: Brassinazole resistant 1
BES1: BRI1-Ems suppressor 1
BKI1: BRI1 Kinase inhibitor 1
BSK1: BR-Signaling kinase 1
BSU1: BRI1 Suppressor 1
BZR: Brassinazole resistant
APX: Ascorbate peroxidase
SOD: Superoxide dismutase
POD: Peroxidase
CAT: Catalase
GR: Glutathione reductase
PPX: Polyphosphate Kinase
GCR1: G-protein coupled receptor 1
GPA1: G-protein α-subunit 1
AOX: Alternative oxidase
PPO: Polyphenol oxidase
AsA: Ascorbic acid
H_2O_2: Hydrogen peroxide
CO_2: Carbon dioxide
POX: Polyphenol oxidase
MDA: Malondialdehyde

DOI: 10.1201/9781003110651-7

ROS: Reactive oxygen species
PAL: Phenylalanine ammonia-lyase
***RCA*:** Rubisco carboxylase activase gene
ARF2: Auxin-response factor
SAUR15: Small auxin upregulated15
GA: Gibberellic acid
ABA: Abscisic acid
Cr: Chromium
NI: Nickel
Cd: Cadmium

7.1 Introduction

Brassinosteroids (BRs) are plant-specific steroid hormones that are characterized by a polyhydroxylated sterol structure and have significant growth promoting activities (Verma et al., 2012; Bajguz and Piotrowska-Niczyporuk, 2014 a; Fang et al., 2019; Peres et al., 2019). In 1979, the first isolated and purified BR from *Brassica napus* pollen was brassinolide (BL; Grove et al., 1979). BRs have been isolated from 64 plant species (53 angiosperms, six gymnosperms, *Equisetum arvense* [a pteridophyte], *Marchantia polymorpha* [a bryophyte], and three algae, *Chlorella vulgaris*, *Cystoseira myrica*, and *Hydrodictyon reticulatum*; Bajguz, 2011). BRs can be classified as C_{27}, C_{28}, or C_{29} according to the number of carbons in their structure (Vardhini, 2013). BRs have been detected in all examined plant organs, including leaves, stems, roots, flowers, grains, seeds, pollen, and anthers as mentioned by Bajguz (2011) and Bajguz and Piotrowska-Niczyporuk (2014a). BRs are detected at a higher level in young growing tissue as compared with older vegetative tissue. The highest levels were found in pollen and immature seeds (Bajguz and Tretyn, 2003). Moreover, Bajguz (2011) stated that pollen and seeds are the richest sources of BRs in the range of 1–100 ng g^{-1} FW, while shoots and leaves usually have the lowest content, 0.01–0.1 ng g^{-1} FW. Notably, BRs may be applied to different plant species at different growth stages of their life cycle viz., vegetative stage (Vardhini and Rao, 1998), flowering stage (Vardhini, 2013), meiosis stage (Saka et al., 2003), anthesis stage (Liu et al., 2006), as a seed treatment (Zhang et al., 2007), root application (Song et al., 2006), and a foliar spray (Vardhini et al., 2008).

It is worthy to mention that BR application at very low concentrations significantly increased plant growth, while higher levels exerted adverse effects on growth and growth-related parameters at early stages (Choudhary et al., 2012 a). BRs at lower doses enhanced cotton fiber development by promoting fiber cell elongation (Sun, 2005). In contrast, higher doses of BRs can inhibit cell elongation, especially in primary root extension and lateral root formation (Sasse, 1994). The optimum concentration of the applied BRs on different types of plants ranged from 0.01 to 100 ppm, and it can be applied with other agrochemicals such as other plant hormones, herbicides, fertilizers, insecticides, and other adjuvants to integrate diverse processes required for plant growth (Mori, 1984; Adam and Schneider, 1999; Khripach et al., 2000; Schneider, 2002; Mussig, 2005).

BRs exhibited high physiological activity even at low concentrations (Bartwal et al., 2013; Sun et al., 2015). BRs are considered as important phytohormones due to their versatile roles in plant growth and development (Vardhini, 2016; Yusuf et al., 2017), and they may be involved in the regulation of many vital physiological activities such as seed germination, cell division, cell expansion, vegetative growth, reproduction, senescence and stress tolerance as mentioned by Clouse (2002); Krishna (2003); Bhardwaj et al. (2006), Oklestkova et al.(2015), Lozano-Durán and Zipfel (2015), and Nolan et al. (2017). Moreover, BRs can activate the cell cycle during seed germination (Zadvornova et al., 2005), control progression of the cell cycle (González-Garcia et al., 2011), induce exaggerated growth in hydroponically grown plants (Arteca and Arteca, 2001), and control the proliferation of leaf cells (Nakaya et al., 2002). In addition, BRs might play prominent roles in various physiological processes such as vascular differentiation, pollen tube growth, proton-pump activation, membrane polarization, source/sink relationships, reproductive development, ion uptake, regulation of gene expression, nucleic acid and protein synthesis, enzyme activation, and photosynthesis (Bajguz 2000; Sasse 2003; Nolan et al., 2019; Praveena et al., 2020). Moreover, BR signaling varies among different cells and tissues, which can be manipulated to improve plant growth and stress responses (Fàbregas et al., 2018). It is worthy to mention that BRs are efficiently used in plants as immune modulators. Praveena et al. (2020) concluded that BRs increased plant tolerance to abiotic and biotic stresses by activation or suppression of key enzymatic reactions, production of chemical defense compounds, induction of protein and ethylene biosynthesis, regulation of the antioxidant system, and nitrogen metabolism.

Although the application of BRs promoted plant growth, they were not widely recognized as a novel phytohormone until the 1990s when several genes involved in BR biosynthesis were identified, including deetiolated2 (DET2), constitutive photomorphogenesis and dwarfism (CPD), and brassinosteroid insensitive 1 (BRI1). The mutation of these genes usually leads to severe growth disorders, including short hypocotyl, dwarfism of seedlings and mature plants, short petioles, dark green leaves, delayed flowering, and reduced male fertility (Li et al., 1996; Li and Chory, 1997; Bishop and Koncz, 2002). Notably, inhibitors of BR synthesis such as uniconazole and brassinazole (BRZ) prevented xylem differentiation in *Zinnia* cell cultures, which can be restored by treatment with BL or 28-homoBL (Iwasaki and Shibaoka, 1991). It was mentioned by Nagata et al. (2001) that the inhibition of growth and secondary xylem development of cress by BRZ was reversed by the exogenous application of BL, emphasizing the necessity of BRs for normal plant growth.

7.1.1 Brassinosteroid Analogs

BR analogs showed positive effects on plant growth, development, and crop yield. Its effects depend on the brassinosteroid type and the concentration and interaction with other hormones as well as plant species and growth stage. The biological effects of synthesized BR analogs from various compounds may be similar to the biological effect of natural BRs (Kowalski et al., 2003; Salgado et al., 2008). These analogs are generally characterized by certain structural groups present in natural

BRs (Yokota, 1997). Brassinolide (BL), 28-homobrassinolide (28-HomoBL), and 24-epibrassinolide (24-EpiBL) are the three bioactive BRs that are widely used in most physiological and experimental studies (Vardhini et al., 2006). The first BR analog was BL, which has the most physiologically active form of BRs. Brassinolide is a C_{28} steroid that exists widely in the plant kingdom. Brassinolide synthesis has been achieved by starting the reaction sequence with pregnenolone, hyodeoxycholic acid, and a sterol mixture containing crinosterol, especially stigmasterol. The second BR analog is 28-homobrasinólida (28-HomoBL), which has been synthesized from stigmasterol and is derived from soybean oil, which is relatively inexpensive and available in sufficient quantities (Khripach et al., 1999). The third BR analog is 24-epibrassinolide (24-EpiBL), which is synthesized from ergosterol, and is used in practical applications on a wide scale because it has a biologically potent activity similar to BL (Salgado et al., 2008; Liu et al., 2014) and plays a critical role in developmental processes (Talaat,2013; Liu et al., 2014; Yuan et al., 2015), regulated cell division and elongation (Bergonci et al., 2014), and gene expression and vascular differentiation (Choudhary et al., 2012 a). BR analogs have been applied to plants grown in soil contaminated with metals, saline soils, and high-temperature conditions to reduce the deleterious effects of stress (Salgado et al., 2008). The application of BR analogs to different plants improved stress resistance and increased crop yield (Ramraj et al., 1997; Vardhini and Rao, 1998; Hasan et al., 2008). It was noted that 5 ppm of BL increased the yielded rice grains/plant by 31.5 % (Lim, 1987). Moreover, BL promoted potato tuber development, inhibited its germination during storage, and increased resistance to infections by *Phytophthora infestans* and *Fusarium sulphureum* (Kazakova et al., 1991). The application of BL on orange trees during flowering increased fruit setting, while when applied during fruit growth it decreased the physiological drop of fruits and increased the number of fruits per plant accompanied by an increase in the average fruit weight. BL retarded fruit abscission in *Citrus madurensis* Lour (Iwaori et al., 1990). BL, 28-HomoBL, and 24-EpiBL treatments increased barley grains weight per ear, the weight of 1000 grains, and the crop yield, as well as increased plant tolerance to lodging (Prusakova et al., 1995). The application of 24-EpiBL or 22,23,24-triepibrassinolide on wheat plant increased panicle weight by 25 (33%) and grain weight by four (37%) and decreased the sterile portion of the ear by 25 (62%), respectively (Takematsu et al., 1988). Application of 24-EpiBL at 1 ppb increased root growth of chickpea by 25% (Singh et al., 1993). The application of 24-EpiBL to three different chickpea cultivars at the flowering stage increased seed yield, crop index, and 100 seeds dry weight, as well as protein and soluble sugars of the yielded seeds (Ramos et al., 1997). Foliar application of BL improved the yield of wheat, mustard, rice, corn, and tobacco (Braun and Wild, 1984; Yokota and Takahashi, 1986); 28-HomoBL and 24-EpiBL treatments significantly increased the yield of potato, mustard, rice, cotton (Ramraj et al., 1997), and *Vigna radiate* (Fariduddin et al., 2003).

7.1.2 Mode of Action of Brassinosteroids

The molecular mechanism of BR action is uncertain, although one might argue from structural considerations that they are likely to work by a mechanism similar to that of animal steroid hormones, which generally act via a soluble receptor–ligand complex that binds to nuclear sites to regulate the expression of specific genes (Mangelsdorf et al., 1995). BRs are perceived by a cell-surface receptor family of leucine-rich repeat receptor kinases BRASSINOSTEROID INSENSITIVE 1 (BRI1), which interacts with co-receptor BRI1 ASSOCIATED RECEPTOR KINASE 1 (BAK1) and undergoes a series of phosphorylation and dephosphorylation events to transduce information to the nucleus that results in the regulation of the expression of several hundred genes involved in diverse physiological functions (Sharma et al., 2013; Belkhadir and Jaillais, 2015; Nakamura et al., 2017).

Earlier, molecular genetics and biochemical studies of BR signaling in *Arabidopsis thaliana* have illustrated a BR signal transduction pathway from ligand perception on the cell surface to gene expression in the nucleus (Wang et al., 2006). In *Arabidopsis*, BRs are perceived by the BRI1 plasma-membrane receptor kinase and activation of BRI1/BRI1-ASSOCIATED RECEPTOR KINASE 1 (BAK1) kinase complexes by transphosphorylation. Subsequently, BRASSINOSTEROID INSENSITIVE 2 (BIN2) kinase is dephosphorylated and inactivated, resulting in the accumulation of unphosphorylated BRASSINAZOLE RESISTANT (BZR) transcription factors in the nucleus (Li and Chory, 1997; Wang et al., 2001; Kim and Wang, 2010).

Recently, Anwar et al. (2018) illustrated the pathway of signal transduction of BRs within plants that play an important role in plant growth and enhanced plant stress tolerance. The signal transduction pathway showed that BR was perceived by the BRI1 receptor kinase at the cell surface and activated BRASSINAZOLE RESISTANT 1 (BZR1) and BRI1-EMS SUPPRESSOR 1 (BES1) transcription factors to induce stress tolerance. Exogenously applied BR bonded with BRI1 to induce an association with BAK1 and disassociation of BRI1 KINASE INHIBITOR 1 (BKI1). Sequential transphosphorylation between BRI1 and BAK1 was required to activate BRI1 and furthermore to phosphorylate BR-SIGNALING KINASE 1 (BSK1) and enhanced BRI1 SUPPRESSOR 1 (BSU1) activity. The active BSU1 inhibited BIN2 through dephosphorylation of the phospho-tyrosine residue of BIN2, which allows the accumulation of unphosphorylated BZR1 and BZR2/BES1 transcription factors. The dephosphorylated BZR1 and BES1 enter the nucleus and act on regulating BR-targeted genes (Li et al., 2016) to enhance plant stress tolerance by increasing the capacity of antioxidant enzymes (Takeuchi et al., 1996; Li et al., 2009; Vardhini and Anjum, 2015) and regulating the accumulation of endogenous hormones (Wei et al., 2015; Wu et al., 2017). Likewise, FERONIA (a receptor-like kinase required for normal pollen tube development and cell elongation) is involved in BR signaling and ethylene signaling to control hypocotyl elongation during seedling growth (Deslauriers and Larsen,2010).

Genetic studies have confirmed that these processes are severely retarded in most BR-deficient and BR-insensitive mutants (Clouse et al., 1996). BR-deficient mutants display severe leaf phenotypic perturbations, including the production of small, round leaves and short petioles (Szekeres et al., 1996; Li et al., 2001), decreasing cell size, and leaf epidermal and mesophyll cell number (Hong et al., 2002). Feeding these mutants with BRs restored the normal leaf size by increasing both the size and number of epidermal and mesophyll cells, implying that BRs promoted leaf growth by enhancing both cell expansion and cell division (Nakaya et al., 2002; Caño-Delgado et al., 2004).

In plants, BR deficiency or BR perception inability results in low seed germination, dwarfism, delay in flowering and senescence, decreased male fertility, and de-etiolation in the dark (Clouse 2015). On the other hand, overexpression of BR biosynthetic genes increased endogenous BR levels leading to an increase in crop yield and enhanced stress tolerance (Divi and Krishna 2009; Xia et al., 2018). In rice, overexpression of a gene that encodes sterol C-22 hydroxylases increased endogenous BR levels, leading to increased grain size and yield up to 40% (Wu et al., 2008).

7.2 Physiological Roles of Brassinosteroids

7.2.1 Regulatory Roles of Brassinosteroids on Plant Growth and Development

The effects of BRs on plants are similar to animal steroid hormones that facilitate processes starting from embryonic development to adult homeostasis through a complex signal transduction pathway (Bergonci et al., 2014). Earlier studies revealed that BRs regulated diverse physiological and developmental processes in plants such as leaf expansion, photomorphogenesis, flower developmental processes, male sterility, stomatal developmental processes, and resistance to stress (Catterou et al., 2001; Hayat et al., 2001a;b; Xia et al., 2009; Choudhary and Yu, 2012; Oh et al., 2012; Tao et al., 2015; Li et al., 2016). BRs regulated plant growth by enhancing cell expansion (Hacham et al., 2011; Zhiponova et al., 2013), regulating seed germination, photosynthesis, xylem differentiation, photomorphogenesis, proton-pump activity, and ethylene biosynthesis (Sasse, 1997; Clouse and Sasse, 1998; Yu et al., 2004; Vriet et al., 2012; Zhang et al., 2014). At cellular levels, BRs regulated cell division and differentiation as well as cell elongation. At whole-plant levels, BRs regulated root and shoot development and fertility (Sasse, 2003; Yang et al., 2011; Kvasnica et al., 2014). BRs are known to induce a broad spectrum of responses, including the stimulation of longitudinal growth of young tissues via cell elongation and cell division (Zurek et al., 1994; Hu et al., 2000) and vascular differentiation, which is a developmental critical process for plant growth. Verma et al. (2012) stated that BRs effectively stimulated the elongation and formation of lateral shoots and shoot buds. Likewise, BRs exhibited a broad spectrum of physiological effects including a marked promotion of reproductive development, vascular differentiation, flowering, fruit set in plants (Yu et al., 2004; Cao et al., 2005; Montoya et al.,2005), senescence, leaf abscission, and nitrogen fixation (Clouse,1996; Clouse and Sasse, 1998; Nakashita et al., 2003; Bajguz and Hayat, 2009).

BR application showed a positive effect on stem elongation, including the promotion of epicotyl, hypocotyl, and peduncle elongation in dicots and enhanced growth of coleoptiles and mesocotyls of monocots (Mandava, 1988). BRs have been shown to stimulate ATPase activity, alter the orientation of cortical microtubules, regulate the timing of flowering (Yu et al., 2008; Domagalska et al., 2010), organ differentiation (Gonzalez-Garcia et al., 2011; Hacham et al., 2011), pollen development (Ye et al., 2010), and expression of specific genes (Zurek and Clouse, 1994). The mechanisms by which BRs regulate stem development are mainly involved in promoting the expression of genes responsible for cell elongation and wall extensibility (Zurek et al., 1994; Horvath et al., 2003).

BRs may also indirectly affect the flowering time by influencing the circadian clock and the photoperiod flowering pathway, as BR application shortened circadian rhythms (Hanano et al., 2006). Defects in BR biosynthesis or signaling pathways also reduced male fertility due to shortening of the stamen and defects in pollen development. These developmental defects correlated with the reduced expression of several key genes involved in anther and pollen development (Ye et al., 2010). BRs control sex determination by promoting stamen and repressing pistil development in tassels in maize (Hartwig et al., 2011). *In vitro*, exogenous application of 24-epiBL stimulated *Arabidopsis* pollen germination, pollen tube growth rates, and increased final pollen tube lengths (Vogler et al., 2014). BRs may also influence branching and flower formation through the modulation of metabolic pathways, nutrient allocation patterns, and promoted filament and pollen growth (Mussig, 2005). In addition, fruit size was increased with the exogenously applied BRs or by overexpressing BR synthetic genes in cucumber and tomato, indicating that fruit development is also positively regulated by BRs (Montoya et al., 2005; Fu et al., 2008). Either foliar application or seed priming with BRs significantly enhanced the growth and number of fruits (Zaharah et al., 2012; Thussagunpanit et al., 2015). BRs played positive roles in fruit ripening, fruit growth, and quality (Zaharah et al., 2012). Spraying BR at the flowering stage significantly increased the production of various crops (Vriet et al., 2012). BRs are also found to increase the growth and yield of sugar beet (Schilling et al.,1991), legumes (Kamuro and Takatsuto, 1991), and mustard seed (Hayat et al.,2001a).

7.2.2 Brassinosteroid Is a Promising Phytohormone in Abiotic Stress Amelioration

It was noted that BR effects depend on a number of factors including dose, plant species, growth stage, growth conditions (with or without stress), kinds of stress, duration of stress, and crosstalk with other hormones and signaling molecules (Nolan et al., 2019; Yin et al., 2019). Extensive research over the years has studied the mitigating role of BRs on various abiotic stresses such as high temperature (Janeczko et al., 2011 a), low temperature (Wang et al., 2014), freezing (Janeczko et al., 2009a), drought (Mahesh et al., 2013), flooding (Liang and Liang, 2009), salinity (Abbas et al., 2013), heavy metals (Bajguz, 2010), and newly reclaimed sandy soil (Ahmed and Shalaby, 2013). BRs reduced the harmful effect of stress by activating a plant defense system (antioxidants), thus significantly increasing growth, yield, and yield components (Meudt et al., 1983; 1984). It has been observed that BRs lead to the modification of cell wall architecture and adjustment of the membrane system, thus providing the first line of plant defense against environmental stresses (Clouse, 1996). BRs actively participate in chlorophyll synthesis, accumulation of osmolyte, antioxidant activity, nutrient management, nitrogen metabolism, and the plant–water relationship either under normal or stressful conditions (Ali et al.,2008 a; b; Ahmad et al.,2017). BRs regulate the defense mechanism and exert a significant role in modulating oxidative damage of reactive oxygen species, regulating antioxidant defense systems, osmoprotectants, production of secondary compounds, biosynthesis of other plant growth regulators, and expression of genes involved in defense responses (Bajguz and Hayat,2009; Yang et al., 2011; Zhao et al., 2016; Ahanger et al., 2018). Recently, there has been a prominent increase in BR application in agricultural processes to improve different crop productivity under stressed conditions. This may be due to the integration of BR signals in many other signaling networks linked to stress mitigation (Sharma et al., 2015; Divi et al., 2016; Marakli and Gozukirmizi, 2016; Liu et al., 2017; Hegazi et al., 2017; Wang et al., 2017). BRs play an important role in a variety of plant physiological processes and adaptation to different kinds of abiotic stresses (Li et al., 2013; Wei et al., 2015) via a sequence of biochemical reactions such as induction of protein synthesis, activation or suppression of key enzymatic reactions, and the production of various defensive chemical compounds (Bajguz and Hayat, 2009). Moreover, BRs improve the plant defense system to tolerate various stresses by increasing chlorophyll content, which ultimately increases photosynthetic capacity, enhancing antioxidant system activities and regulating stress response genes (Yuan et al., 2012; Shu et al., 2016).

BRs enhanced the germination rate and ultimately increased seedling growth under stress conditions (Özdemir et al., 2004; Mahesh et al., 2013; Zhang, 2012). Shu et al. (2016) concluded that BR treatments enhanced tomato seed germination and seedling growth by regulating plant antioxidant enzyme activities (APX, SOD, POD, CAT, and GR) and nitrogen metabolism, as well as increasing chlorophyll, nutrient content, and antioxidant genes expression. BRs control key genes—G-protein coupled receptor 1 (GCR1) and G-protein α-subunit 1 (GPA1)—that are responsible for seed germination. GCR1 is a putative G protein-coupled cell surface receptor and GPA1 is a subunit of the heterotrimeric G protein (Lapik and Kaufman, 2003). GPA1 is a key gene that regulates various biological processes, including biotic and abiotic stresses, growth and developmental processes, and biosynthesis of flavonoids, as well as activating transcription factors and nutrient transporters (Chakraborty et al., 2015).

The application of 24-EpiBL enhanced nitrogen metabolism, the activity of nitrite reductase, nitrate reductase, glutamate synthase, glutamine synthetase, and glutamate dehydrogenase enzymes, and induced photosynthetic characteristics of tomato seedlings under low temperature and weak light stress (Shu et al.,2016). Furthermore, exogenous BRs application increased H^{+}- ATPase and Ca^{2+}- ATPase activities in root and leaf (Steber and Mccourt,2001), which are responsible for establishing an electrochemical potential gradient to maintain ion balance in plants and enhance plant tolerance. BRs alleviated the adverse effect of different stress conditions and regulated the defense system in cucumber by regulating transcription levels of defense genes (Li et al., 2013). Various studies have revealed that BRs integrated with environmental signals to regulate stomatal aperture, one of the most important factors in drought and salinity stress acclimatization, preventing loss of excessive water (Gudesblat et al., 2012; Daszkowska-Golec and Szarejko, 2013). BRs alleviated the adverse effect of different stress conditions and regulated the defense system by regulating transcription levels of defense genes (Li et al., 2013).

7.2.3 Protective Effect of Brassinosteroids on Photosynthesis under Abiotic Stress

Exogenous BR application increased CO_2 fixation and RUBISCO activity in wheat plants (Braun and Wild, 1984) and alleviated photo-inhibition by significantly enhancing the photochemical efficiency of PSII, the quantum efficiency of PSII photochemistry, and the photochemical quenching coefficient (Xia et al., 2009; Ahammed et al., 2015). BRs also significantly enhanced the net photosynthetic rate, intercellular CO_2 concentration, transpiration rate, and stomatal conductance under stress (Farooq et al., 2010; Hayat et al., 2012). BRs enhanced chlorophyll content and reduced the activity of chlorophyllase responsible for the catabolism of chlorophyll pigment under abiotic stress (Bajguz, 2011; Hayat et al., 2012; Sharma et al., 2015). BRs induced the recovery of the photosynthetic apparatus of plants from cold stress by eliciting the enzymes of the Calvin cycle and the antioxidant defense system (Jiang et al., 2013). BR application enhanced alternative oxidase (AOX) that contributed to the balance between chloroplast-mitochondria electron transfer by dissipation of excess photosynthetic reductant leading to a decrease in ROS accumulation and increased protection of photosystems (Deng et al., 2015). Moreover, exogenous BRs significantly increased plant tolerance to various kinds of abiotic stress (Anuradha and Rao, 2001;Yuan et al., 2012; Li et al., 2016) by enhancing photosynthetic capacity, antioxidant system, and chloroplast ultrastructure (Saygideger and Deniz, 2008; Bajguz and Hayat, 2009; Choudhary et al., 2012a; Li et al., 2012; Wang et al., 2012; Yuan et al., 2012; Li et al., 2016; Niu et al., 2016; Shu et al., 2016). Moreover, BRs can regulate the combination of chlorophyll molecules with membrane protein and maintain stability of thylakoid membranes (Bajguz and Hayat, 2009). Moreover, BRs regulate the Rubisco activase gene, which plays a key role in photosynthesis under drought and heat stress in wheat. In addition, BRs significantly increased the antioxidant enzyme activities and the photosynthesis process (Zhao et al., 2017). BR-treated seedlings increased CO_2 assimilation and quantum yield of photosystem II (PSII), ribulose-1, 5-bisphosphate carboxylase/oxygenase (Rubisco) activities, and expression of Rubisco large subunit (*rbcL*) and Rubisco small subunit (*rbcS*) genes to increase photosynthetic capacity (Xia et al.,2009). Thus, BRs promoted the accumulation of chlorophylls and photosynthetic capacity by regulating a variety of enzymes such as chlorophyllase and Rubisco; transcript levels of encoded genes involved chlorophyll and photosynthesis under various stresses (Xia et al.,2009; Hasan et al., 2011; Li et al.,2016; Zhao et al.,2017). BRs regulated the Rubisco carboxylase activase (*RCA*) gene, which played a key role in photosynthesis under drought and temperature stress in wheat and significantly increased the activities of antioxidant enzymes and the process of photosynthesis. Many researchers reported that total chlorophyll contents were increased in the leaves of various crops by application of 24-EpiBL and 28-HomoBL (Hayat et al., 2001a,b; Fariduddin et al., 2009).

7.2.4 Regulatory Roles of Brassinosteroids on Crop Quality

7.2.4.1 Chemical Composition

BR application on different crops increased the quantity and quality of the crops (Salgado et al., 2008). Considerably fewer studies are devoted to the influence of BRs on the quality of the yield of different plant species (Vardhini and Rao, 1998; Janeczko et al., 2009a,b; Janeczko et al., 2011a,b). BR application significantly increased leaf area, chlorophyll content, and photosynthesis, which are the key factors for plant growth under normal or stressed conditions (Yuan et al., 2012; Wei et al., 2015). Application of BRs increased the total chlorophyll content and net photosynthetic rate in *Brassica juncea* (Alam et al., 2007; Fariduddin et al., 2009); soybean seedlings (Cevahir et al., 2008); rice (Farooq et al.,2009); and wheat (Sairam et al.,2005). The positive impact of BRs on the metabolism of carbohydrates is quite well known in plants (Yu et al., 2004; Vardhini et al., 2011). The increased carbohydrate content of tomato pericarp and rice was observed under the influence of BRs as reported by Vardhini and Rao (2002) and Wu et al. (2008), respectively. Vardhini and Rao (2002) revealed the ability of BRs to accelerate tomato fruit maturation and correlated it with the increased content of carbohydrates. This stimulation of the production of carbohydrates might be caused by an enhanced photosynthetic capacity of plants under the influence of BRs. In fact, an increase in CO_2 fixation and levels of reducing sugars was also reported in wheat and mustard plants after the application of BL (Braun and Wild, 1984). It has also been found that these compounds regulated the distribution of assimilates (Fuji and Saka, 2001). Numerous studies have shown that BRs stimulate protein synthesis, soluble protein, and enzymes in the leaves of plants growing either under normal or stressed conditions (Bajguz, 2000; Anuradha and Rao, 2003; Fariduddin et al., 2004; Arora et al., 2008; Shahbaz et al., 2008; Behnamnia et al., 2009) because of the effect of BRs on the transcription and translation processes (Bajguz, 2000) and enhancing the activity of RNA and DNA polymerases (Kalinich et al., 1986). Vardhini and Rao (1998) reported that spraying BL and 24-EpiBL substantially increased the growth of the groundnut plant which was associated with enhanced levels of DNA, RNA, soluble proteins, and carbohydrate. They added that exogenous application of BL and 24-EpiBL elevated the oil content of *Arachis hypogaea* L seeds by 30%. Janeczko et al. (2009a) observed slight changes in the molar percentage of particular fatty acids in soybean, wheat, and oilseed rape seeds after application of 24-EpiBL. Biesaga-Kościelniak et al. (2014) mentioned that BR application increased the amount of total oil content in pea and lupine plants.

7.2.4.2 Antioxidant Enzymes

Numerous studies indicated that BRs—as plant steroid hormones with antistress properties—stimulated the activity of antioxidant enzymes (Mazorra et al., 2002; Hasan et al., 2008). BRs also provided tolerance to plants against abiotic stressors by modulating the activity of enzymatic and non-enzymatic antioxidants (Bartwal and Arora, 2020; Chi et al., 2020; Rattan et al., 2020; Chen et al., 2021; Kour et al., 2021). BR application increased CAT activity in rice (Nunez et al., 2003), tomato (Behnamnia et al., 2009), and *Brassica juncea* L. (Sirhindi et al., 2009;Arora et al., 2010). An increase in polyphenol oxidase (POX) activity in response to the application of BRs has also been reported by Sharma et al. (2007), Sirhindi et al. (2009), and Arora et al. (2010). POX has an important role in the synthesis of lignin and other phenolic polymers. Sirhindi et al. (2009) observed an increase in the activity of PPX after exogenous application of 28-HomoBL on *Brassica juncea* L seedlings. Application of 24-EpiBL (Mazorra et al., 2002; Behnamnia et al., 2009; Arora et al., 2010), 28- HomoBL (Kartal et al., 2009), and BL analog (Nunez et al., 2003) stimulated APX activity. Hayat and Ahmad (2003) found that 28-homoBL treatment increased catalase, amylase, and peroxidase levels of wheat seedlings. It has been found that BRs can induce the expression of some antioxidant genes and enhance the activities of antioxidant enzymes such as SOD, POD, CAT, and APX (Marakli and Gozukirmizi, 2016). The enhanced activities of antioxidative enzymes as a result of BR applications may occur with increasing de novo synthesis or activation of enzymes that are mediated through transcription and/or translation of specific genes to gain tolerance (Bajguz, 2000).

7.2.4.3 Non-Enzymatic Antioxidants

Less is known about the impact of BRs on non-enzymatic antioxidants such as ascorbic acid, tocopherols, β-carotene, and so on. However, exogenous application of 24-epiBl on wheat and soybean plants significantly increased β-carotene and tocopherol contents (Janeczko et al., 2009 b, 2011b). Vardhini and Rao (2002) found that application of 24-epiBl and 28-homoBl on tomato increased lycopene, carbohydrate contents, and ethylene production. Biesaga-Kościelniak et al. (2014) mentioned that 24-epiBl application increased γ-tocopherol content by 9–15% and the ascorbic acid content by 18–52% in the yielded pea seeds depending on the dose of 24-epiBl and method of application. BR application increased ascorbic acid content in the leaves of *Raphanus sativus* L. under the effect of Cr stress (Choudhary et al., 2011). Whereas BR application on tomato roots at the seedling stage decreased ascorbic acid content and increased lycopene and β-carotene content of the developed fruits (Ali et al., 2006). Application of 24-epiBl increased the amount of β-carotene in soybean and oilseed rape (Janeczko et al., 2009b). Meanwhile, Vardhini and Rao (2002) revealed that the administration of 28-homoBl and 24-epiBl to pericarp discs of tomato decreased the levels of ascorbic acid.

7.3 Crosstalk of Brassinosteroids with Phytohormones under Abiotic Stress

Abiotic stress affects the biosynthesis and metabolism of phytohormonal (Maruyama et al., 2014; Saini et al., 2015). Phytohormones played a central role in various plant processes involved in the adaptation to environmental stress by regulating growth, nutrient allocation, and source/sink interactions (Peleg and Blumwald, 2011; Tiwari et al., 2017). Crosstalk between BRs and auxins, gibberellins, cytokinins, abscisic acid (Domagalska et al., 2010), ethylene (Manzano et al., 2011), jasmonic acid (Peng et al., 2011), and salicylic acid (SA; Divi et al., 2010) promoted plant growth, development, and tolerance to abiotic stress by changing the expression of several genes involved in different physiological processes (Divi et al., 2010; Saini et al., 2013; Zheng et al., 2016; Zhu et al., 2016; Yusuf et al., 2017; Ahmad et al., 2018; Bechtold and Field, 2018; Gruszka, 2018). In addition, Divi et al. (2010) reported that 24-EpiBL treatment increased the expression of various hormone marker genes in both wild-type and mutant *Arabidopsis* seedlings. Wu et al. (2017) mentioned that exogenous application of BRs enhanced endogenous levels of salicylic acid, jasmonic acid, and ethylene and concluded that crosstalk occurred between BRs and other plant hormones in the signaling pathway under BR-induced stress tolerance. Therefore, BR signal transduction pathways are involved in many transcriptional activities and metabolic activities, which lead to significant increases in plant tolerance.

7.3.1 Crosstalk of Brassinosteroids with Auxin

Crosstalk between BR and auxin regulates plant growth and developmental process (Hao et al., 2013; Saini et al., 2013; Liu et al., 2014; Chaiwanon and Wang, 2015). Both auxin and BRs promoted cell elongation but their kinetics are different: auxin promoted cell elongation within 10–15 min between the application and the beginning of elongation, while BRs showed at least 45 min to several hours to show the same effect (Taiz, 1984; Clouse, 1996). Microarray analyses showed that BRs and auxin activated overlapping sets of genes, while genetic and physiological studies showed synergistic and interdependent interactions between the two hormones in a wide range of developmental contexts such as hypocotyl elongation and vascular bundle patterning (Nemhauser et al., 2004, Ibañes et al., 2009; Depuydt and Hardtke, 2011). Katsumi (1985) showed that treating cucumber hypocotyl sections with 28-homoBl followed by IAA resulted in synergistic enhancement of auxin-induced elongation, but when the order of treatment was reversed, 28-homoBl was inactive, suggesting that BRs modulate the capacity of stem tissues to respond to IAA. Genetic and physiological studies have shown that BRs and auxin interact antagonistically in roots to maintain the spatiotemporal balance between stem cell division and differentiation required for optimal root growth (Chaiwanon and Wang, 2015). Interaction of BRs and auxin are involved in hypocotyl and root development besides regulating stress responses (Kissoudis et al., 2014). Mechanisms of synergistic interaction between BR and auxin may be explained as follows. First, BZR1 regulates many genes involved in auxin synthesis, transport, and signaling (Sun et al., 2010). Second, auxin activates BR biosynthetic genes and increases BR levels (Mouchel et al., 2006; Chung et al., 2011; Yoshimitsu et al., 2011). Third, BIN2 phosphorylates an auxin-response factor (ARF2; Vert et al., 2008). Fourth, BZR2 and ARF5 bind to the same BR/auxin-response promoter (*SMALL AUXIN UPREGULATED15* [*SAUR15*]; Walcher and Nemhauser, 2012). Fifth, BR and auxin responses are integrated through the actin cytoskeleton, which is regulated by BR and auxin, besides feedback regulated auxin transport and BRs signaling (Lanza et al., 2012).

7.3.2 Crosstalk of Brassinosteroids with Gibberellic Acid

BRs and gibberellic acid (GA) are growth-promoting hormones, having similar effects on various developmental processes throughout the life cycle of plants. The effect of interaction between BRs and GA has been considered to be additive (Depuydt and Hardtke, 2011). Their signaling pathways have been proposed to act on non-overlapping transcriptional responses (Nemhauser et al., 2006). Li et al. (2012) mentioned that BRs interacted with GA to coordinate different physiological processes. Moreover, some studies demonstrated an interdependent relationship and a direct interaction between the BR and GA signaling pathways (Bai et al., 2012; Gallego-Bartolomé et al., 2012; Li et al., 2012). GA is unable to increase hypocotyl elongation in BR-deficient and BR-insensitive mutants, whereas BRs and the dominant gain-of-function *bzr1-1D* mutation can increase cell elongation in GA-deficient mutants. BL has an additive effect with gibberellins (Mayumi and Shiboaka, 1995) and a synergistic effect with auxin that stimulates cell elongation and stem segment elongation (Katsumi, 1985; Salgado et al., 2008; Gudesblat and Russinova, 2011;Tong et al., 2014;Unterholzner et al., 2015). Application of either BL and GA, or BL and auxin, resulted in a synergistic increase in hypocotyl elongation in intact *Arabidopsis* seedlings (Tanaka et al., 2003). Tong et al. (2014) reported that crosstalk between BR and GA has been established in regulating plant cell elongation in rice. They suggested that BRs promoted GA accumulation by inducing the expression of *D18/GA3ox-2*, one of the GA biosynthetic genes. However, exogenous application of a high concentration of BRs led to the activation of *GA2ox-3*, a GA inactivation gene, resulting in inhibition of cell elongation. Moreover, GA inhibited BR signaling as well as its biosynthesis in a feedback inhibiting loop but facilitated cell elongation through activating the primary BR signaling pathway upon applying exogenous high GA concentration, indicating BR–GA crosstalk in regulating cell elongation.

7.3.3 Crosstalk of Brassinosteroids with Cytokinin

Cytokinin stimulates the accumulation of endogenous BRs, suggesting a synergistic interaction between BR and cytokinin in *Chlorella vulgaris*. Upon exogenous treatment of 10 μM trans-zeatin (*t*Z) to the *Chlorella vulgaris* culture, there was a considerable increase in the level of all endogenous BRs by 27–46%. The application of both BL and trans-Zeatin (*t*Z) led to the highest stimulation in the number of *Chlorella vulgaris* cells and the endogenous accumulation of proteins, chlorophylls, and monosaccharides (Bajguz and Piotrowska-Niczyporuk,2014b). Exogenous application of BRs on wheat seedlings regulated the encoded gene expression of cytokinin oxidase and significantly increased the cytokinin level (Yuldashev et al.,2012).

7.3.4 Crosstalk of Brassinosteroids with Abscisic Acid

Physiological and genetic studies have confirmed that BRs and abscisic acid (ABA) can co-regulate many developmental processes (Steber and McCourt 2001; Finkelstein et al., 2008). About 35% of BR-regulated genes are also regulated by ABA, indicating that ABA may regulate BR signaling (Goda et al., 2008; Nemhauser et al., 2006). ABA may act as a signal molecule in response to stressed and unstressed plants where the interaction of BRs and ABA regulated the expression of different genes to enhance stress tolerance (Werner et al.,2001; Zhang et al., 2009; Werner et al.,2010; Nishiyama et al.,2011; Zhu et al., 2016). Furthermore, BRs induced accumulation of ABA content in *Brassica napus* plants and *Chlorella vulgaris* cells exposed to hyperthermal stress as reported by Kurepin et al. (2008) and Bajguz (2009), respectively. BRs promoted seed germination, indicating the antagonistic interaction with ABA (Steber and McCourt, 2001; Finkelstein et al., 2008). Genetic, physiological, and biochemical studies have revealed that BRs and ABA can regulate the expression of genes (Nemhauser et al, 2006; Zhang et al, 2009). Zhou et al. (2014) observed that there is a synergistic correlation between BRs and ABA in inducing various responses such as H_2O_2 production, respiratory burst oxidase homologl (RBOH1) gene expression, NADPH oxidase activity, and mediating heat and oxidative stress tolerance. They suggested that ABA biosynthesis played a key role in sustaining stress tolerance in BR-induced pathways in plants.

7.3.5 Crosstalk of Brassinosteroids with Ethylene

BRs play a key role in the biosynthesis of ethylene in plants (Zhu et al., 2016). Methionine, the starting point of the ethylene biosynthesis pathway, is converted into SAM (*S*-adenosylmethionine) with the help of methionine adenosyltransferases. SAM is converted into 1-aminocyclopropane-1-carboxylic acid (ACC) with the help of ACC synthase and then converted into ethylene by enrichment of the BR signal pathway (Choudhary et al., 2012b). Ethylene accumulation and biosynthesis are induced by BR treatment in tomato (Zhu et al., 2016). These results suggested crosstalk between BRs and ethylene in response to stress tolerance. Likewise, BRs are also known to mitigate stress-caused oxidative damage by inducing ethylene biosynthesis (Wei et al., 2015). Ethylene acts synergistically with BRs to enhance the level of AOX resulting in efficient ROS scavenging for improved stress tolerance. Similar evidence has been obtained where combined treatment of BRs with other hormones like ABA, salicylic acid, and polyamines resulted in a synergistic increase in stress protective effects as compared with individual hormone treatment (Choudhary et al., 2012a; Hayat et al., 2012). BRs and ethylene crosstalk regulate plant growth and developmental processes. BR has been identified as a negative regulator of shoot gravitropism, whereas ethylene has been shown to promote gravitropic reorientation in light-grown seedlings (Vandenbussche et al., 2013). Guo et al. (2008) suggested that BR and ethylene interact indirectly in regulating shoot gravitropic responses through involving auxin signaling genes.

7.3.6 Crosstalk of Brassinosteroids with Salicylic Acid

Interaction between BR and salicylic acid revealed an important role in alleviating biotic and abiotic stresses. The interplay between BR and SA may be due to the NPR1 gene (non-expressor of pathogenesis-related genes 1) that stimulates the expression of the SA-related genes involved in plant defense (Ohri et al., 2015). NPR1 also regulates the BR signaling genes BIN2 and BZRI, which induce thermo- and salinity tolerance in *Arabidopsis thaliana* (Divi et al., 2010).

7.4 Effect of Brassinosteroids on Plant Tolerance to Abiotic Stress

BR application significantly increases plant tolerance against different abiotic stresses by enhancing chlorophyll content, maintaining photosynthetic activities, regulating carbohydrate metabolism, inducing changes in defense enzymes, reducing ion toxicity, activating gene expression, and signal transduction pathways, as well as increasing levels of endogenous plant hormones (Li et al., 2015; Jie et al., 2015; Eremina et al., 2016; Li et al., 2016; Shu et al.,2016). BRs have been shown to induce the accumulation of compatible solutes under various stress conditions that are often associated with improved plant tolerance (Sharma et al., 2012, 2013; Kumar et al., 2013).

7.4.1 Drought

BR application reduced the deleterious effects of drought stress on different plant species by enhancing chlorophyll accumulation, stomatal conductance, photosynthesis, total protein contents, amylase activity, membrane stability, and antioxidant enzyme activities (Singh et al., 1993; Li et al., 2008; Li et al., 2012; Mahesh et al., 2013; Behnamnia, 2015; Talaat and Shawky, 2016). Application of 28-HomoBL stimulated the growth of wheat plant grown under drought stress by increasing relative water content, chlorophyll content, photosynthesis rate, and nitrate reductase activity (Sairam, 1994). Furthermore, BL treatment increased superoxide dismutase, catalase, ascorbate peroxidase, ascorbic acid, and carotenoid content in maize seedlings subjected to water stress (Li et al., 2011). Morillon et al. (2001) proposed that exogenously applied BRs enhanced plant resistance of *Arabidopsis thaliana* to drought by increasing osmotic permeability. Upreti and Murti (2004) reported that the application of either 24-EpiBL or 28-HomoBL increased plant growth including nitrogenase activity and root nodulation under drought. It has also been reported that application of 24-EpiBL enhanced drought tolerance by upregulating transcription factors and regulating the expression of the drought-responsive element in *Arabidopsis thaliana* and *Brassica napus* (Kagale et al., 2007). Furthermore, Fariduddin et al. (2009) noted that the application of 0.01 μM 28-HomoBL to drought-stressed *Brassica juncea* plants at two different developmental stages enhanced the photosynthetic rate, stomatal conductance, and proline content. It has been reported that exogenous BR application reduced reactive oxygen species and protected membrane lipids via the regulation of antioxidant enzyme activity and levels of non-enzymatic antioxidants (Li et al., 2012; Shahana et al., 2015; Ahmad et al., 2017). Exogenous application of BL alleviated the detrimental effects of drought on *Zea mays* by enhancing antioxidant enzyme activities, relative leaf water, proline, and protein content (Anjum et al., 2011). Behnamnia (2015) noted that application of 24-EpiBL decreased oxidative damage on tomato plants during drought stress by downregulating lipoxygenase activity and upregulating the antioxidant defense system. Exogenous application of BRs maintained tissue–water status (Farooq et al., 2009) by stimulating proton pumping (Khripach et al., 2003), activating nucleic acid and protein synthesis (Bajguz, 2000), and regulating gene expressions (Felner, 2003). BR treatment increased the contents of proline and protein under water stress. Zhang et al. (2008) indicated that BR treatment promoted the accumulation of osmoprotectants such as soluble sugars and proline, which may be due to the fact that BRs activated the enzymes of proline biosynthesis (Fariduddin et al.,2009).

7.4.2 Salinity

The application of BR-regulated plant tolerance mechanisms to salinity in several crops included rice (Anuradha and Rao, 2003), *Brassica* species (Sirhindi et al., 2017), and *Vigna sinensis* (El-Mashad and Mohamed, 2012). BR treatments significantly increased plant tolerance to salinity stress (Takeuchi et al., 1996; Ali et al., 2008; De Campos et al., 2009; Abbas et al., 2013; Talaat 2013; Liu et al., 2017; Wu et al., 2017) via increasing photosynthetic rate, chlorophyll content, levels of endogenous hormones, total amino acid contents, activities of antioxidant enzymes, and reducing ion toxicity, reactive oxygen species, and malondialdehyde content (Chanho et al., 2010; Li et al., 2009; Xue-Xia et al., 2011; Wu et al., 2017). BR treatment reduced the inhibitory effect of salinity stress on the germination of rice grain, which may be attributed to the enhanced level of nucleic acids and soluble proteins (Anuradha and Rao, 2001). BR application enhanced plant tolerance to salinity stress during germination and seedling growth of kidney bean and barley by increasing the content of glutathione and betaine (Ali and Abdel-Fattah, 2006). Moreover, exogenous application of BRs regulated the ion balance by enhancing uptake of K^+ and Ca^{2+} and reducing Na^+ content in wheat leaves (Qayyum et al., 2007; Shahbaz et al., 2008). Foliar application 24-EpiBL enhanced chlorophyll fluorescence and membrane stability of pepper leaves under saline conditions (Houimli et al., 2008, 2010). Wu et al. (2017) mentioned that 24-EpiBL treatment improved plant tolerance to salinity stress by increasing the level of hormone and antioxidant enzyme (SOD and CAT) activity, promoting the accumulation of proline and ions (K^+, Ca^{2+}, and Mg^{2+}) in perennial ryegrass. El-Mashad and Mohamed (2012) stated that 0.05 ppm BL as foliar spray treatment mitigated the impact of salinity stress on the cowpea plant by inducing antioxidant enzymes activities such as SOD, POD, PPO, and GR and the contents of AsA; 24-EpiBL treatment decreased the adverse effects of salinity stress on two varieties of pepper by increasing the contents of proline, anthocyanin, minerals, and antioxidative enzyme activities (Abbas et al., 2013). Song et al. (2016) noted that the inclusion of 24-EpiBL in salt-stressed peanut caused enhancement in growth by upregulating oxidative enzyme activity concomitant with reduced electrolyte leakage and malondialdehyde content. Shahbaz et al. (2008) reported that the foliar application of 24-EpiBL reduced the adverse effects of salinity on wheat by increasing the oxidative activity of peroxidase and catalase. BRs increased essential inorganic ions, decreased toxic ions, and promoted ion homeostasis, especially in leaves, roots, and the epicotyl of canola under salinity stress (Liu et al., 2014).

7.4.3 Temperature

The application of BRs on different crops mitigated the adverse effects of either high temperature (Yadava et al., 2016) or low temperature (Aghdam et al., 2012). The thermotolerance attained by the plants in response to BR application may be due to the synthesis of heat shock proteins (Dhaubhadel et al., 1999). Kulaeva (1991) reported that BR treatments increased plant tolerance to low temperature by maintaining chlorophyll content while promoting tolerance to high temperature by maintaining protein synthesis. Regarding high temperature, treating tomato and rape seedlings with 24-EpiBL increased thermotolerance, which is associated with a higher synthesis of heat shock proteins (Dhaubhadel et al., 2002; Kagale et al., 2007). Heat shock proteins have been extensively studied in plants for their protective role in high-temperature tolerance. BRs mitigated the heat-induced inhibition of photosynthetic capability by enhancing carboxylation efficiency and the antioxidative enzyme system in *Lycopersicon esculentum* (Ogweno et al., 2008). Foliar treatment with 24-epiBL mitigated the negative effects of high temperature-induced inhibition of photosynthesis in two cultivars of melon seedlings (Zhang et al., 2013). BRs mitigated high-temperature injury in *Ficus concinna* seedlings by enhancing antioxidative defense mechanisms and improving glyoxalase systems (Jin et al., 2015). Liu et al. (2018) reported that BRs improved lipid productivity and enhanced tolerance of *Chlorella* cells to high temperature. BR application increased rice tolerance to high temperature by maintaining essential plant activities including photosynthesis rate and stomatal conductance by protecting photosynthetic machinery and alleviating photoinhibition (Thussagunpanit et al., 2015). Further, Zhao et al. (2017) reported that application of 24-epiBL mitigated a combination of drought and heat stress in *Triticum aestivum* L. seedlings by increasing the rate of photosynthesis and Rubisco activase gene expression. Lee et al. (2019) reported that the application of BRs improved heat stress tolerance in Kimchi cabbage after three days of heat stress via increasing catalase and peroxidase enzyme activities by 1.76 to 2.08-fold as compared with the control.

Regarding chilling stress, Kumar et al. (2010) reported that foliar application of 24-epiBL on *Brassica juncea* L. seedlings grown under 4°C of chilling stress exhibited reduced H_2O_2 concentration by enhancing antioxidant defense systems (CAT, APX, and SOD). Exogenous application of 24-EpiBL on cucumber alleviated the chilling-induced inhibition of photosynthesis by reducing the generation of reactive oxygen species and increasing the activities of SOD and APX (Hu et al.,

2010). Aghdam et al. (2012) stated that application of 0, 3, and 6 µM BRs to tomato fruits stored at 1°C for 21 days decreased chilling injury, electrolyte leakage, and MDA content while increasing proline levels, total phenolics, and phenylalanine ammonia-lyase (PAL) activity and maintained membrane integrity. Xi et al. (2013) noted that the application of BRs in grapevines reduced the deleterious effect of cold by stabilizing membrane integrity and improvement in antioxidant and osmoregulatory components. Jiang et al. (2013) mentioned that BR treatments protected the photosynthetic apparatus from cold-induced damage in *Cucumis sativus* plants by enhancing the activities of Calvin cycle enzymes and enhancing the antioxidative system, which, in turn, resulted in mitigation of the photo-oxidative stress. Wang et al. (2013) reported that BRs (5, 10, and 15 µM) efficiently decreased the chilling injury of pepper fruit during 18-day storage at 3°C by decreasing electrolyte leakage and MDA content and enhancing antioxidative enzyme activities (CAT, POD, APX, and GR). Likewise, 24-epiBL treatment enhanced photosynthesis and antioxidant defenses, and protected eggplant seedlings from chilling stress (Wu et al., 2015). Further, Filek et al. (2017) stated that BRs mitigated low temperature stress in winter wheat seedlings by regulating their membrane structure. Kaur et al. (2018) observed that application of 28-homoBL regulated antioxidant enzyme activities and gene expression in response to temperature-induced oxidative stress in *Brassica juncea.* Tavallali (2018) mentioned that 24-epiBL treatment significantly delayed chlorophyll degradation and maintained the quality of lime fruit during cold storage, thus increasing its shelf life.

7.4.4 Heavy Metals

BR played a pivotal role in overcoming heavy metal toxicity and improving plant growth (Ali et al., 2008a,b; Bajguz, 2010; Gao et al., 2012; Sharma et al., 2014; Vardhini, 2016). Application of BRs to plants showed lower levels of heavy metals than untreated plants (Sharma et al., 2007; Ali et al., 2008a,b; Bajguz 2010; Vardhini 2016). BRs significantly reduced the toxic effect of Cd in *Helianthus tuberosus* via enhancing chlorophyll content, photosynthetic rate, and antioxidant enzymatic activities (Bajguz, 2000; 2010). Similarly, BRs significantly improved the performance of *Vigna radiate* and *Cicer arietinum* under aluminum and cadmium stresses (Ali et al., 2008a; Hasan et al., 2008), respectively. Similarly, under heavy metal stress, BR treatments enhanced antioxidant enzyme activities (CAT, POD, SOD, GR), expression of defense genes, and proline content, and reduced reactive oxygen species and malondialdehyde content (Ali et al., 2008a,b; Bajguz, 2010; Gao et al., 2012). Foliar application of 28-HomoBL improved Cd- tolerance in *Brassica juncea* through increasing activity of antioxidative enzymes (CAT, POD, SOD) and the content of osmolyte (proline; Hayat et al., 2007). Sharma et al. (2007) reported that BR application regulated antioxidant enzymes and mitigated the toxic effects of zinc in *Brassica juncea*. Hayat et al. (2007) observed that *Brassica juncea* plants grown under Cd stress exhibited a decline in growth, chlorophyll content, activity of nitrate reductase, activity of carbonic anhydrase, nitrate, and sugar content. However, application of 28-HomoBL reduced the toxic effect of Cd stress via enhancing proline accumulation and oxidative enzyme activities. Ali et al. (2008a) studied the application of either 24-EpiBL or 28-HomoBL in reducing the deleterious effect of aluminum stress on *Vigna radiata.* The increase in the aluminum resistance conferred by BR application was reflected in the improvement of plant growth and photosynthesis efficiency. Bajguz and Hayat (2009) noted that BRs have the ability to minimize the toxic effects of heavy metals. Application of 24-EpiBL (5 µM) on *Phaseolus vulgaris* grown under Cd stress, increased activity of antioxidative enzymes and proline content and subsequent improvements in the membrane stability index and relative leaf water content (Rady, 2011). Application of 24-EpiBL ameliorated the toxic effect of Ni stress on *Brassica juncea* by enhancing the activity of antioxidant enzymes (Kanwar et al., 2013); 24-EpiBL treatment regulated antioxidant enzyme activity and increased the level of proline (osmolyte), thereby increasing Ni-plant tolerance of *Vigna radiata* and improving growth, nodulation, and yield attributes (Yusuf et al., 2012). Yusuf et al. (2014) reported that BR application improved the antioxidant defense system and nitrogen metabolism in two cultivars of *Vigna radiata* under different levels of Ni. Kaya et al. (2020) reported that spraying 0.5 µM 24-EpiBL on pepper plants for ten days enhanced the defense mechanism against Cd stress; 24-EpiBL reduced leaf Cd^{2+} content and oxidative stress, enhanced plant growth, regulated water relations, increased proline content, ascorbate-glutathione (AsA-GSH) cycle enzyme activities, the antioxidant defense system, nitric reductase activity, and endogenous nitric oxide content.

7.5 Conclusion

Based on extensive research, BRs have the ability to improve the quantity and quality of horticultural crops and protect plants against many stresses that can be present in the local environment. With the many advances in technology dealing with the synthesis of more stable synthetic analogs and the genetic manipulation of cellular BRs activity, using BRs in the production of horticultural crops has become a more practical and hopeful strategy for improving crop yields and success. Exogenous application of BRs regulated abiotic stress by interactions with other stress-related hormones. Molecular, biochemical, structural, and genomic approaches have increased our understanding of the BR signaling cascade. It is becoming increasingly evident that the BR pathway does not represent a linear signaling pathway that operates in isolation, but rather that BRs undergo crosstalk with multiple other hormones and stress responses. Moreover, BR signaling varies among different cells and tissues, which can be manipulated to improve plant growth and stress responses. It is evident that BRs interact with other phytohormones such as auxin, cytokinin, ethylene, gibberellin, jasmonic acid, abscisic acid, and salicylic acid in regulating a wide range of physiological and developmental processes in plants. A recent study indicated that in response to various intrinsic and extrinsic factors, the signaling components of BR crosstalk with key genes and transcription factors of other phytohormones, thereby regulating multiple functions in plants. The unraveling of these complicated mechanisms of BR signaling and its collaboration with other molecular networks will be of great importance in improving modern agriculture.

REFERENCES

Abbas et al. *Pak J Bot.* 2013; 45:1273–1284.

Adam and Schneider, eds. *Brassinosteroids - Steroidal Plant Hormones*, Springer Tokyo, Japan, 1999; pp. 113–136.

Aghdam et al. *Sci Hort.* 2012; 144:116–120. doi: 10.1016/j.scienta.2012.07.008.

Ahammed et al. *Environ Pollut.* 2015; 201:58–66. doi: 10.1016/j.envpol.2015.02.024.

Ahanger et al. *J Plant Growth Regul.* 2018; 37:1007–1024.

Ahmad et al. *J Plant Inter.* 2017; 12:429–437.

Ahmad et al. *J Appl Biol Biotechnol.* 2018; 6(1):56–62.

Ahmed and Shalaby. *J Appl Sci Res.* 2013; 9:6387–6393.

Alam et al. *Photosynthetica.* 2007; 45(1):139–142.

Ali and Abdel-Fattah. *J Agron.* 2006; 5:167–174.

Ali et al. *Sci Hortic.* 2006; 110:267–273.

Ali et al. *Environ Exp Bot.* 2008a; 62:153–159.

Ali et al. *Chemosphere.* 2008b; 72:1387–1392.

Anjum et al. *J Agron Crop Sci.* 2011; 197:177–185. doi: 10.1111/j.1439-037X.2010.00459.x.

Anuradha and Rao. *Plant Growth Regul.* 2003; 40(1): 29–32.

Anuradha et al. *Plant Growth Regul.* 2001; 33:151–153.

Anwar et al. *Biol Res.* 2018; 51:46. doi: 10.1186/s40659-018-0195-2.

Arora et al. *Can J Pure Appl Sci.* 2(3):503–509.

Arora et al. *Physiol Mol Biol Plants.* 2010; 16(3):285–293.

Arteca and Arteca. *Physiol Plant.* 2001; 112:104–112. doi: 10.1034/j.1399-3054.2001.1120114.x.

Bai et al. *Nat Cell Biol.* 2012; 14:810–817.

Bajguz. *Plant Physiol Biochem.* 2000; 38:209–215.

Bajguz. *J Plant Physiol.* 2009; 166:882–886. doi: 10.1016/j.jplph.2008.10.004.

Bajguz. *Environ Exp Bot.* 2010; 68:175–179. doi: 10.1016/j.envexpbot.2009.11.003.

Bajguz. eds. S. Hayat and A. Ahmad, Springer, Dordrecht, 2011; pp. 1–27.

Bajguz and Hayat. *Plant Physiol Bioch.* 2009; 47:1–8.

Bajguz and Piotrowska-Niczyporuk, eds. 2014a; pp. 163–190. Springer, New York.

Bajguz and Piotrowska-Niczyporuk. *Plant Physiol Biochem.* 2014b; 80C:176–183. doi: 10.1016/j.plaphy.2014.04.009.

Bajguz and Tretyn. *Phytochem.* 2003; 6:1027–1046.

Bartwal and Arora. *Co-Evol Sec Metabol.* 2020; 869–895. doi: 10.1007/978-3-319-96397-6_18.

Bartwal et al. *J Plant Growth Regul.* 2013; 32:216–232. doi: 10.1007/s00344-012-9272-x.

Bechtold and Field. *J Exp Bot.* 2018; 69:2753–2758.

Behnamnia. *Int J Agric Crop Sci.* 2015; 8:552–59. http://ijagcs.com/wp-content/uploads/2015/04/455-462.pdf.

Behnamnia et al. *Gen Appl Plant Phys.* 2009; 35(1–2):22–34.

Belkhadir and Jaillais. *New Phytol.* 2015; 206:522–540. doi: 10.1111/nph.13269.

Bergonci et al. *J Exp Bot.* 2014; 65:2219–2230.

Bhardwaj et al. 2006; pp. 58–84. Aaviskar Publisher, Jaipur.

Biesaga-Kościelniak et al. *Australian J Crop Sci.* 2014; 8(3):378–388.

Bishop and Koncz. *Plant Cell.* 2002; 14 (Suppl.): S97–S110.

Braun and Wild. *J Plant Physiol.* 1984; 116:189–196.

Caño-Delgado et al. *Development.* 2004; 131:5341–51.

Cao et al. *Physi Plant.* 2005; 123(1): 57–66.

Catterou et al. *Plant.* 2001; 212:659–672.

Cevahir et al. *J Appl Biol Sci.* 2008; 2(1): 23–28.

Chaiwanon and Wang. *Curr Biol.* 2015; 20:1031–1042. doi: 10.1016/j.cub.2015.02.046.

Chakraborty et al. *Plant Mol Biol.* 2015; 89:559–576.

Chanho et al. *Phytochemistry.* 2010; 71:380–387.

Chen et al. *Crop J.* 2021; 9:109–119.

Chi et al. *J Exp Bot.* 2020; 71:1092–1106. doi: 10.1093/jxb/erz466.

Choudhary et al. *Chemosphere.* 2011; 84:592–600.

Choudhary et al. *Trends Plant Sci.* 2012a; 17:594–605.

Choudhary et al. *J Exp Bot.* 2012b; 63:5659–5675. doi: 10.1093/jxb/ ers219.

Chung et al. *Plant J.* 2011; 66:564–578.

Clouse. *Curr Biol.* 1996; 6(6):658–661.

Clouse. *Mol Cell.* 2002; 10:973–982.

Clouse. *J Plant Growth Regul.* 2015; 34(4):828–844. doi: 10.1007/s00344-015-9540-7.

Clouse and Sasse. *Ann Rev Plant Physiol Plant Mol Biol.* 1998; 49:427–451.

Clouse et al. *Plant Physiol.* 1996; 111:671–678.

Daszkowska-Golec and Szarejko. *Front Plant Sci.* 2013; 4:138. doi: 10.3389/fpls.2013.00138.

De Campos et al. *Rev Ceres.* 2009; 56:74–79. http://www.ceres.ufv.br.

Deng et al. *J Exp Bot.* 2015; 66:6219–6232. doi: 10.1093/jxb/erv328.

Depuydt and Hardtke. *Curr Biol.* 2011; 21:R365–R373.

Deslauriers and Larsen. *Mol Plant.* 2010; 3:626–640.

Dhaubhadel et al. *Plant Mol Biol.* 1999; 40:333–342.

Dhaubhadel et al. *Plant J.* 2002; 29:681–691.

Divi and Krishna. *New Biotechnol.* 2009; 26(3–4):131–136. doi: 10.1016/j.nbt.2009.07.006.

Divi et al. *BMC Plant Biol.* 2010; 10:151. doi: 10.1186/1471-2229-10-151.

Divi et al. *Plant Biotechnol J.* 2016; 14:419–432. doi: 10.1111/pbi.12396.

Domagalska et al. *Plos One.* 2010; 5:e14012. doi: 10.1371/journal.pone.0014012.

El-Mashad and Mohamed. *Protoplasma.* 2012; 24:625–635. doi: 10.1007/s00709-011-0300-7.

Eremina et al. *Proc Natl Acad USA.* 2016; 113:E5982.

Fàbregas et al. *Nat Commun.* 2018; 9:4680.

Fang et al. *Plant Physiol.* 2019; 180(4): 2061–2076. doi: 10.1104/pp.19.00088.

Fariduddin et al. *Acta Bot Croat.* 2003; 65:19–23.

Fariduddin et al. *J Plant Biol.* 2004; 48:465–468.

Fariduddin et al. *Acta Physiol Plant.* 2009; 31:889–897. doi: 10.1007/ s11738-009-0302-7.

Farooq et al. *J Agron Crop Sci.* 2009; 195(4):262–269. doi: 10.1111/j.1439-037X.2009.00368.x.

Farooq et al. *J Agron Crop Sci.* 2010; 196:336–345. doi: 10.1111/j.1439-037X.2010.00422.x.

Felner. In: Hayat and Ahmad, eds. *Brassinosteroids: Bioactivity and Crop Productivity*, Kluwer Academic Publishers, Dordrecht, 2003; pp. 69–86.

Filek et al. *Steroids.* 2017; 128:37–45. doi: 10.1016/j.steroids.2017.10.002.

Finkelstein et al. *Ann Rev Plant Biol.* 2008; 59:387–415.

Fu et al. *J Exp Bot.* 2008; 59:2299–2308.
Fujii et al. *Plant Prod Sci.* 2001; 4:136–144.
Gallego-Bartolome et al. *Proc Natl Acad Sci USA.* 2012; 109:13446–13451.
Gao et al. *Acta Ecol Sin.* 2012; 33:1935–43.
Goda et al. *Plant J.* 2008; 55:526–542.
Gonzalez-Garcia et al. *Development.* 2011; 138:849–859. doi: 10.1242/dev.057331.
Grove. *Nature.* 1979; 281:216–217.
Gruszka. *Int J Mol Sci.* 2018; 19:2675.
Gudesblat and Russinova. *Curr Op Plant Biol.* 2011; 14(5): 530–537.
Gudesblat et al. *Nat Cell Biol.* 2012; 14:548–554. doi: 10.1038/ncb2471.
Guo et al. *Plant Mol Biol.* 2008; 66:345–360.
Hacham et al. *Development.* 2011; 138:839–848. doi: 10.1242/dev.061804.
Hanano et al. *Genes Cells.* 2006; 11:1381–1392.
Hao et al. *Plant Cell Rep.* 2013; 32:1017–1030.
Hartwig et al. *Proc Natl Acad. Sci USA* 2011; 108:19814–19819.
Hasan et al. *Environ Pollut.* 2008; 151:60–66.
Hasan et al. *Chemosphere.* 2011; 84(10):1446–51.
Hayat and Ahmad. *Ann Appl Biol.* 2003; 143:121–124.
Hayat et al. *Acta Physiol Plant.* 2001b; 23:27–30.
Hayat et al. *Photosynthetica.* 2001a; 38:469–471.
Hayat et al. *Environ Exp Bot.* 2007; 60:33–41. doi: 10.1016/j.envexpbot.2006.06.002.
Hayat et al. *Saudi J Biol Sci.* 2012; 19:325–335. doi: 10.1016/j.sjbs.2012.03.005.
Hegazi et al. *Gesunde Pflanzen.* 2017; 2:91–102. doi: 10.1007/s10343-017-0393-9.
Hong et al. *Plant J.* 2002; 32:495–508.
Horvath et al. *Plant J.* 2003; 34:125–134.
Houimli et al. *EurAsian J BioSci.* 2008; 2:83–90.
Houimli et al. *EurAsian J BioSci.* 2010; 4:96–104.
Hu et al. *Plant J.* 2000; 24(5): 693–701.
Hu et al. *Photosynthetica.* 2010; 48(4): 537–544. doi: 10.1007/s11099-010-0071-y.
Ibanes et al. *Proc Natl Acad Sci USA.* 2009; 106:13630–13635.
Iwaori et al. *Plant Growth Regul.* 1990; 9:119–126.
Iwasaki and Shibaoka. *Plant Cell Physiol.* 1991; 32:1007–1014.
Janeczko et al. *Seed Sci Technol.* 2009b; 37:625–637.
Janeczko et al. *Acta Physiol Plant.* 2009a; 31:71–79. doi: 10.1007/s11738-008-0202-2.
Janeczko et al. *Acta Physiol Plant.* 2011a; 33:1249–1259. doi: 10.1007/s11738-010-0655-y.
Janeczko et al. *Acta Sci Pol.* (*Agric.*) 2011b; 10:33–50.
Jiang et al. *Physiol Plant.* 2013; 148:133–145. doi: 10.1111/j.1399-3054.2012.01696.x.
Jie et al. *J Nucl Agric Sci.* 2015; 29:1001–1008. doi: 10.11869/j.issn.100-8551.2015.05.1001.
Jin et al. *AoB Plants.* 2015; 7:plv009. doi: 10.1093/aobpla/plv009.
Kagale et al. *Planta* 2007; 225:353–364. doi: 10.1007/s00425-006-0361-6.
Kalinich et al. *J Plant Physiol.* 1986; 125:345–353.
Kamuro et al. 1991; pp. 292–297. American Chemical Society, Washington.
Kanwar et al. *Acta Physiol. Plant.* 2013; 3:1351–1362. doi: 10.1007/s11738-012-1175-8.
Kartal et al. *Plant Growth Regul.* 2009; 58 (3):261–267.
Katsumi. *Plant Cell Physiol.* 1985; 26:615–626.
Kaur et al. *Scientific Reports.* 2018; 8:8735. doi: 10.1038/s41598-018-27032-w.
Kaya et al. *Ecotoxicol Environ Safety.* 2020; 196:110483.
Kazakova et al. Izvestiia Timiryazevskoi sel´skokhoziaistvennoi Akademii. 1991:82–88 [C.F. *Biological Abstracts* 94(8):85021].
Khripach et al. Academic Press, San Diego, 1999; p. 456.
Khripach et al. *Netherlands Annal Botany.* 2000; 86:441–447.
Khripach et al. In: Hayat, S., Ahmad, A., eds. *Brassinosteroids; Bioactivity and Crop Productivity*, Kluwer Academic Publisher, Dordrecht, 2003; pp. 189–230.
Kim and Wang. *Ann Rev Plant Biol.* 2010; 61:681–704.
Kissoudis et al. *Front Plant Sci.* 2014, 5:207.
Kour et al. *Front Plant Sci.* 2021; 12:608061. doi: 10.3389/fpls.2021.608061.
Kowalski et al. *Biotecnol Veg.* 2003; 3(2):115–117.
Krishna. *J Plant Growth Regul.* 2003; 22:289–297.
Kulaeva, ed. *ACS Symposium, Ser.* 474, American Chemical Society, Washington, 1991; pp. 141–155.
Kumar et al. *Indian J Biochem Biophys.* 2010; 47(6): 378–382.
Kumar et al. *Rice.* 2013; 6:27. doi: 10.1186/1939-8433-6-27.
Kurepin et al. *Plant Growth Regul.* 2008; 55:165–167. doi: 10.1007/s10725-008-9276-5.
Kvasnica et al. *Steroids.* 2014; 85:58–64.
Lanza et al. *Developmental Cell.* 2012; 22(6):1275–1285. doi: 10.1016/j.devcel.2012.04.008.
Lapik and Kaufman. *Plant Cell.* 2003; 15:1578–1590.
Lee et al. *Hortic Environ and Biotech.* 2019; 60:841–851.
Li and Chory. *Cell.* 1997; 90:929–938.
Li et al. *Science.* 1996; 272:398–401.
Li et al. *Plant Physiol.* 2001; 127:14–22.
Li et al. *New Forotests.* 2008; 35(3):255–266. doi: 10.1007/s11056-007-9075-2.
Li et al. *Plant J.* 2009; 58:275–86. doi: 10.1111/j.1365-313X.2008.03778.x.
Li et al. *Acta Physiol Plant.* 2011; 33:1293–300. doi: 10.1007/s11738-010-0661-0.
Li et al. *Biol Plant.* 2012; 56:192–196.
Li et al. *Plos One.* 2013; 8(6):e66582.
Li et al. *Acta Physiol Plant.* 2015; 37:1–11. doi: 10.1007/s11738-014-1746-y.
Li et al. *Front Plant Sci.* 2016; 7:1–16. doi: 10.3389/fpls.2016.01281.
Liang et al. *Xinan Shifan Daxue Xuebao, ZiranKexueban.* 2009;34:58–62.
Lim. *Agri Res Seoul National University.* 1987; 12:9–14.
Liu et al. *Acta Agron Sinica.* 2006; 32:924–930.
Liu et al. *Plant Biol.* 2014; 16:440–450.
Liu et al. *Steroids.* 2017; 124:1–17. doi: 10.1016/j.steroids.2017.05.005.
Liu et al. *J Appl Phycol.* 2018; 30(1): 253–260. doi: 10.1007/s10811-017-1223-2.
Lozano-Durán et al. *Trends Plant Sci.* 2015; 20:12–19.
Mahesh et al. *Am J Plant Sci.* 2013; 4:2305–2313. doi: 10.4236/ajps.2013.412285.
Mandava. *Ann Rev Plant Physiol Plant Mol Biol.* 1988; 39:23–52.
Mangelsdorf et al. *Cell.* 1995; 83:835–839.
Manzano et al. *Plant Growth Regul.* 2011; 65:213–221. doi: 10.1007/s10725-011-9589-7.
Marakli and Gozukirmizi. In: Chapter 16. eds. A. Shanker and C. Shanker. 2016; pp. 374–394. InTech. doi: 10.5772/61336.

Maruyama et al. *Plant Physiol.* 2014; 164:1759–1771. doi: 10.1104/pp.113.231720.

Mayumi and Shiboaka. *Plant Cell Phys.* 1995; 36:173–181.

Mazorra et al. *Plant Biol.* 2002; 45:593–596.

Meudt et al. *Proc Plant Growth Reg Soc. Am.* 1983; 10:312–318.

Meudt et al. Canadian Patent No. 1173659. Assigned to USA. 1984; pp. 11.

Montoya et al. *Plant J.* 2005; 42(2): 262–269.

Mori. US Patent. 1984; 4:457,967.

Morillon et al. *Planta.* 2001; 212:199–204.

Mouchel et al. *Nature.* 2006; 443:458–461.

Mussig. *Plant Biol.* 2005; 7:110–117.

Nagata et al. *Plant Cell Biol.* 2001; 42:1006–1011. doi: 10.1093/pcp/ pce122.

Nakamura et al. *Plos One.* 2017; 12:e0174015. doi: 10.1371/journal.pone.0174015.

Nakashita et al. *Plant J.* 2003; 33:887–898.

Nakaya et al. *Plant Cell Physiol.* 2002; 43:239–244. doi: 10.1093/pcp/pcf024.

Nemhauser et al. *PLoS Biol.* 2004; 2(9):e258. doi: 10.1371/journal.pbio.0020258.

Nemhauser et al. *Cell.* 2006; 126:467–475.

Nishiyama et al. *Plant Cell.* 2011; 23:2169–2183.

Niu et al. *Chil J Agric Res.* 2016; 76:27–33.

Nolan et al. *Biochem J.* 2017; 474:2641–2661.

Nolan et al. *Plant Cell.* 2019; 32(2):295–318. doi: 10.1105/tpc.19.00335.

Nunez et al. *Plant Biol.* 2003; 47:67–70.

Ogweno et al. *J Plant Growth Regul.* 2008; 27(1): 49–57. doi: 10.1007/s00344-007-9030-7.

Oh et al. *Nat Cell Biol.* 2012; 14:802–809.

Ohri et al. *Curr Prot Pep Sci.* 2015; 16:369–388. doi: 10.2174/138920371666150330141922.

Oklestkova et al. *Phytochem Rev.* 2015; 14:1053–1072. doi: 10.1007/s11101-015-9446-9.

Özdemir et al. *Plant Growth Regul.* 2004; 42:203–211.

Peleg et al. *Curr Opin Plant Biol.* 2011; 14:290–295. doi: 10.1016/j.pbi.2011.02.001.

Peng et al. *J Integ Plant Biol.* 2011; 53:632–640. doi: 10.1111/j.1744-7909.2011.01042.x.

Peres et al. *Int J Mol Sci.* 2019; 20(2):331. doi: 10.3390/ijms20020331.

Praveena et al. *Curr J Appl Sci Technol.* 2020; 39(25): 174–196.

Prusakova et al. *Sel'skokhozyaistvennaya Biologiya.* 1995; 93–97 [*Biological Abstracts* 100:94463, 1995].

Qayyum et al. *Int J Agri Biol.* 2007; 9:584–589.

Rady. *Sci Hortic.* 2011; 129:232–237. doi: 10.1016/j.scienta.2011.03.035.

Ramos et al. Campinas, Brazil. 1997; p. 108.

Ramraj et al. *J Agr Sci.* 1997; 128:405–413.

Rattan et al. *J Plant Growth Reg.* 2020; 1–11.

Saini et al. *Plant Cell Rep.* 2013; 32:741–757. doi: 10.1007/s00299-013-1430-5.

Saini et al. *Front Plant Sci.* 2015; 6:950. doi: 10.3389/fpls.2015.00950.

Sairam. *Plant Growth Regul.* 1994; 14(2):173–181.

Sairam et al. *Biologia Plantarum.* 2005; 49(1):85–91.

Saka et al. *Plant Prod. Sci.* 2003; 6:36–42. doi: 10.1626/pps.6.36.

Salgado et al. *Biológicas.* 2008; 10:18–27.

Sasse. *Proceedings Plant Growth Regul Society of America.* 1994; 21:228–232.

Sasse. *Physiol Plant.* 1997; 100:696–701.

Sasse. *J Plant Growth Regul.* 2003; 22(4):276–288.

Saygideger, Deniz. *Plant Growth Regul.* 2008; 56:219–223.

Schilling et al. In: Cutler HG, Yokota T, Adam G (eds.), *Brassinosteroids: Chemistry, Bioactivity and Applications.* 1991; 208–219.

Schneider, ed. *Progress in Botany*, Springer, Berlin-Heidelberg, Germany. 2002; 63:pp. 286–306.

Shahana et al. *Int J Multi Curr Res.* 2015; 3:905–911.

Shahbaz et al. *Plant Growth Regul.* 2008; 55(1):51–64.

Sharma and Bhardwaj. *Gen Appl Plant Physiol.* 2007; 33:59–73.

Sharma et al. *Brazil J Plant Phys.* 2007; 19(3):203–210.

Sharma et al. *Ecotoxicol Environ Saf.* 2012; 85:72–81. doi: 10.1016/j.ecoenv.2012.07.003.

Sharma et al. *Pest Biochem Physiol.* 2013; 105:144–153. doi: 10.1016/j.pestbp.2013.01.004.

Sharma et al. *Toxicol Int.* 2014; 21:44–50.

Sharma et al. *J Plant Growth Regul.* 2015; 34:509–518. doi: 10.1007/s00344-015-9486-9.

Shu et al. *Plant Physiol Biochem.* 2016; 107:344–353.

Singh et al. *Indian J Agri Sci.* 1993; 63:395–397.

Sirhindi et al. *Phys Mole Biol Plants.* 2009; 15(4): 335–341.

Sirhindi et al. *Acta Physiol Plant.* 2017; 39:228–235.

Song et al. *Yingyong Shengtai Xuebao.* 2006; 17:1871–1876.

Song et al. *Biol Plant.* 2016; 60:329–342.

Steber and McCourt. *Plant Physiol.* 2001; 125:763–769.

Sun. *Plant Cell Physiol.* 2005; 46:1384–1391.

Sun et al. *Dev Cell.* 2010. 19:765–777.

Sun et al. *Hort Sci.* 2015; 50:1518–1523.

Szekeres et al. *Cell.* 1996; 85:171–182.

Taiz. *Ann Rev Plant Physiol.* 1984; 35:585–657.

Takematsu et al. 1988; US Patent 4,767,442.

Takeuchi et al. *Proc Jpn Soc Chem Regul Plants.* 1996; 31:100–101.

Talaat. *Acta Physiol Plant.* 2013; 35:729–740.

Talaat and Shawky. *J Plant Growth Regul.* 2016; 35:518–533.

Tanaka et al. *J Plant Growth Regul.* 2003; 22:259–271.

Tao et al. *Sci Hortic.* 2015; 195:116–123.

Tavallali. *Acta Sci Pol Hortorum Cultus.* 2018; 17(1): 35–48. doi: 10.24326/asphc.2018.1.4.

Thussagunpanit et al. *J Plant Growth Regul.* 2015; 34:320–331.

Tiwari et al. *Curr Genomics.* 2017; 18:469–482.

Tong et al. *Plant Cell.* 2014; 26(11):4376–4393.

Unterholzner et al. *Plant Cell.* 2015; 27(8):2261–2272.

Upreti and Murti. *Biol Plant.* 2004; 48:407–411.

Vandenbussche et al. *Am J Bot.* 2013; 100:215–225.

Vardhini. *Proc and Hra Pradesh Akad Sci.* 2013; 15:75–79.

Vardhini. eds. M. M. Azooz and P. Ahmad. John Wiley & Sons, New York, 2016. doi: 10.1002/9781119081005.ch7.

Vardhini and Anjum. *Front Environ Sci.* 2015; 2:67.

Vardhini et al. *Phytochemistry.* 1998; 48:927–930.

Vardhini et al. *Phytochemistry.* 2002; 61(7):843–847.

Vardhini et al. *Indian J Plant Physiol.* 2006; 11:1–12.

Vardhini et al. eds. S. K. Ashwani Kumar and I. K. Sopory. International Publishing House Ltd., NewDelhi, 2008; pp. 133–139.

Vardhini et al. *Asian Australas J Plant Sci Biotechnol.* 2011; 5:27–30.

Verma et al. International Scholarly Research Network. *ISRN Agron.* 2012; 356485:1–8. doi: 10.5402/2012/356485.

Vert et al. *Proc Natl Acad Sci USA.* 2008; 105:9829–9834.

Vogler et al. *Plant Reprod.* 2014; 27(3):153–167.

Vriet et al. *Plant Cell.* 2012; 24:842–857. doi: 10.1105/tpc.111.094912.
Walcher and Nemhauser. *Plant Physiol.* 2012; 158:273–282.
Wang et al. *Sci Hortic. (Amsterdam).* 2013; 144:195–200. doi: 10.1016/j.scienta.2012.07.018.
Wang et al. *Nature.* 2001; 410:380–383.
Wang et al. *Cell Res.* 2006; 16:427–434.
Wang et al. *Plant Growth Regul.* 2012; 66:87–93.
Wang et al. *Acta Agri Jiangxi.* 2014; 26:36–38.
Wang et al. *Sci Total Environ.* 2017; 592:68–77.
Wei et al. *Front Plant Sci.* 2015; 6:982. doi: 10.3389/fpls.2015.00982.
Werner et al. *Proc Natl Acad Sci USA.* 2001; 98:10487–10492.
Werner et al. *Plant Cell.* 2010; 22:3905–20. doi: 10.1105/tpc.109.072694.
Wu et al. *Plant Cell.* 2008; 20(8):2130–2145.
Wu et al. *J Hortic Sci Biotechnol.* 2015; 90:217–225. doi: 10.1080/14620316.2015.11513175.
Wu et al. *Front Plant Sci.* 2017; 8:1017. doi: 10.3389/fpls.2017.01017.
Xi et al. *Plant Growth Regul.* 2013; 71:57–65.
Xia et al. *Planta.* 2009; 230:1185. doi: 10.1007/s0042 5-009-1016-1.
Xia et al. *Plant Cell Environ.* 2018; 41(5):1052–1064. doi: 10.1111/pce.13052.
Xue-Xia et al. *Plant Physiol J.* 2011; 47:607–612.
Yadava et al. *Nat Sci.* 2016; 8:171–179.
Yang et al. *Mol Plant.* 2011; 4:588–600.
Ye et al. *Proc Natl Acad Sci USA.* 2010; 107:6100–6105.
Yin et al. *Crop J.* 2019; 7(3):411–418. doi: 10.1016/j.cj.2018.10.003.
Yokota. *Trends Plant Sci.* 1997; 2(4):137–143.
Yokota and Takahashi, eds. *Plant Growth Subs.* 1986; 129–138.
Yoshimitsu et al. *Plos One.* 2011; 6: e23851.
Yu et al. *J Exp Botany.* 2004; (55)399:1135–1143.
Yu et al. *Proc Natl Acad Sci USA.* 2008; 105:7618–7623.
Yuan et al. *Sci Hortic.* 2015; 26:103–108.
Yuan et al. *Photosynth Res.* 2012; 112:205–214.
Yuldashev et al. *Plant Physiol Biochem.* 2012; 55:1–6.
Yusuf et al. *Plant Physiol Biochem.* 2012; 57:143–153. doi: 10.1016/j.plaphy.2012.05.004.
Yusuf et al. *Mol Biol Plant.* 2014; 20:449–460. doi: 10.1007/s12298-014-0259-x.
Yusuf et al. Springer International Publishing, Berlin, 2017; pp. 241–260.
Zadvornova et al. *Doklady Natsional'noi Akademii Nauk Belarusi.* 2005; 49:70–73.
Zaharah et al. *J Plant Growth Regul.* 2012; 31:363–372.
Zhang. *Nucl Agric Sci.* 2012. http://lib.cqvip .com/qk/95508X/201201/40981031.html. (in Chinese).
Zhang et al. *Aust J Agric Res.* 2007; 58:811–815. doi: 10.1071/AR06253.
Zhang et al. *Plant Growth Regul.* 2008; 56:257–64. doi: 10.1007/s10725-008-9305-4.
Zhang et al. *Proc Natl Acad Sci USA.* 2009; 106:4543–4548.
Zhang et al. *Photosynthetica.* 2013; 51(3): 341–349. doi: 10.1007/s11099-013-0031-4.
Zhang et al. *Plant Cell Rep.* 2014; 33:683–696. doi: 10.1007/s00299-014-1578-7.
Zhao et al. *Biol Plant.* 2016; 60:563–571.
Zhao et al. *Plant Growth Regul.* 2017; 81(3): 377–384. doi: 10.1007/s10725-016-0214-7.
Zheng et al. *Plant Soil.* 2016; 400:147–164.
Zhiponova et al. *New Phytol.* 2013; 197:490–502. doi: 10.1111/nph.12036.
Zhou et al. *J Exp Bot.* 2014; 65:4371–4383.
Zhu et al. *Sci Rep.* 2016; 6:35392.
Zurek et al. *Plant Physiol.* 1994a; 104:161–170.
Zurek et al. *Plant Physiol.* 1994b; 104:505–513.

8

Jasmonic Acid: Crosstalk with Phytohormones in Growth and Development

G. Kaur and B. Asthir

CONTENTS

Abbreviations

(JA): Jasmonic acid
(GA): Gibberellic acid
(BR): Brassinosteroid
(MeJA): Methyl jasmonate
(JA-Ile): Jasmonoyl-isoleucine
(SL): Strigolactone

8.1 Introduction

Plant growth and development is a complex process mediated by small molecular signals such as phytohormones that act either in the vicinity or are transported to other parts of the plant. Jasmonic acid and its conjugates, such as methyl jasmonate (MeJA) and jasmonoyl-isoleucine (JA-Ile), collectively known as jasmonates. (JAs), serve as natural plant growth regulators that are ubiquitous in the plant kingdom (Ghasemi Pirbalouti et al., 2014). In recent decades, JA biosynthesis has been extensively investigated in both monocots and dicots, especially in *Arabidopsis*. In *Arabidopsis*, at least two pathways are responsible for JA biosynthesis, namely, the α-linolenic acid (18:3) initial octadecane pathway and the hexadecatrienoic acid (16:3) initial hexadecane pathway (Engelberth et al., 2001; Wasternack, 2014; Kazan, 2015). The initial steps of JA biosynthesis are initiated in chloroplasts by the lipase mediated release of α-linolenic acid (α-LeA) from galactolipids until the formation of intermediate cis-12-oxophytodienoic acid (OPDA), via enzymes galactolipase, 13-lipoxygenase (LOX), allene oxide synthase (AOS), and allene oxide cyclase (AOC) (Figure 8.1). Conversely, the enzyme OPDA reductase (OPR3), involved in the reduction of the cyclopentenone ring and shortening of the carboxylic acid side chain, is localized in peroxisomes. These cyclic intermediates are processed by the peroxisomal fatty acid via the β-oxidation pathway, producing JA, which is then exported to the cytosol, where JAR1 leads to the formation of bioactive compound (+)-7-*iso*-jasmonoyl-L-isoleucine (JA-Ile; Wasternack and Strnad, 2016). The import of OPDA into peroxisomes has been shown to be mediated by COMATOSE (CTS), an ABC transporter of the peroxisomal membrane. However, the export of OPDA from chloroplasts

DOI: 10.1201/9781003110651-8

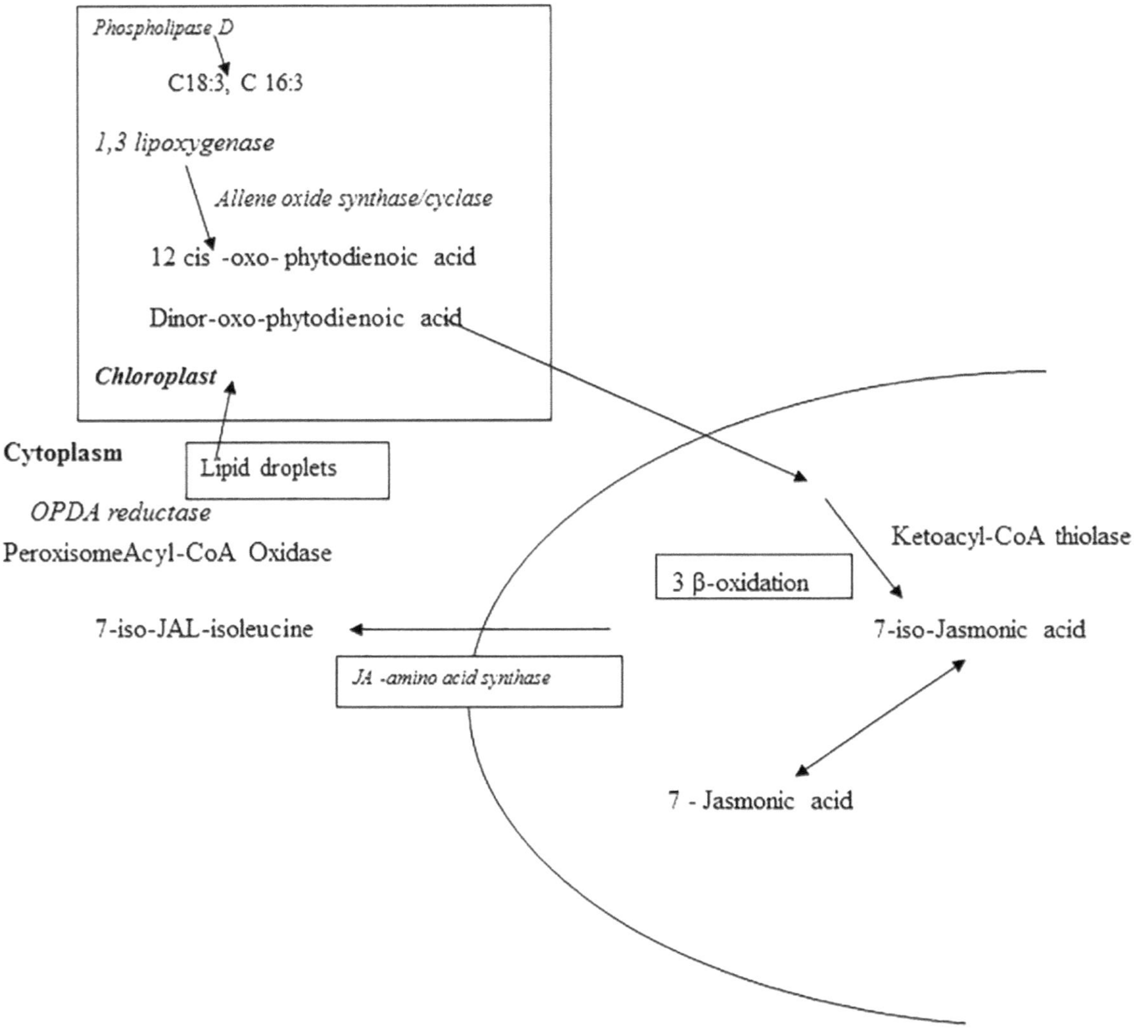

FIGURE 8.1 Biosynthesis of Jasmonic acid.

remained unfolded for a long time, however, recently, Guan et al. (2019) described the role of a protein called JASSY in this transport, which is located in the outer envelope of chloroplast membranes. JASSY contains a steroid acute regulatory protein-related lipid transfer (START) domain, known to be pivotal in the binding and transport of hydrophobic molecules.

Jasmonic acid has been found to play a vital role in plant growth, development, and defense processes (Pauwels et al., 2011; Seo et al., 2011, Du et al., 2013, Huang et al., 2017; Per et al., 2018). To epitomize, it is actively involved in plant reproduction (Song et al., 2011, Xue and Zhang, 2007, Wasternack, 2007; Browse, 2009), floral development, growth inhibition, fruit ripening, tendril coiling, potato tuberization, trichome formation, fungi arbuscular mycorrhizal association, stomatal opening, inhibiting Rubisco biosynthesis, nitrogen and phosphorus uptake, and the transport of organic matter such as glucose (Browse, 2005; Balbi and Devoto, 2008; Reinbothe et al., 2009; Yoshida et al., 2009). In particular, as a signaling molecule, JAs effectively mediate plant responses against pathogens and environmental stresses by inducing a series of genes expression, thereby inducing resistance gene expression (Ballare 2011).

8.2 Crosstalk with Other Phytohormones

JA does not work independently but acts in a complex signaling network combined with other plant hormone signaling pathways (Figure 8.2; Kazan, 2015; Ahmad et al., 2016; Wasternack and Strnad, 2016; Hu et al., 2017).

8.2.1 JA–Auxin Crosstalk

JA regulates root growth in many aspects, including inhibition of the primary root (Chen et al., 2011), promoting lateral root formation (Cai et al., 2014), negatively regulating adventitious roots (Gutierrez et al., 2012; Lakehal et al., 2019), and inducing root regeneration (Ye et al., 2019; Zhang G. et al., 2019) via cross-talking with auxin. JA inhibits root elongation and development through auxin-dependency by reducing both cell counts and dimension, suggesting that JA-induced primary root growth inhibition is a complicated process involving diverse cellular processes in different root tissues (Chen et al., 2011, 2012; Wasternack and Hause, 2013). In addition to this, JA activates *MYC2*, leading to

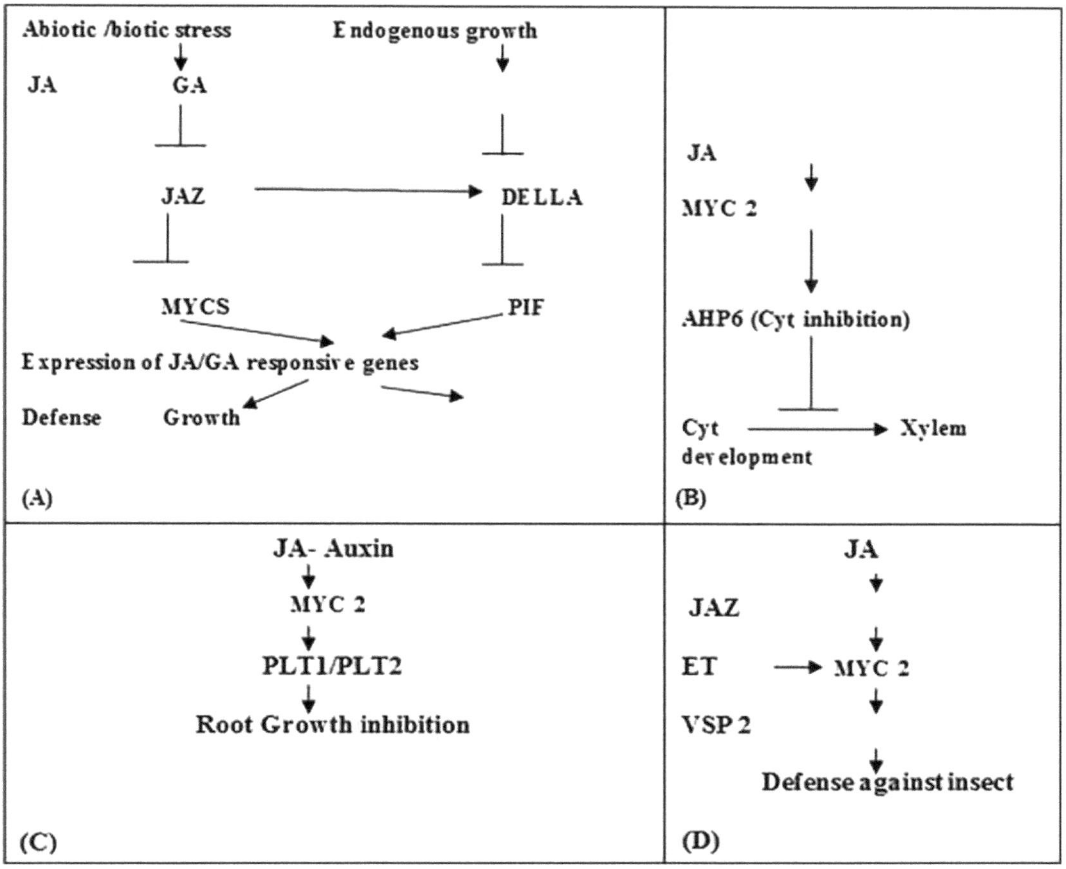

FIGURE 8.2 Interaction of JA with other phytohormones. (A) Crosstalk between JA and GA; (B) crosstalk between JA and Cytokinin; (C) crosstalk between JA and Auxin; and (D) crosstalk between JA and ethylene. Arrows and T bars represent positive and negative regulation.

the repression of *PLT1* (*PLETHORA1*) and *PLT2* (which encodes members of the AP2/EREBP transcription factor family), thus acting as key effectors for the establishment of the stem cell niche during embryonic pattern formation (Chen et al., 2011). This response occurs due to auxin accumulation and in response to auxin-responsive TFs. Therefore, PLTs serve as a key node for JA-auxin crosstalk in regulating the maintenance of the stem cell niche in roots (Chen et al., 2011; Figure 8.3).

In the root, MeJA has been found to activate the transcription of *ASA1* and several other auxin biosynthesis-related genes such as *YUCCA2* (Cheng et al., 2006), *ASB1* (Stepanova et al., 2005), and *NITRILASE 3* (*NIT3*) (Kutz et al., 2002). JA failed to increase lateral root initiation in mutants with disrupted auxin signaling, like the *slr1* (*iaa14*) and *arf7/19* double mutant (Sun et al., 2009), which strongly supports that JA-induced lateral root formation is auxin-dependent. Activated expression of the transcription regulator *HOMEODOMAIN GLABROUS11* (*HDG11*) increased the level of JA in the roots via upregulating the expression of several genes encoding JA biosynthetic enzymes, resulting in enhanced auxin signaling and lateral root formation (Cai et al., 2015). MeJA also induced the expression of *YUC8* and *YUC9* and thus participated in auxin-mediated primary root growth and lateral root initiation (Hentrich et al., 2013).

8.2.2 JA–GA Crosstalk

JA and GA coordinately and antagonistically interact to regulate plant growth and defense (Jang et al., 2017, Huot et al., 2014, Hu et al., 2017). Extensive studies on JAZ JA signaling repressor proteins and DELLA GA signaling repressor proteins revealed that direct interaction between JAZs and DELLAs mediates the antagonistic interaction between JA and GA. A quadruple *della* mutant (which lacks four of the five *Arabidopsis* DELLA proteins) was shown to be partially insensitive to gene induction by JA, whereas the constitutively active dominant DELLA mutant *gai* was found to be sensitized for JA-responsive gene induction, implicating DELLAs in JA signaling and/or perception (Navarro et al., 2008). In the absence of GA, stable DELLA interacts with JAZ to release MYC2, resulting in the activation of MYC2 downstream genes. At the same time, DELLA interacts with JAZ to inhibit the expression of JA biosynthetic genes (*DAD1* and *LOX*) and further inhibits JA biosynthesis as well as the activities of MYB21 and MYB24, thereby regulating stamen development (Song et al., 2011). Cheng et al. (2004) found that GA promotes JA biosynthesis, thereby inducing the expression of *MYB21*, *MYB24*, and *MYB57* to promote stamen filament elongation. Most recently, it has been shown that DELLA repressors promote JA signaling through physically interacting with JAZ1 (Hou et al., 2010),

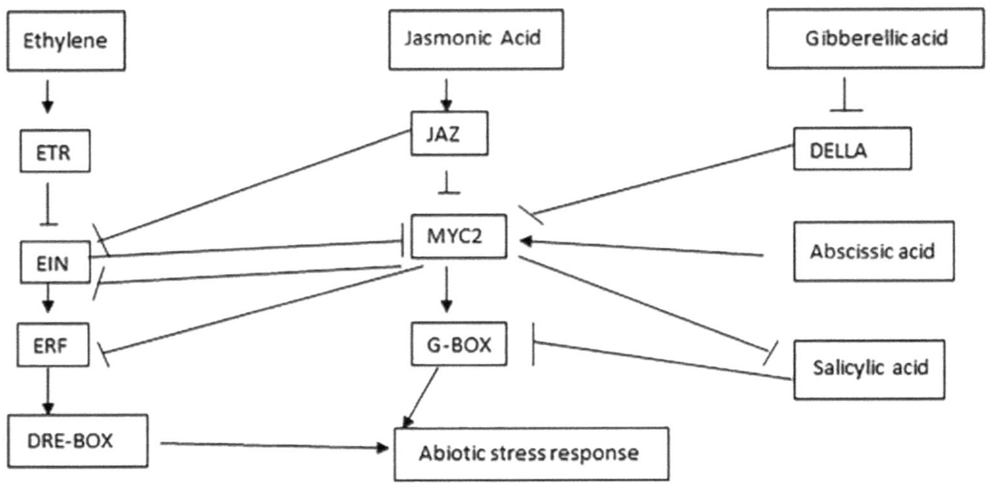

FIGURE 8.3 Crosstalk of JA with other plant hormone signaling pathways. MYC2 is the major component involved in interactions between JA and gibberellin (GA). DELLAs interact with JAZ repressors, relieving MYC2 from JAZ repression, and facilitating JA-mediated defense responses by the activation of MYC2. MYC2 is also positively regulated by ABA. Conversely, MYC2 inhibits salicylic acid (SA) regulation of abiotic stress response genes. The JAZ inhibition of EIN mediates JA and ET signaling synergy in plant resistance, whereas the reciprocal counteraction between MYC2 and EIN mediates JA and ethylene (ET) signaling antagonism. Positive and negative regulatory actions are indicated by arrows and lines with bars, respectively.

suggesting a mechanism for GA-mediated downregulation of JA defense responses.

8.2.3 JA–Cytokinin Crosstalk

JA and cytokinin differently regulate the expression of genes involved in chlorophyll development, indicating the existence of an antagonistic interaction between JA and cytokinin. A recent study revealed that xylem differentiation is regulated by JA in *Arabidopsis* roots, and an antagonistic interaction between JA and cytokinin is fundamentally important for JA-dependent xylem development (Jang et al., 2017). In *Arabidopsis*, an exogenous cytokinin treatment inhibits xylem development and *wooden leg* mutants with defects in cytokinin signaling strongly exhibit an all-xylem phenotype and lack procambial cells in their roots. In addition to this, mutants lacking transcription of Type-B *ARRs*, such as *ARR1*, *ARR10*, and *ARR12*, or transgenic plants overexpressing *AHP6*, a negative regulator of cytokinin signaling, form extra xylem (Jang et al., 2017, Yokoyama et al., 2007). A time-course experiment showed that suppression of cytokinin responses by JA does not occur rapidly, but the JA-mediated xylem phenotype is tightly linked to the suppression of the cytokinin response. Further analysis of arabidopsis histidine phosphotransfer protein6-1 and myc2-3 mutants revealed that the JA-responsive transcription factor MYC2 regulates the expression of AHP6 in response to JA and the expression of AHP6 is involved in the JA-mediated xylem phenotype (Jang et al., 2017).

8.2.4 JA–Ethylene Crosstalk

JA and ET antagonize or coordinately regulate plant stress response (Zhu, 2014; Zhu and Lee, 2015). On the one hand, exogenous JA triggers the degradation of JAZ, and the release of MYC2 regulates the expression of ORA59/ERF1 and wound responsive gene *VSP2*, so as to resist herbivorous insects. On the other hand, JAZ inhibits the transcriptional activity of EIL2/EIN3 in the ET signaling pathway and activates downstream ORA59/ERF1 that targets the promoter of *PLANT DEFENSIN 1.2* (*PDF1.2*) and induces its expression, thereby resisting the infection of necrotrophic pathogens and hemibiotrophic pathogens (Zhu et al., 2011). The GCC box in the promoter of *PDF1.2* is targeted by ET response factor (ERF) proteins, such as ERF1 and ORA59, which confers JA responsiveness and synergy between JA and ET (Brown et al., 2003; Pre et al., 2008). ETHYLENE-INSENSITIVE3 (EIN3) and its closest homolog ETHYLENE INSENSITIVE 3-likel (EIL1) are two primary transcription factors downstream of EIN2 (Chao et al., 1997; Guo and Ecker, 2013). Zhu et al. (2011) confirmed that JA-Zim (JAZ) domain proteins directly interact with EIN3/EIL1 and repress its transcriptional activity. JA-induced EIN3/EIL1 activation and ET-induced EIN3/EIL1 stabilization underlie the synergistic crosstalk between JA and ET in response to necrotrophic fungi in *Arabidopsis* (Zhu et al., 2011). However, in responses to chewing insects, JA and ET act antagonistically in *Arabidopsis*, which might be mediated by MYC2 and EIN3 (Memelink, 2009; Verhage et al., 2011). EIN3 interacts with and represses MYC2 to inhibit the JA-induced expression of herbivory-inducible genes and attenuate JA-regulated plant defenses (Song et al., 2014a, 2014b; Zhang et al., 2014). However, some of the conclusions in *Arabidopsis*, which is a dicotyledonous plant, cannot be verified in the monocotyledonous plant rice (De Vleesschauwer et al., 2014). For example, JA or ET biosynthesis- and signaling-related genes are all critical for the positive regulation of rice resistance to *Magnaporthe oryzae*, which is another destructive disease of rice (Nasir et al., 2018).

8.2.5 JA–ABA Crosstalk

As a major phytohormone regulating abiotic stress responses, ABA interacts with the JA signaling pathway to induce plant physiological responses to mitigate the effects of various abiotic stress factors (Chen et al., 2011). ABA receptor PYRABACTIN RESISTANCE 1-like proteins (PYLs) regulate metabolic reprogramming in *Arabidopsis thaliana* and tobacco through the JA signaling pathway (Per et al., 2018). The bHLH transcription factor, MYC2, a major regulator of the core JA signaling mechanism, participates in the ABA signaling pathway in response to drought stress (Abe et al., 2002; Liu et al., 2014). This transcription factor also regulates AOC1, a key JA biosynthesis gene, at signal intersections of the ABA/JA signaling pathway (Browse et al., 2009, Kazan and Manners, 2013). OsbHLH148 transcript levels increased rapidly following treatment with MeJA or ABA, and abiotic stress factors included drought, salinity, and cold stress. Expression profiling analyses of transgenic versus wild-type rice have identified OsDREB and OsJAZ genes that are upregulated by OsbHLH148 overexpression (Seo et al., 2011). Proline accumulation under water stress appeared to be an ABA-independent response, as both JA-deficient (jarl-1) and JA-insensitive (jail) in *A. thaliana* lines accumulate similar levels (De Ollas et al., 2015). In addition to this, ABA receptor PYL forms a complex with JAZ, which activates the transcriptional activity of MYC2. On the one hand, MYC2 activates the expression of JA-responsive gene *VSP2* under the mediation of MED25 to resist herbivorous insect feeding. On the other hand, MYC2 inhibits the expressions of *PTL1* and *PTL2* as well as root growth. Furthermore, ABA also initiates the degradation of JAZ12, which plays a specific role in the crosstalk between JA and ABA signaling pathways (Pauwels et al., 2015). Li et al. (2018) reported that low temperature increased JA and ABA concentrations, while also inducing the upregulation of ZjCBF, ZjLEA, and ZjDREB1 in *Zostera japonica* leaves. Overall, it indicates that the JA and ABA signaling pathways coordinate to regulate each other's responses to various biotic and abiotic stresses.

8.2.6 JA–Strigolactone Crosstalk

The role of strigolactone (SL) and its crosstalk with jasmonate and other phytohormones under biotic stress cannot be described in detail due to limited data. However, the available data points to a connection between SLs and jasmonates, as depicted in a study conducted on an SL-deficient tomato mutant (*Sl-ccd8*), where reduced contents of jasmonic acid (JA), ABA, and SA, and lower expression of the jasmonate-dependent gene, *PINII* (responsible for tomato resistance to *Botrytis cinerea*), were shown (Torres-Vera et al., 2014). Further alluding to this link is the fact that methyl jasmonate, a plant defense signaling molecule, exerts some influence on *Nicotiana tabacum PDR6* (*Nt-PDR6*), which is an orthologue of the SL transporter gene, *Ph-PDR1* (Xie et al., 2014). In another study, the expression of SL biosynthetic D27 and CCD8 and SA, JA, and ABA marker genes was found to be upregulated in response to the parasitic plant *Phelipanche ramosa* (Torres-Vera et al., 2016). A rather inconsistent observation was reported recently in which GR24 supply did not affect JA accumulation in WT plants, *At-max1*, and *At-max2*, nor did inoculation with *Mucor* sp. (Rozpądek et al., 2018). Therefore, there is not adequate data to draw any valid conclusions about the nature of SL–jasmonate crosstalk, but the fact that both hormones elicit responses in similar developmental processes such as mesocotyl elongation, senescence, and plant-microbe interactions offer some indications that SL–jasmonate crosstalk is likely to feature actively in these processes and cannot be completely neglected.

8.3 JA–Brassinosteroid Crosstalk

A link between JA and BR signaling in rice was established by performing a functional characterization of OsGSK2, a glycogen synthase kinase 3(GSK3)-like kinase in rice (*Oryza Sativa*; He et al., 2020). In plants, GSK3s have long been known to participate in BR signaling, but recent functional studies have identified their involvement in other pathways (Youn and Kim, 2015). OsGSK2 is the rice homolog of *Arabidopsis* (*Arabidopsis thaliana*) BRASSINOSTEROID INSENSITIVE 2, which plays a key role in BR signaling (Tong et al., 2012). In another study, the JA-inducible transcription of *PDF1.2a* and *PDF1.2b* was significantly reduced in the *BRASSINOSTEROID INSENSITIVE1-ETHYL METHANESULFONATE-SUPPRESSOR1* (*BES1*) gain-of-function mutant *besl-D*, highly susceptible to *Spodoptera exigua* and *B. cinerea*. BES1 directly targeted the terminator regions of *PDF1.2a/PDF1.2b* and suppressed their expression. *PDF1.2a* overexpression diminished the enhanced susceptibility of *besl-D* to *B. cinerea* but did not improve the resistance of *besl-D* to *S. exigua*. In response to *S. exigua* herbivory, BES1 inhibited biosynthesis of the JA-induced insect defense-related metabolite indolic glucosinolate by interacting with transcription factors MYB DOMAIN PROTEIN34 (MYB34), MYB51, and MYB122, and suppressing expression of genes encoding CYTOCHROME P450 FAMILY79 SUBFAMILY B POLYPEPTIDE3 (CYP79B3) and UDP-GLUCOSYL TRANSFERASE 74B1 (UGT74B1; Liao et al., 2020). Thus, BR contributes to the growth-defense tradeoff by suppressing the expression of defensin and glucosinolate biosynthesis genes.

8.4 Jasmonate in Plant Growth and Development

8.4.1 Seed Germination

The exogenous application of JAs inhibits various aspects of seedling growth, including primary root growth, leaf expansion, and hypocotyl elongation (Kim et al., 2015; Song et al., 2014b; Wasternack and Hause, 2013, Huang et al., 2017). However, the suppression of the inhibitory effect of JAs in particular on primary root growth was observed in the overexpression of NINJA or JAZ proteins carrying a deletion, mutation, or variation in the Jas domain, but enhanced by the abolishment of NINJA/TPL or combined mutations in JAZ7, JAZ8, JAZ10, and JAZ13 (Huang et al., 2017). JA-induced inhibition of primary root growth is mediated by MYC2, MYC3, and MYC4, which are distributed in different layers of the primary root apex (Fernández-Calvo et al., 2011; Gasperini et al., 2015). Additionally, JA-mediated root growth inhibition was also found to be affected by PLANT U-BOX PROTEIN10 (PUB10)—mediated MYC2 ubiquitination and 3-MAPK 6-mediated phosphorylation of MYC2 (Chico et al., 2014; Sethi et al., 2014). In addition, JA application delays the ABA-mediated inhibition of seed germination in *Arabidopsis*; however, *jar1* and *coi1-16* show increased sensitivity to the inhibition of seed germination by ABA (Ellis and Turner, 2002; Staswick et al., 1992). OPDA inhibits seed germination in *Arabidopsis* in a COI1-independent manner (Dave et al., 2011). During the cold-stimulated germination of wheat (*Triticum aestivum*) seeds, JA biosynthesis-related gene expression and JA biosynthesis increase rapidly in the dormant embryos after transfer to room temperature, and JA suppresses ABA biosynthesis to promote cold-stimulated germination (Xu et al., 2016). JA also inhibits coleoptile growth and plant height in rice and represses ear shoot growth in maize (*Zea mays*) (Biswas et al., 2003; Riemann et al., 2013; Riemann et al., 2008; Yan et al., 2012; Yang et al., 2012). JA-Ile was found to activate defense responses by triggering the degradation of JASMONATE ZIM DOMAIN (JAZ) transcriptional repressor proteins. The combined mutations within the 13-member *Arabidopsis JAZ* gene family resulted in elucidating the effects of chronic JAZ deficiency on growth, defense, and reproductive output. A higher-order mutant (*jaz* decuple, *jaz D*) defective in 10 *JAZ* genes (*JAZ1-7*, *-9*, *-10*, and *-13*) exhibited robust resistance to insect herbivores and fungal pathogens, which was accompanied by slow vegetative growth and poor reproductive performance (Guo et al., 2018).

8.4.2 Leaf Senescence

The relationship between JA and leaf senescence has been extensively studied (Jung et al., 2007; Seltmann et al., 2010; Breeze et al., 2011; Qi et al., 2015; Yu et al., 2016). Higher genetic markers transcript of developmental senescence (such as ERD1and SEN4), and the senescence-associated gene 21(SAG21) were reported as a result of MeJA (Jung et al., 2007). Additionally, an increased JA level was observed in leaves as they senesce developmentally (He et al., 2002; Seltmann et al., 2010; Breeze et al., 2011). In *A. thaliana*, a novel model for MeJA-regulated COI1-dependent leaf growth retardation has been proposed (Noir et al., 2013). TFs such as MYC2, MYC3, and MYC4 can upregulate the expression of genes associated with senescence (e.g., SENESCENCE-ASSOCIATED GENE29 [SAG29]) and chlorophyll catabolic enzyme (CCGs; e.g., pheophorbide A oxygenase) and thereby control JA/dark-induced leaf senescence (Qi et al., 2015; Zhu et al., 2015). JA-induced leaf senescence can also be negatively regulated by the outcome of WRKY57, which interacts with JAZ4/8 (Jiang et al., 2014). Recently, interaction between JAZs with axial regulators YABBY1 and YABBY3 was also reported to promote chlorophyll degradation (Boter et al., 2015).

8.4.3 Reproductive Development

Jasmonates serve essential roles in reproductive processes such as the development of petals and stamens in plants. In maize, JA-deficient mutant *opr7 opr8* showed outgrowth of multiple female reproductive buds and extreme elongation of ear shanks, indicating JA is a crucial signal for female organ growth (Yan et al., 2012). In *Arabidopsis*, JA biosynthesis mutants such as *fad3/7/8*, *aos*, *opr3*, and JA signaling mutant *coi1* are male sterile, strongly supporting JA as an essential signal for the development of the male organ of bisexual flowers (Browse, 2009). Reeves et al. (2012) showed an interaction between auxin-responsive transcription factors and JA biosynthesis in flower development and maturation. The R2R3 MYB transcription factors MYB21 and MYB24 are key regulators of petal and stamen growth, and the auxin-responsive transcription factors ARF6 and ARF8 regulate the expression of JA-responsive MYB21 and MYB24 by controlling JA biosynthesis, indicating that auxin interacts with JA to regulate the development of floral organs (Huang et al., 2017). GA suppresses DELLA repressor proteins to activate expression of the JA biosynthetic gene *DAD1* and JA biosynthesis, and triggers *MYB* expression to promote filament elongation (Cheng et al., 2009). Auxin signaling components, including TRANSPORT INHIBITOR RESPONSE1(TIR1)/AUXIN SIGNALING F-BOX PROTEINs (AFBs), INDOLE-3-ACETIC ACID 8 (IAA8), and AUXIN RESPONSE FACTOR6 (ARF6)/ARF8, regulate JA biosynthesis and MYB expression to modulate filament elongation and anther dehiscence (Cecchetti et al., 2008; Nagpal et al., 2005; Wang et al., 2013). The homeotic protein AGAMOUS (Ito et al., 2007), E3 ligase DAD1-ACTIVATING FACTOR (Peng et al., 2013), and the peroxisomal membrane protein DAYU (Li et al., 2014) activate the expression of several JA biosynthetic genes and JA biosynthesis to influence anther dehiscence and pollen germination, while the NO APICAL MERISTEM/ARABIDOPSIS TRANSCRIPTION ACTIVATION FACTOR/CUP-SHAPED COTYLEDON (NAC) TF ANTHER INDEHISCENCE FACTOR (Shih et al.,2014) and JINGUBANG (Ju et al., 2016) repress anther dehiscence and pollen maturation, respectively, in *Arabidopsis* by suppressing JA biosynthesis.

The *Arabidopsis* JA-deficient mutants *aos* and *opr3* and the JA perception mutant *coi1* exhibit larger petals than wild-type plants at anthesis, suggesting that JA restricts petal expansion (Brioudes et al., 2009; Reeves et al., 2012). MYB21 and MYB24 are required for petal expansion (Reeves et al., 2012). In *Arabidopsis*, from the final stage of flower bud opening to the final spreading of the petals after anthesis, JA represses the expression of *MYB21* in petals, resulting in restricted petal growth; in contrast, the expression level of *MYB21* is increased in the petals of *aos* and *coi1* plants, leading to persistent petal expansion and large petals (Reeves et al.,2012).

The rice JA-deficient mutants coleoptile photomorphogenesis (cpm)1 (Biswas et al., 2003), cpm2/hebiba (Riemann et al., 2013), osjar1 (Xiao et al., 2014), and extra glume (eg)1 (Li et al., 2009); the JA signaling mutant eg2-D (with a dominant mutation in OsJAZ1; Cai et al., 2014a); RNAi lines of OsCOI1a and OsCOI1b (Yang et al., 2012); transgenic lines of truncated JAZs (Hori et al., 2014); and rice transgenic lines expressing an *Arabidopsis* jasmonic acid carboxyl methyltransferase gene (Kim et al., 2009) exhibit complete or partial male sterility due to abnormalities in spikelet organs, including abnormal or reduced stamens, reiterative glume-like structures, stigma-like organs, and impaired anther dehiscence.

8.4.4 Seed and Embryo Development

Embryo development was found to delayed and increased programmed cell death in the tomato JA-deficient mutant *prosystemin-mediated responses2*, which carries a mutation in FAD7 (Goetz et al., 2012). The *jasmonic acidinsensitivel* (*jai1*) mutant, which exhibits a loss of function of the tomato homolog of COI1, cannot set viable seeds (Li et al., 2004). Interestingly, a wound-induced endogenous rise in OPDA was found to restore seed development in *jai1*, suggesting a role for OPDA signaling in the maternal control of seed development (Goetz et al., 2012). However, *SiOPR3*, a tomato transgenic line in which the *OPR3* gene is silenced, contains a similar amount of OPDA to wild type and sets only a few viable seeds while methyl-JA treatment restored the seed-setting of *SiOPR3* (Scalschi et al.,2015), suggesting that JA but not OPDA has a major role in the maternal control of seed development. In apples (*Malus sylvestris*) and sweet cherries (*Prunus avium*), endogenous JA accumulated in the early ripening stage of the fruit and seeds, indicating an important role of JA in fruit/seed development (Kondo et al., 2000).

8.4.5 Trichome Development

Trichomes are branching structures or hair-like appendages differentiated from epidermal cells in the aerial part of the plant, which function as barriers by acting as sensors or barriers, or by releasing volatile compounds (Ishida et al., 2008). Trichome formation is initiated by different endogenous developmental signals, including phytohormones such as jasmonate (Traw and Bergelson, 2003; Li et al., 2004; Yoshida et al., 2009), gibberellin (Perazza et al., 1998), ethylene (Plett et al., 2009), and salicylic acid (Traw and Bergelson, 2003). In the absence of JA, *Arabidopsis* JAZ proteins interact with WD-repeat/bHLH/MYB complexes, repressing their formation and transcriptional activity (Qi et al., 2011). Following wounding or an insect attack, however, JA biosynthesis is triggered, leading to the turnover of JAZ proteins; under these conditions, WD-repeat/bHLH/MYB complexes are able to promote trichome formation. Subgroup IIId bHLH TFs antagonize WD-repeat/bHLH/MYB complexes by binding competitively to the promoters of their mutual target genes to inhibit trichome formation (Nakata et al., 2013; Song et al., 2013a). JA and GA synergistically enhance trichome formation. DELLA repressors in the GA signaling pathway also directly target and inhibit WD-repeat/bHLH/MYB complexes (Qi et al.,2014).

8.4.6 Sex Determination

In maize, the formation of the tassel (male inflorescence) and ear (female inflorescence) results from the abortion of pistil and stamen development, respectively. In maize mutants, *tasselseed1* (*ts1*) (the mutant of a LOX), *ts2*, and *opr7 opr8*, the tassels are converted to fertile ears that are able to set seeds; JA treatment can restore tassel development, demonstrating that JA controls sex determination in maize (Acosta et al., 2009; Yan et al., 2012). *TS1* encodes a 13-lipoxygenase (i.e., *LOX8*), disruption of which causes JA-deficiency locally in the tassel meristem. *opr7 opr8* is a double mutant of OPR isoforms required for JA biosynthesis, mutation of which results in JA depletion systemically in the plant. Several studies have shown that gibberellin (GA) is involved in ear formation. GA biosynthesis mutants such as *an1*, *d1*, *d2*, *d3*, and *d5*, and GA perception mutants *D8* and *D9*, all showed dwarfism and masculinized ears (i.e., male florets are produced in ears), indicating GA is another important phytohormone for sex determination in maize (Chuck, 2010). Thus, studies of JA and GA hypothesize that JA and GA act antagonistically in male and female flowers, respectively, in the maize sex determination process.

8.4.7 Flower and Fruit Development

The presence of jasmonate and related volatile fatty acid derivatives may be involved in insect attraction related to pollen dispersal. Other aspects of flower, fruit, and seed development that can be modulated by jasmonate include fruit ripening, fruit carotenoid composition, and the expression of genes encoding seed and vegetative storage proteins. Jasmonate-stimulated tomato and apple fruit ripening most likely occurs through activation of EFE and the production of ethylene (Czapski and Saniewski, 1992). It is possible that jasmonate levels gradually increase in developing fruit, leading to the enhanced synthesis of ethylene and subsequent fruit ripening. The application of JA to tomato fruit inhibited the accumulation of lycopene and stimulated the accumulation of b-carotene (Saniewski and Czapski, 1983).

8.5 Conclusion

Plant hormone JA has its own specific biosynthetic and signaling pathways, but their roles in plant development and physiology overlap. This suggests that JA modulates plant growth and physiology through interactions with other hormones, and the extensive interplay between auxin, GA, cytokinin, ethylene, ABA, SL, and BR in the regulation of all aspects of plant growth and development supports this idea. JA mediates the plant response and its underlying molecular mechanisms have been well reported in previous studies. JA also modulates plant development, such as root, stamen, hypocotyl, chloroplast, and xylem development, and increasing evidence of JA-dependent modulation of plant growth and development. Many studies have revealed that the crosstalk between phytohormones is mediated through regulatory proteins controlling phytohormone metabolic and signaling pathways. This chapter briefly described the metabolism and signaling pathways of the phytohormones GA, cytokinin, and auxin that interact with JA in the modulation of plant growth and development, and discussed the recent findings on JA crosstalk, focusing on the JA–GA, JA–cytokinin, and JA–auxin interactions.

REFERENCES

Abe et al. *Plant Cell* 2002, 15:63–78.

Acosta et al. *Science* 323:262–265.

Ahmad et al. *Frontiers in Plant Science* 7:813. doi: 10.3389/fpls.2016.00813.

Balbi et al. *New Phytologist* 177:301–318.
Ballare´. *Trends in Plant Science* 2011, 16:249–257.
Biswas et al. *Plant and Cell Physiology* 2003, 44:242–254.
Boter et al. *The Plant Cell* 2015, 27:3160–3174.
Breeze et al. *The Plant Cell* 2011, 23:873–894.
Brioudes et al. *The Plant Journal* 2009, 60:1070–1080.
Brown et al. *Plant Physiology* 2003, 132:1020–1032.
Browse. *Vitamins and Hormones* 2005, 72:431–456.
Browse. *Phytochemistry* 2009a, 70:1539–1546.
Browse. *Annual Review of Plant Biology* 2009b, 60:183–205.
Cai et al. *Journal of Integrative Plant Biology* 2015, 57:1017–1030. doi: 10.1111/jipb.12347.
Cai et al. *Nature Communications* 2004, 5:3476.
Cecchetti et al. *The Plant Cell* 2008, 20:1760–1774.
Chao et al. *Cell* 1997, 89:1133–1144. doi: 10.1016/S0092-8674(00)80300-1.
Chen et al. *Plant Cell* 2011, 23:3335–3352. doi: 10.1105/tpc.111.089870.
Chen et al. *Plant Cell* 2012, 24:2898–2916.
Cheng et al. *Development*. 2004, 131(5):1055–1064. doi: 10.1242/dev.00992.
Cheng et al. *Genes and Development* 2006, 20:1790–1799.
Cheng et al. *PLoS Genetics* 2009, 5:e1000440.
Chico et al. *The Plant Cell* 2014, 26:1967–1980.
Chuck. In: *Advances Botanical Research*. Eds. K. Jean-Claude and D. Michel. Academic Press, 2010, 54:53–83.
Czapski and Saniewski. *Journal of Plant Physiology* 1992, 139:265–268.
Dave et al. *The Plant Cell* 2011, 23:583–599.
De Ollas et al. *Plant Signaling & Behavior* 2015, 10:e1078953.
De Vleesschauwer et al. *Frontiers in Plant Science* 2014, 5:e611.
Du et al. *Frontiers in Plant Science* 2013, 4:397.
Ellis et al. *Planta* 215:549–556.
Engelberth et al. *Plant Physiology* 2001, 125:369–377.
Fernández-Calvo et al. *The Plant Cell* 2011, 23:701–715.
Gasperini et al. *PLoS Genetics* 2015, 11:e1005300.
Ghasemi Pirbalouti et al. *Archiv der Pharmazie - Chemistry in Life Sciences* 2014, 347:229–239.
Goetz et al. *Plant Physiology* 2012, 158:1715–1727.
Guan et al. *Proceedings of the National Academy of Sciences of the United States of America* 2019, 116(21):10568–10575.
Guo and Ecker. *Cell* 2013, 115:667–677. doi: 10.1016/S0092-8674(03)00969-3.
Guo et al. *Current Opinion in Plant Biology* 2018, 44:72–81. doi: 10.1016/j.pbi.2018.02.009.
Gutierrez et al. *Plant Cell* 2012, 24:2515–2527.
He et al. *Plant Cell* 2020, 32:2806–2822.
Hentrich et al. *Plant Journal* 2013, 74:626–637. doi: 10.1111/tpj.12152.
Hori et al. *Plant Signal Behavior* 2014, 9:e970414.
Hou et al. *Developmental Cell* 2010, 19(6):884–94.
Hu et al. *Journal of Experimental Botany* 2017, 68(6):1361–1369.
Huang et al. *Journal of Experimental Botany* 2017, 68:1349–1359.
Huot et al. *Molecular Plant* 2014, 7(8):1267–1287.
Ishida et al. *Annual Review of Plant Biology* 2008, 59:365–386.
Ito et al. *The Plant Cell* 2007, 19:3516–3529.
Jang et al. *Scientific Reports* 2017, 7(1):10212.
Jiang et al. *Plant Cell* 2014, 26 (1):230–245.
Ju et al. *The Plant Cell* 2016, 28:2131–2146.
Jung et al. *Plant Cell Reports* 2007, 26:1053–1063.
Kazan. *Trends in Plant Science* 2015, 20:219–229.
Kazan et al. *Molecular Plant* 2013, 6:686–703.
Kim. *Plant Physiology* 149:1751–1760.
Kim et al. *Frontiers in Plant Science* 2015, 6:20.
Kutz et al. *Plant Journal* 2002, 30:95–106.
Lakehal et al. *International Journal of Molecular Sciences* 20:4428.
Li et al. *The Plant Cell* 2004,16:126–143.
Li et al. *The Plant Cell* 2014, 26:619–635.
Li et al. *PLoS ONE* 2018, 13:e0198885.
Liao et al. *Plant Physiology*, 2020, 182:1066–1082.
Liu et al. *Nucleic Acids Research* 2014, 42:5556–5566.
Memelink. *Phytochemistry* 2009, 70:1560–1570.
Nagpal et al. *Development* 2005, 132:4107–4118.
Nakata et al. *The Plant Cell* 2013, 25:1641–1656.
Nasir et al. *Seminars in Cell & Developmental Biology* 2018, 83:95–105.
Navarro et al. *Current Biology* 2008, 18:650–655.
Noir et al. *Plant Physiology* 2013, 161:1930–1951.
Pauwels and Goossens. *Plant Cell* 2011, 23:3089–3100.
Pauwels et al. *Plant Physiology* 2015, 169:1405–1417.
Peng et al. *The Plant Journal* 2013, 74:310–327.
Per et al. *Frontiers in Plant Science* 2016, 7:.
Per et al. *Environmental and Experimental Botany* 2018, 145:104–120. doi: 10.1016/j.envexpbot.2017.11.004.
Perazza et al. *Plant Physiology* 1998, 117:375–383.
Pieterse et al. *Annual Review of Cell and Developmental Biology* 2012, 28:489–521. doi: 10.1146/annurev-cellbio-092910-154055.
Plett et al. *Journal of Experimental Botany* 2009, 60:3923–3933.
Pré et al. *Plant Physiology* 2008, 147:1347–1357.
Qi et al. *Plant Cell* 2011, 23:1795–1814.
Qi et al. *The Plant Cell* 2014, 26:1118–1133.
Qi et al. *Plant Cell* 2015, 27:1634–1649.
Reeves et al. *PLoS Genetics* 2012, 8:e1002506.
Reinbothe et al. *FEBS Journal* 2009, 276:4666–4681.
Riemann et al. *Plant, Cell and Environment* 2008, 31:783–792.
Riemann et al. *The Plant Journal* 2013, 74:226–238.
Rozpądek et al. *Frontiers in Microbiology* 2018, 9:441.
Saniewski and Czapski. *Experientia* 1983, 39:1373–1374.
Saniewski et al. *Acta Physiologiae Plantarum* 2002, 24:211. doi: 10.1007/s11738-002-0013-9.
Scalschi et al. *The Plant Journal* 2015, 81:304–315.
Seltmann et al. *Plant Physiology* 2010, 152:1940–1950.
Seo et al. *Plant Journal* 2011, 65:907–921.
Sethi et al.*The Plant Cell* 2014, 26:3343–3357.
Shih et al. *Journal of Experimental Botany* 2014, 65:621–639.
Song et al. *Plant Cell* 2011, 23:1000–1013. doi: 10.1105/tpc.111.083089.
Song et al. *PLoS Genetics* 2013, 9:e1003653.
Song et al. *Plant Cell* 2014a, 26:263–279.
Song et al. *Current Opinion in Plant Biology* 2014b, 21:112–119.
Staswick et al. *Proceedings of the National Academy of Sciences of the United States of America* 1992, 89:6837–6840.
Stepanova et al. *Plant Cell* 2005, 17:2230–2242.
Sun et al. *Plant Cell* 2009, 21:1495–1511. doi: 10.1105/tpc.108.064303.
Tong et al. *Plant Cell* 2012, 24:2562–2577.
Torres-Vera et al. *Molecular Plant Pathology* 2014, 15:211–6. doi: 10.1111/mpp.12074.
Torres-Vera et al. *Physiological and Molecular Plant Pathology* 2016, 94:100–107.

Traw et al. *Plant Physiology* 2003, 133:1367–1375.
Verhage et al. *Frontiers in Plant Science* 2011, 2:e47.
Wang et al. *Plant Molecular Biology* 2013, 82:71–83.
Wasternack. *Annals of Botany* 2007, 100:681–697.
Wasternack. *Biotechnology Advances* 2014, 32:31–39.
Wasternack and Hause. *Annals of Botany* 2013, 111:1021–1068.
Wasternack and Strnad. *New Biotechnology* 2016, 33:604–613.
Xiao et al. *Plant Molecular Biology* 2014, 86:19–33.
Xie et al. *Physiologia Plantarum* 2014, 153:299–306.
Xu et al. *Journal of Experimental Botany* 2016, 67:3497–3508. doi: 10.1093/jxb/erw172.
Xue et al. *Journal of Genetics and Genomics* 2007, 34:339–346.
Yan et al. *The Plant Cell* 2012, 24:1420–1436.
Yang et al. *Proceedings of the National Academy of Sciences of the United States of America* 2012, 109:1192–1200.
Ye et al. *Plant Cell*, 2019, 31:687–698.
Yokoyama et al. *Plant and Cell Physiology* 2007, 48:84–96.
Yoshida et al. *Development* 2009, 136:1039–1048.
Youn et al. *Molecullar Plant* 2015, 8:552–565.
Yu et al. *Journal of Experimental Botany* 2016, 67:751–762.
Zhang et al. *Plant Cell* 2014, 26:1105–1117.
Zhang et al. *Nature Plants* 2019, 5:491–497.
Zhang et al. *New Phytologist* 2019, 224:761–774.
Zhu. *Journal of Experimental Botany* 2014, 65:5743–5748. doi: 10.1093/jxb/eru349.
Zhu and Lee. *Plant and Cell Physiology* 2015, 56:414. doi: 10.1093/pcp/pcu171.
Zhu et al. *Proceedings of the National Academy of Sciences of the United States of America* 2011, 108:12539–12544.
Zhu et al. *Plant Journal* 2015, 84:597–610.

9

Bioscience of Jasmonates in Harmonizing Plant Stress Conditions

Shruti Kaushik, Anmol Sidhu, Anil Kumar Singh, and Geetika Sirhindi

CONTENTS

DOI: 10.1201/9781003110651-9

Abbreviations: *ERF*, ethylene response factor; *WRKYs*, Transcription factors containing highly conserved amino acid sequence WRKYGQK; *PAP1*, production of anthocyanin pigment 1; *ORCAs*, octadecanoid-responsive *Arabidopsis* AP2/ERF-domain protein; *ZCTs*, zinc finger *Catharanthus* transcription factor; *MAPKs*, mitogen-activated protein kinases; *COI1*, *CORONATINE INSENSITIVE1*; *LOX*, lipoxygenase; *JAZ*, jasmonate ZIM domain; SCF, Skp-Cullin-F-box E3 ubiquitin ligase complex; bHLH, basic helix–loop–helix; *MYCs*, myelocytomatosis oncogenes transcription factors; *CYP*, cytochrome P450 transcription factor.

9.1 Introduction

The sedentary nature of plants obliges them to compensate for their reduced mobility with high plasticity in order to hastily and efficiently regulate their metabolism and maintain their growth performance in a broad range of environmental conditions (Walters et al., 2005). Acclimation of plants to various environmental challenges involves a variety of physiological and metabolic changes that allow them to cope with new environments. Various environmental stresses hamper agricultural crop productivity, revealing a large variety of responses happening in different time ranges from hasty biophysical adjustments to slower changes in gene expression (Kouřil et al., 2013).

A striking phenomenon associated with this acclimation process was the production of volatile and non-volatile compounds including phytohormones that help them to acclimatize to a changing environment (Ashraf et al., 2010). An important phytohormone, JA (jasmonic acid) and its methyl ester, methyl jasmonates (MeJAs), are the derivatives of fatty acid metabolism (Jalalpour et al., 2014). JA is ubiquitously present in plant species, where their levels are elevated in the reproductive tissues and flowers, but low in the mature leaves and roots (Pirbalouti et al., 2014). Jasmine oil, also called ethereal oil of *Jasminum grandiflorum*, was used for the first time to isolate MeJA (Avanci et al., 2010) and JA was isolated from the culture filtrates of the fungus *Lasiodiplodia theobromae*, which is an imperative member of the JAs (Tsukada et al., 2010). Apart from JA and MeJA, other JAs, mainly *cis*-jasmone, jasmonoyl ACC (JA-ACC), and jasmonoyl isoleucine (JA-Ile), displaying numerous biological functions have been reported (Wasternack and Kombrick, 2010; Koo and Howe, 2012).

Jasmonate and MeJA are believed to modulate various crucial processes that are involved in plant growth and development such as vegetative growth, cell stamen and trichome development, fruit ripening, senescence, biotic and abiotic stress tolerance, sex determination, storage organ formation, and root elongation (Camposa et al., 2014). In addition, the biosynthesis and proper accretion of secondary metabolites are also induced by JAs along with various transcription factors (TF) involved in the synthesis of these secondary metabolites. A common face underlying JA-mediated transcriptional control of biosynthesis of secondary metabolites involves *JAZ* proteins, the *SCFCOI1* complex, and *MYC2* together with additional components, such as *ERFs*, *WRKYs*, *PAP1*, *ORCAs*, *MYBs*, and *ZCTs*, and all of them are active in diverse pathways (De Geyter et al., 2012). As a signaling molecule, JA regulates the expression of various genes in response to biotic and abiotic stresses and promotes definite protective mechanisms (Li et al., 2018).

JAs have been advocated to significantly sustain a balance between growth and defense mechanisms in plants' regulatory systems but do not act independently; rather, they operate in complex networks with crosstalk to other phytohormonal signaling pathways including gibberellin (GA), auxin (indole-3-acetic acid, IAA), cytokinin (CK), brassinosteroids (BRs), abscisic acid (ABA), ethylene (ET), jasmonic acid (JA), and salicylic acid (SA; Wasternack and Hause, 2013). An intensive synergistic or antagonistic crosstalk among the signaling pathways of JAs with other phytohormones has been established to modulate their biosynthesis and responses (Kazan, 2015). The crosstalk among the phytohormones can also reprogram the genetic machinery, affect defense reactions, and thereby improve plant stress tolerance.

Given these recent developments, there is a promising need to harmonize the earlier updates on jasmonate (Wasternack, 2007). Taking the latest information and fundamental breakthroughs into contemplation, we will discuss here in parallel the miscellaneous roles of jasmonates in plant stress responses and development.

9.2 JA Biosynthesis and Metabolism

Jasmonic acid is a lipid-derived hormonal signal that controls the expression of defense genes in response to various stress responses such as wounding and other environmental stresses. To express wound and stress-induced genes, the formation of jasmonates in a lipid-based signaling pathway via octadecanoid seems to be a common principle for many plant species (Wasternack and Parthier, 1997; Creelman and Mullet, 1997).

Biosynthesis of JA is initiated in the chloroplast thylakoid membrane where JA and its numerous derivatives emanate from chloroplast membrane lipids, preferentially α-linolenic acid. In this octadecanoid pathway, the lipid intermediates translocate from chloroplast membranes into the cytoplasm and then into peroxisomes (León, 2013).

9.2.1 Scheme of JA Biosynthesis

9.2.1.1 Production of Linolenic Acid from Linoleic Acid

α-linolenic acid (18:3), precursor of JA, is produced from linoleic acid (18:2) via a desaturation reaction catalyzed by the *fatty acid desaturase* (*FAD*) enzyme. Three *ω-3 fatty acid desaturases* (*FAD3*, *FAD7*, and *FAD8*) have been identified in the *Arabidopsis* genome which functions to convert dienoic fatty acids (18:2) to their trienoic forms (18:3). *fad3*, *fad7*, and *fad8* triple mutant of *Arabidopsis* is defective in JA biosynthesis because of no detectable production of 18:3 (McConn and Browse, 1996).

9.2.1.2 Release of Linolenic Acid from Galactolipids Involved in JA Biosynthesis

The JA precursor α-linolenic acid (18:3) is released from galactolipids of chloroplast membranes and it is generally accepted that a *phospholipase 1* (*PLA_1*) releasing α-linolenic acid from the *sn-1* position of galactolipids is responsible for the generation of JA substrate, whereas the large family of phospholipases is not involved in JA biosynthesis. However, it is a matter of debate as to which of the *PLA_1s* are involved in JA biosynthesis. Initially, *DEFECTIVE IN ANTHER DEHISCENCE 1* (*DAD1*) was considered to be responsible for JA synthesis as the mutant form dad1 showed decreased JA levels, especially in flowers and male sterility (Ishiguro et al., 2001). *DONGLE* (*DGL*), *PLA_1* extracted from *Arabidopsis thaliana* was also believed to be involved in basal and wound-induced JA biosynthesis (Hyung et al., 2008; Yang et al., 2007). But a later comparison of RNAi lines of both *DAD1*and *DGL* with wild type in early wound response (Ellinger et al., 2010) suggested that none of these enzymes are involved in JA synthesis. More recently, a study of *PLASTID LIPASE* (*PLIP1*) in the export of acyl groups from plastids for seed oil biosynthesis has clarified this long-term puzzle in which numerous lipases, preferentially *galactolipases* are involved in JA biosynthesis (Wang et al., 2017). Homologs of *PLIP2* and *PLIP3* were found to have *glycerolipase A_1* activity.

9.2.1.3 Oxygenation of α-Linolenic Acid by 13-LOX

After its release from galactolipids, free α-linolenic acid (18:3) is oxidized to *13-HPOT* in a reaction catalyzed by *13-LOX* enzymes. Oxygenation of α-linolenic acid is considered as the initial step in JA biosynthesis. In this reaction, oxygen is inserted by lipoxygenase enzyme (*LOX*) at the C-13 position of α-linolenic acid. Among six *LOXs* reported in *Arabidopsis*, the *LOX2* is mainly responsible for wound-induced JA production (Bell et al., 1995; Schommer et al., 2008; Glauser et al., 2009). Whereas *LOX3* and *LOX4* are mainly involved in the development of fertile flowers (Caldelari et al., 2011), further genetic and physiological studies have denoted that *LOX6* is involved in wound- and drought-induced JA synthesis (Chauvin et al., 2013; Grebner et al., 2013). The crystal structure of several *LOXs* has been elucidated, which allowed an improved mechanistic explanation of their catalytic mechanism (Newcomer and Brash, 2015).

9.2.1.4 Dehydration of 13-HPOT *by* AOS

This step is catalyzed by a plastid localized *13-allene oxide synthetase* (*AOS*) enzyme, where *13-HPOT* is dehydrated to an unstable *12,13(S)-epoxy octadecatrienoic acid* (*12,13-EOT*). There is a single *AOS* gene reported in *Arabidopsis*, whereas tomato has two (Howe et al.,2000; Park et al., 2002) that are known to be induced by wounding thus, suggesting its role in wound-induced JA biosynthesis (Howe et al., 2000; Park et al., 2002). The *aos* mutant of *Arabidopsis* is defective in wound and basal JA biosynthesis, as it disrupts the expression of the *AOS* gene, thus providing direct evidence of the importance of *AOS* for JA biosynthesis (Park et al., 2002). In addition, the *aos* mutant exhibits a male sterile phenotype which did not recover during development but was rescued by exogenous MeJA spraying (Park et al., 2002). These results thus demonstrated that *AOS*-dependent JA biosynthesis is important not only for wound responses but for reproductive development as well.

9.2.1.5 Synthesis of OPDA *by* AOC

Hydrolysis of unstable allylic epoxides leads to the formation of α- and ɤ-ketols. Allene oxide cyclase (*AOC*) is directed to cyclize allylic epoxides to form 12-oxo-phytodienoic acid (*12-OPDA*) which is usually a racemic mixture of (+) and (–) enantiomers (Brash et al., 1988). The stereo-specific cyclization of allene oxide into (+) enantiomer OPDA is done exclusively by *AOC*, indicating its role in the establishment of the enantiomeric structure of the JA precursor (Schaller and Stintzi, 2009). *OPDA* is the end product of the first half of JA biosynthesis localized in plastids.

9.2.1.6 Export of OPDA from Chloroplast to Peroxisome

After completion of the first half of JA biogenesis in the chloroplast, *OPDA* is further exported to the cytosol and then into peroxisome for subsequent conversion into the active forms of JA. Recently, a protein named *JASSY* was acknowledged to play a prominent role in the export activity of OPDA from the chloroplast to the cytosol (Guan et al., 2019; Figure 9.1). *JASSY*, a chloroplast outer envelope (OE) protein, was found to be co-expressed with numerous other genes involved in JA response (Hu et al., 2013). This protein is testified to contain a steroid acute regulatory protein-related lipid transfer (START) domain, which suggests a role in binding/transport of hydrophobic molecules (Guan et al., 2019). Thus, *JASSY* binds with *OPDA* and functions as a membrane channel. The import of OPDA from the cytosol to peroxisome is further supported by

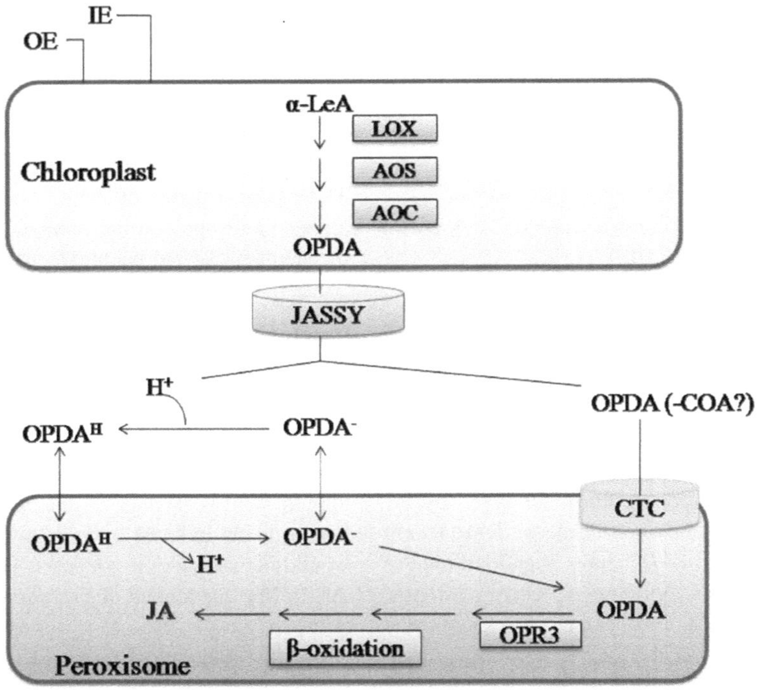

FIGURE 9.1 Recent model of the biosynthetic pathway of JA in chloroplasts and peroxisomes showing the export of *OPDA* from the chloroplast to the cytosol via the *JASSY* protein.

the activity of *COMATOSE* (*CTS*), an ABC transporter of peroxisomal membrane.

9.2.1.7 *Action of OPDA Reductase* (OPR3) *on OPDA*

The second half of JA biosynthesis occurs in peroxisomes upon transport of *OPDA* across two membranes. In peroxisome, the reduction of the cyclopentanone ring of *OPDA* is catalyzed by *OPDA* reductase (OPR) to produce *OPC-8:0* (Schaller and Stintz, 2009).

9.2.1.8 *β- Oxidation of Carboxylic Acid Side Chain* (ACX, MFP, KAT)

The carboxylic acid side chain of 3-oxo-2-(2'(z)-pentenyl)-cyclopentane-1-octanoic acid (*OPC-8:0*) is shortened by fatty acid β-oxidation machinery (Hu et al., 2012). The shortening of the octanoic acid side chain of *OPC* involves three rounds of β-oxidation to yield (+)-7-*iso*-JA. These final steps of JA biosynthesis are catalyzed by *acyl-CoA oxidase* (*ACX*), multifunctional protein (*MFP*), and *3-ketoacyl-CoA thiokinase* (*KAT*), known as core enzymes of β-oxidation.

9.2.2 OPR3-Independent Pathway: A Bypass in JA Biogenesis

For the first time in 25 years, a new pathway for JA biosynthesis was identified. Lately, a new mutant allele, *opr3-3*, was identified, which exhibited a complete absence of OPR3 activity (Chini et al., 2018). As production of JA is primarily exhibited by a reduction of *OPDA* via *OPRs*, thus a loss of functional allele *opr3-3* unveiled a new bypass pathway of JA biosynthesis. In this bypass pathway, peroxisomal OPDA was reported to be metabolized to dinor-OPDA (dn-OPDA), tetranor-OPDA (tn-OPDA), and 4,5-didehydr-JA, which is further reduced to JA via *OPR2* (Chini et al., 2018).

9.2.3 Metabolism of JA Compounds for Active Homeostasis

After biosynthesis, JA is subjected to diverse modifications to form a series of metabolites. JA can be converted to active, partially active, and certain inactive compounds. So far, about 12 different metabolic pathways are known (Wasternack and Strnad, 2018). Among them, the most important derivatives and metabolites are a result of the following: conjugations with amino acids (Staswick and Tiryaki, 2004; Suza and Staswick, 2008); hydroxylation (Gidda et al., 2003); carboxylation, decarboxylation (Koch et al.,1997); methylation; esterification; sulfation; O-glycosylation; and lactone formation of 12-OH-JA derivatives. JA is known to be metabolized by alteration of the carboxyl group, C-12 carbon oxidation or the reduction of C-6 ketone. These derivatives are known to vary in their biological functions. The majority of JA signaling is attributed to homeostasis among different derivatives of JA-Ile but some reactions lead to compounds with activity in specific reactions of stress responses and developments such as leaf movement. Some of the common metabolic pathways of JA are summarized in the following sections.

9.2.3.1 *Conjugation*

JA is subjected to an amide-linked conjugation with isoleucine (Ile) and some other amino acids to yield JA-Ile and other amino acid conjugates. The (3R,7S) diastereoisomer of JA-Ile is recognized as the most active biological form of JA (Fonseca et al., 2009), which is synthesized by the action of a conjugating enzyme *JAR1*, a member of the Gretchen Hagen 3 (GH3) gene family (Suza and Staswick, 2008). Recently, apart from JA-Ile (+)-7-iso-JA-Ala, (+)-7-iso-JA-Val., (+)-7-iso-JA-Leu, and (+)-7-iso-JAMetwere were also identified as bioactive conjugates of JA (Yan et al., 2016). Many evident studies revealed that JA-Ile, but not free JA, promotes the interaction between JAZ proteins and F-box protein COI1 (Fonseca et al., 2009; Sheard et al., 2010). Furthermore, the homeostasis of JA-Ile is structured by diverse environmental and developmental processes of JA responses.

9.2.3.2 *Hydroxylation*

Hydroxylation at C-11 or C-12 carbon of the pentenyl side chain of jasmonic acid yields 12-hydroxy-jasmonic acid (12-OH-JA), commonly known as tuberonic acid (Sembdner and Parthier, 1993). However, the direct hydroxylation of JA had not been reported so far. Several members of the *CYP94* gene family are known to hydroxylate JA-Ile (Heitz et al., 2012). Hydroxylation of jasmonic acid attributes to JA catabolism via the ω-oxidation pathway (Miersch et al., 1992). Furthermore, hydroxylated jasmonic acid is in turn known to yield sulfonated and glucosylated derivatives of JA by the addition of SO_3 and glucose to 12-OH-JA derivatives. Several consequent research approaches have concluded that metabolic conversion of JA by hydroxylation and sulfation leads to a fine-tuning of the pattern of gene expression, including switch-off in JA signaling for a subset of genes (Miersch et al., 2008).

9.2.3.3 *Carboxylation*

Carboxylation is akin to hydroxylation (Wasternack and Song, 2017) in a view that instead of direct modification of JA, certain specific genes of the *CYP94* gene family exhibit stepwise oxidative activity to yield carboxylated 12-OH-JA-lle. The major consequence of hydroxylation and carboxylation of JA compounds is switch-off in JA and JA-lle signaling owing to sustained homeostasis among the active and inactive derivatives of jasmonic acid. Thus, in nature, this homeostasis amid different derivatives of JA may allow plants to reciprocate flexibly toward different herbivores (Wasternack and Strnad, 2018).

9.2.3.4 *Decarboxylation*

Decarboxylation of JA yields a volatile compound known as *cis*-jasmone, which is a component of the floral bouquet in many flowering plants involved in enticing insects for pollination (Wasternack and Strnad, 2018). Apart from this, wounding is also known to form active *cis*-jasmone. It has even been described that treatment of *cis*-jasmone expresses a distinct set of genes different from that of JA (Matthes et al., 2010).

9.2.3.5 *Methyl Ester of JA*

Jasmonic acid is converted to a volatile derivative MeJA by a SAM-dependent carboxyl methyltransferase. MeJA is alleged to act as an airborne signaling interplant communication, being released in response to various abiotic stresses as well as wounding (Wasternack and Strnad, 2016). The methylation of JA in *Arabidopsis* is catalyzed by S-adenosyl-L-methionine:jasmonic acid carboxyl methyltransferase (JMT) to form MeJA (Seo et al., 2001). The overexpression of *JMT* is reported to hoist MeJA level, thereby increasing the expression of JA responsive genes leading to increased disease resistance (Seo et al., 2001). Thus, *JMT*-mediated methylation forms a crucial phase of JA-regulated plant immunity.

9.3 JA Signaling Network versus OPDA Signaling

9.3.1 Instigation of Jasmonic Acid Signaling

Perception of stress cues and their further circulation to switch on adaptive responses is a key step toward plant stress tolerance. Various abiotic and biotic stress signals are first perceived by plants, which in their course activate JA signaling, starting with the initiation of JA biosynthesis. Abiotic stress is perceived in plants by certain primary sensory mechanisms which translate the physical and chemical environment, such as water availability, ion concentration, and temperature, into a biological signal to further initiate many downstream cellular responses (Lamers et al., 2020). In cases of biotic stress or mechanical wounding, the prosystemin is hydrolyzed to systemin, which is transported to other cells via the apoplastic route and combines with cell surface receptors (Li et al., 2003). As a result of certain traumatic stress signals during abiotic or biotic stresses, biosynthesis of phytohormones is initiated to generate active defense responses. After effective biosynthesis of JA, the signal is further transmitted via long (vascular bundle or air transmission) or short (adjacent cells) distance transmission to regulate environmental adaptation of plants (Ruan et al., 2019).

9.3.2 JA Signal Perception and Induction of Response

This pathway transmogrifies plant responses regulated by JAs from transcription to translation (Wasternack and Hause, 2002; Santino et al., 2013; Wasternack and Hause, 2013). JA-lle is rapidly synthesized and perceived by the *COI1F-box* protein in response to developmental or environmental signals, which further recruits *JAZ* repressors for its ubiquitination and degradation through *26S proteasome* and thus modulating expression of the corresponding target genes, which regulate JA-induced plant development and defense responses (Figure 9.2).

Coronatine-insensitive1 (*Coi1*) was a pioneer JA deficient mutant observed in *Arabidopsis* by Kazan and Manners (2008). Analysis of *Arabidopsis coronatine insensitive-1* mutant, which is insensitive to JA enormously added to our understanding of JA signaling. Various JA-dependent functions, for example, pest resistance, pathogen resistance, wound response; secondary metabolite biosynthesis, and fertility are defective in *Coi1* mutants. Protein, having an *F-box* motif and 16 leucine-rich repeats, is coded by the *Coi1* gene (Browse, 2009), which forms multiprotein structures that act as receptors (Avandia et al., 2010). Coi1 is also known to participate

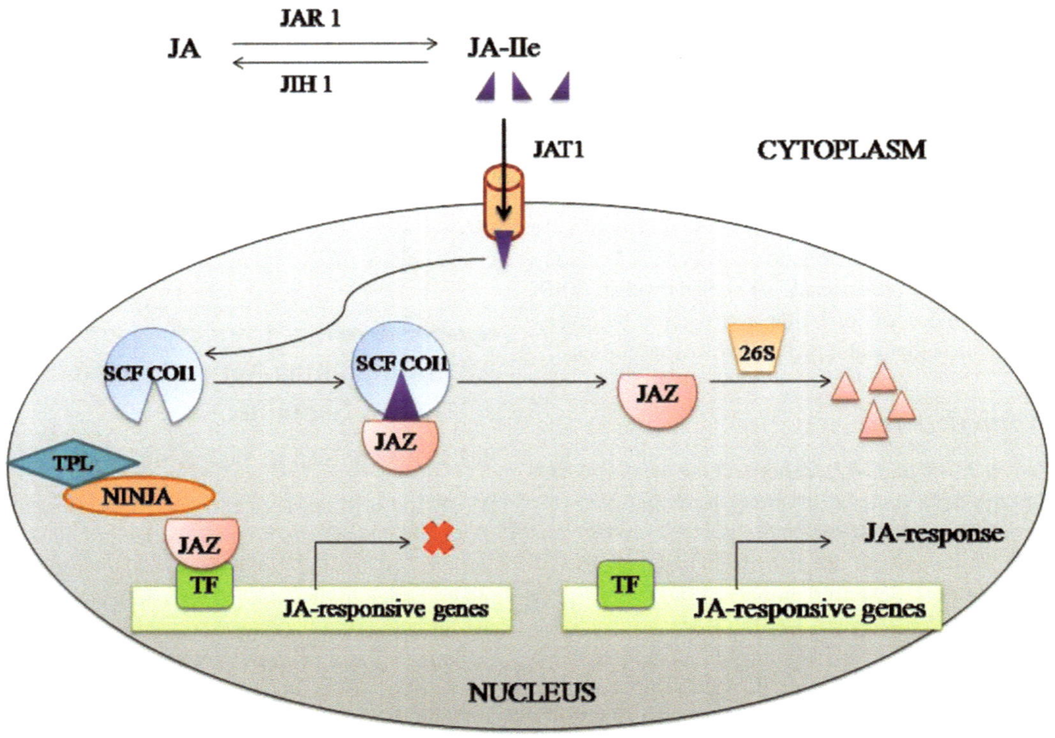

FIGURE 9.2 Model of perception and signal transduction of jasmonates under abiotic stress.

in removing repressors of JA transduction (Xie et al., 1998) mediated by *SCF* complexes found in all eukaryotes (Jiang et al., 2013; Wasternack and Hause, 2013). A proteinaceous *SCF* complex composed of *SKP1*, *F-box*, and *Cullin* protein is reported to regulate the cell cycle by activating ubiquitination and degradation of cell cycle proteins in *26S proteasome*-dependent degradation (Kazan and Manners, 2008). The dynamic equilibrium between SCF^{COI1} mediated stabilization and *26S proteasome* mediated abasement of *COI1* maintains this protein at a suitable level, which is essential for exercising its biological function (Yan et al., 2013).

Examination of *jar1* and *jai1* mutants of *Arabidopsis* revealed that *JASMONATE INSENSITIVE1* (*JA1*) and *JASMONATE RESISTANT1* (*JAR1*) quash the effectiveness of externally applied JAs. Mutants of *Jar1* are highly sensitive to pathogens (Rosahl and Feussners, 2005). JA amino acid synthetase encoded by JAR1 is known to facilitate JA links to isoleucine, thus presenting its role as a signaling molecule by suppressing root growth (Staswick, 2008; Wasternack and Kombrink, 2010).

JASMONATE INSENSITIVE1 (*Jai1*) resistant to JA is a sterile mutant of tomato (Li et al., 2004). As *Jai1* showed 68% amino acid identity with *coi1* (Li et al., 2004), it is often considered a tomato homolog of *coi1*. *Jai1* mutant plants exhibited loss of seed maturation, appearance of glandular trichomes on leaves, unripened fruits, and sepals stipulating the JA-signaling pathway in glandular trichomes of tomato toward its defense responses (Li et al., 2004).

Despite all these reports, there is no information available regarding the repressor that can provide evidence about the relationship between the entire transcriptome and the SCF^{Coi1} complex regulated by JA (Browse, 2009). After treatment of *Arabidopsis opr3* mutant with exogenous JA, upregulation of 31 genes over the control was found (Thines et al., 2007). Hence, it was proposed that interaction of *Coi1* and *JAZ* proteins is enabled by JAs and consequently they aid *JAZ* proteins to cause degradation by ubiquitination through *26S proteasome* (Koo et al., 2011). Chini et al. (2007) identified the *JASMONATE INSENSITIVE3* (*JAI3*) protein, a member of the JAZ protein family which is recognized as a primary repressor of gene expression in JA signaling. *JAI3*and other *JAZ* proteins are reportedtointeractwith*Coi1*. The proteasome *26S* activity is degraded by JA treatment due to the activity of *Coi1* (Chung et al., 2010). Robson et al. (2010) described that JA signaling in wounding and shade is interconnected with *JAZ1* stability in *Arabidopsis*. Therefore, it can be advocated that *JAI3*, *JAZ3*, and *JAZ1* are repressors of JA signaling. Additionally, for activation of JA responsive genes, SCF^{Coi1} dependent ubiquitination is required (Chini et al., 2007; Kazan and Manners, 2008).

9.3.3 JA Signaling versus *OPDA* Signaling

A manifest exception was noted in the binding assays of JA compounds when the fundamental concept of JA/JA-lle perception was first established in 2007, where JA precursor *OPDA* was not an active ligand in *COI1-JAZ* degradation assays (Thines et al., 2007). Sheard et al. (2010) studied the crystal structure of *COI1-JAZ* coreceptor complex and formulated that *OPDA* does not fit in the binding pocket of JA-lle. Accordingly, there were an increasing number of examples elucidating the JA/JA-lle independent roles of *OPDA* (Wasternack et al., 2013). Many evident studies reported a major role of OPDA in tendril coiling (Blechert et al., 1999), hypocotyl growth inhibition (Bru¨x et al., 2008), and COI1-independent defense signaling via ARF2 (Stotz et al., 2011), suggesting at least a role of JA in these activities. Thus, *OPDA* signaling tends to differ from that of JA. However, not much is known regarding *OPDA* perception during OPDA-specific responses (Farmer and Davoine, 2007). Lately, individual signaling properties of *OPDA-Ile* have been reported, but further studies are required for the analysis of an SCF^{COI1}-JAZ co-receptor-independent route of signaling (Wasternack and Hause, 2016).

9.4 JA Signaling Network Amid Abiotic Stress

Plant growth, development, and survival rely on certain complex biological networks coupled with catabolic and anabolic pathways, which are disrupted by abiotic stresses. As a result of abiotic stress regulated uncoupling of functional metabolic pathways, toxic by-products and ROS are generated which might affect cell longevity (Demine et al., 2014). Jasmonates have a well-documented role in orchestrating genome-wide transcriptional changes in response to abiotic stress signals in plants. During abiotic stress signaling, JA is chiefly involved in physiological and molecular responses. Some of the major roles of JA in assuaging abiotic stress in plants are discussed in the following sections.

9.4.1 Cold Stress/Freezing Stress

Low-temperature stress is a significant limiting factor for the growth and development of plants, which causes ice crystal formation and cellular dehydration. There are two types of low-temperature stress viz. chilling(>0°C) and freezing (<0°C) against which plants have evolved a nexus of complex tolerance mechanisms including expression of hormone-linked genes and accumulation of low temperature-induced stress-related amino acids, proteins, and soluble sugars to counterpoise cell membrane damage (Hincha and Zuther, 2014).

JA signaling plays a prominent role against such stress factors by downstream regulation of certain cold-responsive genes and thereby enhancing cold tolerance (Hu et al., 2017). In a study conducted on *Musa acuminate* under cold storage, the expression of *MYC* transcription factors and several cold-responsive genes such as *CBF1*, *CBF2*, *COR1*, and *RD2* was reported (Zhao et al., 2013). Alleviation of cold stress in guava (González-Aguilar et al., 2004), tomato (Zhang et al., 2012), mango (González-Aguilar et al., 2000), and so on by an increased synthesis of antioxidants and activation of certain defense related compounds like phenols and heat shock proteins (*HSPs*) is a well-reported function of MeJA signaling. Thus, such findings demonstrate that under cold stress conditions, the activation of the *MYC* gene family is attributed to MeJA/JA signaling and amplification in the plant body (Ruan et al., 2019), which ultimately mitigates cold injury by the promotion of defense-related compounds and the antioxidant system.

9.4.2 Drought Stress

Water deficit or drought stress is one of the main reasons for reduced crop yield in many agricultural regions globally. Drought stress increases ion toxicity, decreases turgor pressure, and inhibits photosynthesis. JA signaling pathways are known to alleviate drought stress by enhancing endogenous JA levels and subsequently evolving the ability to tolerate stress conditions.

Immediately after perception of drought stress, a rapid and transient increase in the endogenous content of JA was reported in *Arabidopsis thaliana* (Balbi et al., 2008) and *Citrus* sp. (De Ollas et al., 2013). The exogenous application of MeJA improves drought resistance by activating the signaling pathways attributed to increasing total carbohydrates, total sugars, free amino acids, total proline, and protein contents, as well as elevating the activities of certain enzymatic (*CAT, POD, SOD*) and non-enzymatic (proline and soluble sugar) antioxidant systems (Abdelgawad et al., 2014; Wu et al., 2012). On the other hand, MeJA is also reported to improve drought resistance in some plants by adjusting their metabolism (Fu et al., 2017; Mohamed and Latif, 2017). Apart from this, MeJA also improves water deficit conditions in some plants by regulating their stomatal closure by modulating K^+ channels (Evans, 2003). Therefore, jasmonates could be well documented as crucial players of drought tolerance that retaliate by increasing organic osmoprotectants and antioxidative enzyme activity.

9.4.3 Salt Stress

Salt stress has both osmotic and cytotoxic effects on the growth and development of plants, interfering with their metabolism and leading to membrane disorders and oxidative stress (Kadri et al., 2014). Both endogenous and exogenous JA is well versed to enhance plant salt stress tolerance (Domenico et al., 2019).

Jasmonates were reported to enhance salt tolerance in tomato plant by maintaining homeostasis among reactive oxygen species (ROS; Abouelsaad and Renault, 2018). On the other hand, some studies reported that exogenous treatment of JA reduced salt-induced damage by increasing photosynthetic rates, proline content (Bandurska et al., 2003), and antioxidant enzyme activity (Walia et al., 2007) or by reducing Na^+ accumulation rates in shoots (Khan et al., 2012). Also, exogenous JA was reported to improve Na^+ exclusion in roots by decreasing Na^+ uptake and thereby facilitating salt stress tolerance (Shahzad et al., 2015). Thus, this line of evidence indicates that JA alleviates salt stress by enhancing endogenous hormones and the antioxidant defense system (ADS).

9.4.4 Heavy Metal Stress

Heavy metals impair plant growth and development by mimicking essential mineral nutrients that eventually generate ROS. Many studies revealed that JA signaling pathways are allied with heavy metal toxicity. Exogenous application of MeJA to plants exposed to heavy metal toxicity could ameliorate their damage by enhancing the activities of antioxidants like *SOD, POD, CAT,* and so on (Yan et al., 2015; Yan et al., 2013). Also, MeJA was reported to mitigate the toxicity of boron in sweet wormwood (*Artemisia annua*) by plummeting the amount of lipid-peroxidation and stimulating synthesis of antioxidative enzymes (Aftab et al., 2011). Under heavy metal stress, MeJA minimizes oxidative stress by induction of gene expression encoding antioxidants and secondary metabolites

(Farooq et al., 2016). Thus, JA regulates plant responses toward heavy metal toxicity by regulating their ADS.

9.4.5 Light Stress

Light is a key regulator of JA biosynthesis and signal transduction in plants (Kazan and Manners, 2011). JA biosynthesis induced by UV-B treatment, initiated JA signaling pathways in *Brassica* and *Nicotiana* sp., which further increased plant defense mechanisms (Svyatyna and Riemann, 2012). Chen et al. (2018) reported that exogenous MeJA treatment enhances the interaction between FR-insensitive 219 (FIN219) and the C terminus of cryptochrome 1 (CCT1) under blue light, suggesting that JA signaling and blue light signaling pathways play a vital role in stress responses. Furthermore, JA pre-treatment was reported to reduce the adverse effects of UV-B light on PS-ll function by significantly increasing PS-ll quantum yield, the photosynthetic electron transport rate, and the capture efficiency of excitation energy in the reaction center by decreasing the non-photochemical quenching in wheat seedlings (Liu et al., 2012). Thus, exogenous JA could effectively counteract the negative effects of UV-B stress in plants.

9.5 JA Signaling Network to Regulate Biotic Stress

Due to their sedentary habitat, plants struggle with numerous biotic stresses caused by different living organisms like bacteria, fungi, viruses, nematodes, insects, and so on, which tend to cause various diseases and infections in plant tissues and ultimately reduce crop productivity. The class of lipid-derived phytohormones known as jasmonates act as major plant immunity hormones that promote plant defense against mechanical wounding, chewing insects, and various necrotrophic pathogens.

JA and JA-lle are rapidly induced in response to various biotic stress factors (Chauvin et al., 2013; Fragoso et al., 2014), which subsequently induce almost all major secondary metabolites and protein expression involved in defense responses, including terpenoids, alkaloids, amino acid derivatives, anti-nutritional proteins, and certain pathogen-related proteins (De Geyter et al., 2012; De Vleesschauer et al., 2013).

Due to tissue injury, two types of responses are shown by plants viz. local and systemic response, and JA signaling plays a significant role in these responses. The local response during tissue damage activates various damage-associated plant defense signals and attacker-derived signals, which are either chemical or physical in nature and are further recognized by PRR receptors of the cell membrane. The recognition step is followed by activation of *de novo* synthesis of JA/JA-lle, which further activates *SCF^{COI1}/26 proteasome* complex to degrade *JAZ* repressor proteins and expresses transcription factors (TFs) involved in defense-related responses (Abraham, 2018). On the other hand, systemic responses are controlled by two discrete pathways involving JA, viz. cell-nonautonomous pathway (slow pathway) and cell autonomous pathway (fast pathway). JA is produced upon leaf damage in the cell-nonautonomous pathway, which is further translocated to undamaged leaves where it triggers JA responses in target cells. Whereas in the case of the cell autonomous pathway, production by a wound-induced mobile signal other than JA activates synthesis of JA/JA-lle and their subsequent defense-related responses in distal tissues (Abraham, 2018).

9.5.1 JA Signaling during Plant–Insect Interactions

JA signaling in response to insect attack is known to be very specific in nature. This is supported by the finding that when *Arabidopsis* sp. is attacked by a specialist insect herbivore *Pieris rapae*, only one branch of the downstream JA signaling is activated, which is regulated by *MYC2* transcription factor, inducing the expression of the *VEGETATIVE STORAGE PROTEIN 2 (VSP2)* gene (Verhage et al., 2010). Mechanical wounding or damage instigated by herbivore feeding triggers a rapid accumulation of JA at the wounding site (Glauser et al., 2008). Also, a steady surge of JA was noted in the entire plant body as a result of successive leaf feeding by herbivores (Reymond et al., 2004).

In response to insect injury, the first step of retaliation against the stress factors includes upregulation of various biosynthesis (*LOX2*, *LOX4*, *LOX3*, *OPR3*, *AOC3*, etc.) and signaling genes (*MYC2*, *JAZ1*, *JAZ2*, etc.) of JA (Chung et al., 2008). Furthermore, insect feeding or wounding induces a large number of JA-dependent genes involved in pathogenesis, glucosinolate metabolism, and detoxification (Reymond et al., 2004), thereby activating synthesis of certain anti-insect substances like proteinase inhibitors (PIs; Farmer et al.,1992), artemisinin (De Geyter et al., 2012), and some poisonous secondary metabolites like glucosinates and camalexin. Apart from JA, its precursor *OPDA* is also known to actively participate against plant insect attacks (Stintzi et al., 2001). Also, JA signaling under insect attack often operates in combination with complex networks of other phytohormones like SA, ABA, and ET. Durrant and Dong (2004) reported that during plant–pathogen interactions, JA signaling interacts both synergistically and antagonistically with that of salicylic acid.

Apart from numerous defense related JA signaling pathways, a "memory regulated" JA-mediated stress response has also been studied in plant–pathogen interactions, where a "memory" of the stress responds more efficiently after a second stress experience (Conrath et al., 2006; Pastor et al., 2012). This process is known as "priming," a horizontal multicomponent phenomenon that involves many different signaling pathways involving activities of ROS and mitogen-activated protein kinases (*MAPKs*) that activate JA-dependent signals (Frost et al., 2008; Pastor et al., 2012).

9.5.2 JA Signaling during Plant–Pathogen Interactions

JA is regarded as an essential phytohormone against a wide range of pathogens, either alone or in combination with other plant hormones like ethylene, salicylic acid, auxin, and so on, where it may interact synergistically or antagonistically with these hormones during plant–pathogen interactions (Adie et al., 2007). For instance, the interaction of rice–*Xanthomonas oryzae*, is known to activate *MITOGEN-ACTIVATED PROTEIN*

KINASE 6 (MPK6), which results in accumulation of both JA and SA that further induces expression of JA and SA responsive genes (Santino et al., 2013). In *Arabidopsis* sp., it was reported that both SA- and JA-mediated signaling was activated to initiate systemic acquired resistance against pathogens. But a systemic induction of genes involved in JA biosynthesis was detected prior to SA induction (Truman et al., 2007). Thus, JA actively participates in long-distance information transmission.

For disease resistance against oomycetes and fungal pathogens, COI1 is acknowledged to have a central role in JA signaling (Glazebrook, 2005; Adie et al.,2007). However, in many other events of biotic interactions, it may also depend upon specific plant–pathogen interactions. For instance, *COI1* promotes susceptibility toward hemibiotrophic root fungus *Fusarium oxysporum* (Thatcher et al., 2009) and the soil-borne bacteria *Ralstonia solanacearum* (Hernández-Blanco et al., 2007) but, on the contrary, the *coi1* mutant shows resistance toward these pathogens and fails to develop symptoms of chlorotic lesions caused by bacterial effector coronatine (COR). The bacterial phytotoxin *CORONATINE* (*COR*) is known to act as a structural mimic of JA-Ile (Fonseca et al., 2009), which interferes with JA signaling, thereby suppressing the basal defense mechanism against pathogen invasions.

Activation of transcription factors acts as a key step in mediating immune response in plants, which controls regulation of specific defense-related genes. *MYC2* is a major transcription factor in JA signaling (Lorenzo et al., 2004), and is known to act as a key factor in ABA/JA signaling and thus regulates different defense-responsive genes in both the signaling pathways against pathogen interaction.

Thus, JA signaling is known to play a vital role in defense responses against microbial, necrotrophic, and viral pathogens either individually or in combination with other phytohormones by regulating activity of JA biosynthetic defense-responsive genes.

9.6 Physiological Responses of JA in Stress Conditions

For adequate growth and development of an organism, appropriate signal perception, amalgamation, and transduction are mandatory. In the case of plants, a structurally diverse group of phytohormones are known to regulate cell division, gene expression, transcription levels, and growth (Went and Thiman, 1937). Endogenous integration of phytohormones like auxin, gibberellins, ethylene, abscisic acid, brassinosteroids, and jasmonates are known to regulate plant growth, development, and defense. A large number of studies on lipid-derived stress hormone JAs are available in the literature, which are well acclaimed as regulators of both biotic and abiotic stresses and exhibited their remarkable role in a number of plant developmental events including reproductive development, leaf senescence, and primary root growth.

9.6.1 Seed Germination

JA was described as an inhibitor of seed germination. However, lately biochemical and genetic evidence has shown that *OPDA* is an inhibitory compound that acts together with ABA in a *COI1*-independent manner (Dave et al., 2011). JA and MeJA inhibit germination of non-dormant seeds and instigate germination of dormant seeds (Creelman and Mullet, 1997). Application of JA delays ABA-mediated seed germination inhibition in *Arabidopsis*; however, *coi1-16* and *jar1* show increased sensitivity to inhibition of seed germination by ABA (Ellis and Turner, 2002; Staswick et al., 1992).

9.6.2 Regulation of Embryo/Seed Development

In *prosystemin-mediated responses2*, a JA-deficient mutant of tomato that carries a mutation in *FAD7*, embryo development is delayed or arrested owing to increased programmed cell death (Goetz et al., 2012). Also, in the *jai1* mutant of tomato, due to a loss of function of *COI1*, viable seed production is inhibited (Li et al., 2004). Moreover, a viable seed set is noted in tomato mutant *acx1a*, which produces *OPDA* and a residual amount of JA. But in *SiOPR3*, a tomato transgenic line with a silenced *OPR3* gene, a similar amount of *OPDA* compared with wild type is known to set only a few viable seeds (Scalschi et al., 2010). Thus, the role of JA and *OPDA* in seed development requires further study.

9.6.3 Fruit/Seed Ripening

The *Jar1* mutant of tomato, which is incapable of JA perception, bears much smaller fruits as compared with wild-type plants. Also, the young seeds of mutant fruits are known to suffer from a high rate of abortion (>97%), thus indicating that JA perception and signal transduction is a vital step in early fruit development and seed maturation in tomato (Li et al., 2004). Endogenous JA accumulation during early ripening stage in apples (*Malus sylvestris*) and sweet cherries (*Prunus avium*) indicated that JA plays an important role in fruit/seed development (Kondo et al., 2004).

9.6.4 Root Growth Inhibition by JA

Growth inhibition and promotion of senescence were the first physiological responses described under JA (Ueda and Kato, 1980; Dathe et al., 1981). JA needs *COI1* for root growth inhibition, as it is indicated by JA unresponsiveness of the *coi1* mutant. JA-induced root growth inhibition was strongly supported by short root phenotype mutants such as *cev1*, which showed constitutive elevation of JA levels (Ellis et al., 2002). *ETHYLENE INSENSITIVE* (*EIN*)*3* and *EIN3-LIKE1* (*EIL1*) TFs of *Arabidopsis*, in ethylene signaling, interrelate with *JAZ* proteins and positively mediate both JA-induced root hair formation and JA-dependent primary root growth inhibition (Zhu et al., 2011). Thus, root growth inhibition induced by JA needs to be examined in relation to other factors controlling complex processes of root growth (Petricka et al., 2012). Crosstalk with other hormones involved in JA perception and signaling indicated JA involvement in root growth. Taken together, the root growth inhibition by JA seems to occur preferentially via modulation of auxin effects in root growth and development, as biosynthesis of auxin, a key player in root growth, is regulated by JA. JA-mediated gene expression and root growth inhibition were induced by conversion of *OPDA*, which is a primary mobile compound relocating from roots to bioactive hormones. Thus, shoot-to-root translocation of the JA precursor, *OPDA*, is known to coordinate plant growth responses (Schulze et al., 2012).

9.6.5 Lateral Root Formation

Various crosstalks between JA and auxin clearly suggest a role for JA in lateral root formation. It had been shown that lateral root formation is induced by auxin and inhibited by a conjugate of JA and tryptophan (Staswick, 2009). Most of the lateral root mutants of *Arabidopsis* are affected in auxin signaling, homeostasis, and transport, thereby indicating the dormant role of auxin in lateral root formation (Petricka et al., 2012).

9.6.6 Adventitious Root Formation

Adventitious roots are either formed naturally or by some environmental stimuli in aerial organs. It is a complex process regulated by hormones and environmental factors. Auxin acts as a positive regulator that arbitrates by *ARF6* and *ARF8*, which are targets of *miR167* (Gutierrez et al., 2012). Interestingly, there is a negative *COI1* and *MYC2* mediated regulation via altered *JA/JA-Ile* homeostasis in downstream auxin-induced adventitious root formation. Mutants like *coi1-16*, *myc2*, *myc3*, *myc4*, and *jar1*, impaired in JA perception and signaling, form far more adventitious roots than their wild type (Gutierrez et al., 2012). This data provides instances of auxin–JA crosstalk that occurs during adventitious root formation.

9.6.7 JA Regulates Vegetative Growth

JA defense signaling activation during abiotic or biotic stress depletes available resources and severely affects plant growth. It is well established that JAs act as growth inhibitors in the roots and shoots of plants (Staswick et al., 1992). Accumulation of endogenous JA during wound responses strongly suppresses the growth of roots and shoots by inhibition of cell mitosis (Zhang and turner, 2008). The JA signal integrates other phytohormones like GA, IAA, and ET to modulate defense responses and growth processes (Nagpal et al., 2005).

9.6.8 Tuber Formation

Various pathways are involved in tuber formation (Sarkar, 2008). For a long time, JA regulated tuber induction by 12-OH-JA and its glucoside (*TAG*) had been suggested (Wasternack and Hause, 2002). In *Solanum tuberosum*, *LOX-1* is known to be involved in tuber formation and tuber yield (Kolomiets et al., 2001), while its derived metabolites like JA and *TAG* are known to accumulate at low tuber inducing temperatures (Nam et al., 2012). However, the occurrence of 12-OH-JA in various non-tuber bearing plant species questions any specific role of JA in tuber formation (Miersch et al., 2008). Cell expansion in stolons and changes in microtubule orientation could possibly be involved in tuber induction because JA biosynthesis occurs in developing stolons (Cenzano et al., 2007).

9.6.9 JA in Trichome Development

Hair-like appendages or branching structures differentiated from epidermal cells in aerial parts of plants are known as trichomes that protect plants from insects, herbivores, and other abiotic damage (Ishida et al., 2008). The formation of trichomes is initiated by various environmental cues like wounding and insect attack (Yoshida et al., 2009) as well as by different endogenous developmental signals of phytohormones like jasmonates (Yoshida et al., 2009). A JA -ensitive mutant of tomato, *jar1*, is known to be absent in surface trichomes in young fruits and has significantly less on stem and leaf surfaces (Li et al., 2004). Also, the perception mutant *coi1* and biosynthesis mutant *aos* of JA are known to form fewer trichomes than their wild forms, which were shown to be increased by treatment of MeJA in *aos* but not in *coi1*, indicating that the JA signal acts as a positive regulator of trichome development in *Arabidopsis* (Yoshida et al., 2009).

9.6.10 JA Induced Leaf Senescence

Senescence of a leaf is a complex developmental program that relies on nutrients, light/dark conditions, biotic or abiotic stresses, and numerous hormones including JA. Leaf senescence is promoted by JA in a *COI1*-dependent manner (Qi et al., 2015), while *JAZ7* is known to repress JA-regulated dark-induced leaf senescence (Yu et al., 2016). High-resolution transcript profiling of senescing leaves recognized a distinct group of *TFs*, which relates to leaf development, metabolic pathways, and senescence (Breeze et al., 2011). JA-linked *TFs* viz. *WRKY70* and *WRKY54* (Besseau et al., 2012), *WRKY53* (Miao and Zentgraf, 2007), and *ORE1* (Balazadeh et al., 2010) are identified to be active in leaf senescence.

JA-induced chlorophyll breakdown is known to promote leaf senescence, and chlorophyll degradation is shown to be regulated by the interaction of *YABBY1* and *YABBY3* with *JAZs* (Boter et al., 2015). Moreover, a mechanistic explanation regarding the role of JA in senescence was studied by JA-mediated downregulation of *Rubisco activase* in a *COI1*-dependent manner (Shan et al., 2012).

9.6.11 JA in Reproductive Organ Development

Fertility can be restored in JA-biosynthesis impaired mutants by treatment with JA during floral development stages (Mandaokar et al., 2006). Stamen and JA-specific *mRNA* regulating genes involved in metabolic pathways of synthesis of terpenoids, wax, and pollen constituents were detected by transcript profiling of JA-treated stamens of *opr3* plants (Mandaokar et al., 2006). It is well established that *ARF6* and *ARF8* regulate JA biosynthesis in anther filaments (Reeves et al., 2012). In *Arabidopsis*, besides regulating stamen development, JA is also known to regulate petal growth (Brioudes et al., 2009). JA plays a crucial role in sex determination in maize (Acosta et al., 2009). Maternally inherited cytoplasmic male sterile phenotype, which leads to pollen abortion, is known to be associated with JA biosynthesis in rice plants (Liu et al., 2012).

9.7 JA-Mediated Secondary Metabolites

Plants produce a wide range of low molecular weight organic compounds that are generally classified into two large classes: primary and secondary metabolites. Primary metabolites are vital for plant growth and development, whereas secondary metabolites act as defense molecules and protect plants from various adverse conditions. Biosynthesis and accumulation of secondary metabolites are strictly prohibited in a spatial and temporal manner and are altered by a number of biotic and abiotic factors (Verpoorte et al., 2000). In addition, plant secondary metabolites are also stimulated by plant hormones under normal and stress conditions (Zhao et al., 2005). As one of the most representative hormones allied with plant defense responses, MeJA activates plant defense mechanisms in environmental stresses such as chilling, UV light, wounding, and insect attack by generating the particular enzymes that are responsible for the production of defensive chemicals and pathogenesis-related (PR) proteins (Koo et al., 2009). JAs act as ubiquitous and conserved elicitors for the generation of secondary metabolites across the plant kingdom, from gymnosperms to angiosperms. The promising illustration of transcriptional regulation of secondary metabolism in response to JAs suggests that TFs form a dynamic regulatory network to fine-tune the timing, amplitude, and tissue-specific expression of gene pathways and subsequent accretion of secondary metabolites (Vom et al., 2002). Some of the major secondary metabolites that are transcriptionally regulated by jasmonates are illustrated in the following sections.

9.7.1 Terpenoid Indole Alkaloids

Catharanthus (*Vinca*) produces a large array of terpenoid indole alkaloids (TIAs) such as vincolin, vinblastine, or vincristine that have a large array of pharmaceutical applications. The biosynthesis of TIAs has been preferentially studied in the cell suspension cultures of *C. roseus* and characterized into two principal groups containing three pathways: Two of these pathways, the MEP pathway leading to isopentenyl pyrophosphate pursued by the iridoid pathway leading to organic acid. Enzymes required for these two pathways are located in the internal phloem-associated parenchyma cells (Goossens et al., 2017). The third pathway is the shikimate pathway that is located in the epidermal cells and aids in the formation of tryptamine, converge with iridoid branch during the formation of strictosidine, and is converted to catharanthine and vinblastine/vincristine. During the formation of vinblastine, the JA-responsive TFs were first identified in *C. roseus* and these TFs were called *OCTADECANOID RESPONSIVE CATHARANTHUS ALKALOID 2* and *3* (*ORCA2* and *ORCA3*). These are characterized as members of the *ERF* subfamily within the AP2/ERF TF superfamily (Memelink et al., 2001). Interaction of *ORCA2* and *ORCA3* was shown on a specific sequence including GCC boxes of target promoters of genes that encode all enzymes of the TIA pathway (Paul et al., 2017). Meanwhile, an *ORCA* gene cluster, *CrORCA*,3,4,5 plays an important role based on the fact that all the branches of TIA pathways are under the control of *CrORCA3*, and overexpression of this did not show any elevated level of TIA synthesis, suggesting the involvement of other TFs (Van der et al., 2000). The *MYC2*-regulated *CrORCA* gene cluster *CrORCA*,3,4,5 shows the functional overlap with *CrORCA3* (Paul et al., 2017). Both *CrORCAs* and *CrMYC2* are coregulated and activate different branches of TIA pathways such as *TRYPTOPHAN DECARBOXYLASE* (*TDC*), some steps of the iridoid pathway, and the strictosidine synthase (*STR*) gene (Paul et al., 2017). Finally, biosynthesis of TIA is regulated by two JA-inducible bHLH TFs of the sub-clade IVa, *bHLH IRIDOID SYNTHESIS1* and 2 (BIS 1, BIS2; Van Moerkercke et al., 2016). Overexpression of *CrBIS1* or *CrBIS2* leads to amplified expression of iridoid pathway genes and previous MEP pathway genes as well as improved accumulation of TIAs (Van Moerkercke et al., 2016). At last, *ORCA3* regulates most but not all the steps of TIA synthesis (Suttipanta et al., 2011) and negative regulation of *CrMYC2* by *JAZ* was observed in several JA-dependent pathways that remain to be elucidated.

9.7.2 Nicotine

Nicotine belongs to the Solanaceae family and is one of the most imperative research models in plants. The most important secondary metabolites present in *Nicotiana* species are tobacco alkaloids (Wang et al., 2015). Various structural genes encoding enzymes are involved in the nicotine biosynthetic pathway and these genes are transcriptionally regulated by JA (Shoji et al., 2008). In response to JA, JA receptor *NtCOI1* causes degradation of Nt*JAZs* protein (a transcriptional repressor) and induces the expression of *putrescine N-methyltransferase* (*PMT*) that is involved in the formation of the pyrrolidine ring in *N. tabacum* (Shoji et al., 2010). JA-inducible *ORC1* also known as *ERF221* and *ERF189* TFs come together in the NIC2 locus and are upregulated by JA elicitation. Both ORC1 and *ERF189* have overlapping but non-redundant roles in regulating nicotine biosynthetic genes. *ORC1* is a homolog of *C. roseus ORCA3*, and its overexpression stimulates alkaloid biosynthesis (De Boer et al., 2011). In the case of *N. benthamiana*, two bHLH TFs such as bHLH1 and bHLH2, which are homologs of *MYC2*, were identified to be active in nicotine biosynthesis by binding to the PMT promoter (Todd et al., 2010). In addition, JA-induced phosphorylation cascade has also been revealed to play an important role in nicotine biosynthesis such as activity of both *ORC1* and *MYC*-type bHLH proteins that can be post-translationally upgraded by a JA-modulated phosphorylation cascade. Conversely, in tobacco, the JA-inducible *MYBJS1* (an *R2R3-MYB* TF) was shown to stimulate phenylpropanoid biosynthetic genes that accumulate polyamine conjugates in stress conditions (Galis et al., 2006). The extensively distributed stimulation of biosynthesis of secondary metabolites by JA may signify an evolutionary advantage of a recognized regulatory module.

9.7.3 Artemisinin

Artemisinin is a sesquiterpene lactone endoperoxide that accumulates at a high level in the glandular trichomes of *Artemisia annua*. Its biosynthesis takes place by chloroplast MEP pathway and cytosolic MVA pathway through farnesyl pyrophosphate and amorphadiene, which generates artemisinic and dihydroartemisinic acid. During the formation of dihydroartemisinic acid, its encoding genes are activated by the JA-dependent pathway by using *AaMYC2* and *AaGSW1* (*GLANDULAR TRICHOME-SPECIFIC WRKY1*), and finally, dihydroartemisinic acid is photo-oxidatively converted into artemisinin (Geyter et al., 2012). Jasmonate-responsive two *AP2/ERF* TFs like *AaERF1* and *AaERF2* regulate the transcription of ADS and CYP sesquiterpene oxidase (*CYP71AV1*) and activate the expression of both genes. *AaERF1* or *AaERF2* overexpression led to improved accumulation of artemisinin and artemisinic acids (Yu et al., 2012). In *A. annua*, JA-inducive *WRKY1* transactivates the promoter of the *ADS* gene that is involved in the regulation of artemisinin biosynthesis (Ma et al., 2009). In some cases, *AP2/ERF* TF-*AaORA1* regulates the same enzymes as *AaWRKY1*— those directing the pathway toward artemisinin formation. AabHLH1 (bHLH TF from *A. annua*) also positively regulate the transcription of ADS and CYP71AV1 by binding to the E boxes promoter and transient the expression of AabHLH1 TF in *A. annua* leaves. The prior study indicates that in *C. roseus*, JA-responsive expression of alkaloid biosynthesis genes is controlled by TF cascades consisting of *CrMYC2* and bHLH protein that regulates the gene expression *AP2/ERF*-domain transcription factor *ORCA* (*ORCA2* and *ORCA3*; Li et al., 2013).

9.7.4 Taxol

Taxol is a complex diterpene that belongs to the genus *Taxus* and is one of the most promising anticancer agents (Wilson et al., 2014). Taxus is present in small quantities in plants (0.01%–0.03% of the dry weight of the bark of *Taxus*) so the production of Taxol remains a challenging problem due to the limited resources of *Taxus* sp. (Lenka et al., 2015). Similar to other secondary metabolites, JAs significantly stimulate the production of Taxol and related taxanes in *Taxus* sp. (Cusido et al., 2014). For the biosynthesis of Taxus, two regulatory steps exist: (i) formation of a taxane ring which is upregulated by Jas; (ii) acylation at C-13 position. Li et al. (2013) identified a *WRKY* TF, *TcWRKY1* from *Taxus chinensis*, which acts as a regulator of *10-deacetylbaccatin III-10 β-o-acetyl transferase* (*DBAT*) which is a rate limiting step in the biosynthesis of Taxol (Walker et al., 2000). In *T. chinensis* suspension cells, the gene expression of *TcWRKY1* was exclusively induced by JAs, which further stimulates the expression of DBAT but the transcript levels of DBAT are reduced by *RNA interference (RNAi)*. It means that *TcWRKY1* participates in Taxol biosynthesis regulation and *DBAT* is a target gene of this *WRKY* TF. Recently, Lenka et al. (2015) identified three JA-inducible MYC TFs, *TcJAMYC1*, *TcJAMYC2*, and *TcJAMYC4*, in *Taxus cuspidata*. These *MYC* TFs show the similarity in sequence and expression with *Arabidopsis MYC2* and these similarities suggest a conserved response to JAs despite important divergence between the gymnosperm and angiosperm lineages.

9.7.5 Ginsenoside

GS is present in species of order Brassicales, such as broccoli, cabbage, horseradish, and mustard. The main active component of ginseng is ginsenoside and triterpenoid saponin. To date, more than 40 ginsenosides have been characterized and divided into three categories based on the position of side chains. These three types are protopanaxatriol (PPT), protopanaxadiol (PPD), and oleanolic acid (Yang et al., 2015). Ginsenosides are synthesized via the mevalonate pathway by cyclization of 2,3-oxidosqualene. A large number of genes encoding enzymes in glucosinolate biosynthesis are JA-inducible, such as *anthranilate synthase* (*ASA1*), *C-S-lyase SUR1*, *cytochrome P450 enzymes CYP79F1/2*, *CYP79B2/3*, *sulfotransferases SOT16/17/18*, and *glucosyl transferase UGT74B1* unfolding JA-regulation of GS biosynthesis via *MYC2*, *MYC3*, and *MYC4*. Thus, *MYB*-type TFs are the targets of JA-Ile-mediated *COI1/MYC2/JAZ* signaling modules that are active in glucosinolate biosynthesis. These *MYB*-type TFs are *MYB28* and *MYB29* which activate *CYP79F1/2*, *MAM1-3*, *SOT17/18*, and *C-S lyase* (*SUR1*), while *MYB51*, *MYB34*, and *MYB122* activate *CYP79B2/3*, *TSB1*, *UGT74B1*, *CYP83B1*, and *SOT16* (Schweizer et al., 2013). Leading to the formation of methionine and tryptophan within the chloroplast, various oxidative reactions of glucosinolate core pathway downstream the aldoxime in the ER–cytosol interface and subsequent transport within the phloem or sequestration in the vacuole (Grubb et al., 2006). Among the final products, the indole glucosinolates (IG) and aliphatic glucosinolates (AG) are phytoalexins, brassinin, cyclobrassinin, and camalexin (Kim et al., 2017).

9.7.6 Anthocyanin

Anthocyanins are accumulated in flowers and are involved in the attraction of pollinators. Anthocyanins act as an antioxidant defense system against reactive oxygen species in leaves. Any stress conditions induce the formation of JA, which leads to anthocyanin accumulation in epidermal cells of the leaves and shoots (Chen et al., 2006). Anthocyanin biosynthesis is the general phenylpropanoid pathway via the shikimate acid pathway and leads to the formation of flavonons such as C-glycosylated flavonons and naringenin O-glycosylated flavonons (Jiang et al., 2016). Most of the steps in anthocyanin formation are JA inducible. *Phenyl alanine ammonium lyase (PAL)* is one of the first detected JA-inducible enzymes during secondary compound formation and is most sensitive to any type of stress.

Biosynthesis of anthocyanin is mainly regulated by WD-repeat/bHLH/*MYB* transcriptional complexes (Goossens et al., 2017). Members of such complexes mainly include group III, subgroup f bHLH TFs, *GLABRA3* (*GL3*), *ENHANCER OF GLABRA3* (*EGL3*), *TRANSPARENT TESTA8* (*TT8*), *WD-repeat TF TRANSPARENT TESTA GLABRA1* (*TTG1*), and *MYB TF GLABRA1* (*GL1*). All TFs act as a positive regulator in different steps of anthocyanin biosynthesis and are targets of JAZ proteins (Qi et al., 2011). Generally, the interaction of JAZ protein with TFs takes place at the N terminal of the JAZ-interacting domain (JID) and is shown for two subgroups, bHLHd and bHLHe. Although, in the case of subgroup bHLHf, interaction takes place at the C-terminal part together with the bHLH domain despite the presence of the JID domain (Qi et al., 2011). Members of the former, such as *MYC2* act as a positive regulator via interaction with *MED25* (Chen et al., 2012), whereas other members act as a negative regulator, for example, *JA ASSOCIATED MYC2-LIKE* (*JAM*) TFs. *JAM1*, *JAM2*, and *JAM3* take the JID domain for the interaction with *JAZ*. Both JAMs and MYCs are competitors to cis-regulator element binding, which leads to a balance between the activation and suppression of anthocyanin biosynthesis and JA-dependent processes (Qi et al., 2015).

Another proof of a direct link between anthocyanin and JA formation has been obtained with the thylakoid formation mutant thf1 (Gan et al., 2014). Mutant thf1 showed elevated levels of *α-LeA*, *OPDA*, and JA, *COI1*-dependent elevation of anthocyanin biosynthetic and regulatory genes, and upregulation in anthocyanin accumulation.

9.8 Crosstalk of JA with Other Phytohormones to Mitigate Plant Stress Conditions

Interaction between plant hormones is the core of plant stress responses (He et al., 2017). JA does not have a self-regulatory role but acts within a complex signaling network with other plant hormone signaling pathways like those of ET, ABA, GA, SA, BR, and IAA (Aleman et al., 2016). Conspicuously, hormonal crosstalk involving both positive and negative feedback can influence synthesis, transport, and signaling of other phytochromes (Hu et al., 2017). The following discussion looks at current studies on the mechanisms by which JAs and other plant hormones respond to both biotic and abiotic stress.

9.8.1 JA–Auxin Crosstalk

Interaction between JA and auxin coordinately regulates plant development and physiology processes such as tendril coiling, cell elongation, and the production of secondary metabolites (Saniewski et al., 2002). COI1, MYC2, and JAZ are the main core components that participate in the crosstalk of JA and auxin signaling pathways. Auxin/indole-3-acetic acid can induce the expression of JA signaling repressor JAZ1/TIFY10A (Grunewald et al., 2009). JA also inhibits auxin stimulated apical growth of roots; JA-treated wild-type plants have much shorter roots as compared with untreated wild-type plants. Even mutant plants having defects in JA signaling form roots with similar lengths to the roots of wild-type plants that are treated with JA (Jang et al., 2017). By contrast, auxin is imperative for root growth but due to auxin deficiency or signaling mutants, as (*trp2-12*) and *auxin resistant 3* (*arx3-1*), develops very short roots in comparison with wild-type plants (Zhang et al., 2019). It means that JA-stimulate inhibition of root growth by suppressing the expression of auxin-responsive transcription factors *PLETHORA*s (*PLTs*) that are responsible for the maintenance of stem cell niche and cell proliferation (Mahonen et al., 2014). Notably, *PLT* expression levels did not suppress in JA-signaling mutants such as *coi1-1* and *myc2* but COI1-dependent JA signaling intervenes in the JA-induced root phenotype and MYC2 transcription factor suppresses the expression of *PLTs* by directly binding to the promoter region (Chen et al., 2011).

In addition, the interaction between JA and auxin regulates the development of floral organs. In the auxin signaling pathway, ARF6/ARF8 are key regulators of petal and stamen growth through modulation in the endogenous level of JA, and MYB transcription factors, such as MYB21/MYB24, downstream the JAZ in JA signaling pathways also synchronal regulate petal and stamen growth (Reeves et al., 2012). JA and auxin also regulate leaf senescence. JAZ7 suppresses dark-stimulated leaf senescence, while MYCs such as MYC2, promote senescence by stimulating the expression of chlorophyll degradation-related genes and senescence-associated genes, which indicate that JA activates leaf senescence through the COI1-dependent JA signaling pathway (Qi et al., 2015). In JA-dependent leaf senescence, the repressors of JA signaling like JAZ4, JAZ8, and WRKY57 also function as negative regulators while auxin signaling repressor IAA29 functions as a positive regulator. More imperatively, the interaction between WRKY57–JAZ4/8 and WRKY57–IAA29 mediates JA-dependent leaf senescence, suggesting an antagonistic interaction between JA and auxin is involved in leaf senescence (Jiang et al., 2014).

9.8.2 JA-ABA Interaction

JA and ABA signaling pathways coordinately induce plant physiological responses to mitigate the effects of various stress factors (Gomez-Cadenas et al., 2015). The findings of numerous studies suggest that ABA and JA signal transduction pathways show both synergistic as well as antagonistic regulation characteristics (Lackman et al., 2011). Jasmonic acid was initially described as an activator of the ABA signaling pathway (Abe et al., 2003). In JA signaling, the involvement of ABA receptor PYRABACTIN RESISTANCE1-Like proteins (PYL4) was found to regulate metabolic reprogramming in tobacco and *Arabidopsis* (Lackman et al., 2011). ABA receptor PYLs form a complex with JAZ (jasmonate ZIM-domain protein) and activate the transcriptional activity of the basic helix–loop–helix (bHLH) protein transcription factor, MYC2. MYC2 acts as a major regulator of the core JA signaling mechanism and participates in the ABA signaling pathway in response to drought stress (Liu et al., 2014). Besides, MYC2 transcription factor regulates a key gene of JA biosynthesis, *AOC1*, during signal intersections of ABA/JA signaling pathways (Kazan et al., 2013). On the other hand, MYC2 transcription factor also activates the expression of the JA responsive gene *VSP2* under the mediation of MED25 to resist herbivorous insect feeding and inhibits the expressions of *PTL1* and *PTL2* as well as root growth (Pauwels et al., 2015).

Additionally, JA interacts with ABA and polyamines to tolerate chilling stress in apple (Yoshikawa et al., 2007). Several key genes such as *LEA*, *CBF*, and *DREB*, have been shown to respond under low-temperature stress in JA and ABA signal transduction pathways (Hu et al., 2013). Li et al. (2018) observed that low-temperature stress increases the concentration of JA and ABA, while also inducing the upregulation of *ZjLEA*, *ZjCBF*, and *ZjDREB1* in *Z. japonica* leaves.

Under salinity, drought, and cold stress, MeJA and ABA treatment increased the transcript levels of OsbHLH148, which helps in the upregulation of *OsDREB* and *OsJAZ* genes (Seo et al., 2011). When *A. thaliana* is under water stress condition, octadecanoid-responsive AP2/ERF-(APETALA2/ethylene responsive factor) domain transcription factor 47 (*ORA47*) acts as a gene target in JA and ABA biosynthesis (Chen et al., 2016). In ABA-independent as well as both JA-insensitive (*jai1*) and JA-deficient (*jar1-1*), *A. thaliana* lines accumulate proline under water stress. Many studies have reported that JA and ABA are also involved in salt stress responses in plants. Under salt stress, JA treatment further increased ABA concentration (Seo et al., 2005). Furthermore, Kim et al. (2009) reported that the application of JA after NaCl treatment, rather than before, could enhance endogenous ABA concentrations. According to Wang et al. (2001), during plant growth, the concentrations of JA and ABA normally increased while SA and IAA concentrations declined under salt stress. Brossa et al. (2011) observed that during the regulation of antioxidant status, the interaction between JA and ABA signaling is responsible for acclimating osmotic stress in plants. This indicates that both JA and ABA signaling coordinate to regulate the responses produced under stress conditions.

Besides this, crosstalk between ABA and JA might be possible through other signaling proteins such as MAP kinase MPK6 (Takahashi et al., 2007; Xing et al., 2008) and AUXIN RESISTANT 1, a subunit of RUB1-activating enzyme that regulates the protein degradation activity of Skp1–Cullin–F-box complexes (Tiryaki and Staswick, 2002; Tiryaki, 2007). Previously, an antagonistic interaction of ABA with JA signaling pathways were noted in the jasmonic acid insensitive4 (jin4) and jasmonic acid resistant1 (jar1) mutants showing hypersensitivity to ABA inhibition of germination (Staswick et al., 2004). In *Oryza sativa*, the ABA and JA-mediated antagonistic regulation of the expression of salt stress-inducible transcripts were also noted (Moons et al., 1997). Thus, crosstalk between JA and ABA signaling helps to coordinate the balance between plant growth and defense resistance.

9.8.3 JA–Cytokinin Interaction

Cytokinin affects various aspects of plant growth and development by regulating stem cell identity and cell proliferation (Kieber et al., 2018). The genes involved in cytokinin responses are largely affected by JA or JA-dependent stress responses (Argueso et al., 2009). Moreover, cytokinin transgenic and deficient mutant plants showed increased tolerance to stresses under JA responses (Qiu et al., 2014). These findings suggested that cytokinin response is essential to JA-dependent stress responses and growth modulation. In *Arabidopsis*, three histidine kinases such as AHK2, AHK3, and AHK4/WOODEN LEG function as cytokinin receptors, and these receptors are affected by JA or environmental stresses (Xie et al., 2018). JA maintains an optimum level of cytokinin concentrations that are required for plant growth by modulating gene expression as well as the activity of cytokinin oxidase. Avalbaev et al. (2016) observed that exogenous application of MeJA regulates cytokinin content by modulating cytokinin oxidase (cytokinin oxidase/dehydrogenase) activity in wheat seedlings under salinity stress.

Previous findings have proposed that JA also antagonistically interacts with cytokinin in several aspects of plant development. JA reduces soybean (*Glycine max*) callus growth induced by cytokinin (Ueda et al., 1982), and also abolishes the effect of cytokinin on chlorophyll development by regulating the expression of genes involved in chlorophyll development (Liu et al., 2016). In *Arabidopsis* roots, an antagonistic interaction between JA and cytokinin is fundamentally important to regulate xylem differentiation and development (Jang et al., 2017). Cytokinin maintains the stem cell identity and function as a negative regulator of xylem differentiation was confirmed by showing that JA reduces the expression of the cytokinin-responsive *PIN-FORMED* 7 (*PIN7*) gene, which is responsible for xylem development, whereas in *wooden leg* mutants, having defects in cytokinin signaling strongly showed all-xylem phenotypes and a lack pro-cambial cells in their roots (Jang et al., 2017). Additionally, in transgenic plants overexpressing *AHP6* or mutants that lack transcription of Type-B *ARRs*, such as *ARR1*, *ARR10*, and *ARR12*, a negative regulator of cytokinin signaling forms extra xylem (Yokoyama et al., 2007). Similarly, JA-deficient *OPDA reductase 3* (*opr3*) mutants or wild-type plants treated with exogenously JA showed an extra xylem phenotype but JA signaling mutants, *jasmonate resistant 1* (*jar1*) and *coi1* did not (Jang et al., 2017). Together, these results show that JA represses the procambium-specific cytokinin response and the effect of JA on extra xylem formation is inhibited by cytokinin.

Moreover, the *myc2* mutant did not form extra xylem under exogenous JA response, and also reduced the expression of *AHP6* encoding a cytokinin signaling inhibitor. It suggests that the JA-responsive MYC2 transcription factor intercedes this process by promoting *AHP6* expression. It is likely that an antagonistic interaction between cytokinin and JA is also involved in the regulation of JA-dependent stress responses. Recently, Nitschke et al. (2016) observed that plants with defective cytokinin signaling or reduced cytokinin levels showed a JA-dependent cell death phenotype in response to circadian stress. Liu et al. (2015) observed that exogenous application of JA and cytokinin can antagonistically also regulate leaf senescence-associated gene expression in *O. sativa*. It was also proved that leaf senescence not only depends on the level of JA or cytokinin but also on the balance between JA and cytokinin.

9.8.4 JA–ET Interaction

JAs and ET can antagonize or act coordinately in controlling stress responses or development. JA and ET synergistically stimulate the expression of pathogen-responsive genes, like PLANT DEFENSIN 1.2 (PDF1.2), to support plant tolerance against infections (Zhu et al., 2011). Two APETALA 2/ETHYLENE RESPONSE FACTOR (AP2/ERF) domain transcription factors, OCTADECANOID-RESPONSIVE ARABIDOPSIS AP2/ERF 59 (ORA59) and ETHYLENE RESPONSE FACTOR 1 (ERF1) are required for the generation of PDF1.2 expression by direct association with the GCC box in the PDF1.2 promoter (Zarei et al., 2011). It has been suggested that ORA59 and ERF1 are important molecules for combining JA and ET signaling and controlling the expression of pathogen-responsive genes (Pre et al., 2008).

Zhu et al. (2011) tested whether or not EIN3/EIL1 are the direct links between ET and JA signaling and found that double mutants of ein3 eil1 are insensitive to ET and JA treatment for root hair development and induction of pathogen-responsive gene expression. In addition, it was shown that EIN3/EIL1 is physically interacting with three JAZ members (JAZ1/3/9), so EIN3/EIL1 is another branch of JAZ interacting transcription factors. Chromatin immunoprecipitation (ChIP) assay illustrates that JA or ET treatment elevates EIN3 DNA binding ability.

Next, it was revealed that HISTONE DEACETYLASE 6 (HDA6) interacts with JAZs and EIN3/EIL1 as a co-repressor, even JA treatment diminished HDA6–EIN3 interaction to release its inhibition. Consistent with this result, it was noticed that the histone acetylation levels on the ERF1 promoter are improved after JA treatment (Zhu et al., 2011). On the whole, this work expressed that JA through a derepression mechanism directly activates EIN3/EIL1, while ET activates EIN3/EIL1 through reported post-transcriptional regulation. These two different activation steps are required for fully activating EIN3/EIL1.

Additionally, the mitogen-activated protein kinase (MAPK) cascade has been widely studied in plant hormone biosynthesis and signaling pathways (Schweighofer and Meskiene, 2008). It was observed that JA activates the MKK3–MPK6 cascade to positively regulate MYC2-governed wounding-related gene expression and pathogen-responsive gene expression and negatively regulate root elongation. More amusingly, it has been exposed that either ein2 or ein3 can suppress the JA- or ET-induced higher PDF1.2 expression in the MKK3 overexpression background (Takahashi et al., 2007). It suggests that the MKK3–MPK6 cascade might be playing a positive role in modulating EIN3/EIL1 activities.

Antagonistically, both JA and ET mutually inhibit each other's functions. ET-induced apical hook formation by the induction of HOOKLESS1 (HLS1) expression that encodes a protein similar to N-acetyltransferase (Raz and Koornneef, 2001). It is hypothesized that JA antagonizes ET's positive role in hook development by modulation in HLS1. Song et al. (2014) experimental results show that the hls1 mutant is insensitive to JA treatment because JA downregulates HLS1 mRNA expression even in the presence of ET.Song et al. (2014) focused on MYC2, the most prominent JAZ-interacting transcription factor that regulates various JA responses (Kazan and Manners, 2013). It was observed that MYC2 and its closest homologs, MYC3 and MYC4, function redundantly in the inhibition of ET-induced hook formation. HLS1 expression level in myc234 (referring to the myc2 myc3 myc4 triple mutant) is higher than in the wild type and JA could not repress this expression because MYC2 is a negative regulator of HLS1 expression and JA negatively regulates the expression of HLS1 through MYC2 family proteins.

Zhang et al. (2014) found that MYC2 interacts with EIN3/EIL1 and further reported that MYC2 invalidates EIN3 DNA binding capacity in an *in vitro* electrophoretic mobility shift assay (EMSA). In support of these *in vitro* studies, it was reported that even MYC2 interacts with EIN3, MYC2 could not associate with the EIN3 binding site (EBS) on the HLS1 promoter, which further confirms that the interaction between MYC2 and EIN3 blocks the association between the EIN3 protein and the HLS1 promoter (Zhang et al., 2014). Besides this, the repression of HLS1, MYC2 also represses the expression of ERF1 through inhibition of EIN3/EIL1 (Song et al., 2014). On this basis, these results demonstrate that JA-activated MYC2 interacts with EIN3/EIL1 and represses their DNA binding ability and reduces the expression of EIN3/EIL1 target genes, such as HLS1 and ERF1.

9.8.5 JA-GA Interaction

The crosstalk between JA and GA signaling pathways also synergistically and antagonistically regulate the plant growth and defense response; however, the plant defense response is wielded at the cost of inhibiting growth (Wasternack and Hause, 2013). In the "relief of repression" model, the interaction between JAZ–DELLA debilitates the functions of JAZs and DELLAs as signaling repressors. In the absence of GA, stable DELLA interacts with JAZ to release MYC2, which, in turn, activates the expression of JA-responsive genes. In the presence of GA, JAZs are liberated from the DELLA–JAZ complex by the degradation of DELLAs, and free JAZs vitiate the JA response by inhibiting expression of JA biosynthetic genes such as *DAD1* and *LOX* via direct interaction with MYC2 (Song et al., 2011).

Besides, in the GA pathway, JAZ interacts with DELLA and represses to activate bHLH factor PHYTOCHROME INTERACTING FACTOR 3 (PIF3), although JA signals stimulate JAZs degradation and release DELLAs to repress the PIF3 and reduce GA-enhanced hypocotyl elongation (Yang et al., 2012). On the other hand, DELLAs interact with JAZs and repress to release MYC2, which positively regulates JA-mediated root growth inhibition and susceptibility to Pst DC3000 (Wild et al., 2012). In rice, OsJAZ9 proteins directly interact with DELLA protein SLENDER RICE 1 (SLR1), and this interaction confirmed the antagonistic association between JA and GA by showing that overexpression of *OsJAZ9* promotes the GA response while knockout of *OsJAZ9* reduces the GA response (Daviere et al., 2013).

JA and GA synergistically regulate trichome initiation, sesquiterpene biosynthesis, and stamen development. Both JAZs and DELLAs interact with the same downstream transcription factors, such as WD-repeat/bHLH/MYB and MYC2, to alter the JA and GA synergy in regulating trichome formation and sesquiterpene biosynthesis. Under JA and GA signal responses, JAZs and DELLAs are degraded and WD-repeat/bHLH/MYB complexes or MYC2 are released to instigate trichome formation (Qi et al., 2014) or sesquiterpene biosynthesis (Hong et al., 2012). Additionally, GA was regulating JA biosynthesis through DELLAs that suppress the expression of LIPOXYGENASE (*LOX*) and DELAYED ANTHER DEHISCENCE1 (*DAD1*; Cheng et al., 2009). As a result, *MYB21*, *MYB24*, and *MYB57* become activated for stamen development (Mandaokar et al., 2006). It remains to be clarified whether DELLAs also correlated with MYB21, MYB24, and MYB57 for suppression of stamen development.

9.8.6 JA–SA Interaction

One of the most fascinating examples of hormonal crosstalk occurs between the JA and SA signaling pathways (Khan et al., 2012). The first suggestion for JA–SA crosstalk came from the study of tomato, which indicated that SA and its acetylated form aspirin are very effective suppressors of JA-dependent wound responses (Doherty et al., 1988). After tomato, antagonism interaction between JA and SA signaling pathways was demonstrated in other plant species (Spoel et al., 2003). SA targets JA by suppressing JA biosynthetic enzymes like LOX2 and AOS (Laudert and Weiler, 1998). In addition, many points downstream of JA biosynthesis also are targeted by SA (Leon-Reyes et al., 2010). Mitogen-activated protein kinase 4 (MAPK4) was also reported to mediate the antagonism between JA and SA mediated signaling pathways (Petersen et al., 2000). In *Arabidopsis*, MAP KINASE4 (MPK4) was identified as a positive regulator of JA signaling and a negative regulator of SA signaling, as mutant mpk4 plants demonstrated, elevated levels of SA and constitutive expression of SA-responsive PR genes fail to stimulate JA defense marker genes (Petersen et al., 2000). The antagonism interaction between JA- and SA-mediated signaling can also be affected by glutaredoxin, GRX480 (Meyer, 2008). GRX480 was confirmed to be involved in thiol disulfide reductions and redox regulation of protein activities involved in a large number of cellular processes (Meyer, 2008). SA responsive PR genes are activated by the interaction of GRX480 with TGA TFs (Ndamukong et al., 2007). Conversely, GRX480 also inhibited the expression of the JA-responsive PDF1.2 gene. It is remarkable to note that GRX480 can be activated by SA-induced NPR1, which, in turn, can form a complex with TGA factors that inhibit the expression of JA-responsive genes. In the case of *Arabidopsis*, the specific time frame in which SA inhibits JA-responsive gene expression corresponds with a transient elevated level of glutathione (Koornneef et al., 2008). Glutathione biosynthesis inhibitor L-buthionine sulfoximine (BSO) also impacts this suppressive effect on SA, suggesting that SA-mediated modulation of the cellular redox state is an imperative trigger for the attenuation of the JA pathway (Koornneef et al., 2008).

Synergistic interaction between JA and SA is also reported in many plant species. Górnik et al. (2014) demonstrated that treatment of seeds with JA or SA could improve seedling resistance to chilling, although Ilyas et al. (2017) reported that exogenous application of JA and SA in wheat could increase drought stress tolerance, while JA was more efficient than SA. In cases of combined application, JA and SA did not considerably influence plant growth. Sayyari et al. (2010) found that treatment with both MeJA and methyl salicylate (Me-Sa) increased the hydrophilic total antioxidant activity in pomegranates fruits and no significant changes were reported in lipophilic total antioxidant activity. JA and SA protect plants from salt stress via induction of protein-coding gene expression (Khan et al., 2012). Besides, Farhangi-Abriz et al. (2018) observed that exogenous application of JA and SA decreased the concentration of Na^+ in soybean under different salt stress levels and JA has a greater effect on Na^+ reduction than SA. In response to light stress, MAPK4 is a positive regulator in the JA signaling pathway but is a negative regulator in the SA signaling pathway in *Nicotiana attenuate* (Meldau et al., 2012).

Indeed, trade-offs between JA-dependent defense against insect herbivores or necrotrophs and SA-dependent resistance to biotrophs have been reported (Verhage et al., 2010). For example, induction of the SA pathway by avirulent *P. syringae* inhibits JA signaling and rendered infected leaves of *Arabidopsis* more susceptible to the necrotrophic fungus *Alternaria brassicicola* (Spoel et al., 2007). In the same way, prior inoculation with SA-inducing biotrophic pathogen *Hyaloperonospora arabidopsidis* inhibited JA mediated defenses, which were activated by caterpillars feeding on imported cabbageworm, *Pieris rapae* (Koornneef et al.,2008). In lima bean, SA stimulating the phloem-feeding sweet potato whitefly (*Bemisia tabaci*) negatively affected JA biosynthesis and JA-dependent indirect plant defenses that were triggered by a two-spotted spider mite (*Tetranychus urticae*) on the same plant (Zhang et al., 2009). In the absence of whitefly, spider mite-infested plants produce volatile blends that attract predatory mites that kill spider mites. Therefore, in whitefly infested plants induced indirect defense was suppressed, which resulted in reduced attractiveness to predatory mites and this antagonistic effect could be mimicked by SA. This indicates that the effect of JA and SA crosstalk extends plant immune responses against organisms from different trophic levels (Brooks et al., 2005).

9.8.7 JA–BR Interaction

Another hormonal crosstalk that plays an important role in plant development and stress responses is JA and brassinosteroids (BRs; Campos et al., 2009). A mutually antagonistic interaction between JA and BR pathways was reported in *O. sativa* (Nahar et al., 2013). In an antagonistic manner, JA was reported to suppress BR biosynthesis in roots by negatively regulating BR biosynthetic genes such as *OsD11* and *OsDWARF*, whereas BRs were also found to exert a negative impact on JA biosynthesis by downregulating expression of *OsAOS2* (Nahar et al., 2013). Similarly, in the case of *Arabidopsis*, JA was reported to negatively regulate the expression of DWARF4 in a COI1-dependent manner and BR also inhibited JA-dependent gene induction and root inhibition (Kim et al., 2013). BR signaling cascades including BR receptor BRI1 and BR-related kinase BAK1 play a well-characterized role in BR signaling (Yang et al., 2011). It was argued that JA application extracts a high level of trypsin proteinase inhibitors in *NaBAK1*-silenced plants as compared with empty vector plants and can elevate the accumulation of thiol protein inhibitors (TPIs; Yang et al., 2011).

9.9 Conclusion

JAs have received attention in recent years owing to their striking involvement in regulating plant development and defense. JA responses are harshly regulated by various positive or negative signaling components. These components act spatially and temporally and allow a fine-tuning in JA-dependent regulation. JAs show synergistic and antagonistic actions with other phytohormones to provide a powerful buffer system that is decisive for plants in dealing with continuously fluctuating environments. This chapter shows the modification of hormone biosynthetic pathways for the production of stress tolerant plants as well as improved crop productivity for the coming decades. Moreover, an extensive approach should be used to reveal further mechanisms of the upregulation of JA biosynthesis genes by stresses and hormone homeostasis of JAs and their biosynthetic pathway at the genetic level to get a deeper understanding of plant responses under stress conditions. Hence, future studies that can unravel major insights into the role and regulation of JAs in combined stress can yield promising outcomes.

REFERENCES

Abdelgawad et al. *Agric. Sci.* 2014; 5:1077–1088.
Abe et al. *Plant Cell.* 2003; 15:63–78.
Abouelsaad et al. *J. Plant Physiol.* 2018; 226:136–144.
Abraham et al. *Phytochem. Rev.* 2018; 17:51–80.
Acosta et al. *Science.* 2009; 323:262–265.
Adie et al. *Plant Cell.* 2007; 19:1665–1681.
Aftab et al. *Protoplasma.* 2011; 248:601–612.
Aleman et al. *Sci. Rep.* 2016; 6:28941.
Argueso et al. *Plant Cell Environ.* 2009; 32:1147–1160.
Ashraf et al. *Crit. Rev. Plant Sci.* 2010; 29:162–190.
Avalbaev et al. *J. Plant Physiol.* 2016; 191:101–110.
Avanci et al. *Genet. Mol. Res.* 2010; 9:484–505.
Balazadeh et al. *The Plant J.* 2010; 62:250–264.
Balbi et al. *New Phytol.* 2008; 177:301–318.
Bandurska et al. *Acta Physiol. Plant.* 2003; 25:279–285.
Bell et al. *Proc Natl Acad Sci USA* 1995; 92:8675–8679.
Besseau et al. *J. Exp. Bot.* 2012; 63:2667–2679.
Blechert et al. *Planta.* 1999; 207:470–479.
Boer et al. *Plant J.* 2011; 66:1053–1065.
Boter et al.*The Plant Cell.* 2015; 27:3160–3174.
Brash et al. *Proc. Natl. Acad. Sci.* 1988; 85:3382–3386.
Breeze et al. *Plant Cell.* 2011; 23:873–894.
Brioudes et al. *The Plant Journal.* 2009; 60:1070–1080.
Brooks et al. *Mol. Plant Pathol.* 2005; 6:629–39.
Brossa et al. *J. Plant Growth Regul.* 2011; 30:322–333.
Browse et al. *Ann. Rev. Plant Biol.* 2009; 60:183–205.
Bru¨x et al. *The Plant Cell.* 2008; 20:1088–1100.
Caldelari et al. *Plant Mol. Biol.* 2011; 75:25–33.
Campos et al. *J. Exp. Bot.* 2009; 60:4347–61.
Camposa et al. *J. Chem. Ecol.* 2014; 40:657–675.
Cenzano et al. *J. Plant Physiol.* 2007; 164:1449–1456.
Chauvin et al. *New Phytol.* 2013; 197(2):566–575.
Chen et al. *FEBS Lett.* 2006; 580:2540–2546.
Chen et al. *Plant Cell.* 2011; 23(9):3335–3352.
Chen et al. *The Plant Cell.* 2012; 24:2898–2916.
Chen et al. *New Phytol.* 2016; 211:599–613.
Chen et al. *PLoS Genet.* 2018; 14:e1007248.
Cheng et al. *PLoS Genet.* 2009; 5:e1000440.
Chini et al. *Nature.* 2007; 448:666–671.
Chini et al. *Nat. Chem. Biol.* 2018; 14:171–178.
Chung et al. *Plant Physiol.* 2008; 146:952–964.
Chung et al. *Plant J.* 2010; 63:613–622.
Conrath et al. *Mol. Plant Microbe Interact.* 2006; 19:1062–1071.
Creelman et al. *Rev. Plant Biol.* 1997; 48(1):355–381.
Cusido et al. *Biotech Adv.* 2014; 32:1157–1167.
Dathe et al. *Planta.* 1981; 153:530–535.
Dave et al. *The Plant Cell.* 2011; 23:583–599.
Daviere et al. *Development* 2013; 140:1147–1151.
De Geyter et al. *Trends Plant Sci.* 2012; 17(6):349–359.
De Ollas et al. *Physiol. Plant.* 2013; 147:296–306.
De Vleesschauwer et al. *Trends Plant Sci.* 2013; 18(10):555–565.
Demine et al. *Metabolites.* 2014; 4:831–878.
Doherty et al. *Physiol. Mol. Plant Pathol.* 1988; 33:377–84.
Domenico et al. *Physiol. Plant.* 2019; 165:198–208.
Durrant et al. *Ann. Rev. Phytopathol.* 2004; 42(503):185–209.
Ellinger et al. *Plant Physiol.* 2010; 153:114–127.
Ellis et al. *Planta.* 2002; 215:549–556.
Evans et al. *Plant Physiol.* 2003; 131:8–11.
Farhangi-Abriz et al. *Ecotox. Environ. Saf.* 2018; 147:1010–1016.
Farmer et al. *Plant Physiol.* 1992; 98:995–1002.
Farmer et al. *Curr. Opin. Plant Biol.* 2007; 10:380–386.
Farooq et al. *Front. Plant Sci.* 2016; 7:1–16.
Fonseca et al. *Nat. Chem. Biol.* 2009; 5:344–350.
Fragoso et al. *New Phytol.* 2014; 202(4):1335–1345.
Frost et al. *Plant Physiol.* 2008; 146:818–824.
Fu et al. *Front. Plant Sci.* 2017; 8:2108.
Gális et al. *Plant J.* 2006; 46:573–592.
Gan et al. *J. Integr. Plant Biol.* 2014; 56:916–927.
Gidda et al. *J. Biol. Chem.* 2003; 278:17895–17900.
Glauser et al. *J. Biol. Chem.* 2008; 283:16400–16407.
Glauser et al. *J. Biol. Chem.* 2009; 284:34506–34513.
Glazebrook et al. *Ann. Rev. Phytopathol.* 2005; 43:205–227.
Goetz et al. *Plant Physiol.* 2012; 158:1715–1727.
Gomez-Cadenas et al. *Curr. Protein Pept. Sci.* 2015; 16:413–434.
González-Aguilar et al. *J. Agric. Food Chem.* 2000; 48:515–519.
González-Aguilar et al. *Res. Commun.* 2004; 313:694–701.
Goossens et al. *J. Exp. Bot.* 2017; 68:1333–1347.
Górnik et al. *Acta Physiol. Plant.* 2014; 36:2547–2554.
Grebner et al. *Plant Physiol.* 2013; 161(4):2159–2170.
Grubb et al. *Trends Plant Sci.* 2006; 11:89–100.
Grunewald et al. *EMBO Rep.* 2009; 10:923–928.
Guan et al. *Proc. Natl. Acad. Sci.* 2019; 116:10568–10575.
Gutierrez et al. *The Plant Cell.* 2012; 24:2515–2527.
Heitz et al. *J. Biol. Chem.* 2012; 287:6296–6306.
Hincha et al. New York: Springer. 2014; 1166.
Hong et al. *Plant Cell.* 2012; 24:2635–2648.
Howe et al. *Plant Physiol.* 2000; 123:711–724.
Hu et al. *Plant Cell.* 2012; 24:2279–2303.
Hu et al. *Plant Cell.* 2013; 25:2907–2924.
Hu et al. *J. Exp. Bot.* 2017; 68:1361–1369.
Hyung et al. *Dev. Cell.* 2008; 14:183–192.
Ilyas et al. *Plant Anal.* 2017; 48:2715–2723.
Ishida et al. *Ann. Rev. Plant Biol.* 2008; 59(1):365–386.
Ishiguro et al. *Plant Cell.* 2001; 13:2191–2209.
Jalalpour et al. *Turk. J. Biol.* 2014; 38:76–82.
Jang et al. *Sci. Rep.* 2017; 7:10212.
Jiang et al. *Nat. Biotechnol.* 2013; 31:233–239.
Jiang et al. *Plant Cell.* 2014; 26:230–245.
Jiang et al. *Plants.* 2016; 5:27.
Kadri et al. *Biochem. Genet.* 2014; 52:106–115.
Kazan et al. *Plant Physiol.* 2008; 146:1459–1468.
Kazan et al. *J. Exp. Bot.* 2011; 62:4087–4100.
Kazan et al. *Mol. Plant.* 2013; 6:686–703.
Kazan et al. *Trends Plant Sci.* 2015; 20(4): 219–229.
Khan et al. Springer, Berlin/Heidelberg. 2012; 277–300.
Kieber et al. *Development.* 2018; 145:149344.
Kim et al. *J. Environ. Biol.* 2009; 30:333–338.
Kim et al. *PLoS One.* 2013; 8:e60505.
Kim et al. *Int. J. Mol. Sci.* 2017; 18:1004.
Koch et al. *Helv. Chim. Acta.* 1997; 80:838–850.
Kolomiets et al. *The Plant Cell.* 2001; 13:613–626.
Kondo et al. *J. Am. Soc. Hortic. Sci.* 2004; 129(2):152–157.

Koo et al. *Phytochemistry*. 2009; 70:1571–1580.
Koo et al. *Proc. Natl. Acad. Sci. U.S.A.* 2011; 108:9298–9303.
Koo et al. *Front. Plant Sci.* 2012; 3:19.
Koornneef et al. *Plant Physiol.* 2008; 147:1358–68.
Kouřil et al. *Biochim. Biophys. Acta*. 2013; 1827:411–419.
Lackman et al. *Proc. Natl. Acad. Sci. U.S.A.* 2011; 108:5891–5896.
Lamers et al. *Plant Physiol.* 2020; 182:1624–1635.
Laudert et al. *Plant J.* 1998; 15:675–84.
Lenka et al. *Front. Plant Sci.*; 6: 1152015.
León et al. *Subcell. Biochem.* 2013; 69:299–313.
Leon-Reyes et al. *Mol. Plant Microbe Interac.* 2010; 23:187–197.
Li et al. *Plant Cell*. 2003; 15:1646–1661.
Li et al. *The Plant Cell*. 2004; 16:126–143.
Li et al. *BMC Plant Biol.* 2013; 13:155–172.
Li et al. *Plant Biol.* 2013; 15:19–26.
Li et al. *Plant J.* 2018; 95:444–457.
Li et al. *PLoS ONE*. 2018; 13:e0198885.
Liu et al. *Plant Growth Regul.* 2012; 31:436–447.
Liu et al. *J. Biol. Chem.* 2012; 287:40051–40060.
Liu et al. *Nucleic Acids Res.* 2014; 42:5556–5566.
Liu et al. *J. Plant Growth Regul.* 2015; 1–11.
Liu et al. *J. Plant Growth Regul.* 2016; 35:366–376.
Lorenzo et al. *Plant Cell*. 2004; 16:1938–1950.
Ma et al. *Plant Cell Phys.* 2009; 50:2146–2161.
Mähönen et al. *Nature*. 2014; 515:125–129.
Mandaokar et al. *Plant J.* 2006; 46:984–1008.
Matthes et al. *Planta*. 2010; 232:1163–1180.
McConn et al. *Plant Cell*. 1996; 8:403–416.
Meldau et al. *BMC Plant Biol.* 2012; 12:213.
Memelink et al. *Trends Plant Sci.* 2001; 6:212–9.
Meyer et al. *J. Plant Physiol.* 2008; 165:1390–1403.
Miao et al. *The Plant Cell*. 2007; 19:819–830.
Miersch et al. *Nat. Prod. Letters*. 1992; 2:293–299.
Miersch et al. *New Phytologist*. 2008; 177(1):114–127.
Mohamed et al. *Physiol. Mol. Biol. Plants*. 2017; 23:545–556.
Moons et al. *Gene*. 1997; 191(2):197–204.
Nagpal et al. *Development*. 2005; 132:4107–4118.
Nahar et al. *Mol. Plant Microbe Interac.* 2013; 26:106–115.
Nam et al. *Cell*. 2012; 110:203–212.
Ndamukong et al. *Plant J.* 2007; 50:128–139.
Newcomer et al. *Protein Sci.* 2015; 24:298–309.
Nitschke et al. *Plant Cell*. 2016; 28:1616–1639.
Park et al. *Plant J.* 2002; 31:1–12.
Pastor et al. *Env. Exp. Bot.* 2012; 94: 46–56.
Paul et al. *New Phytol.* 2017; 213:1107–1123.
Pauwels et al. *Plant Physiol.* 2015; 169(2):1405–1417.
Petersen et al. *Cell*. 2000; 103:1111–1120.
Petricka et al. *Ann. Rev. Plant Biol.* 2012; 63:563–590.
Pirbalouti et al. *Asian Pac. J. Trop. Biomed.* 2014; 4:505–509.
Pre et al. *Plant Physiol.* 2008; 147:1347–1357.
Qi et al. *The Plant Cell*. 2011; 23:1795–1814.
Qi et al. *Plant Cell*. 2014; 26:1118–1133.
Qi et al. *The Plant Cell*. 2015a; 27:1620–1633.
Qi et al. *The Plant Cell*. 2015b; 27:1634–1649.
Qiu et al. *Environ. Saf.* 2014; 104:202–208.
Raz et al. *Plant Physiol.* 2001; 125:219–226.
Reeves et al. *PLoS Genet.* 2012; 8:e1002506.
Reymond et al. *Plant Cell*. 2004; 16:3132–3147.
Robson et al. *Plant Cell*. 2010; 22:1143–1160.
Rosahl et al. Oxford: Blackwell Publishing. 2005; 329–354.
Ruan et al. *Int. J. Mol. Sci.* 2019; 20:2479.
Saniewski et al. *Acta Physiol. Plant.* 2002; 24:211.
Santino et al. *Plant Cell*. 2013; 32:1085–1098.
Sarkar et al. *Plant Cell Reports*. 2008; 27:1–8.
Sayyari et al. *Food Chem.* 2010; 124:964–970.
Scalschi et al. *The Plant J.* 2010; 81:304–315.
Schaller et al. *Phytochemistry*. 2009; 70:1532–1538.
Schommer et al. *PLoS Biol.* 2008; 6(9):e230.
Schulze et al. *BMC Genom.* 2012; 13:524.
Schweighofer et al. *Mol BioSys.* 2008; 4:799–803.
Schweizer et al. *Plant Cell*. 2013; 25:3117–32.
Sembdner et al. *Annu. Rev. Plant Physiol. Plant Mol. Biol.* 1993; 44:569–589.
Seo et al. *Proc. Natl. Acad. Sci. U.S.A.* 2001; 98:4788–4793.
Seo et al. *Biol. Plant.* 2005; 49:447–450.
Seo et al. *Plant J.* 2011; 65:907–921.
Shahzad et al. *J. Agron. Crop Sci.* 2015; 201:443–451.
Shan et al. *Curr. Opin. Plant Biol.* 2012; 15:84–91.
Sheard et al. *Nature*. 2010; 468:400–405.
Shoji et al. *Plant and Cell Physiol.* 2008; 49:1003–1012.
Shoji et al. *The Plant Cell*. 2010; 22:3390–3409.
Song et al. *Plant Cell*. 2011; 23(3):1000–1013.
Song et al. *The Plant Cell*. 2014; 26:263–279.
Spoel et al. *Plant Cell*. 2003; 15:760–770.
Spoel et al. *Proc. Natl. Acad. Sci. USA*. 2007; 104:18842–47.
Staswick et al. *PNAS*. 1992; 89:6837–6840.
Staswick et al. *Plant Cell*. 2004; 16:2117–2127.
Staswick et al. *Trends Plant Sci.* 2008; 13:66–71.
Staswick et al. *Plant Cell*. 2009; 16:2117–2127.
Stintzi et al. *Proc. Natl. Acad. Sci. U.S.A.* 2001; 98:12837–12842.
Stotz et al. *Plant Cell Physiol.* 2011; 52:1941–1956.
Suttipanta et al. *Plant Physiol.* 2011; 157:2081–2093.
Suza et al. *Planta*. 2008; 227:1221–1232.
Svyatyna et al. *Protoplasma*. 2012; 249 Supplement 2:137–145.
Takahashi et al. *The Plant Cell*. 2007; 19:805–818.
Thatcher et al. *The Plant Journal*. 2009; 58(6): 927–939.
Thines et al. *Nature*. 2007; 448:661–665.
Tiryaki et al. *Plant Physiol.* 2002; 130:887–894.
Tiryaki et al. *J. Mol. Cell Biol.* 2007; 6:189–195.
Todd et al. *The Plant J.* 2010; 62:589–600.
Tsukada et al. *Phytochemistry*. 2010; 71:2019–2023.
Ueda et al. *Plant Physiol.* 1980; 66:246–249.
Truman et al. *PNAS USA*. 2007; 104: 1075–1080.
Ueda et al. *Physiol. Plant.* 1982; 54:249–252.
Van der Fits et al. *Science*. 2000; 289:295–297.
Van Moerkercke et al. *Plant J.* 2016; 88:3–12.
Verhage et al. *Plant Physiol.* 2010; 154:536–540.
Verpoorte et al. *Trans. Res.* 2000; 9:323–343.
Vom Endt et al. *Phytochemistry*. 2002; 61:107–114.
Walia et al. *Plant Cell Environ.* 2007; 30:410–421.
Walker et al. *Proc. Natl. Acad. Sci. U.S.A.* 2000; 97:583–587.
Walters et al. *J. Exp. Bot.* 2005; 56:435–447.
Wang et al. *J. Chem. Ecol.* 2001; 27:327–342.
Wang et al. *Plant Cell*. 2017; 29:1678–1696.
Wang et al. *Mol Genet Genomics*.2015; 290: 1–11.
Wasternack et al. *Trends Plant Sci.* 1997; 2:302–307.
Wasternack et al. New York: Academic Press. 2002; 165–222.
Wasternack et al. *Annals of botany*. 2007; 100(4):681–697
Wasternack et al. *ACS Chem. Biol.* 2010; 5:63–77.
Wasternack et al. *Ann. Bot.* 2013; 111:1021–1058.
Wasternack et al. *Plant Signal Behav.* 2016; 11(11): e1253646.
Wasternack et al. *New Biotechnol.* 2016; 33(5):604–613.

Wasternack et al. *J. Exp. Bot.* 2017; 68(6):1303–1321.
Wasternack et al. *Int. J. Mol. Sci.* 2018; 19, 2529.
Went et al. New York: Macmillan. 1937.
Wild et al. *Plant Cell.* 2012; 24:3307–3319.
Wilson et al. *Curr. Opin. Biotech.* 2014; 26:174–182.
Wu et al. *J. Plant Growth Regul.* 2012; 31:113–123.
Xie et al. *Science.* 1998; 280:1091–1094.
Xie et al. *Nat. Commun.* 2018; 9:1604.
Xing et al. *Plant J.* 2008; 54:440–451.
Yan et al. *Plant Cell* 2013; 16:393–442.
Yan et al. *Ecotoxicol. Environ. Saf.* 2013; 98:203–209.
Yan et al. *Biol. Plant.* 2015; 59:373–381.
Yan et al. *Plant Physiol.* 2016; 172:2154–2164.
Yang et al. *J. Chem. Biol.* 2007; 282:18116–18128.
Yang et al. *Mol. Plant.* 2011; 4:588–600.
Yang et al. *Proc. Natl. Acad. Sci. USA* 2012; 109:E1192–E1200.
Yang et al. *Molecules.* 2015; 20:1293–1303.
Yokoyama et al. *Plant Cell Physiol.* 2007; 48:84–96.
Yoshida et al. *Curr. Opin. Plant Biol.* 2009; 21:133–139.
Yoshikawa et al. *Plant Growth Regul.* 2007; 52:199–206.
Yu et al. *Mol. Plant.* 2012; 5:353–365.
Yu et al. *Mol. Plant.* 2016; 7(10):1505–1507.
Zarei et al. *Plant Mol. Biol.* 2011; 75:321–331.
Zhang et al. *PLoS One.* 2008; 3: e3699.
Zhang et al. *Proc. Natl. Acad. Sci. U.S.A.* 2009; 106:21202–7.
Zhang et al. *Postharvest Biol.* 2012; 64:160–167.
Zhang et al. *The Plant Cell.* 2014; 26:1105–1117.
Zhang et al. *New Phytol.* 2019; 224:761–774.
Zhao et al. *Biotechnol. Adv.* 2005; 23:283–333.
Zhao et al. *Plant Cell Environ.* 2013; 36:30–51.
Zhu et al. *PNAS.* 2011; 108:12539–12544.

10

Jasmonic Acid in Root Patterning Mechanisms: Wound Healing, Regeneration, and Cell Fate Decisions

Javier Raya-González and José López Bucio

CONTENTS

10.1 Introduction

Across their developmental transitions, plants encounter biotic and abiotic stressors that modulate organogenesis and adaptive behaviors. The jasmonates, including jasmonic acid (JA), methyl jasmonate (MeJA), and JA-isoleucine (JA-Ile), are rapidly biosynthesized as a local response to cellular injury, activate immunity in distant tissues, and may inform neighbors of the presence of attackers acting as volatile info-chemicals (Chen et al., 2011; Raya-González et al., 2012; Campos et al., 2014). More recently, the role of jasmonates in plant growth and the readjustment of root architecture became clear given their critical functions in cell division, elongation, and differentiation (Raya-González et al., 2012). Molecular elements of a canonical JA signal transduction pathway, including the CORONATINE INSENSITIVE 1 (COI1) receptor and the transcriptional regulator MYC2 direct the crosstalk with other phytohormones, peptides, and second messengers, noteworthy are glutamate and calcium (Mousavi et al., 2013; Toyota et al., 2018; Schulze et al., 2019). In the last few years, information has been emerging that has pointed out the influence of JA on stem cell activation, wound healing, and regeneration after tissue damage (Zhang et al., 2019; Zhou et al., 2019). This chapter updates the roles and functioning of jasmonates on plant organogenesis and cellular programs for tissue reconstruction.

10.2 Jasmonic Acid in Wound Signaling

The plant oxylipin jasmonic acid has been traditionally considered the master regulator orchestrating the cellular response to damage and a critical element in the signaling network behind immunity. Its role as a phytohormone was established in the 1990s following a strong effect on the induction of protease inhibitors in tomato and the discoveries that its biosynthesis in injured tissues conferred resistance to herbivores through influencing many processes including, but not limited to, the synthesis of metabolites that are toxic to attackers, reinforcement of mechanical barriers, and readjustments of growth and development (Koo and Howe, 2009; Campos et al., 2014).

Endogenous levels of JA are typically low (~5, 10, and 15 ng/g-FW for *Arabidopsis*, rice, and tobacco leaves, respectively [Du et al., 2013; Caarls et al., 2017; Xu et al., 2020]), but rise as a response to wounding, mechanical damage, or tissue injury caused by herbivores (Howe, 2004). The signaling pathway underlying JA recognition has been thoroughly investigated and yielded basic information about the molecular players involved. In unstressed plants that produce low amounts of JA, wound and stress-induced genes remain silent because their responsive transcription factors are repressed by JASMONATE ZIM-domain (JAZ) proteins (Chini et al., 2007; Thines et al., 2007). When endogenous cellular concentrations of JA rise upon wounding or leaf damage, the JAZ repressors bind to COI1, which is the F-box protein component of the E3 ubiquitin ligase SCFCOI1 that acts as a JA receptor (Thines et al., 2007; Katsir et al., 2008). COI1-JAZ interaction triggers JAZ degradation via the ubiquitin/26S proteasome pathway, thereby releasing transcription factors to activate defense-related gene expression (Durand et al., 2016). Thus, activation of cellular responses to JA implies de-repression of the transcriptional machinery, which may be important to its proposed function in orchestrating the tight balance between growth and defense required for plant survival and adaptation (Figure 10.1).

DOI: 10.1201/9781003110651-10

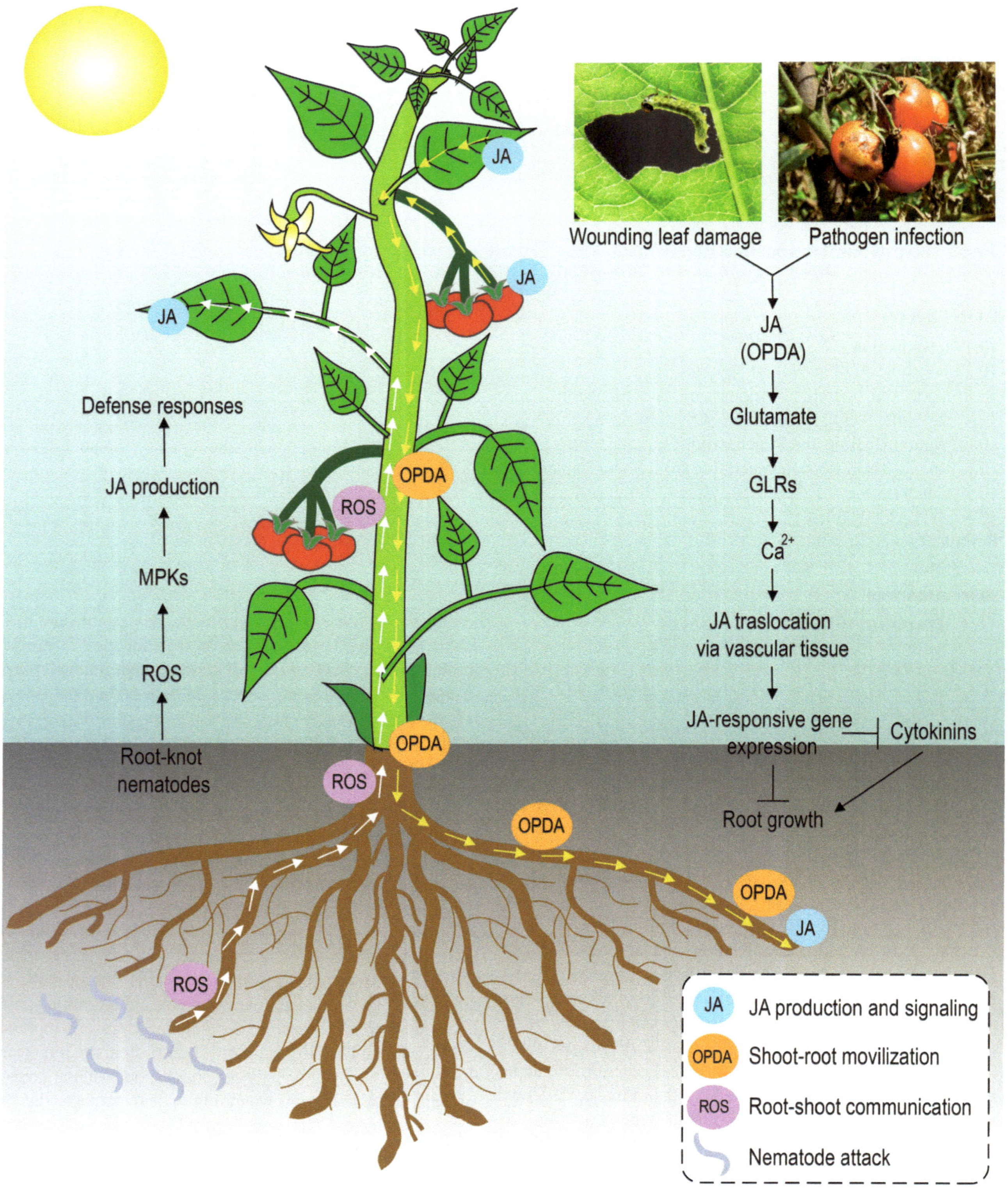

FIGURE 10.1 Jasmonic acid acts as a versatile molecule for long-distance communication within plant tissues. Wounds and leaf damage trigger the rapid release of jasmonic acid that locally or systemically modulates growth-defense balance via several interactors and second messengers, including glutamate, cytokinins, reactive oxygen species, and calcium. In roots, nematode attacks or colonization by fungal symbionts activate jasmonic acid biosynthesis through MAPK signaling that improves immunity in aerial tissues.

Over the years, the progress made toward understanding the JA response used leaves and aerial tissues exposed to mechanical damage or herbivores. It appears, however, that the JA being produced at the wound sites can move to distant organs, including unwounded roots (Gasperini et al., 2015a, b). Recently, a JA transporter (AtABCG16/JAT1) has been identified that acts as a dual permease, moving the hormone from the cytosol to the apoplast and exporting JA-Ile from the cytosol to the nucleus (Li et al., 2017). Although the systemic wound response is important to orchestrate effective protection against defoliation by herbivores, knowledge about how JA induces signaling in roots to influence immunity, growth, and recovery after damage is just starting to emerge.

10.3 Local and Systemic Responses to Jasmonic Acid

Because of their sessile lifestyle, plants have evolved highly efficient and dynamic responses to the perception of environmental challenges in a local and systemic manner through cell-to-cell and/or organ-to-organ communication. This allows communication between aboveground and belowground systems through mobile signals as a response to environmental inputs. Long-distance signals are transmitted in both distal (shoot to root; root to shoot) and radial (cell-to-cell) manners and the overall plant adaptation involves changes in levels of endogenous phytohormones, peptides, and second messengers (Gasperini et al., 2015a; Mousavi et al., 2013; Chen et al., 2016; Ohkubo et al., 2017; Toyota et al., 2018). When plants are damaged by herbivores, insect attacks, or mechanical injury, JA biosynthesis is rapidly induced to reinforce immunity through complex transduction networks and fine changes in metabolism, growth, and development (Koo and Howe, 2009; Larrieu et al., 2015). These responses are regulated by sophisticated local and systemic signals, which allow plant adaptation to multiple environmental conditions. For systemic responses, JA precursors are transported to specific distal sites through the phloem (Schulze et al., 2019). To study this phenomenon, different approaches have been developed including the use of biosensors Jas9-VENUS and GCaMP3 that include fluorescent proteins to monitor JA distribution and rises in cytosolic Ca^{2+} concentration (Larrieu et al., 2015; Toyota et al., 2018).

Shoot-to-root communication is critical for plant adaptation to biotic and abiotic stress (Gasperini et al., 2015a; Schulze et al., 2019). In response to wounding, through mechanisms yet to be defined, JA triggers glutamate (GLU) biosynthesis and accumulation, which is recognized by the GLUTAMATE RECEPTOR LIKE (GLR) family and turns on intracellular calcium levels (Mousavi et al., 2013; Toyota et al., 2018). GLR genes (*GLR3.1, 3.2, 3.3*, and *3.6*) are expressed in xylem and phloem tissues, which indicates that GLU signaling is transmitted through plant vasculature to improve fitness (Acosta et al., 2013; Mousavi et al., 2013; Gasperini et al., 2015a; Toyota et al., 2018; Schulze et al., 2019).

The role of JA in the reconfiguration of root system architecture via long-distance signaling has just started to emerge by using pharmacological, genetic, and reciprocal micro-grafting experiments in model plants and crops using wild-type (WT) and JA-related mutants, JA-responsive gene markers, and tissue-specific JA application (Gasperini et al., 2015b; Vázquez-Chimalhua et al., 2019; Schulze et al., 2019). In response to wounding, JA-dependent signaling is activated and amplified to unwound roots, mediating translocation of JA and JA precursors, including *cis*-12-oxo-phytodienoic acid (OPDA; Larrieu et al., 2015; Gasperini et al., 2015a; Schulze et al., 2019). Once in the root, OPDA is converted to JA-Ile, which is transported to the nucleus through JASMONATE TRANSPORTER 1 (JAT1) and activates the *Novel Interactor of JAZ* (*NINJA*) and the key transcription factor MYC2 in COI-dependent signaling. Thus, JA is mobile, and upon activation of defense circuits in leaves, it goes to the roots to modulate the cell cycle in meristems and the cell elongation program (Acosta et al., 2013; Gasperini et al., 2015a; Schulze et al., 2019).

JA mimics the effects that mechanic damage has on root elongation and this depends on its crosstalk with cytokinins (CK). Vázquez-Chimalhua et al. (2019) employed a system in which only the shoots of *Arabidopsis* seedlings were exposed to a medium supplemented with different JA concentrations. They found that JA application to the shoot triggers expression of lipoxygenase enzyme *LOX2* and *JAZ1* in leaves that correlated with inhibition of the primary root growth. Wounding of cotyledons also repressed growth depending on the inhibition of the cytokinin response; consistently, exogenously applied cytokinins restored root growth of JA-treated seedlings. This report demonstrates an antagonist interaction with JA/cytokinin in regulating root growth (Vázquez-Chimalhua et al., 2019).

JA biosynthesis and signaling are activated in response to multiple signals, including volatile compounds emitted by plants and/or rhizobacteria. *Arthrobacter agilis* UMCV2, a plant growth-promoting rhizobacterium (PGPR), has phytostimulating and protective functions in plants, through the production of the volatile compound dimethyl-hexa-decylamine (DMHDA). DMHDA regulates root system architecture by repressing cell division and elongation through JA and CK signaling in *Arabidopsis* (Raya-González et al., 2017; Vázquez-Chimalhua et al., 2019). The alkamides are secondary metabolites from plants and quorum-sensing modulators in bacteria. While investigating the genetic mechanisms by which *Arabidopsis* roots react to the alkamide *N*-isobutyl decanamide, a JA and NO-interaction was discovered, which modulates defense responses in leaves and the formation of lateral roots (Méndez-Bravo et al., 2010; Morquecho-Contreras et al., 2010). JA long-distance responses also go from root to shoot, indicating mutual regulation loops in coordinating organogenesis (Nalam et al., 2012; Wang et al., 2019). Root damage leads to translocation of root-derived 9-LOX oxylipin and triggers ROS signaling and interaction with mitogen-activated protein kinase (MPK) signaling to activate JA production and defense responses in leaves (Nalam et al., 2012; Wang et al., 2019). Thus, local and long-distance signaling enables JA to act in cross-kingdom communication which is important for plant defense and morphogenesis.

10.4 Jasmonic Acid Modulates Root System Architecture

Initially, jasmonic acid was associated with the defensive plant reactions to pathogen challenges (Koo and Howe, 2009). However, the last ten years demonstrated its critical functions in root system remodeling (Figure 10.2; Sun et al., 2009; Raya-González et al., 2012). Jasmonic acid application represses primary root growth in a dose-dependent manner and this phenotype is a very good visual marker to isolate and identify JA-related mutants. Treatment of *Arabidopsis* roots with methyl jasmonate (MeJA), as well as the bacterial phytotoxin coronatine, which interacts with the JA receptor COI1, unveiled distinct JA signaling components, including JASMONIC ACID INSENSITIVE 1 (JIN1/MYC2), JASMONIC ACID RESISTANCE 1 (JAR1), and CORONATINE INSENSITIVE 1 (COI1) (Staswick et al., 1992; Feys et al., 1994; Berger et al., 1996).

JA regulates root growth through the control of cell division and elongation (Chen et al., 2011; Raya-González et al., 2012). In the meristem, it alters quiescent center (QC) activity and leads to the differentiation of columella stem cells modulating the expression of the AP2-domain transcription factors PLETHORA 1 (PLT1) and PLT2, through the MYC2 transcription factor (Chen et al., 2011). Auxin signaling, via auxin response factors (ARFs), directly controls PLT1 and PLT2 expression to regulate QC specification and stem cell niche activity (Aida et al., 2004). Downstream of JA signaling, a number of second messengers transduce the underlying message to orchestrate gene expression. Nitric oxide (NO), a ubiquitous and highly reactive molecule, mediates the effects of JA in primary root growth inhibition and in promoting lateral root formation (Schlicht et al., 2013; Barrera-Ortiz et al., 2018). In *Arabidopsis* seedlings exposed to JA, NO accumulation is triggered at the root apical region in wild-type seedlings, but not in JA-related mutants, whereas application of the NO donor, sodium nitroprusside (SNP), induces expression of JAZ1 and JAZ10, two JA signaling repressors.

Analysis of several ethylene signaling mutants in response to JA and SNP revealed that ethylene through ETHYLENE INSENSITIVE 2 (EIN2) acts as a JA and NO response integrator to modulate root system configuration (Barrera-Ortíz et al., 2018). Like NO, reactive oxygen species (ROS) influence plant development and defense in a tissue-specific manner. JA interacts with multiple hormone pathways for control

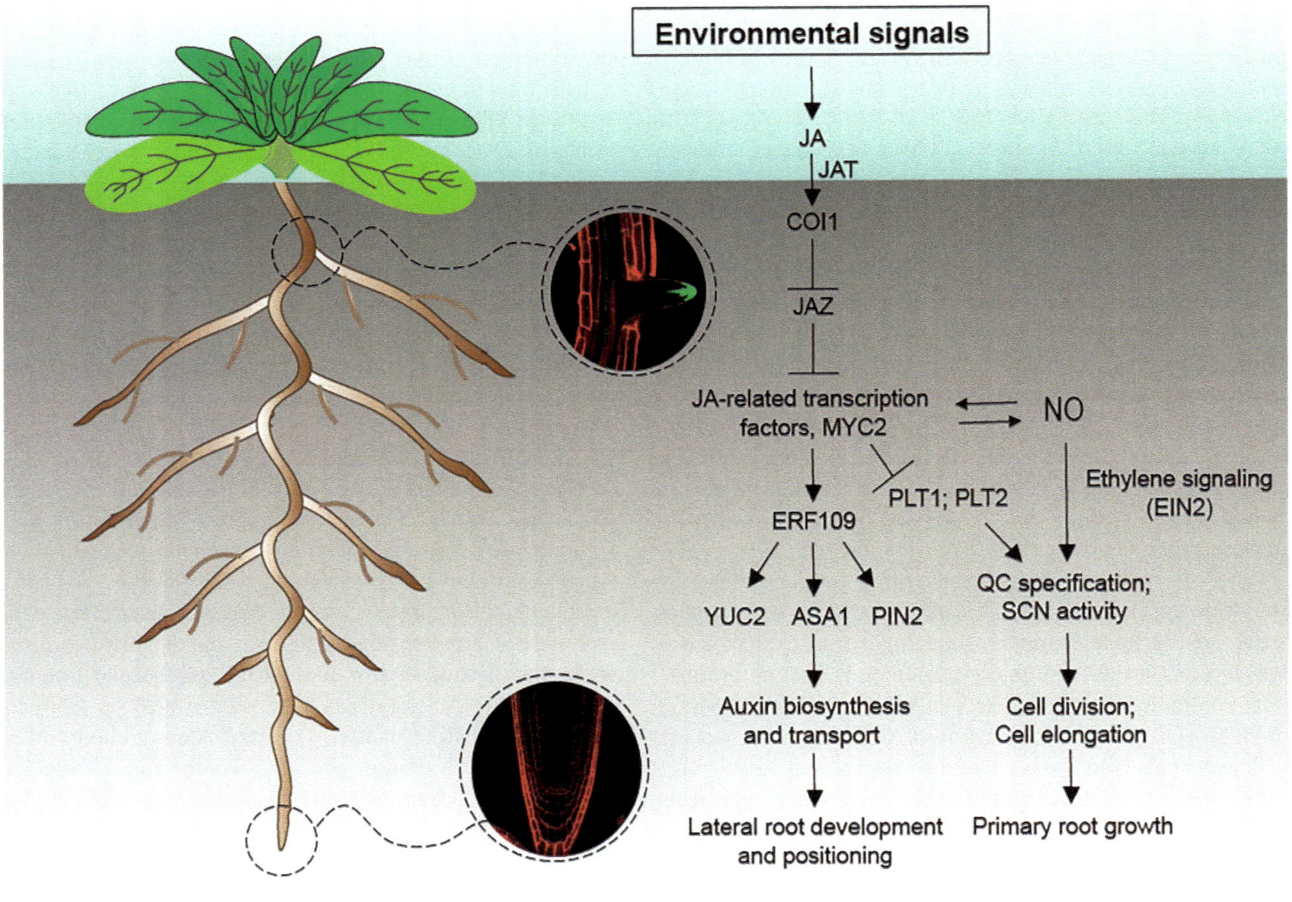

FIGURE 10.2 Jasmonic acid is a master regulator for the reconfiguration of root architecture. The root lies belowground where it supports aerial growth, soil exploration, and nutrition. The growth zones of main roots and their branches are the main sites where jasmonic acid-related stimuli are perceived and influence cell division through the PLETHORA transcription factors and auxin-ethylene crosstalk.

of root elongation, including serotonin, brassinosteroids (BRs), and abscisic acid (ABA). Serotonin (5-hidroxytryptamine), a neurotransmitter in vertebrates and an emerging bioactive molecule of plants, represses primary root elongation via ROS accumulation and JA-ethylene signaling (Pelagio-Flores et al., 2016). However, BRs negatively regulate JA responses to root elongation and defensive reactions, which depend on the DWARF4 (DWF4) enzyme (Ren et al., 2009; Huang et al 2010; Gan et al., 2015; Liao et al., 2020). Several reports have shown that JA and ABA interact in processes of plant development and stress responses, including seed germination, shoot and root growth, and herbivory through JA biosynthesis involving JA-induced ABA receptor PYL4 and SAPK10-bZIP72-AOC module in *Arabidopsis* and rice seedlings, respectively (Lackman et al., 2011; de Ollas et al., 2015; Wang et al., 2020). JA and gibberellins show antagonistic effects on root development. The DELLA and JAZ repressors physically interact, and GA application triggers DELLA degradation, which allows JAZ repressors to bind MYC2 and suppress JA-responsive genes (Hou et al., 2010).

Lateral and adventitious roots are formed through de novo organogenesis from the parent root, and these branches increase soil exploration and plant interaction with the rhizosphere. Two reports evidenced that JA promotes lateral root formation and coordinates their positioning along the main root axis (Sun et al., 2009; Raya-González et al., 2012). JA activates pericycle cells to induce lateral root primordium formation and its subsequent maturation inducing auxin biosynthesis, transport, and signaling. JA directly activates the gene encoding ANTHRANILATE SYNTHASE α1 (ASA1), an enzyme involved in auxin biosynthesis as well as the accumulation of the auxin transporter PIN-FORMED 2 (PIN2) in cell membranes of auxin transporting tissues. Consistently, auxin-related mutants are resistant to JA effects on lateral root induction, indicating JA–auxin crosstalk that triggers lateral root development (Sun et al., 2009; Raya-González et al., 2012; Hsu et al., 2013).

Lateral root formation can be induced by distinct abiotic signals, including sensing of obstacles. When the primary root encounters an obstacle, it curves to form waves and coils, in such situations, auxin accumulation spreads toward the concave side of the curvature and lateral root initiation is then triggered, which leads to a functional lateral root. This is a physiological response dependent of JA signaling via JA-receptor COI1 (Ditengou et al., 2008; Raya-González et al., 2012). In an effort to unravel how JA interacts with auxin, Cai et al. (2014) identified the ETHYLENE RESPONSE FACTOR 109 (ERF109), a transcription factor highly inducible by MeJA. JA-induced lateral root development is under the control of ERF109, which regulates auxin biosynthesis by ASA1 and YUC2 proteins, indicating that ERF109 is an integrator between JA and auxin signaling in controlling root system architecture. Indeed, JA requires an intact ethylene signaling pathway to activate lateral root development in *Arabidopsis*, since an ethylene-insensitive mutant (*ein2*), fails to respond to JA-induced lateral root formation (Barrera-Ortiz et al., 2018). Thus, the available evidence points to a link between JA, auxin, and ethylene for root system architecture remodeling in plants. How JA induces NO and ROS production/accumulation to influence root organogenesis is still an open question.

10.5 Wound Healing and Regeneration

During growth in the soil, plants encounter obstacles, uneven patches of nutrients or pollutants, and may be attacked by pathogens, and all these stimuli not only cause damage to external cell layers but also to inner tissues (Figure 10.3). Particularly sensitive to stress are the highly dividing cells within meristems, which upon experiencing osmotic, salt, or metal stress die via an apoptotic-like program (Fulcher and Sablowski, 2009; Sakamoto et al., 2018). Apoptosis is an important adaptive strategy for an organism to avoid the accumulation of mutations but implies continuous damage to the organ and vulnerability to further stress. In addition, herbivores, nematodes, and microorganisms may lead to wound induction since they depend on roots for their activities and may colonize differentiation zones and/or lateral root initiation sites (Zhou et al., 2020; Lee and Belkhadir, 2020). Fungal and bacterial species also release, as a competitive strategy, antibiotics, secondary metabolites, and virulence factors that may damage the genetic material and act as genotoxins, triggering damage to root meristems (Ortiz-Castro et al., 2014; Garnica-Vergara et al., 2016).

Inner organ destruction caused by abiotic stress (i.e., salt, cold, or heat stress), as well as wounding, triggers a strong defense reaction and, at the same time, the cells located in the vicinity of damaged tissue change their fates to activate the regeneration of the missing tissue, this enables plants to stay viable, survive, and reproduce even under very hard situations (Ikeuchi et al., 2016; Sang et al., 2018; Zhou et al., 2020).

In *Arabidopsis* and maize plants, removal of the root tip induces the regeneration of the missing part. This enabled critical advances toward understanding wounding recovery (Lim et al., 2000; Heyman et al., 2013, 2016). The gathered data show that wound healing in plants is possible by oriented cell divisions and changes in cell identity, aspects of great interest not only in the biology of plants but also in animals and humans. Cutting of the most apical portion of the root, comprising at least part of the meristem, showed that regeneration after injury correlates with global changes in gene expression and has been typified using molecular markers and reporter genes as well as using mutants and transgenic lines in the model plant *Arabidopsis thaliana* (Heyman et al., 2013, 2016; Efroni et al., 2016; Marhava et al., 2019; Zhou et al., 2019). Following damage, surrounding cells proliferate and produce a daughter lineage for tissue recovery that ultimately fills the wound. The proliferating cells, which have lost their previous identities, acquire new phenotypes prior to the acquisition of embryonic stem cell properties (Efroni et al., 2016). Deletion of the root tip causes the removal of the stem cell niche including the quiescent center and initial cells that are located within root tissues, thus neighbor cells change their fates not only to develop a strong defense response (Zhou et al., 2020) but also to re-organize a new quiescent center and meristem initials in order to resume growth (Efroni et al., 2016). Moreover, under combined stresses imposed by mutation of the MED18

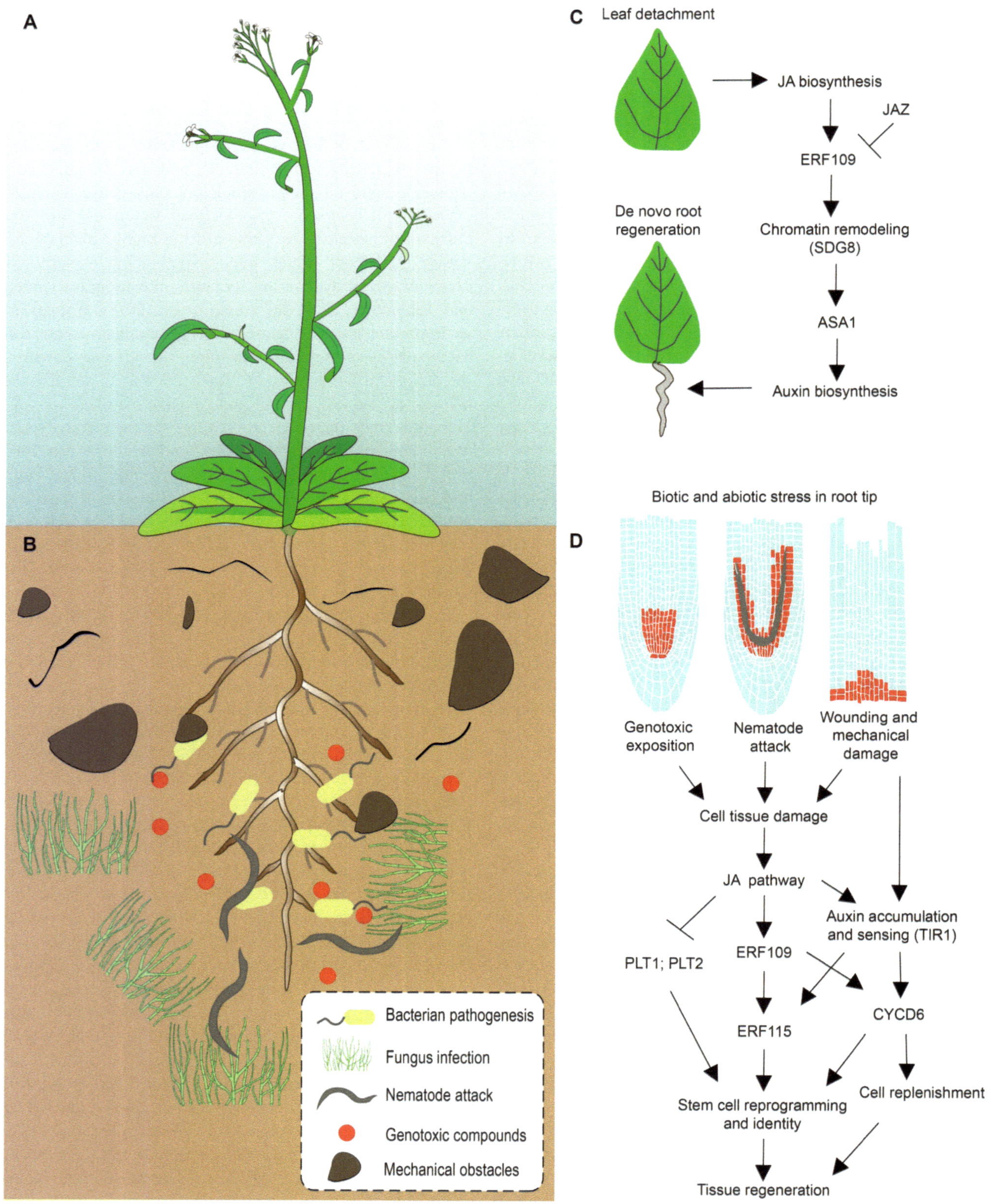

FIGURE 10.3 Jasmonic acid helps plants to recover from wounding and mechanical damage. Tissue damage in roots and shoots elicits cell reprogramming to recover the missing parts. Jasmonic acid orchestrates the regeneration of roots from leaf explants and restores tissue integrity upon genotoxic or biotic stress, wounding and mechanical damage, which operates through ethylene response factors (ERF) 115 and 119, and auxin transport and signaling for cell replenishment and organ regeneration.

subunit of the transcriptional MEDIATOR complex, the entire root meristem splits, giving rise to dichotomous branches each attaining their own root meristem and acquiring independent apical dominance to grow (Raya-González et al., 2018; Ruiz-Aguilar et al., 2020).

Recent research uncovered the emerging role of JA as an integrator of both the defense and regeneration response of roots to wounding, herbivory, and genotoxic stress. JA accumulates very rapidly after wounding, in just seconds or a few minutes (Glauser et al., 2009; Heyer et al., 2018), and triggers a signaling cascade through the COI1 receptor and the master transcription factor MYC2 that influences gene expression (Katsir et al., 2008; Kazan and Manners, 2013). Restoration of damaged tissue occurs via two possible manners: (1) inducing the IAA biosynthetic proteins from the YUCCA family (Xu et al., 2017), or (2) promoting local auxin accumulation within the damaged area, where cell death patches disrupt auxin transport (Canher et al., 2020). Another link involves AUXIN RESISTANT 1 (AXR1), which encodes a ubiquitin-activating enzyme E1 that orchestrates both defense and developmental responses, mediated by auxin and jasmonic acid, respectively (Tyriaki and Staswick, 2002; Martinez-García et al., 2020).

Laser ablation techniques enable the elimination of single cells to study wounding and regeneration *in vivo* and the accompanying patterning mechanisms in real time (Hoermayer and Friml, 2019; Shanmukhan et al., 2020). Either ablation of single root cells, treatment of whole root tips with bleomycin, a DNA damaging agent that kills pro-vasculature cells, or resection of the root tip triggers the growth of adjacent cells, promotes cell division, and activates the expression of *ETHYLENE RESPONSE FACTOR 115* (*ERF115*) for restorative patterning, allowing replacement of dead cells in a process orchestrated by auxin through the *PLETHORA 2* transcription factor (Durgaprasad et al., 2019; Marhava et al., 2019; Hoermayer et al., 2020, Canher et al., 2020). Jasmonate and auxin cooperatively act to regenerate damaged tissues upon laser ablation of quiescent center cells in root tissues during soil penetration and in response to nematode herbivory (Zhou et al., 2019). Wounding causes the rapid elevation of JA in *Arabidopsis* root meristems, which induced the division of quiescent center cells through the activation of ERF109 and ERF115 via the JA receptor COI1. The finding that JA signaling mediates root regeneration competence after damage reveals a link with developmental regulators to orchestrate both immunity and tissue repair mechanisms. Certainly, through an adequate balance of growth-defense trade-offs, plants successfully adapt to the environment.

10.6 Role of JA in Root Regeneration from Shoot Explants

The formation of a root system from shoot explants is critical for plant propagation, particularly in economically important varieties. In tissue culture technologies, control of the auxin-cytokinin ratio is important to achieve efficient organogenesis, with auxin playing an inductive role in root formation and cytokinin promoting shoot proliferation, such hormonal balance is important for cell fate determinations and even lateral root primordia can be transformed into shoot meristems following appropriate cytokinin treatments (Rosspopoff et al., 2017). Roots can be regenerated from detached leaves and stems as well as from calli and the overall cellular reprogramming is conserved in the plant kingdom (Figure 10.3; Birnbaum et al., 2016; Xu et al., 2018).

Wounding causes JA accumulation in leaf explants and activates ERF109, which in turn upregulates the gene encoding ASA1 (Zhang et al., 2019, Ye et al., 2020). During root initiation from hypocotyl cuttings or from callus induced by auxin/cytokinin treatment, cell fate reprogramming relies on changes in the chromatin states that might ensure the correct spatiotemporal expression pattern of the key regulators (Zhang et al., 2019). Jasmonic acid interactors involved in wound healing in primary roots may also mediate root branching from hypocotyl explants after wounding.

10.7 Conclusion

In the 1990s, the oxylipin jasmonic acid emerged as a phytohormone coordinating the defensive reaction on herbivory and leaf mechanical wounding. The last few years demonstrated its critical function for the reconfiguration of root organogenesis, modulating primary root, lateral root formation, and wound healing caused by mechanical damage and root nematodes, which suggest a direct link with root immunity that operates via local and long distal signaling from shoot to root and vice-versa (Zhou et al., 2019; Canher et al., 2020). The molecular players mediating JA-dependent root tissue regeneration still remain to be identified and this is a major goal to be achieved. Transcriptomic, proteomic, and metabolomics approaches in WT, mutants, and overexpressor lines for selected genes in model plants (i.e., *Arabidopsis thaliana*) and in crops (i.e., maize) will help to unravel the phytohormone networks by which JA accomplish these functions.

Jasmonic acid is not alone in orchestrating the wounding and stress signaling response from leaves to roots, the inverse reaction is also true since root colonization by the beneficial fungus *Trichoderma* triggers the production and translocation to the shoot of two additional oxylipins, 12-oxo-phytodienoic acid (12-OPDA) and α-ketol of octadecadienoic acid (KODA), which mediate the induced systemic response against necrotrophic pathogens (Wang et al., 2020). A major advance toward understanding the root immune system was done by Zhou et al. (2020), who indicate that differentiated root cells are immune-competent and mount a strong defense reaction through combined stimuli that include wounding and inoculation with the probiotic bacteria *Pseudomonas protegens* CHA0 (CHA0) that unlocks cell immunity. From these reports, it appears that the root developmental window is important to boost the immune response not only against pathogens but also the priming reaction unchained by probiotic microbial symbionts.

A few volatiles and secondary metabolites from fungi and bacteria have been found to elicit JA biosynthesis and transport from shoot to root (Vázquez-Chimalhua et al., 2019). These molecules may act as alarm stimuli for plants to mount strong adaptive reactions, which directly or indirectly promote nutrient uptake (Hernández-Calderón et al., 2018). How

JA interacts with the cellular machinery to induce transcription of iron transporters and promote exudation of sugars and other attractants to rhizospheric microorganisms remains to be investigated. However, the classic and emergent roles played by JA in plant growth and development ensure many possible applications in agriculture beyond plant nutrition, which may be helpful in the search for more effective and sustainable strategies to combat pathogens and pests, heal leaves and roots, and promote root organogenesis from shoot explants.

10.8 Acknowledgements

The writing of this chapter was possible by the financial support made by CONACYT, México (grant A1-S-34768).

REFERENCES

Acosta et al. *Proceedings of the National Academy of Sciences USA* 110 (2013): 15473–15478.
Aida et al. *Cell* 119 (2004): 109–120.
Barrera-Ortíz et al. *Journal of Plant Growth Regulation* 37 (2018): 438–451.
Berger et al. *Plant Physiology* 111, no. 2 (1996): 525–531.
Birnbaum. *Current Opinion in Plant Biology* 34 (2016): 61–67.
Caarls et al. *Proceedings of the National Academy of Sciences* 114 (2017): 6388–6393.
Cai et al. *Nature Communications* 5 (2014): 5833.
Campos et al. *Journal of Chemical Ecology* 40 (2014): 657–675.
Canher et al. *Proceedings of the National Academy of Sciences USA* (2020): doi: 10.1073/pnas.2006620117.
Chen et al. *The Plant Cell* 23, no. 9 (2011): 3335–3352.
Chen et al. *Current Biology* 26, no. 5 (2016): 640–646.
Chini et al. *Nature* 448 (2007): 666–671.
de Ollas et al. *Plant, Cell & Environment* 38 (2015): 2157–2170.
Ditengou et al. *Proceedings of the National Academy of Sciences USA* 105 (2008): 18818–18823.
Du et al. *Frontiers in Plant Science* 4 (2013): 397.
Durand et al. *Plants (Basel)* 5 (2016): 6.
Durgaprasad et al. *Cell Reports* 29 (2019): 453–463.
Efroni et al. *Cell* 165 (2016): 1721–1733.
Feys et al. *The Plant Cell* 6, no. 5 (1994): 751–759.
Fulcher et al. *Proceedings of the National Academy of Sciences USA* 106 (2009): 20985–20988.
Gan et al. *Plant Science* 241 (2015): 238–245.
Garnica-Vergara et al. *New Phytologist* 209 (2016): 1496–1512.
Gasperini et al. *Plant Physiology* 169 (2015): 2244–2254.
Gasperini et al. *Plos Genetics* 11, no. 6 (2015): e1005300.
Glauser et al. *The Journal Biological Chemistry* 284 (2009): 34506–34513.
Hernández-Calderón et al. *Plant Molecular Biology* 96 (2018): 291–304.
Heyer et al. *Frontiers in Plant Science* 9 (2018): 1569.
Heyman et al. *Science* 342 (2013): 860–863.
Heyman et al. *Nature Plants* 2 (2016): 16165.
Hoermayer et al. *Current Opinion in Plant Biology* 52 (2019): 124–130.
Hoermayer et al. *Proceedings of the National Academy of Sciences USA* 117 (2020): 15322–15331.
Hou et al. *Developmental Cell* 19 (2010): 884–894.
Howe et al. *Journal of Plant Growth Regulation* 23 (2004): 223–237.
Hsu et al. *Journal of Plant Physiology* 170 (2013): 63–69.
Huang et al. *Plant Signaling & Behavior* 5 (2010): 140–142.
Ikeuchi et al. *Development* 143 (2016): 1442–1451.
Katsir et al. *Proceedings National of Academy of Sciences USA* 105 (2008): 7100–7105.
Kazan and Manners. *Molecular Plant* 6 (2013): 686–703.
Koo and Howe. *Phytochemistry* 70 (2009): 1571–1580.
Lackman et al. *Proceedings of the National Academy of Sciences* 108 (2011): 5891–5896.
Larrieu et al. *Nature Communications* 6 (2015): 1–9.
Lee et al. *Cell Host and Microbe* 27 (2020): 308–310.
Li et al. *Molecular Plant* 10 (2017): 695–708.
Liao et al. *Plant Physiology* 182 (2020): 1066–1082.
Lim et al. *Plant Cell* 12 (2000): 1307–1318.
Marhava et al. *Cell* 177 (2019): 957–969.
Martinez-Garcia et al. *Science Reports* 10 (2020): 8860.
Méndez-Bravo et al. *Plant and Cell Physiology* 51 (2010): 1612–1626.
Morquecho-Contreras et al. *Plant Physiology* 152 (2010): 1659–1673.
Mousavi et al. *Nature* 500 (2013): 422–426.
Nalam et al. *Plant Cell* 24 (2012): 1643–1653.
Ohkubo et al. *Nature Plants* 3, no. 4 (2017): 1–6.
Ortiz-Castro et al. *Molecular Plant-Microbe Interactions* 27 (2014): 364–378.
Pelagio-Flores et al. *Physiologia Plantarum* 158 (2016): 92–105.
Raya-González et al. *Plant Journal* 96 (2018): 895–909.
Raya-González et al. *Journal of Plant Physiology* 169, no. 14 (2012): 1348–1358.
Raya-González et al. *Protoplasma* 254 (2017): 1399–1410.
Ren et al. *Plant Physiology* 151 (2009): 1412–1420.
Rosspopoff et al. *Development* 144 (2017): 1187–1200.
Ruiz-Aguilar et al. *Plant Cell and Environment* 43 (2020): 1989–1999.
Sakamoto et al. *Nature Communications* 9 (2018): 5285.
Sang et al. *New Phytologist* 218 (2018): 1334–1339.
Schlicht et al. *New Phytologist* 200 (2013): 473–482.
Schulze et al. *Molecular Plant* 12, no. 10 (2019): 1383–1394.
Shanmukhan et al. *Current Opinion in Plant Biology* 53 (2020): 117–127.
Staswick et al. *Proceedings of the National Academy of Sciences* 89, no. 15 (1992): 6837–6840.
Sun et al. *The Plant Cell* 21, no. 5 (2009): 1495–1511.
Thines et al. *Nature* 448 (2007): 661–665.
Tiryaki and Staswick. *Plant Physiology* 130 (2002): 887–894.
Toyota et al. Science 361 (2018): 1112–1115.
Vázquez-Chimalhua et al. *Protoplasma* 256 (2019): 643–654.
Wang et al. *Current Biology* 29 (2019): 1–9.
Wang et al. *Plant Cell* 32 (2020): 166–185.
Wang et al. *New Phytologist* 228 (2020): 1336.
Xu et al. *Plant and Cell Physiology* 58 (2017): 1710–1723.
Xu et al. *Current Opinion in Plant Biology* 41 (2018): 39–45.
Xu et al. *Frontiers in Physiology* 11 (2020): 847.
Ye et al. *Plant Cell* 32 (2020): 226–241.
Zhang et al. *Nature Plants* 5 (2019): 491–497.
Zhou et al. *Cell* 177 (2019): 942–956.
Zhou et al. *Cell* 180 (2020): 440–453.

11

Understanding the Role of Jasmonic Acid in Growth, Development, and Stress Regulation in Plants

Pooja Jha, Ritu Sharaya, Punam Kundu, Ashmita Chhikara, Shruti Kaushik, Anmol Sidhu, Geetika Sirhindi, M. Naeem, Ritu Gill, and Sarvajeet Singh Gill

CONTENTS

11.1 Introduction

Jasmonic acid (JA), with its various metabolic derivatives, forms a vital group of jasmonates (JAs) to help plants retaliate against various environmental challenges in the form of abiotic and biotic stress factors. JA is an efficient regulator of plant growth and development and JA signaling plays a crucial role in stress management. JA is activated on receiving stress-related stimuli and is thereafter known to activate transcription factors (TFs), which, in turn, upregulate stress-responsive genes. JA basically acts as a signal transducer and modulates the activities of antioxidant enzymes to combat stress-induced oxidative burden on plants. JA makes its contribution in various signaling cascades to defend plants from harsh environmental conditions (Raza et al., 2020). JA also interacts with other phytohormones to accelerate complex signaling responses under various abiotic and biotic stresses. The literature revealed that JA signaling has been extensively studied in model plants *Oryza sativa*, *Arabidopsis thaliana*, *Nicotiana benthamiana*, and so on, which revealed its crucial role in the day-to-day affairs of plants with the environment. Extensive research revealed that different bioactive compounds of JA and its conjugative derivatives isolated from a variety of plants, such as

DOI: 10.1201/9781003110651-11

methyl jasmonate (MeJA), 12-hydroxyJA sulfate (12-HSO4-JA), jasmonyl-isoleucine (JA-Ile), 12-O-glucosyl-JA-Ile, JA-glucosyl ester, JA-Ile methyl ester, jasmonoyl-amino acid, cis-jasmone, 12-carboxy-JA-Ile, 12-O-glucosyl-JA, and JA-Ile glucosyl ester, play an important role and are collectively called jasmonates (JAs; Sheard et al., 2010; Ghorbel et al., 2021). Jasmonates and their biologically active derivatives are known as lipid hormones. Among these JA, isoleucine-jasmonate (Ile-JA) and methyl jasmonate are the basic compounds (Wasternack and Strnad 2018; Wang et al., 2020). When a plant faces harsh climatic conditions, it makes more of an effort for its survival to maintain equilibrium between growth and developmental processes, which directly influence the quality and quantity of agronomically important crops (Harfouche et al., 2019; Raza et al., 2020). JAs consist of oxylipin signaling molecules, which together act as jasmonate and function in a juxtaposed signaling manner between the defensive and development mechanisms of plants. MYC2 is a circuit controller that regulates the crosstalk between various signaling cascades of JA and other phytohormones like auxin, ethylene, ABA, SA, and so on. The crosstalk switch on molecular mechanisms counteracts the unfavorable circumstances through interactions with the phytochrome-circadian clock.

JA is one of the important beneficial phytohormones. It is involved in the physiological, morphological, and developmental progression of plants by activating stress-responsive proteins/enzymes against several abiotic and biotic stresses. Different abiotic stress responses are adapted by plants such as tolerance to cold (Mustafa et al., 2018), heat (Degu et al., 2016; Balfagon et al., 2019), drought (Parmoon et al., 2019; Ghaffari et al., 2020), salinity (Farhangi- Abriz et al., 2019; Alisofi et al., 2020), heavy metal (Ahmad et al., 2017; Ali et al., 2018), waterlogging (Kamal and Komatsu 2016; Ouli-Jun et al., 2017), ozone stress (Tuominen et al., 2004; Cui et al., 2016), and UV radiation (Liu et al., 2012). Antagonistic signaling interactions of JA with SA/ET/BR/ABA are principal mediators in plant defense mechanisms for systematic acquired resistance (SAR). In *A. thaliana*, JA signaling is well studied and the JA receptor CORONATINE INSENSITIVE (COI1) interacts with MED 25 and regulates the activation of MYC2, which forms a complex with MYC3 and MYC4 for the management of defense mechanisms in infected plants (An et al., 2017). The CRISPR/Cas9 approach is used to identify a single copy gene *Allene oxide* cyclase mutant involved in JA biosynthesis in rice. CRISPR-Cas9 genome editing technology is utilized *in vivo* and *in vitro* for crop improvement where it improves nutritional value and develops resistance toward unfavorable conditions by using the approach of target gene editing tools. Researchers developed a better environment for plants using a plant genetic engineering approach by developing transgenic plants with health benefits and innate immunity and reducing the time consumption resource in comparison with conventional efforts (Wahab and James, 2013).

11.2 Biosynthesis of JA

JA is a vital phytohormone involved in growth, photo morphogenesis, and stress modulation in plants under abiotic and biotic stress conditions. JA biosynthesis occurs from polyunsaturated fatty acid by a variety of enzymes through the octadecanoid pathway. This pathway is regulated at multiple stages by distinct enzymes and cellular compartments including plastids and peroxisomes. The precursor of JA, α-linolenic acid, is released from galactolipids through a lipase-mediated process (Waternack and Hause, 2019). The first step of the JA biosynthesis process takes place in the plastid where α-linolenic acid is transformed into 13-hydroperoxy-octadecatrienoic acid (13-HPOT) by a 13-lipoxygenase enzyme. Further, 13-HPOT directs the formation of 12-oxo-phytodienoic acid (OPDA), catalyzed by allene oxide synthase (AOS) and allene oxide cyclase (AOC; Figure 11.1). The AOS enzyme is a member of the P450 cytochrome family and a catalyst for the biosynthesis of JA (Ahmad et al., 2016). In this pathway, AOC is formed by the cyclization of the AOS, which further forms the stable intermediate of JA biosynthesis, that is, OPDA (precursor of JA and a cis-enantiomer; Yan et al., 2013). This OPDA is mobilized into peroxisomes with the help of the COMATOSE transporter. Here, the reduction of OPDA takes place. It is reduced into 3-oxo-2(2)[Z]-pentenylcyclopentanel octanoic acid using OPDA reductase (Li et al., 2005). In the last step, three cycles of β-oxidation are carried out via enzymes: acyl-CoA oxidase (AOX), a multifunctional protein, and 3-ketoacyl CoA thiolase (KAT; Warernack and Strnad, 2018). Jasmonyl-CoA is the final product formed after β-oxidation of OPDA. Thioesterase cleaves it into cis 7-iso-jasmonic acid. JA carboxyl methyltransferase further catalyzes the methylation of cis-7-iso-JA to form MeJA, the methyl ester of JA. The reversible reaction between JA and jasmonyl-isoleucine is mediated using jasmonate amino acid synthetase (Sharma and Laxmi, 2015). Recently, Guan et al. (2019) reported that JASSY, a protein localized to the outer chloroplast envelope, facilitates the export of OPDA from the chloroplast for the biosynthesis of jasmonates and loss-of-function mutants further confirmed the essential role of JASSY.

11.3 Vital Growth Activities Performed by JA

JA, with its derivatives, is an important cellular growth regulator. It is involved in many developmental and morphological processes such as seed germination, flowering, fertility, fruit ripening, and senescence (Lyons et al., 2013). Microarray analysis reported that exogenous application of jasmonate induces the activation of numerous genes that take part in several metabolic processes. Genes induced by jasmonate (Fu et al., 2020) also play a vital role in the synthesis of secondary metabolites and cell wall formation. However, some enzymes that take part in photosynthetic activity are downregulated by jasmonates. Ribulose biphosphate carboxylase (RUBISCO) is a key enzyme in the photosynthesis process that is downregulated by jasmonates (Koo et al., 2018). JA is a vital biochemical that controls the flow of internal information within plant cell membranes by balancing metabolic signaling (Ye et al., 2019).

11.4 Physiological and Morphological Functions of JA

To understand the importance of JA in growth, development, and stress mitigation, researchers focused on the utilization of biotechnological tools with natural and artificial mutants for the impaired

FIGURE 11.1 Biosynthesis of JA.

functioning of corresponding genes. Furthermore, transgenic plants have also been developed through various approaches: (1) potential roles of JA and their derivatives at the molecular level; (2) enhancement of the regulatory pathways of gene expression for biosynthesis, degradation, and signaling networks of JA; and (3) regulation of gene expression at the level of genome, transcriptome, proteome, metabolome, and enzyme activity profiles.

Environmental conditions directly affect the integration between physiological and morphological functions that govern the overall growth and yield of crop plants. Some of the comprehensive summaries of recent modifications of JA's functions are discussed in the following sections.

11.4.1 Root Growth Development

Two types of root systems are found in plants: the tap root system (embryonic development) composed of primary, lateral, and adventitious roots, and the fibrous root system (asymmetrical divisions of the pericycle), which is mainly composed of adventitious roots present in dicotyledons and monocotyledons plants, respectively (Martinez-de la Cuz et al., 2015). The root developmental process is highly influenced by external environmental factors and soil type. It is a plastic trait that results in the evolution of new molecular mechanisms in fine roots. JA plays an important role in the regulation of these mechanisms during the root developmental process. Under O_2 deficient conditions, JA signaling is also observed in root cells of higher plants (Zhang et al., 2019). In O_2 deficient conditions (also known as hypoxia), a TF, RAP2.12, senses it and induces the repressor (JAZ) of JA signaling (Shukla et al., 2020). In response to environmental stresses, the exogenous application of JAs plays an important role with ethylene and auxin. Phenotype and transcriptomic analysis suggest that the regulation of Al-induced expression of JA's receptor COI1 and MYC2 was upregulated in the root tips (Yang et al., 2017). It was also observed that both JA and auxin interact and regulate the growth of roots in sunflower seedlings. JA negatively regulates the growth of primary and lateral roots in seedlings. Treatment of seedlings with ibuprofen increases the length of primary and lateral roots. Ibuprofen is an inhibitor of JA biosynthesis that blocks JA synthesis and results in increased root length (cell elongation) in seedlings. JA also affects the nutrition value of plants as they interact with iron and melatonin (Pelagio-Flores et al., 2020).

11.4.2 Leaf Expansion

JA negatively regulates the expansion of true leaves and cotyledon. Wounded plants after an injury start the synthesis of JA, which then inhibits cell division, this is called the bonsai effect. This effect is observed in ornamental plants. The same process takes place in plants under stress. In *A. thaliana*, JA treatment represses leaf expansion by inhibiting the activity of the mitotic cyclin CycB1during the cell cycle. Jazmonate zim protein (JAZ) along with COL1 (F-box protein function as E3 ubiquitin ligase) and MYC2 mediate the JA-induced inhibition of cell expansion in *Arabidopsis thaliana* (Wasternack and Hause, 2007; Chehab et al., 2012). The ABA receptor,

PYRABACTIN RESISTANCE LIKE 6 (PYL6) with helix loop, and TF MYC2 regulate ABA and JA signaling and inhibit cotyledon expansion (Aleman et al., 2016). In *A. thaliana*, the wound-induced systemic resistance mechanism was studied. According to this study, wound signal transmission processes are mediated by GLR3, which regulates the propagation of calcium ions. GLR3 also increases JA biosynthesis in distal leaves. Furthermore, GLR3 inactivates the repressor of JA biosynthesis through lipogenase activity in the phloem. Both xylem and phloem interplay with mobile signals in distal leaves. In *Arabidopsis*, distribution of jasmonate and its translocation have been studied in detail. JA synthesized in chloroplast was translocated to distal leaves in a controlled manner. A five-member clade of ATP binding cassette G (ABCG) was identified, which behaves like a transporter of JA (JAT). It performs influx and efflux of JA and JA-ILe: ABCG on plasma membrane efflux JA from cytoplasm and influx of JA-ILe in the nucleus. AtJAT3 and AtJAT4 are responsible for the transport of JA from local to distal leaves. GLR3 and JAT3/4 facilitate long-distance translocation in a self-controlled manner (Wasternack and Hause, 2013).

11.4.3 Hypocotyl Elongation

In the model plant *Arabidopsis*, JA inhibits the elongation of hypocotyls under visible ranges of light when bound to its receptor COI1. JA deficient jarl and COI1 mutant (Insensitive for JA) showed elongation in hypocotyls under red and far-red light (Chen et al., 2013). JA act as an inhibitor for hypocotyl growth (Robson et al., 2010). JA/MYC2 acts as a positive regulator in the inhibition of hypocotyl elongation under red and far-red light but acts as a negative regulator in blue light. Thus, along with JA/MYC2, the presence of compatible light is equally important for the elongation of hypocotyl JA repressed coleoptiles growth, the height of the plant in *Oryza sativa*, and shoot growth in maize plant (Riemann et al., 2013; Yan et al., 2012; Yang et al., 2012).

11.4.4 Petal Expansion

Jasmonates inhibit petal expansion in *Arabidopsis*. JA deficient mutants *aos,* opr3, and JA receptor mutant *coil I* showed larger petals than wild-type plants (Reeves et al., 2012*)*. For petal expansion, both MYB24 and MYB21 are essential factors. JA suppresses its expression, resulting in restricted petal growth after anthesis (the stage where the flower is fully functional; Reeves et al., 2012). Increased level of MYB21 enhances petal expansion in aos and coi1 mutants (Reeves et al., 2012). Jasmonate and its derivatives also play an important role in limiting petal size expansion. bHLH TF BIGPETAL also limits the size of the petal by controlling post-mitotic processes (Brioudes et al., 2009). bHLH TF positively regulates the expansion of petals in opr3 plants; bHLH and BIGPETALp increased and lead to petal expansion.

11.4.5 Apical Hook Formation

Dark-grown etiolated dicotyledons form a hook-like structure known as the apical hook in the early stage of development. It bends downward to protect the shoot from damage. JA prevents the formation of the apical hook through a COI-JAZ-MYC2 mediated chain. In normal conditions, EIN3/EIL4 transcriptionally activate the HOOKLESS1 gene that regulates apical hook formation. In the presence of JA, MYC2/3/4 become functionally active, which activates the EIN3 binding proteosome. It degrades EIN3 so that the HOOKLESS gene remains inactive, resulting in a no apical hook formation (Song et al., 2014; Zhang et al., 2014; Song et al., 2014) (Figure 11.2).

11.5 JA and Its Communicating Response against Abiotic and Biotic Stress Factors

JA modulates several critical mechanisms such as vegetative growth, biosynthesis of anthocyanin, stomata opening, root elongation, cell cycle regulation, and fruit ripening in stressed environments (Ali and Baek, 2020). JA enhances leaf senescence in a COI1-dependent manner and TFs regulate JA-generated leaf senescence via the upregulation of gene expression that is associated with senescence (Lalotra et al., 2020). Jasmonate starts to be deposited in plant leaves in response to various stress factors like drought, cold, osmotic stress, and in contact with pathogens and herbivores (Grebner et al., 2013). In tomato, it promotes AsA-GSH cycle enzymes and reduces the generation of reactive oxygen species (ROS) under Pb (Bali et al., 2018). Exogenous JA increases antioxidant activity and phytohormone levels in tobacco against imazapic (Kaya and Doganlar, 2016). JA also increases the level of betaine in pear leaves when faced with water deficiency (Gao et al., 2004). Metal toxicity also has a negative impact on plant growth and productivity, even in small amounts. It has been found that JA induces a decrease in the levels of malondialdehyde content and H_2O_2 with an increase in the accumulation of proline and glycine betaine. It also restores the pigment amount and leaf relative water content to a specific level under Cd stress in *Vicia* bean. It has also been demonstrated that JA regulates the metabolism of GSH by enhancing glutathione reductase (GR) activity. Exogenous application of JA could help plants develop tolerance against the undesirable impact of ultraviolet-B stress in wheat (Liu et al., 2012). Zhao et al. (2013) observed that MYC2, a transcription factor, controls the action of JA and demonstrated that two MYC2 genes are involved in chilling tolerance associated with methyl jasmonate in banana. Exogenous application of MeJA facilitates the biosynthesis of soyasaponin in *Glycyrrhiza glabra* cultures cells (Hayashi et al., 2003). MeJA also performed as an ozone scavenger in cotton (Grantz and Vu, 2012).

11.6 JA Defenses against Necrotrophic Pathogens and Herbivorous Insects

Plants evolved innate immunity against numerous pathogens. Pathogenic microorganisms invade plant cells and deliver proteins in the cytoplasm. These proteins are called effectors and are responsible for the activation of a number of pathways involved in pathogen and plant interaction. It has been reported that pathogenic or herbivore attack is connected with herbivore-associated

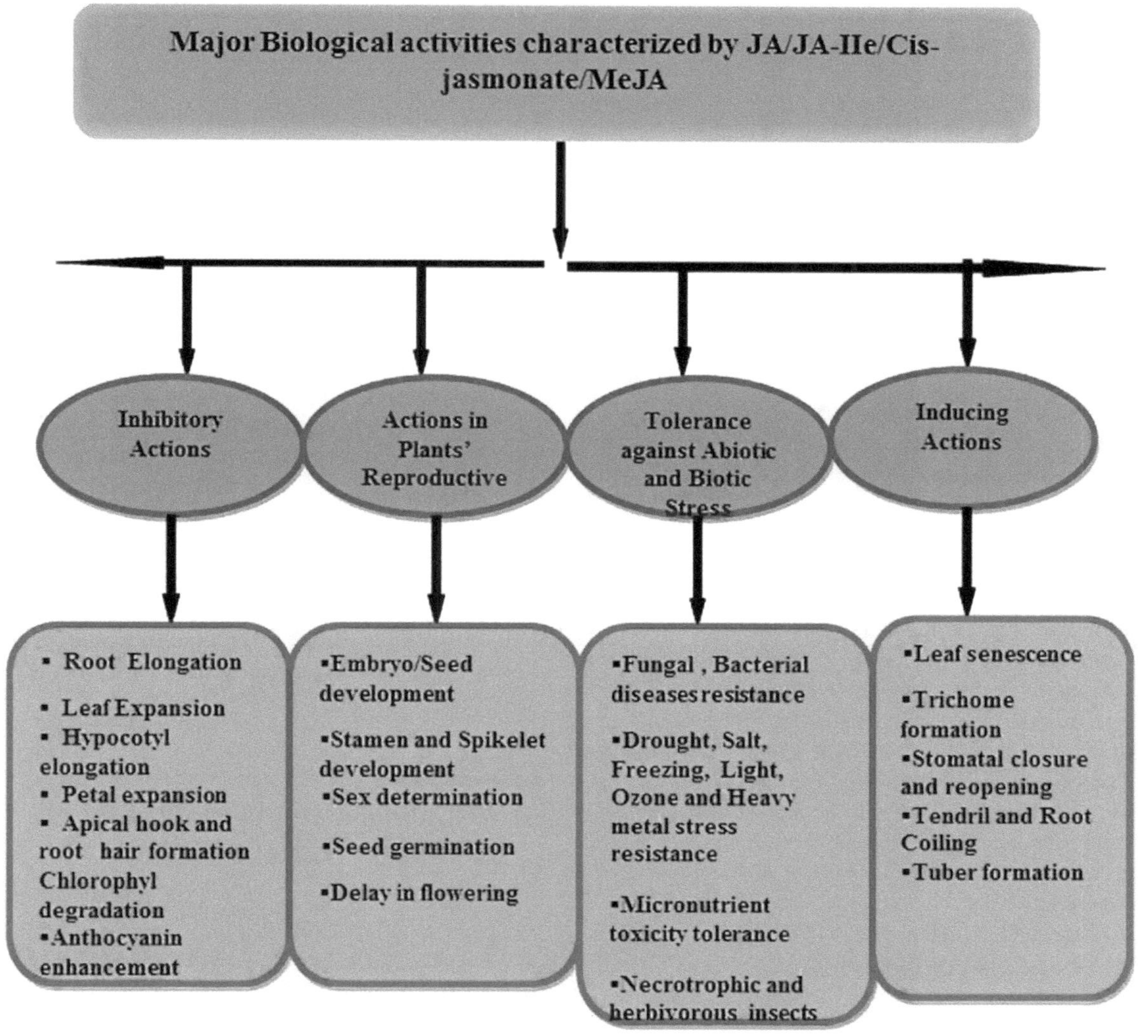

FIGURE 11.2 Vital growth activities such as physiological, morphological, and developmental processes in plants performed by jasmonates and their biologically active compounds in additional responses to several biotic and abiotic stress conditions.

molecular patterns and damage-associated molecular patterns (DAMPs), pathogen-associated molecular patterns (PAMPs), and herbivore-associated molecular patterns (HAMPs). These patterns are recognized by pathogen recognition receptors located on the cell membrane. Studies have shown that JA accumulation is enhanced in response to DAMP, HAMP, and PAMP. However, the exact mechanism of JA biosynthesis and its accumulation in response to pathogenic attack remains unclear (Howe and Jander, 2008; Dodds and Rathjen 2010; Tang et al., 2017). Downstream responses developed after pathogenic attacks partially overlap with ROS, phytohormones, and the biosynthesis of secondary metabolites for the activation of defense systems in plants (Feng and Shan, 2014; Wu et al., 2014). Jasmonate alone and with other phytohormones provide a defense against tissue damaging (wounding), leaf-eating herbivores, and necrotrophs by activating protection mechanisms in plants (Furstenberg–Hagg et al., 2013). Antagonistic signaling interactions of JA with SA, ABA, and ethylene are principal mediators in the plant defense mechanism for developing SAR against phloem-feeding aphids, spider mites, and biotrophic or hemibiotrophic pathogens (Grant and Jones, 2009; Kloth et al., 2016). JA plays a dominant role in helping plant resistance to the hemibiotrophic pathogen *Verticillium dahliae*. JA and SA cascades have important roles in developing tolerance against aphids. Some transcriptional factors like WRKY 22 in *A. thaliana,* suppress JA/SA signaling and promote susceptibility to aphids involving pathogen trigger immunity. The expression of WRKY22 is induced by aphids. Overexpression of WRKY22 showed its role in promoting susceptibility to aphids through the mesophyll mechanism. WRKY22 suppresses JA/SA signaling and promotes cell wall loosening and deterioration in plant growth (Kloth et al., 2016). In plants, SAR is an important mechanism in inducing defense resistance against a broad spectrum of microbes. The effective induction defense-related gene in response to infection of blast fungus is the JA-induced rice gene (JAmyb) that encodes a Myb TF involved in host cell death in rice seedlings.

11.7 JA-Based Defense Responses against Fungal Diseases

Fungus infection damages~125 million tons of economically important and staple food crops every year. It is estimated that fungus is the biggest threat to plants as well as animals. Due to a wide host range, fungus causes quantitative and qualitative yield losses in many important crops. In general, two types of fungus—hemibiotrophic and necrotrophic—harm crop plants (Pandey et al., 2009; Mengiste 2012). Necrotrophic fungus leads to necrosis or cell death while hemibiotrophics behave like obligate parasites, where they take nutrients from the plant in a living condition and produce a reproductive structure in the plant cells. Invasion of fungus inside cells elicits a defensive response by activating *PDF1.2* (plant defense factors). These factors are associated with SA/JA/ethylene signaling. In *Arabidopsis,* it was shown that powdery mildew fungus increases the level of SA and JA. Okada et al. (2015) showed that JA displays defense resistance against fungus. *Magnaporthe grisea* infects wheat, rye, barley, and pearl millet. It is commonly known as rice blast fungus, and after infection, it activates an enzyme antibiotic biosynthetic monooxygenase (Abm), which deteriorates JA signaling and endorses the colonization of *Magnaporthe oryzae* via conversion of JA into 12OH-JA (Patkar et al., 2015; Zhang et al., 2017). The plant's innate system tries to degrade Abm, resulting in the accumulation of MeJA and activation of the downstream signaling network for developing immunity. Recently, a key regulator in the JA signaling network has been reported in the form of bHLH and MYC2 that antagonistically coordinates defense responses by regulating two different strategic resistances for pests and pathogens, respectively (Liu et al., 2019). As a result of triggered beneficial soil microbes, induction of ISR arises by MYC2 expression during effector-mediated suppression of innate immunity in the roots of infected plants (Kazan and Manners, 2013). Receptors of JA function as transcriptional repressors of MYC2/MYC3/MYC4 and regulate the various transcriptional activation complexes with MEDIATOR25 (MED25) and CORONATINE INSENSITIVE for the induction of defense mechanisms in infected plants (An et al., 2017).

11.8 JA-Based Defense Responses against Bacterial Diseases

Pseudomonas syringae is a common example of a pathogenic variant of a hemibiotroph. It produces a phytotoxic coronatine, that is, COR. This toxin mimics iso leucine jasmonate and has an affinity to bind with COI1 (JA receptor), and after binding, it starts JA signaling and suppresses SA signaling via antagonistic conversation. It results in chlorosis and stomatal closure to facilitate the entry of bacteria into the core parts of leaves (Bender et al., 1999; Sheard *et al.*, 2010; Zhang et al., 2015b). Activation of the COR-JAZ mediated complex suppresses the SA signaling pathway to minimize resistance against *Pseudomonas syringae* infection via the JA signaling pathway (Zeng and He, 2010; Zhang et al., 2015b). COR toxin disturbs secondary metabolite synthesis pathways that deposit in the plant cell wall to provide rigidity and defense from the surroundings, hence suppressing the cell wall defense process (Millet et al., 2010; Geng et al., 2012; Yi et al., 2014). Likely COR, JA-Ile boosts the interaction with JAZ proteins through modulation in a COI1-dependent manner in *Arabidopsis* (Zheng et al., 2012; Zhang et al., 2015b; Zhou et al., 2015). *Arabidopsis* protein RPM1-INTERACTING PROTEIN 4 (RIN4) appears to be involved in bacterial infection that interacts with the Avr band and triggers the plasma membrane-localized AHA1, thus, the interaction of AvrB and AHAI modulates JA receptor (COI1) and COI1 and repressor (JAZ) expression to regulate stomatal openings (Zhou et al., 2015).

11.9 Convergence in the JA Signaling Network between Abiotic and Biotic Stress

The junction between signal perception and response is a central part of the protection against a variety of stressors in plants (He et al., 2017). It has been reported that JA-ET and JA-ABA convergence produces obligatory responses for the resistance of necrotrophic pathogens, such as *Botrytis cinerea* and *Erwinia carotovora*, and also responds to several abiotic stresses, such as drought, salt, and cold, as well as plant growth and development (Shinozaki et al., 2003). JA functions in both dependent and independent manners by forming a complex signaling network with other phytohormone signaling pathways. JAZs-MYC2 is the crucial component for the crosstalk between different phytohormone signaling, particularly in the convergence with auxin, ethylene, ABA, SA, and GA (Shinozaki et al., 2003; Figure 11.3).

11.10 Common Molecular Players for JA Crosstalk

SKP1/CULLIN/F-box (SCF) COI1 60 complex-jasmonates receptors CORONTINE form an assembled complex called $SCFE_3$ ubiquitinine ligase complex or SCF^{COI}. It possesses box protein that mediates the interaction of target proteins with coi1 and 26s proteasome for the degradation of target proteins. JAZ interacts with TFs and represses its activity by modulating its structure. It inhibits the action of isoleucine-jasmonate. JAZ contains the ZIM domain which facilitates the binding of two cofactors—NINJA and TOPELESS—within the ZIM domain. Overexpression of JAZ reduces the sensitivity of JA in wound-induced genes, thus decreasing the resistance toward insects and making it susceptible to aphids

MYC TFs belong to the BASIC HELIX–LOOP–HELIX family. It comprises a JAZ binding domain, a transcription activation domain, and one Bhih region, which may or may not be followed by a leucine zipper region. These TFs are mostly expressed in the green parts of plants, especially in leaves. MYC plays a key role in many physiological and molecular processes. In *A. thaliana,* it acts synergistically with MYC2 and MYC3 to regulate leaf senescence, root cell elongation, chlorophyll degradation, fruit dehiscence, pollen wall formation, and so on. MYC TFs activate by JA and regulate the expression of various genes that participate in developing necrotrophic herbivore resistance in plants.

11.11 JA and ABA

JA can alleviate environmental stress either independently or in collaboration with other phytohormones. JA crosstalk

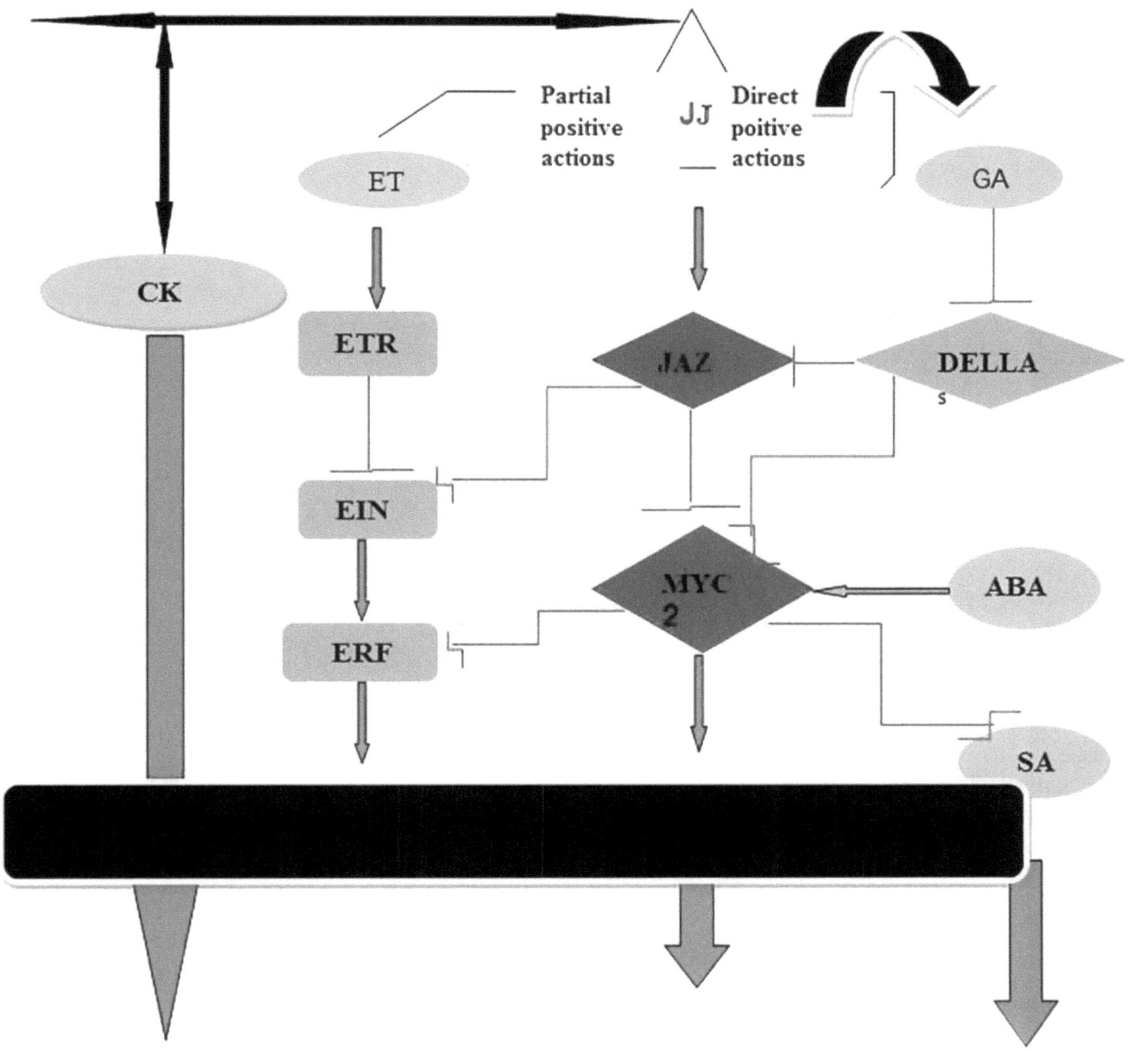

FIGURE 11.3 JA-mediated signaling convergence with other phytohormone signaling pathways, showing that MYC2 is the central crosstalk TF for communication between JA and GA. DELLAs relieve MYC2 from JAZ repression through interaction with JAZ repressors.

with ABA induces various physiological and morphological responses to diminish the harmful effects caused by abiotic and biotic stresses. JA has antagonistic and synergistic interrelationships with ABA in the regulation of plant responses to various stresses. It has also been demonstrated that ABA is used as a switch of herbivore-induced resistance through activation of JA-regulated responses in *Arabidopsis* upon a secondary attack of a herbivore (Wang et al., 2020). Yang et al. (2018) illustrated in their study that JA and ABA play a crucial part in phytochrome A and phytochrome B-mediated salt tolerance. Furthermore, MeJA and ABA application could modulate salt stress-related gene expression and also diminish phytochrome gene expression in plants. The F-box gene, F-Box Stress Induced1 (FBS1), is produced in adverse environmental situations such as osmotic stress and pathogen attack. The expression of the FBS1 gene is increased by treatment with JA and ABA. Several genes respond to JA and ABA and are strongly regulated by FBS1. Therefore, FBS1 maintains a balance between both stress hormones JA and ABA (Gonzalez et al., 2017).

Small G-proteins, Nucleolar GTP Binding Protein (NOG1and NOG2), regulate the stomatal response and nonhost resistance in response to abiotic and bacterial pathogens by JA and ABA dependent pathways (Lee et al., 2017). In *Arabidopsis*, NOG1-2 obstructs the interactions between JAZ9 and COI1 by competing with COI1, which leads to stomata regulation function. NOG1-2 could be a point of crosstalk as it integrates signals for stomata closure between JA and ABA pathways in a stressful environment. NOG1-2 plays an important role in the production of JA-associated genes after treatment with ABA and phytotoxin coronatine COR (Lee et al., 2018). Munoz–Espinoza et al. (2015) indicated that JA levels and expression of OPR3 in flc plants showed that the generation of the OPR3 gene is regulated by ABA. The level of both hormones is regulated in leaves and roots during water stress.

11.12 JA and Ethylene

Rapid gene transcription involves antimicrobial proteins or enzymes encoded for secondary metabolites synthesis and is a vital phase in JA and ET-based defense responses. The signal cascade includes ET and JA response pathways that have a crucial role in plants, specifically in the resistance response to microbial disease (Pre et al., 2008). JA and ET collectively modulate plant responses against the attack of microbes and necrotrophic fungi. The plant defensin (PDF1.2) protein protects plants against pathogens and is activated by both JA and ET. ET response factor (ERF) proteins, like ERF1 and ORA 59, target the GCC box that is present in the PDF1.2 promoter and confers responsiveness of JA and in crosstalk in JA and ST (Ma et al., 2020). The most famous TF for JA signal transduction is the MYC2 switch. On one side, JA activates JAZ degradation exogenously and MYC2 regulates ORA59, ERF1, and VSP2 (wound responsive gene) expression to give resistance to herbivores insects, and on the other side, JAZ inhibits EIL2, EIL3 transcriptional activity in the ET signal cascade, activates ORA59, ERF1 to downstream the target of the PDF1.2 promoter, and generates its expression for providing tolerance against pathogen infection (Zhu et al., 2011, Yang et al., 2019). Downregulation of NRT1.5 is mediated by JA and ET pathways through EIN2, EIL1, and many other components. SINAR promotes resistance to stress and affects plant growth during non-stressed environments via the signaling module of ET and JA (NRT1.5, NRT1.8; Zhang et al., 2014). Several plants induce defense responses involving rapid ET and JA burst after an attack of insects. These hormones mediate the rapid occurrence of a signaling event. This stronger reaction is a direct result of the compounds in the oral secretions of insects that are acknowledged by the plant system. Sunflower leaves show an injury that leads to rapid induction of the expression of HAHB4. MeJA and ET also upregulate HAHB4 expression in plants (Manavella et al., 2008).

11.13 JA and SA

JA and SA-mediated signaling pathways share a cordial and antagonistic relation. A JA-dependent resistance mechanism is reported in necrotrophs and SA-dependent resistance is well documented in biotrophs. Exogenous application of SA on several mutants and wild-type plants established SA-independent methyl-D-erythritol 2,4-cyclodiphosphate (MecPP) regulated JA responsive gene induction with increasing SA level and MecPP, both together and individually (Onkokesung et al, 2019). The JA marker gene is highly induced in the cehl/eds16-1 mutant (SA-deficient) line in comparison with *cehl*. MecPP is a penultimate metabolite in the methyl-D-erythritol 4-phosphate pathway. It regulates the transcription of genes involved in the SA biosynthesis pathway without affecting the JA signaling pathway (Lemos et al., 2016). MecPP also helps in the generation of known marker genes of JA by 12-OPDA and COI1—not in an SA-dependent manner (Lemos et al., 2016). SA targets the major enzyme activity of JA biosynthesis such as AOS, AOC2, LOX2, and OPR3 to repress its downstream signaling. Thus, SA shows its antagonistic effect on its potential target, the JA pathway (Leon-Reyes et al., 2010). Le et al. (2004) reported that endogenous JA rapidly increased in wound stress and it coincides with a decrease in SA level in rice. This study provides strong evidence of an inverse correlation between SA and JA in wound response. WRKY70 plays a crucial role in *Arabidopsis thaliana* in the activation of SA-responsive genes and suppressing JA-responsive genes. JA and ET act together in a synergistic way to mediate the expression of defense-related genes, however, they often behave antagonistically with the SA pathway (Manavella et al., 2008).

11.14 JA with Other Hormones

The cytokinin (CK) response is an essential element of the JA hormone's stress response and growth regulation. JA-dependent stress responses severely affect the CK response involving gene expression. Several histidine kinases such as AHK2, AHK3, AHK4, and WOODEN LEG work as receptors for CK. There is direct intercommunication between CK and histidine kinase receptors which induces the receptor's kinase activity, thus resulting in autophosphorylation on histidine residue. JA alters CK pathway components such as AHPs and ARRs. JA and CK modulate the expression pattern of genes involved in the development of chlorophyll, displaying an antagonistic elationnship between JA and CK (Jang et al., 2020). ERF109, a B-3 group member of the ERF/AP2 super family, mediates crosstalk between JA and auxin in regulating the formation of lateral roots through the induction of biosynthesis genes of auxin (*ASA1* and *YUC2*). Several elements of JA responses are identified in the promoters *ASA1* and *YUC2*, which further highlights the complexity of JA and auxin interplay and the crucial ERF109 role in regulating genes of auxin biosynthesis and JA signals (Cai et al., 2014). Auxin and JA levels rise, which boosts Aux or JAZ recruitment via the SCF E3 ubiquitin ligase complex. IAA and JA-Ile play critical signaling roles with wide consequences (Zhang et al., 2016).

11.15 Genetic Engineering of JA Genes toward Biotic Stress

Transgenic crops covered ~1.7 million hectares of land in 1996; however, herbicide, bacterial, fungal, and viral disease resistance means they now cover more than 195.2 million hectares globally (James, 2019; Tohidfar and Khosravi, 2015). Severe damages in plants are caused by several pathogens that result in biotic stress, which creates losses in agricultural production worldwide, and reducing these effects increases the quality, quantity, and yield. One of the major phytohormones is JA, which promotes plant immunity in response to biotic stress (Ruan et al., 2019; Farmer et al., 2003; Kessler et al., 2004). Among traditional and biotechnological approaches in land practices, traditional breeders are not able to generate resistance to biotic stress. An alternative source of biotechnology application is developing resistance in plants by using actual gene of interest recombination as well as JA (Roy et al., 2011). Researchers developed a better environment for plants by using plant genetic engineering approaches that develop transgenic plants with innate immunity and a reduction in the time consumption resource (Wahab and James, 2013). Advanced genetic modifications in plant signaling molecules, genetic makeup, and JA biosynthesis-related genes give vigorous outcomes for the activation of defense genes and resistance to disease (Berger, 2002). Jas induce the expression of several defense-related genes and initiate phytoalexins—antioxidative substances after infection of pathogens into a plant (Okada et al., 2015). JA biosynthesis is interfered by tetcyclacis and triazole derivatives, leading to growth retardation even when applied to JA endogenous diminution and preventing the induction of pathogen PR1 (Ruen et al., 2019; Chinni et al., 2018; Koo et al., 2009; Larrieu et al., 2016).

11.16 Manipulating Laccase Gene *GhLac1* in Cotton

Laccases, multicopper oxidases, play a crucial regulatory role in several biological mechanisms implicated in the synthesis of flavonoids, lignification, cell elongation, pigmentation, and cotton fiber quality (Lu et al., 2020). Laccase gene *GhLac1* expression is regulated via transgenic manipulation in cotton that provides a strong defense response for various infectious agents. Downregulation of *GhLac1* creates significant enhancement in lignification and is linked with tolerance against *Verticillium dahliae* (fungal pathogen), *Helicoverpa armigera* (cotton bollworm), and *aphis gossypii* (cotton aphid). If the expression of GhLac1 is suppressed, it results in metabolic flux redirection in the phenylpropanoid pathway, leading to JA and secondary metabolite acquisition that provides immunity to *V. dahliae* and bollworm and also enhanced sensitivity against cotton aphids (Hu et al., 2018). *Verticillium dahliae*, a fungal pathogen that creates vascular wilt diseases in cotton, results in a significant decrease in yield (Zhang et al., 2018). Hui et al. (2020) reported that GhLac1 RNAi lines have shown enhancement in fiber initials differentiated from the epidermis and shorter fiber length. GhLac1 expression was suppressed that subsequently led to the hyperaccumulation of JA and flavonoids in fiber and ovule cells. Arcuri et al. (2020) demonstrated in their genome-wide study of *Eucalyptus grandis* that EgrLAC gene expression profiling indicated differential expressions against osmotic and oxidative stress.

11.17 Overexpression of the Laccase Gene in *Verticillium dahliae* Confers Resistance to Pathogens

Verticillium dahliae releases toxins made of acidic protein lipopolysaccharide complex that adversely affect CO_2 fixation and the plant metabolism, which sometimes leads to the death of the plant. Plants enhance their tolerance to *V. dahliae* through modifications in the cell wall, extracellular enzymes, TFs, and the JA-associated signaling pathway (Song et al., 2020). Downregulation of GhLac15 promotes lignification in the cell wall that leads to increments in the lignin amount, G monolignol, and G/S ratio that further enhance the resistance of vascular wilt in *Arabidopsis*. The GhLac15 gene also improved the accumulation of arabinose and xylose in plant cell walls (Zhang et al., 2018). Liu et al. (2020) reported that overexpression of GhLAC1 also provides resistance to *Botrytis cinereal* in tomato, making strong physical defenses, hence, increasing resistance (Liu et al., 2020).

11.18 Enhancing the Expression of *OsAOS2* and *WRKY30* Genes in Rice

OsAOS2, a pathogen inducible gene in rice, encodes AOS, a vital enzyme in the JA biosynthesis pathway. Overexpression of the *OsAOS2* gene leads to an increase in the level of endogenous JA and enhances gene expression and pathogenesis-related PR proteins to increase resistance against the fungus *Magnaporthe grisea* which is a causal organism of rice blast. Basal expression of OsAOS2 is very low in leaves as compared with sheath, culm, and flower because this gene has four domains of *cytochrome P450*, having no signal peptide for targeting it into the chloroplast. It was shown that the expression of the *OsAOS2* gene is enhanced in leaves after *M. grisea* infection. To exploit this feature, transgenic lines of rice are developed to overexpress the *OsAOS2* gene by putting it under a strong promoter—*PBZ1*. This pathogen-induced promoter accumulates a huge amount of *OsAOS2* transcript, which further enhances JA biosynthesis. Insertion of the *PBZ1* promotor also plays a key role in the activation of pathogenesis-related genes PR1a, PR3, and PR5. Overexpression of *OsFBN1* has affected *OsAOS2* transcription levels in the biosynthesis of JA; *OsFBN1* overexpression also induces the formation of plastoglobules and decreases the percent of grain filling and JA levels in response to heat stress in rice (Li et al., 2019). *WRKY* TFs act as an important regulator toward pathogen infection in plants. *WRKY30* overexpression in rice promotes resistance against the fungus *Rhizoctonia solani*, which causes sheath blast and *Magnaporthe grisea*. The enhancement of resistance in transgenic lines of rice is associated with the activation of JA synthesis-linked genes like AOS2 to increase the accumulation of endogenous JA. This study shows that JA has a vital role in defense responses mediated by the *WRKY30* gene against pathogens. *WRKY30* may seem a potential candidate gene to upgrade disease tolerance in rice (Peng et al., 2012).

11.19 Regulating the Expression of the *OPR1* Gene in *Arabidopsis*

OPR1 is a ~7kb gene fragment from *A. thaliana*, which codes for 12-oxo-phytodieonic acid-10,11 reductase. JA upregulates the expression of the OPR1 gene. Promotor deletion analysis suggests that the promoter region of OPR1 contains two regulatory elements: JASE1 and JASE 2. JASE1 contains a motif without any sequence signature and JASE2 consists of A/C box-like motifs. These regulatory elements are necessary for the regulation of *OPR*1 through JA (He et al., 2001). Exogenous JA application causes premature senescence of wild-type *Arabidopsis* leaves, whether attached or detached, but is not able to generate precocious senescence in the insensitive JA mutant coi1 plants, therefore, the JA pathway is necessary for enhancing senescence in leaves (Shan et al., 2011). In addition to AOS, in *Arabidopsis* senescent leaves, several genes encoding the enzyme group of the JA biosynthetic pathway are activated. During leaf senescence, the expression of lipoxygenase 2 is highly reduced, but the expression of the lipoxygenase 1 gene strongly increases the senescence process of leaves by downregulating AOC4, while others are upregulated (*AOC1, AOC2, AOC3*; Mochizuki et al., 2016).

11.20 Overexpression of the TomloxD Gene in Tomato

A chloroplast localized lipoxygenase, TomLoxD is involved in wound-induced biosynthesis of JA, and point mutation in the TomLoxD catalytic domain results in an spr8 mutant (*suppressor of prosystemin-mediated responses8*). spr8 plants include a group of JA-based immunodeficiencies, including the inability to express wound response genes, abnormal development of glandular hairs, and adverse effects on tolerance to *Helicoverpa armigera* and *Botrytis cinerea* (Yan et al., 2013; Li et al., 2014). In transgenic tomatoes, overexpression of TomLoxD leads to enhancement in lipoxygenase activity level and endogenous JA content, which further indicates that TomLoxD makes use of an α-linolenic acid-like substrate to generate 13-HPOT. The TomLoxD gene has shown involvement in the endogenous synthesis of JA and resistance to abiotic and biotic stress factors and applications in the genetic engineering of cropping plants (Zang et al., 2017). The study demonstrated that TomLoxD overexpression results in increasing wound-induced JA synthesis and elevation in wound-responsive gene expression, and therefore, promotes tolerance against herbivores, insect attacks, and necrotrophic pathogen infection (Huang et al., 2010).

11.21 RO-292 Protein Accumulation in Response to Abiotic Stresses

Plant roots have major functions not only for the absorption of nutrients and water but also for sensing abnormal climatic stress in response to drought, salt, temperature, flooding, and freezing tolerance. Phytohormones play a significant role in stress management; one of them is JA, which helps in developing tolerance against stress conditions through enhancing or reducing the functions of stress-responsive proteins in the stressed cell of the plants. RO-292 is one of the stress-responsive proteins found in weak rice seedlings during salt and drought stress. When it is subjected to the above-mentioned stress conditions, it shows compatibility with the known proteins OsPR10a, PBZ1, and OsPR10b previously found in rice. Therefore, JA emits light not only in salt and drought stress but also in rice blast infection due to the accumulation of R0-292 protein, which is why the gene of the RO-292 stress protein is named RSOsPR10 "root-specific rice PR10" (Makoto Hashimoto et al., 2004).

11.22 JA Signaling Gene Mutants Impaired through CRISPR/CAS9

The key player for clusters of regularly interspaced short palindromic repeat (CRISPER) is Cas9, which has a specialized region in DNA that is characterized by the presence of nucleotide repeats and spacers (bits of DNA interspersed between repeated sequences). Compared with ZFN and TALEN, Cas9 nuclease has a new genome editing technology at the target site of a double-strand break. Since Cas9 cleavage is guided by a piece of RNA rather than a protein, this makes it cost-effective to modify specific genes. CRISPR-based biological surveys transform research into more specific and accurate fields of biotechnology and agricultural production. Epigenetic control is generally focused on three gene-editing layers, that is, DNA methylation, histone modification, and non-coding RNAs through which multiple targetable epigenomic-editing tools are recognized for locus-explicit chromatin adjustments that encompass gene regulation and editing. JAs play an important role in the interaction between the environment and the physiological processes of plants. Mutant genes that are involved in the production and JA signaling are very important to analyze the function of a particular phytohormone. These mutants are mainly produced by mutagenesis methods, for example, irradiation, EMS treatment, or T-DNA insertion, and potential undesired mutations could involve other biological processes. Furthermore, the CRISPER/Cas9 technique accurately and efficiently edits the genome by creating DNA modifications at specific loci and limits undesired mutations.

The CRISPR/Cas9 approach has been used to identify a new JA-deficient mutant, OsAOC, which was gene-targeted since it is a single copy gene in the JA biosynthesis pathway in rice. CRISPR-Cas9 genome editing technology is utilized *in vivo* and *in vitro* for the development of crop improvements as an enhancement to nutritional value while developing resistance toward abiotic and biotic stress by using targeting gene-editing tools (Nguyen et al., 2020).

11.23 Conclusion

Frequently changing environmental conditions poses a serious threat to crops in the form of abiotic and biotic stresses, causing significant loss and food insecurity. To overcome the ill effects of abiotic and biotic stress factors, a wide range of mechanisms are activated in plants. In this regard, the perception of stress stimuli, signaling, and activation of stress-responsive genes by phytohormones, such as JA, play a significant role in the regulation of various plant defense responses. Furthermore, JA signaling crosstalk with other phytohormones and signaling components like auxins, CKs, MAP kinase, SA, and so on play a crucial role in overcoming the stress burden on plants. Therefore, such integrations are particularly important for stress responses and plant development. In this respect, genome-wide exploration among genomes of a variety of organisms can help in understanding the ABA signaling network, which can further be exploited for the development of tough crop plants to bear the burden of stresses in open sky agriculture.

Acknowledgments

PJ acknowledges financial support in the form of NET-JRF from the University Grants Commission. RS also acknowledges the support of a university research scholarship from Maharshi Dayanand University, Rohtak. Work on plant abiotic stress tolerance in SSG laboratory was partially supported by University Grants Commission (UGC), Department of Science and Technology (DST), Council of Scientific & Industrial Research (CSIR), Govt. of India. SSG and RG also acknowledge partial support from DBT-BUILDER grant.

REFERENCES

Abdullah et al. *Agric. Water Manag.* 2015; 159:11–18.
Agnihotri et al. *Chemosphere.* 2020; 243:125361.
Ahmad et al. *Front. Plant Sci.* 2016; 7:813.
Ahmad et al. *Arch. Agron. Soil Sci.* 2017; 63(13):1889–1899.
Ali et al. *Int. J. Mol. Sci.* 2020; 21(2):621.
Alscher et al. *Physiologiaplantarum.* 1989; 77(3):457–464.
Anderson et al. *The Plant Cell.* 2004; 16(12):3460–3479.
Bali et al. *Sci. Total Environ.* 2018; 645:1344–1360.
Bidhan et al. *Biotechnology.* 2011; 10(1):1–22.
Cai et al. *Nat. Commun.* 2014; 5:5833.
Caillaud et al. *PLoS Biol.* 2013; 11(12):e1001732.
Carvalho et al. In *Salt Stress Plants*; Springer. 2013; 283–299.
Chini et al. *Nat. Chem. Biol.* 2018; 14:171–178.
Demirci et al. *J. Cell Physiol.* 2018; 233(3):1844–1859.
Ding et al. *BMC Plant Biol.* 2017; 17(1):225.
Feng et al. *Sci. Signal.* 2014; 7(320):pe10.
Fu et al. *Front. Plant Sci.* 2017; 8:2108.
Fu et al. *Plant Cell Rep.* 2020; 39:273–288.
Gaj et al. *Trends Biotech.* 2017; 31 (7):397–405.
Gao et al. *Plant, Cell Environ.* 2004; 27:497–507.
Ghorbel et al. *Plant Cell Rep.* 2021; 40:1471–1494.
Gonzalez et al. *BMC Genomics.* 2017; 18:533.
Grantz et al. *J. Exp. Bot.* 2012; 63(11):4303–4313.
Grebner et al. *Plant Physiol.* 2013; 161(4):2159–2170.
Gregersen et al. *Plant Mol. Biol.* 2013; 82(6):603–622.
Griffiths et al. *Essays Biochem.* 2020; 64(3):501–512.
Guan et al. *Proc. Natl. Acad. Sci. U.S.A.* 2019; 116(21):10568–10575.
Hai et al. *J. Appl. Phycol.* 2020.
Hayashi et al. *Plant Cell Physiol.* 2003; 44(4):404–411.
He et al. *Plant Mol. Biol.* 2001; 47(5):595–605.
He et al. *Plant Physiol.* 2002; 128(3):876–884.
He et al. *Nat. Commun.* 2018; 9(1):460.
Hernandez et al. *Biochem. Biophys. Res. Commun.* 2009; 387(2):300–304.
Hernandez et al. *J. Exp. Bot.* 2015; 66(10):2901–2911.
Hossain et al. Springer. 2020; 649–673.
Hu et al. *Hortic. Res.* 2020; 7(1):1–5.
Huang et al. *BMC Biotechnol.* 2010; 11:1–15.
Huang et al. *Sci. Silvae Sin.* 2017; 119:384–391.
Islam et al. *Plant Cell Physiol.* 2020.
Jang et al. *Int. J. Mol. Sci.* 2020; 21(1):305.
Javed et al. *Photochem. Photobiol.* 2011; 87:1354–1362.
Jia et al. *Nat. Plants.* 2019; 5(2):212–224.
Jiang et al. *Int. J. Mol. Sci.* 2020; 21(3):1018.
Jikumaru et al. *Biosci. Biotechnol. Biochem.* 2004; 68(7):1461–1466.
Kaya et al. *Ecotoxicol. Environ. Saf.* 2016; 124:470–479.
Kazan et al. *Mol. Plant.* 2013; 6(3):686–703.
Kim et al. *Proc. Natl. Acad. Sci. USA* 2018; 115(33):8448–8453.
Koo et al. *Plant J.* 2009; 59:974–986.
Koo et al. *Phytochem.* 2018; 17:51–80.
Koyama et al. *Plant Physiol.* 2017; 175(2):874–885.
Lalotra et al. *Jasmonates: An Emerging Approach in Biotic and Abiotic Stress Tolerance.* 2020. doi: 10.5772/intechopen.84608.
Larrieu et al. *BMC Boil.* 2016; 14:79.
Lee et al. *Biochem. Biophys. Res.* 2004.
Lee et al. *Sci. Rep.* 2017; 7(1):9260.
Lee et al. *Int. J. Mol. Sci.* 2018; 19(7):1922.
Lei et al. *J. Integr. Plant Biol.* 2020; 62(2):218–227.
Lemos et al. *J. Exp. Bot.* 2016; 67(5):1557–1566.
Leon-Reyes et al. *Planta.* 2010; 232(6):1423–1432.
Li et al. *Plant Cell.* 2005; 17(3):971–986.
Li et al. *BMC Genomics.* 2014; 15:1–12.
Li et al. *Plant J.* 2017; 92(4):650–661.
Li et al. *Int. J. Mol. Sci.* 2019; 20(3):671.
Li et al. *Plant Sci.* 2019; 285:230–238.
Li et al. *Mol. Plant.* 2020; 13(10):1485–1498.
Lim et al. *Annu. Rev. Plant Biol.* 2007; 58:115–136.
Lim et al. *Regulatory Mechanisms of Leaf Senescence under Environmental Stresses.* 2020.
Liu et al. *J. Plant Growth Regul.* 2012; 31:436–447.
Liu et al. *Plant, Cell Environ.* 2016; 39(11):2515–2529.
Liu et al. *Plant Cell.* 2019a; 31(2):430–443.
Liu et al. *Plant J.* 2019b; 97(2):368–377.
Liu et al. *The Plant Cell.* 2019c; 31(1):106–127.
Liu et al. *Genes.* 2020a; 11(10):1134.
Liu et al. *J. Genet.* 2020b; 99:23. PMID: 32366734.
Liu et al. *Int. J. Agric. Biol.* 24:221–228.
Lyons et al. *Plant Cell Rep.* 2013; 32:815–827.
Ma et al. *Exp. Bot.* 2018; 145:1–11.
Ma et al. *New Phytol.* 2020; 225(1):474–487.
Manavella et al. *Plant J.* 2008; 56(3):376–388.
Mei et al. *Mol. Plant Microbe Interact.* 2006; 19(10):1127–1137.
Monzón et al. *Plant Growth Regul.* 2012; 66(2):129–136.
Munemasa et al. *Plant and Cell Physiol.* 2019; 60(10):2263–2271.
Munné-bosch et al. *Funct. Plant Biol.* 2004; 31:8808–8818.
Munoz-Espinoza et al. *Front. Plant Sci.* 2015; 6:997.
Nejat et al. *Curr. Issues Mol. Biol.* 2017; 23:1–16.
Nguyen et al. *J. Appl. Phycol.* 2020a; 14:1–3.
Nguyen et al. *Plant Biol.* 2020b; 199–209.
Niwa et al. *Biosci. Biotech. Biochem.* 2018; 82(2):292–303.
Noriega et al. *J. Plant Growth Regul.* 2012; 31:79–89.
Okada et al. *Plant Cell Physiol.* 2015; 56(1):16–27.
Onkokesung et al. *Plant, Cell Environ.* 2019; 42(7):2309–2323.
Pandey et al. *Plant Physiol.* 2009; 150(4):1648–1655.
Parmar et al. *Biotech.* 2017; 7(4):239.
Pelagio-Flores et al. In *Neurotransmitters in Plant Signaling and Communication.* Springer, Cham. 2020; 93–105.
Penfold et al. *J. Exp. Bot.* 2014; 65(14):3859–3873.
Peng et al. *Planta.* 2012; 236(5), 1485–1498.
Pre et al. *Plant Physiol.* 2008; 147(3):1347–1357.
Raza et al. *Plant Cell Reports.* 2020; 1–29.
Rojo et al. *J. Plant Growth Regul.* 2003; 22(1):82–98.
Ruan et al. *Mol. Sci.* 2019; 20:2479.
Ruan et al. *Int. J. Mol. Sci.* 2019; 20(10):2479.
Santisree et al. *Protective Chemical Agents in the Amelioration of Plant Abiotic Stress: Biochemical and Molecular Perspectives.* 2020:342–73.
Schippers et al. *Curr. Opin. Plant Biol.* 2015; 27:77–83.
Schommer et al. *PLoS Biol.* 2008; 6(9):230.
Seltmann et al. *Plant Physiol.* 2010; 152(4):1940–1950.
Shan et al. *Plant Sci.* 2010; 178(2):130–139.
Shan et al. *Plant Signal. Behav.* 2011; 6(4):575–577.
Sharma et al. *Front. Plant Sci.* 2015; 6:1129.
Shinozaki et al. *Curr. Opin. Plant Biol.* 2003; 6(5):410–417.
Shukla et al. *Plants.* 2020; 9(8):1046.
Silva et al. *Curr. Gene Ther.* 2011; 11(1):11–27.
Singh et al. Springer, Cham. 2020:263–304.
Song et al. *Int. J. Mol. Sci.* 2020; 7; 21(3):1120.

Springer et al. *Proc. Natl. Acad. Sci. USA* 2016; 113(12):3383–3388.
Sree et al. Springer. 2015; 611–622.
Staswick et al. *Proc. Natl. Acad. Sci.* 1992; 89(15):6837–6840.
Sun et al. *Nature communications.* 2014; 5(1):1–2.
Sun et al. *Appl. Soil Ecol.* 2017; 119:384–391.
Tamogami et al. *FEBS Lett.* 1997; 401(2–3):239–242.
Tamogami et al. *Adv. Bot. Res.* 2011; 60:309–348.
Thomas et al. *Ann. Rev. Plant Physiol.* 1980; 31:83–111.
Thomma et al. *Curr. Opin. Immunol.* 2001; 13(1):63–68.
Thomma et al. *Biotechnol. Agron. Soc. Environ.* 2015; 19(1):62–70.
Vos et al. *Front. Plant. Sci.* 2013; 4:539.
Wang. *Functions and Mechanisms of Chloroplast Lipid Remodeling in Plants.* Michigan State University; 2017.
Wang et al. *Plant Cell.* 2018; 30(5):1006–1022.
Wang et al. *PLoS Genet.* 2019; 15(4):e1008068.
Wang et al. *Int. J. Mol. Sci.* 2020; 21:1446.
Wang et al. *Crop J.* 2020.
Wasternack. *Trends Plant Sci.* 2019; 24(8):667–9.
Wasternack & Hause. *Ann. Bot.* 2013; 111(6):1021–1058.
Wasternack & Hause. *Nat. Plants.* 2019; 5(8):776–777.
Wasternack & Strnad. *Int. J. Mol. Sci.* 2018; 19(9):2539.
Wei et al. *Plant Physiol.* 2015; 169(2):1371–1381.
Woo et al. *Ann. Rev. Plant Biol.* 2019; 70:347–376.
Xiang & Oliver. *Plant Cell.* 1998; 10(9):1539–1550.
Yan et al. *PLoS Genet.* 2013; 9(12):e1003964.
Yan et al. *Plant Dev. Stress Resp.* 2013. doi: 10.5772/52675.
Yang et al. *Plant Physiol.* 2017; 173(2):1420–1433.
Yang et al. *Plant Cell Physiol.* 2018; 59(11):2381–2393.
Yang et al. *Front. Plant Sci.* 2019; 10:1349.
Yuan et al. *Plant Physiol.* 2017; 173(3):1864–1880.
Yuan et al. *Mol Plant.* 2020; 13(8):1149–1161.
Zeng et al. *Biol. Trace Elem. Res.* 2012; 148:255-63.10.
Zhai et al. *The Plant Cell.* 2015; 27(10):2814–2828.
Zhang et al. *Plant Cell.* 2014; 26(10):3984–3998.
Zhang et al. *J. Exp. Bot.* 2016; 67(7):2107–2120.
Zhang et al. *Sci. Rep.* 2017; 7:2816.
Zhang et al. *Mol. Plant.* 2018; 11(2):326–337.
Zhang et al. *Mol. Plant Pathol.* 2019; 20(3):309–322.
Zhao et al. *Plant Cell Environ.* 2013; 36(1):30–51.
Zhao et al. *BMC Plant Biology.* 2020; 20(1):1–22.
Zhu et al. *Proc. Natl. Acad. Sci.* 2011; 108(30):12539–12544.

12

Jasmonates and Plant Responses Under Metal Stress

M. Reyes-Díaz, J. González-Villagra, C. Figueroa, C. Inostroza-Blancheteau, M. Morales, and L.A. Bravo

CONTENTS

12.1 Introduction

Several factors such as pH, the presence of different elements, and the accumulation of metals in soil cause plant growth reduction, affecting various physiological and molecular activities of plants (Panuccio et al., 2009; Hassan et al., 2017). Metals such as zinc (Zn), copper (Cu), molybdenum (Mo), manganese (Mn), cobalt (Co), and nickel (Ni) are essential for the biological processes of plants (Salla et al., 2011; Shahid et al., 2015). These metals with other highly toxic metals like arsenic (As), lead (Pb), cadmium (Cd), mercury (Hg), chromium (Cr), aluminum (Al), and beryllium (Be) can reduce crop productivity when their concentrations reach supraoptimal values (Xiong et al., 2014; Pierart et al., 2015). It is known that these toxic elements cause morphological anomalies and metabolic disorders, leading to yield reduction in plants (Amari et al., 2017). These metabolic disorders also increase the production of reactive oxygen species (ROS), such as superoxide anion radical (O_2^-), hydrogen peroxide (H_2O_2), and hydroxyl radical (OH^-), causing disturbance of the redox homeostasis of cells (Gill and Tuteja, 2010; Shahid et al., 2015). However, plants possess mechanisms to resist oxidative damage depending on the capacity to prevent ROS under metal stress (Choudhury et al., 2017).

On the other hand, phytohormones play a remarkable role in plant development and response to stresses, regulating several developmental processes and signaling networks in plants under different abiotic stresses. Currently, it has been established that phytohormones have the potential to alleviate the harmful effects of abiotic stresses (Sytar et al., 2019; Wang et al., 2020), which are mediated by three major pathways of the phytohormones: jasmonic acid (JA), salicylic acid (SA), and ethylene (ET; Takahashi et al., 2004). Jasmonates are important cellular regulators involved in a wide range of plant processes, such as growth and development, including defense against biotic and abiotic stresses (Dar et al., 2015; Ali and Baek, 2020). Thus, this chapter provides a critical overview of the biosynthesis, signaling, and transport of JAs and their role in different physiological, biochemical, and molecular responses of plants under metal stresses.

12.2 Function of Jasmonates in Plants

The first discovered jasmonate was methyl jasmonate (MeJA), a volatile jasmonate constituent of essential oils of *Jasminum grandiflorum* (Demole et al., 1962), then, jasmonic acid (JA) was identified from cultures of the fungus *Lasiodiplodia*

DOI: 10.1201/9781003110651-12

theobromae (Aldridge et al., 1971). They are fatty acid-derived compounds. For instance, JA is synthesized from linolenic acid (Zimmerman and Feng, 1978). Even though JA was recognized to have a similar chemical structure to mammals' prostaglandin by these authors, their function in plants was recognized a couple of years later when the senescence promoting activity of MeJA was demonstrated experimentally using MeJA obtained from aqueous acetone extracts from leaves and stems of wormwood *Artemisia absinthium* (Ueda and Kato, 1980). Likewise, the authors used as a bioassay the senescence of cut pieces of *Avena sativa* leaves. MeJA was stronger in promoting senescence than ABA, and even in small concentration was able to abolish the anti-senescence action of kinetin. Senescence is promoted via the degradation of photosynthetic proteins, with increased proteolytic and peroxidase activities and increased cellular respiration (Koetje, 2003).

Later, in the mid-1980s, a series of articles were published about jasmonate promoting plant defense against pathogens. First, mechanical wounding caused by herbivorous, such as the induction of protease inhibitors, which inhibits proteases of insect guts, causing feeding to misfunction, therefore, preventing massive foliage damage of chewing insects, but later it was also found that exogenous elicitors from microbial pathogens (necrotrophs) were sufficient to trigger jasmonate action even in the absence of wounding (Thomma et al., 1998). Here, jasmonates play two major roles: first, the response of the plant against insect attack through the induction of genes involved in herbivore deterrence, wound healing, and defense-related process locally; second, jasmonate also plays a role in a systemic component involved in preconditioning undamaged leaves to repel future insect attacks (Sun et al., 2011). The systemic response required a long-distance traveling signal. For many years it was attributed to systemin, an 18 amino acid peptide that was identified to be produced in the wounding sites of tomato plants (Ryan and Pearce, 1998). Later on, using reciprocal grafting of two tomato mutants, a JA biosynthetic mutant (*spr-2*) and a JA response mutant (*jai-1*), it was evidenced that the graft-transmissible wound signal was JA or at least a related compound derived from the octadecanoid pathway (Li et al., 2002). These authors also proposed that the system has a regulatory but local role in the systemic wound response. One of the major ecophysiological roles of JAs was interplant communication. In the early 1990s, this was found when experiments incubated sagebrush branches (*Artemisia tridentata*), which release MeJA, in proximity in a closed chamber with tomato plants. After two days of incubation, tomato leaves exhibited elevated levels of protease inhibitors (Farmer and Ryan, 1990). Therefore, MejA was proposed as the second volatile hormone after ethylene functioning in interplant communication.

Today, a variety of plant developmental responses seem to be associated with JAs. For instance, the earlier recognized developmental effects additional to senescence were germination and root elongation inhibition in *Helianthus annus* (Corbineau et al., 1988). This effect has been associated with JA inhibition of alfa-amylase (Norastehnia et al., 2007). However, the effect of JA on germination is species-specific; for instance, some reports indicated that JA could break seed dormancy (Yildiz et al., 2008; Xu et al., 2016), particularly associated with low-temperature stratification. Finally, JA has been involved in regulating reproductive development, such as pollen development, stamen elongation, and pollen release timing (Thines et al., 2013). The dehiscence of anthers is delayed in the delayed dehiscence 1 (*dde1*) mutant of *Arabidopsis*, resulting in inefficient fertilization (male sterility). The JA treatment has been found to restore the WT phenotype and help the plant produce seeds.

The JA and MeJA have also been associated with a variety of plant responses to abiotic stress such as salt, drought, chilling, and heavy metal toxicity (Yu et al., 2019). More importantly, JAs also affect a trade-off between primary and secondary metabolism, degrading photosynthetic enzymes and inactivating ribosomes while stimulating terpenoid, phenylpropanoid, and alkaloid biosynthesis. This trade-off has important implications for controlling resource allocations to plant productivity or self-defense against biotic (herbivores, pathogens) and abiotic stress (Koetje, 2003).

Jasmonate association with increased resistance to metal and metalloid toxicity in the plant was recognized long ago. The first evidence of this relationship was the strong association of JA activation of the same set of genes induced by copper and cadmium in *Arabidopsis thaliana*. For instance, JA increased mRNA level and the capacity for glutathione synthesis (Xiang and Oliver, 1998). MeJA applied to Cd stressed *Brassica juncea* enhanced GSH production more prominently in the presence of sulfur and protected photosynthetic functioning of plants (Per et al., 2016). MeJA was also reported to promote phytochelatin production in Cd-exposed *A. thaliana* (Maksymiec et al., 2007).

Additionally, the expression of *type-2 metallothionein* gene (*KoMT2*) in leaves was enhanced in Cd-treated plants, whereas the exogenously applied MeJA significantly restored the expression of *KoMT2* (Chen et al., 2014). There is no direct evidence of JA/MeJA regulating genes involved in the synthesis of phytochelatin (oligomer of glutathione) or metallothioneins (low molecular mass proteins that contain cysteine). Both are important components of heavy metal sequestration in plants. Recent information supports the idea of epigenetic regulation by miRNAs of these important chelating agents (Çelik et al., 2020).

The action of JA in metal and metalloid toxicity is not fully elucidated. The most accepted hypothesis is associated with the control of oxidative stress caused under metal/metalloid toxicity. The exposure to toxic levels of metals/metalloids favors an oxidative burst; this likely promotes JA biosynthesis mediated by ROS signaling. Then, JA provides stress tolerance by modulating major enzymatic and non-enzymatic components of the antioxidant defense system (Xu et al., 2010). Additionally, a variety of plant species treated with MeJA have exhibited a reduction in the root uptake of metals. For instance, JA modulates Cd uptake and translocation genes (Lei et al., 2020). Finally, JAs play multifunctional roles in stress tolerance (Ahmad et al., 2016), crosstalking with almost all other known growth regulators/phytohormones (Per et al., 2018).

12.3 Jasmonates Biosynthesis, Transport, and Signaling

Nowadays, JA and its derivatives—collectively called jasmonates (JAs)—as well as their precursors, the octadecanoids, are a growing class of lipid-derived signaling molecules with multiple functions ranging from the initiation of abiotic/biotic stress responses to plant growth and development regulation (Huang et al., 2017; Wang et al., 2020). This section summarizes the current knowledge about the JA biosynthesis pathway, how these molecules are transported related to specific responses, and discusses their complex signaling—highlighted in metal stress.

12.3.1 Biosynthesis

As mentioned earlier, although the first jasmonate compound—jasmonic acid methyl ester—was isolated from the essential oil of *J. grandiflorum* in the 1960s (Demole et al., 1962), it was two decades later when the JA biosynthetic pathway was elucidated by Vick and Zimmermann (1984). It has been extensively described as the octadecanoic pathway (Feussner and Wasternack, 2002) since numerous studies have extensively characterized other JAs using biochemical and genetic tools in model plants such as *Arabidopsis* and tomato (Wang et al., 2020).

Jasmonic acid, along with its derivates and precursors, comes from oxygenated octadecanoid fatty acids and has a pentacyclic ring structure (Wasternack and Strnad, 2018). Initially, the cyclopentanone JA received the most attention as a plant growth regulator belonging to the JAs family (Vick and Zimmerman, 1984; Wasternack and Xie, 2010), as JA is the best characterized and well known as the most abundant among JAs. It is currently well known that JA also confers abiotic stress tolerance in plants, such as heavy metal stress. So far, the biological activity of plants toward stress stimuli is not limited to JA but extends to various JA-metabolites and JA-conjugates along with its cyclopentenone precursors (Wasternack and Hause, 2013; Ahmad et al., 2016).

The biosynthesis of JAs follows a lipid esterification pathway sequentially and involves various subcellular compartments including chloroplasts, peroxisomes, and the cytosol through the following steps: (1) release of α-linolenic acid (α-LeA) from lipid chloroplast membranes to OPDA production (chloroplast); (2) JA biosynthesis (peroxisomes); to culminate with (3) the formation of different JA-conjugates (cytosol).

In the first step, JAs are lipid-derived compounds that originate from the tri-unsaturated fatty acids α-linolenic acid (18:3) localized in chloroplast membranes (Wasternack and Strnad, 2018). So, first, JA biosynthesis starts by liberation of α-LeA from these membrane lipids by the action of phospholipases (Wasternack and Hause, 2013; Hou et al., 2016). Second, the addition of molecular oxygen to α-LeA is catalyzed by a chloroplast-located 13-lipoxygenase (13-LOX) from the intermediate compound 13-hydroperoxy-9,11,15-octadecatrienoica cid (13-HPOT; Feussner and Wasternack, 2002; Reyes-Díaz et al., 2016). This substrate is then oxidized to the unstable allene oxide by allene oxide synthase (AOS), which is rapidly converted to 12-oxo-phytodienoic acid (12-OPDA) by the enzyme allene oxide cyclase (AOC; Schaller and Stintzi, 2009; Gfeller et al., 2010). These first steps take place in plastids, whereas 12-OPDA metabolism follows their occurrence in peroxisomes.

In the second step, OPDA has several fates, including esterification into plastid lipids and transformation into the 12-carbon prohormone jasmonic acid (JA) by the activity of 12-oxo-phytodienoic reductase (OPR) and three cycles of β-oxidation in the peroxisomes (Delker et al., 2006). This step named the JA biosynthetic pathway as the octadecanoid pathway (Feussner and Wasternack, 2002; Han, 2017).

The third step occurs in the cytosol, where JA is a substrate for further diverse conjugations such as MeJA and JA-isoleucine (JA-Ile), the latter is the major bioactive compound among a growing number of JAs in plants (Fonseca et al., 2009a,b; Wasternack and Strnad, 2018).

Including metal stress to several biotic stimuli such as herbivory and pathogen attack, these are environmental factors that could induce JA and MeJA biosynthesis (Farmer et al., 2003). Metal toxicity is one of the major abiotic stresses leading to hazardous health effects in animals and plants. As a potent abiotic elicitor, the metal stress effect triggers JA accumulation, which has been investigated by Maksymiec et al. (2005). In this study, Cu and Cd *in vivo* conditions induced accumulation of JAs in *Arabidopsis thaliana* mature leaves and young and oldest *Phaseolus coccineus* plants.

12.3.2 Transport

Beyond biotic responses (Tamogami et al., 2012), JAs are the major plant hormones involved in various responses to abiotic stress, such as metal stress (Raza et al., 2020). It means that most knowledge about JA transport is focused on wounding responses; the same mechanisms are followed in their intracellular transport and throughout different tissues and organs of plants.

Current knowledge about the intracellular transport of JAs is limited due to technical difficulties in JA measurements throughout different subcellular compartments (Skalicky et al., 2018). The importance of subcellular distribution on the metabolism and signaling of JAs has been revealed by peroxisomal ATP-binding cassette (ABC) COMATOSE (CTS) transporter identification (Wang et al., 2019). CTS is the key point for the peroxisomal entry of the substrates of β-oxidation, including the precursor, 12-OPDA (Footitt et al., 2007), although OPDA could also be imported into peroxisomes via passive diffusion or a second importer (Theodoulou et al., 2005). Some studies highlight that CTS-mediated distribution of OPDA between the cytosol and peroxisome plays an essential role in the regulation of the homeostasis and signaling between OPDA and JA, as *cts* mutants enhanced OPDA levels in the cytosol, promoting the distinct OPDA signaling pathway (Maynard et al., 2018).

So far, JA biosynthesis occurs by an external stimulus such as damaged leaves that activate systemic wounding responses. After the signaling, JA and JA-Ile are transported from wounded (damaged) to undamaged leaves. A JA-Ile transporter has recently been described, GTR1, which is involved in the translocation of JA and JA-Ile in plants and perhaps contributed to the correct positioning of JA and JA-Ile to attenuate an excessive wound response in undamaged leaves (Tamogami et al., 2012). Also, MeJA is transported to distal leaves via the vascular process, metabolizing itself into JA-Ile and triggering

VOCs emission as defensive metabolites (Tamogami et al., 2012; Ishimaru et al., 2017). JA-Ile are translocated to the nucleus, where JAs have retrograde signaling (Muñoz and Munné-Bosch, 2019). Although it has not been reported, studies about jasmonate transport under plant metal stress could be a similar pathway to a systematic response under soil and water contaminants in order to prevent toxicity and plant death.

12.3.3 Signaling

Originally the word "hormone" comes from the Greek meaning "to stimulate," so it does not require the notion of transport *per se* (Davies, 2010). In this sense, plant hormones can have their site of action far from the place where they are synthesized due to complex signaling cascades. In this section, we will discuss transcription factors and genes involved in the regulation of the JA signaling pathway and the important role in crosstalk in response to abiotic stress by metals.

Under stressful environmental conditions, plants suffer an intracellular imbalance of ROS triggering cell oxidative stress. It could lead to whole-plant death if plants are not able to cope with them by the stimulation of antioxidant mechanisms (Miller et al., 2010). Alternatively, plants can also alter endogenous levels of phytohormones in response to stressful conditions. To date, climate change and anthropogenic activities modulate the carbon balance in ecosystems and contribute to soil pollution and water with metals (Senesil et al., 1999; Vareda et al., 2019). One of the major consequences of metals for plants is enhanced ROS production which usually damages cellular components such as lipid membranes (Gratão et al., 2005). Lipid peroxides formation may be a prolonged consequence of metal-induced oxidative stress and may act as an activation signal for plant defense genes by increasing the octadecanoid pathways to jasmonate biosynthesis (Maksymiec et al., 2005).

JA, OPDA, and JA-amino acid conjugates are well described to act as retrograde signals in plant development processes and plant adaptation to the environment by controlling responses to external abiotic/biotic stressors. According to the abiotic stress response pathway, the JAZ-MYC module (JAZ: JAsmonate ZIM domain) plays a central role in the JA signaling pathway through the integration of regulatory transcription factors (MYC) and related genes (Wang et al., 2020). After hormone perception by CORONATINE INSENSITIVE1 (COI1), JAZ repressors are targeted for proteasome degradation, releasing MYC2 and derepressing transcriptional activation. JAZs are homomeric and heteromeric proteins and have been instrumental in recent advances in the field, such as identifying COI1 as a critical component of the jasmonate receptor and the discovery of the bioactive jasmonate in *Arabidopsis*, (+)-7-iso-JA–Ile. Small changes in the jasmonate structure result in hormone inactivation and might be the key to switching-off signaling for specific responses to stimulus and long-distance signaling events (Fonseca et al., 2009a; b).

Finally, it has been reported that JA is a plant-signaling molecule closely associated with plant resistance to abiotic stress, which usually is involved in physiological and molecular responses. Physiological responses often include a relevant activation of the antioxidant mechanisms, both enzymatic (such as O_2^-, peroxidase, and NADPH-oxidase) and non-enzymatic antioxidants (e.g., ascorbate; Noriega et al., 2012; Zhao et al., 2016; Ahmad et al., 2017). Whereas molecular responses often involve the expression of JA-associated genes (JAZ, AOS1, AOC, LOX2, and COI1), interactions with other plant hormones (ABA, ET, SA, GA, IAA, and BR), and interactions with TFs (MYC2 and bHLH148; Wang et al., 2020; Raza et al., 2020).

Several studies have reported that the exogenous application of MeJA alleviates metal damage by increasing antioxidant enzyme activity and secondary metabolite levels (reviewed by Ali and Baek, 2020). Siddiqui and Husen (2019) reported several studies that demonstrated that JA mitigates the deleterious effects of metal stress-reducing oxidative stress, such as Cd (Chen et al., 2014; Ahmad et al., 2017) and As (Farooq et al., 2018a) in the model *Arabidopsis thaliana* plant. However, further studies are needed since metallic stress responses could be species-specific, as shown by Wiszniewska et al. (2019), in which JAs increased under Pb but not under Cd in the halophyte *Aster tripolium*. Further studies are required to understand these relationships in the context of species tolerance to metal stress.

12.4 Jasmonates and Their Roles in Plants under Metal Stresses

JAs, more known as lipid-derived phytohormones such as jasmonoyl isoleucine (JA-Ile), JA-glucosyl ester, 12-hydroxyjasmonic acid sulfate (12-HSO4-JA), MeJA, jasmonic acid 3-oxo -2-2′-cis-pentenyl-cyclopentane-1-acetic acid (JA), among other compounds derived from cyclopentanones and belonging to the oxidized lipids family more commonly named oxylipins (Ruan et al., 2019; Ali and Baek, 2020). These natural compounds are crucial in plant growth regulation and are involved in several physiological, morphological, and biochemical processes (Ueda and Saniewski, 2006; Huang et al., 2017). For example, vegetative growth, cell cycle regulation, anthocyanin biosynthesis, fruit ripening, stomatal opening, nitrogen, and phosphorus uptake, as well as signaling molecules that regulate abiotic stress like low temperature, drought, salt, micronutrient toxicity, and metals, among others (Yoshida et al., 2009; Keramat et al., 2010; Paeizi et al., 2018). Furthermore, metabolic genes have been associated with glutathione in response to metals and JAs, suggesting that this phytohormone could be involved in signaling for cooper and cadmium, increasing mRNA levels and the capacity for glutathione synthesis under these stress conditions (Xiang and Oliver, 1999). In this sense, reports made by Piotrowska et al. (2009) showed that low JA doses inhibit metal accumulation, restored the growth of *Wolffia arrhizal*, and increased GSH content. With respect to these metals, studies performed by Maskymiec et al. (2005) showed that Cu and Cd stress act as potent elicitors triggering the jasmonate accumulation in leaves of *Arabidopsis thaliana* and *Phaseolus coccineus* plants. The accumulation behavior was carried out in two stages in both species, in the first seven hours, jasmonate accumulated in *A. thaliana* and 14 hours for *P. coccineus*, followed by a rapid decrease during the next seven hours. These findings could indicate that JAs are connected with the strategies of toxic action of these metals in plants. Other works, performed with Cd and As, described which metals may affect root growth in a paddy field in interactions among phytohormones like auxin and JAs during adventitious and lateral root formation (Ronzan et al., 2019). However, little is known about the impact at the physiological, biochemical, or molecular level under these metal stress conditions.

12.4.1 Jasmonates and Cadmium Stress

Cd is a non-essential toxic heavy metal for plants that negatively affects cellular processes, enzymatic activity, and electron transport. Cd stress in plants can indirectly produce an overaccumulation of ROS, leading to oxidative stress that can cause severe damage to lipids, proteins, and DNA (Lei et al., 2020; Singh and Shah, 2014; Yan et al., 2015; Yan et al., 2013).

Jasmonates such as JA and MeJA are phytohormones that regulate many important processes for plant growth and development and can also act as signaling molecules involved in the cellular response to Cd stress (Chen et al., 2014; Singh and Shah, 2014). In plants of *Oryza sativa* treated with 50 μM of $CdCl_2$, MeJA increased roots up to four times and shoots up to six times after days of Cd treatment, possibly due to hormone stimulation accumulation induced by H_2O_2 (Singh and Shah, 2014). Jasmonic acid increased by two times in plants of *Pisum sativum* after 14 days of 50 μM of $CdCl_2$ exposure (Rodriguez-Serrano et al., 2006). Changes in endogenous JA in the leaves of *Capsicum frutescens* were not significant after 24 hours of Cd treatment (at 50 mg L^{-1} $CdCl_2$), but a significant increase in JA was noticed after 48 hours of exposure (Yan et al., 2013). By contrast, in *Kandelia obovata* leaves, 200 μM of Cd caused a decrease in JA concentrations after nine days of treatment (Chen et al., 2014).

Moreover, plants of *Solanum nigrum* after seven days of treatment with Cd did not show any significant effect on leaf endogenous JA content, whereas an increase in MeJA accumulation in the leaves was induced in Cd-stressed plants (Yan et al., 2015). In *Arabidopsis thaliana* treated with 100 μM of Cd, an increase of six times of JAs was observed after seven hours of treatment and three times after five days (Maksymiec et al., 2005). Another study in *A. thaliana* found that the root MeJA concentration was significantly increased after Cd exposure from six hours to three days (Lei et al., 2020). This study also found the overexpression of genes involved in JAs biosynthesis such as *AtLOX3*, *AtLOX4* (*lipoxygenases 3* and *4*), and *AtAOS* (*allene oxide synthase*) after one hour of exposure to Cd, indicating that Cd stress enhances *Arabidopsis* endogenous JA biosynthesis via upregulation of the expression of JA biosynthesis genes (Lei et al., 2020). In the same study, the authors used mutants of *AtAOS* that had less mRNA abundance than a wild type, and both were exposed to Cd stress (at 50 μM), which resulted in more severe chlorosis symptoms and a higher accumulation of Cd in shoots and roots in the mutant line (Lei et al., 2020). Similar results were observed in JA-deficient *Lycopersicon esculentum* seedlings subjected to 50 mg L^{-1} $CdCl_2$, which showed oxidative stress and Cd accumulation symptoms than a wild-type line (Zhao et al., 2016). Both studies indicate that the decrease in endogenous JA synthesis enhances plant sensitivity to Cd, suggesting that JA positively regulates plant response to Cd stress (Lei et al., 2020; Zhao et al., 2016). The increase in the concentration of JAs in plants subjected to stress by Cd may also be due to the increase in the activity of the enzyme lipoxygenase (LOX), a biosynthetic enzyme of the jasmonate pathway, or the stimulation of the octadecenoic pathway induced by Cd that can lead to jasmonate synthesis (Singh and Shah, 2014; Yan et al., 2013). Despite advances and studies in plant molecular biology, further studies are needed to elucidate the involvement of JAs to alleviate Cd stress.

12.4.2 Jasmonates and Arsenic Stress

As is an extremely toxic metal pollutant, which affects human health worldwide, especially in Asian countries such as India and Bangladesh (Zhao et al., 2010; Shri et al., 2019). Arsenic pollution in soil, which is derived from anthropogenic and natural resources, is becoming a serious concern for the human population due to crop plants accumulating As, leading to contamination of the food chain and causing arsenicosis in humans (Babar and Tariq, 2018; Shri et al., 2019). In plants, As is accumulated in roots and the shoot, inducing detrimental effects on growth and plant development, and decreasing crop yields (Zhao et al., 2010; Niazi et al., 2017; Ronzan et al., 2019). At the morphological level, As toxicity reduces leaf area, number of leaves, and shoot growth inhibition; meanwhile, As toxicity reduces the root cortex area, and it induces cell death of root tips and root growth inhibition in several species such as *Brassica napus*, *Brassica juncea*, *Zea mays*, and *Glycine max* (Armendariz et al., 2016; Anjum et al., 2017; Niazi et al., 2017). Besides, it has been reported that As toxicity reduces CO_2 assimilation, stomatal conductance, and chlorophyll concentrations, which is related to plant growth inhibition (Srivastava et al., 2013; Sanglard et al., 2016; Anjum et al., 2017). Arsenic toxicity increases ROS, which causes damage to lipids, proteins, and membranes, triggering oxidative stress (Jin et al., 2010; Nath et al., 2014; Armendariz et al., 2016). Thus, plants have developed complex mechanisms to cope with metal stress, such as As. Among As-induced mechanisms is the biosynthesis of the phytohormones, which regulates several responses in plants (Ali et al., 2014; Ryu and Cho, 2015; Ronzan et al., 2019). Endogenous JAs trigger positive effects on alleviating As stress in plants, modulating jasmonate-responsive genes, and activating antioxidant defense systems (Howe, 2004; Islam et al., 2016; Mousavi et al., 2020). At the morphological level, Ronzan et al. (2019) reported that As toxicity affected adventitious and lateral root formation and increased JA levels in *Oryza sativa* plants, showing that roots accumulated higher As levels than shoots. Likewise, Mousavi et al. (2020) reported that JA levels were higher (about 20%) in *Oryza sativa* plants subjected to 50 μM of As compared with controls. On the other hand, Farooq et al. (2016) reported that endogenous jasmonates significantly increased in *Brassica napus* plants subjected to As toxicity (200 μM) compared with non-As treated plants. The authors also showed that high JAs levels stimulated the phenylpropanoid pathway, which biosynthesizes phenolic compounds in *B. napus* plants. Thus, Dar et al. (2015) suggested that high JA levels in plants mitigate ROS during metal stress throughout the biosynthesis of phenolic compounds.

12.4.3 Jasmonates and Aluminum Stress

Al is the third most common element in the earth's lithosphere (Lindsay, 1979). Despite its high levels, Al is non-toxic to plant growing soils with neutral or near-neutral pH values due to low solubility (Lindsay, 1979; Schneider et al., 2004). However, at the low pH of the soil (acidity), Al solubilizes from non-toxic silicate or oxide forms into phytotoxic Al^{3+} (Tahara et al., 2008). Aluminum toxicity inhibits root growth in many plant species at micromolar concentrations in acidic conditions, suggesting that Al interferes with dividing and expanding root cells (Tamás et al., 2006). This inhibition limits the uptake of water and nutrients and increases plant susceptibility to other stresses in the root zone, that is, low pH and Ca imbalance, among others (Kochian et al., 2015). Despite extensive efforts to decipher the Al phytotoxicity mechanisms, the primary cause remains largely speculative (Rengel, 1996; Barcelo and Poschenrieder, 2002; Kochian et al., 2015; Reyes-Díaz et al., 2015). Currently, it has been reported that JAs could decrease the harmful effect of Al^{3+} stress in plants. However, relatively very few studies relate Al stress tolerance with endogenous concentrations of jasmonates. A study in the root tips of *Cassia tora*, a plant species well adapted to acid soil, showed an increase in JA concentration and lignin synthesis related to Al-induced oxidative stress (Xue et al., 2008). After 48 hours of Al exposure, roots of Nipponbare also increased concentrations of JAs; however, the author did not relate this increase with Al-induced lipid peroxidation, indicating that further studies are necessary for relating the increase of JA concentrations with the enhancement of lignification (Roselló et al., 2015).

In this context, some research showed the relation between WRKY transcription factors, Al^{3+} toxicity, and jasmonic acid-regulated resistance responses (Rabara et al., 2013; Wang et al., 2020). The transcriptions factor WRKYs participate as regulators of Al-activated malate transporter (ALMT) proteins through binding to their promoters and altering malate efflux in plant roots as the main resistance mechanisms of Al toxicity (Ding et al., 2013; Ye et al., 2017). It is known that Al^{3+} toxicity induced inhibition of root growth in several plants, but JAs improve it when plants are exposed to Al^{3+} stress (Delhaize and Ryan, 1995; Kochian et al., 2004; Huang et al., 2017). Recently, Wang et al. (2020) reported the crosslink in *Solanum lycopersicum* roots among jasmonate and Al at the molecular levels; they found that *SlALMT3* is a regulator gene between Al and JAs in root growth inhibition, and six SlWRKY transcription factors could be involved in upstream regulations of the *SlALMT3* gene. Although there is little research that evaluates how JAs and metals influence physiological and molecular traits, such as root growth, GSH content, WRKYs transcription factor, or *ALMT* genes, the understanding of this complex crosslink between these factors in plants is not yet fully elucidated.

12.4.4 Jasmonates and Other Heavy Metals

Nutrients are fundamental in different physiological and biochemical processes in plants, being involved in chlorophyll biosynthesis, photosynthetic processes, DNA synthesis, protein modifications, redox reactions in the mitochondrion, sugar metabolism, and nitrogen fixation, among others. As mentioned previously, some metals occur naturally in the earth's crust at various levels; the problem arises when released in excess into the environment due to natural processes or anthropogenic activity (Singh et al., 2016). Elements such as silicon (Si) are considered beneficial elements, and it has been reported to be associated with the maintenance of structures in some plant species (Epstein, 1999). Zinc is crucial in the activities of many enzymes and transcription factors involved with maintaining membrane integrity, auxin metabolism, and reproduction (Marschner, 1995; Briat et al., 2007; Ricachenevsky et al., 2013). Nonetheless, excess or high concentrations of metals can produce toxicities in different organs or whole plants, and therefore, their uptake and utilization are limited by the plant cells to avoid oxidative damage (Saito et al., 2010; Srivastava et al., 2012; Farias et al., 2013; Millaleo et al., 2013). Studies performed using Si and MeJA in *O. sativa* with the objective to evaluate the effects of combined treatments on vegetative development and genetic stability showed that the use of Si or Si +MeJA promoted an increase in the height, relative chlorophyll index, and biomass, but no difference in the quantity of DNA was observed; whereas MeJA contributed to the vegetative development and did not affect the genetic stability in rice plants (Nascimento et al., 2019).

On the other hand, Zamani et al. (2019) investigated the effect of MeJA, and sodium silicate on the content of some mineral elements in *Solanum lycopersicum* plant leaves under salt stress, suggesting that MeJA and Si can partially mitigate the negative effects of salinity stress, contributing to an increase in the uptake of nutrients under this condition. Apparently, MeJA and Si could contribute to alleviating the negative effects of salinity stress than each one alone; however, more research is necessary to confirm these findings of the combined treatments of MeJA and Si in plants. In other matters, Ye et al. (2013) described that Si enhances JAs signaling pathways, which mediate biotic defense responses in *O. sativa* by increasing resistance to insect herbivores by Si accumulation in leaves, mediated by the jasmonate pathway. Recently, using high atmospheric CO_2 suppresses jasmonate and Si using like defenses without affecting herbivores in *Brachypodium distachyon* grown under ambient (400 ppm) and high (640 ppm) CO_2 concentrations. The results suggested that Si alters the jasmonate response in plants due to Si treatment, which had higher jasmonate levels than without Si plants, and also showed that jasmonate induces Si uptake; however, Si reduces jasmonate response in plants subjected to stress conditions (Hall et al., 2020). Almost all metal pollutes of the environment negatively impact plant growth and development in several plant species. Indeed, in *Suaeda glauca* and *A. thaliana* exposed to high Pb, Ni, Cd, and Mn concentrations, the fresh weight and photosynthetic pigment concentrations decreased in both species (Zhang et al., 2018). Many of these metals have no beneficial functions in plants and could be toxic for plants even at very low levels or in excess conditions (Wang et al., 2020). In this context, Sirhindi et al. (2015), using *Glycine max* plants with applicate JAs before treated $NiCl_2$ toxicity enhanced seeding tolerance to Ni^{2+} stress. Furthermore, the authors revealed that JAs protected the seedlings by regulating the antioxidant machinery, protecting DNA synthesis, and total proteins. Similar studies performed in *Zea mays* under Ni-stress also revealed that exogenous application of JAs ameliorates the adverse effects of oxidative damage, growth, biomass production, and protein concentrations in plants due to enhanced antioxidant enzyme activity (Azeem, 2018).

12.5 Exogenous Jasmonate Application and Resistance Mechanisms in Plants under Metal Stresses

An excess of essential metals such as Cu, Mo, and Mn, among others, induce toxicity in plants, increasing oxidative stress, and therefore, physiological damage (Dhankar and Solanki, 2011). The application of jasmonates (JA or MeJA) enhances the accumulation of osmolytes, antioxidant enzyme concentration, and so on, which prevents plants from receiving damage by excess metal ions (Poonam et al., 2013).

12.5.1 Exogenous Jasmonates and Cadmium Stress

Cd stress affects the growth and development of plants, which is noticed as a reduction in the length, fresh weight, and dry mass of plant shoots and roots (Figure 12.1). In *Vicia faba* plants exposed to Cd ($CdSO_4*8H_2O$; 150 mg L^{-1}), application of JA increased shoot and root length by 49.9% and 30.6%, respectively, in comparison with the Cd-alone treatment (Ahmad et al., 2017; Figure 12.1). In *Oryza sativa* seedlings, the treatment of Cd at 50 μM caused a reduction in the length of the shoots by up to 24% and up to 25% in roots after ten days of exposure to stress by Cd. The growth of these seedlings was almost 100% restored with the application of MeJA at 5 μM after ten days (Singh and Shah, 2014). Comparable results were observed in *Brassica juncea* treated with 50 μM of Cd and 10 μM of MeJA, see Table 12.1 (Per et al., 2016). Likewise, in *Glycine max* treated with 500 μM of Cd, shoot dry weight was dramatically reduced by 52%, but exogenous MeJA (at 0.01 mM) increased by 30% shoot dry weight in comparison with Cd-stressed plants (Keramat et al., 2010). Otherwise, in *Kandelia obovata* plants, there were no significant changes in plant growth after nine days of treatment with Cd (at 200 μM) and Cd+MeJA (200 μM Cd+0.1; 1; 10 μM MeJA) compared with control plants (Chen et al., 2014), whereas application of 0.01 μM MeJA to *Solanum nigrum* plants subjected to 40 mg L^{-1} $CdCl_2$ was effective on reverting root dry mass reduction caused by Cd toxicity (Yan et al., 2015). However, in the same study, Cd stressed *S. nigrum* plants subjected to 1000 μM of MeJA showed deterioration symptoms such as yellow and wilted leaves and blackened roots, suggesting toxicity by a higher dose of MeJA application. Yan et al. (2013) showed that in *Capsicum frutescens* treated with 50 mg L^{-1} $CdCl_2$, application of MeJA at 10 and 1000 μM reduced shoot and dry root mass, while lower concentrations of this hormone (0.1 and 1 μM) increased shoot and root dry mass, alleviating Cd negative effect on plant growth.

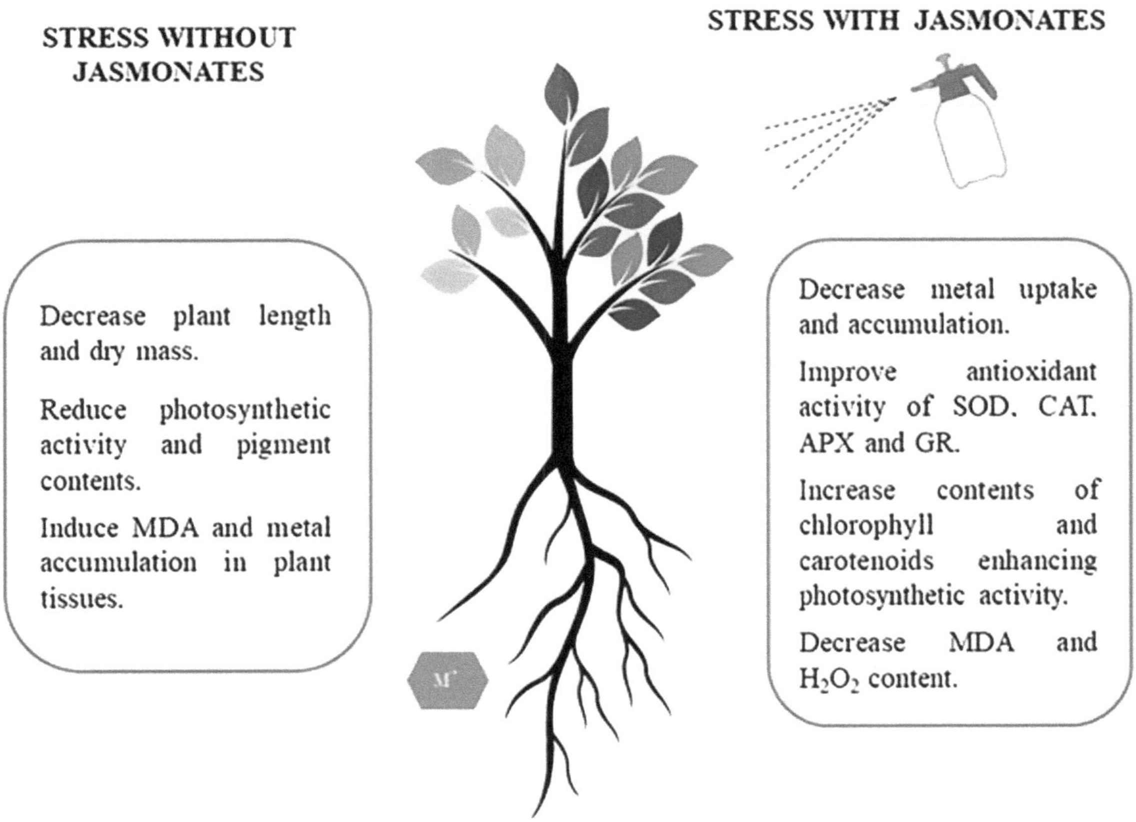

FIGURE 12.1 Impact of jasmonates on physiological and biochemical responses of plants under metal stress conditions.

TABLE 12.1

Effects of Jasmonate Application in Plant Species under Metal Stress

Metal stress/ Dose	Plant Species	Hormone dose	Effects	Reference
Cd 50 μM	*Oryza sativa*	MeJA at 5 μM after 10 days of metal exposure	Recovery by almost 100%; reduction of root and shoot length and increased chlorophyll content by 25% in leaves. Increased levels of GSH in plants and improved GSH/GSSG ratio.	Singh and Shah (2014)
Cd 50 μM	*Brassica juncea*	10 μM of MeJA	MeJA application enhances total chlorophyll content and net photosynthesis and improves root and shoot length.	Per et al. (2016)
Cd 500 μM	*Glycine max*	MeJA at 0.01 mM	Increased shoot dry weight and chlorophyll content. Reduce MDA accumulation.	Keramat et al. (2010)
Cd 200 μM	*Kandelia obovata*	MeJA at 0.1; 1; 10 μM after 9 days of treatment	Inhibited Cd translocation to the leaves and reduced MDA accumulation.	Chen et al. (2014)
Cd 40 mg L^{-1}	*Solanum nigrum*	MeJA 0.01 and 0.1 μM	Reverted root dry mass reduction. Increase GSH content and inhibit Cd uptake.	Yan et al. (2015)
Cd 40 mg L^{-1}	*Solanum nigrum*	MeJA at 1000 μM	Metal and hormone application caused yellow and wilted leaves and blackened roots. Decreased GSH levels in roots and leaves.	Yan et al. (2015)
Cd 50 mg L^{-1}	*Capsicum frutescens*	MeJA at 0.1 and 1 μM	Increase shoot and root dry mass alleviating. Improved SOD, POD, and GPx.	Yan et al. (2013)
Cd 50 mg L^{-1}	*Capsicum frutescens*	MeJA at 10 and 1000 μM	Reduced shoot and root dry mass. MeJA at 1000 uM decreased chlorophyll content. Decreased the activity of CAT and APx.	Yan et al. (2013)
Cd 150 mg L^{-1}	*Vicia faba*	JA at 0.01 mM	Increased levels of chlorophyll α (by 59.8%), chlorophyll β (by 150%), and carotenoids (by 92.9). Hormone application reduced Cd accumulation in different plant tissues. Improved levels of SOD, CAT, APX, and GR.	Ahmad et al. (2017)
Al 100 μM	*Vaccinium corymbosum*	5 μM MeJA	Reduced Al concentration, lipid peroxidation, and H_2O_2 concentrations.	Ulloa-Inostroza et al. (2017)
As 50 μM	*Oryza sativa*	1 μM MeJA	Recovery yield components enhanced the activities of SOD, CAT, APX, and POD enzymes and decreased the gene expression of Lsi1, Lsi2, and Lsi6.	Mousavi et al. (2020)
As 25 μM	*Oryza sativa*	0.25 μM MeJA	Increased root and shoot growth and recovered CO_2 exchange. Decreased Lsi1, Lsi2, and Lsi6 genes, which increased plant growth.	Verma et al. (2020)
As 200 μM	*Brassica napus*	1 μM MeJA	Recovered plant growth.	Farooq et al. (2016)
As 200 μM	*Brassica napus*	1 μM MeJA	Alleviated disruption of the nucleus and nuclear membrane, enhanced the activities of SOD, CAT, APX, and POD enzymes.	Farooq et al. (2018a)
As 200 μM	*Brassica napus*	1 μM MeJA	Mitigated chloroplast damage.	Farooq et al. (2018b)

On the other hand, photosynthetic activity and pigment contents can be reduced in Cd-stressed plants, which is possibly a consequence of inhibition of enzymes responsible for chlorophyll biosynthesis or damage to photosystems (Ahmad et al., 2017; Keramat et al., 2010). Stress-induced by Cd in *V. faba* plants decreased chlorophyll a (chl a), chlorophyll b (chl b), and carotenoids concentrations but treatment with JA 0.01 mM helped increase levels of chl a, chl b, and carotenoids levels by 59.57%, 150%, and 92.85%, respectively, compared with Cd-alone treated plants (Ahmad et al., 2017). Total chlorophyll contents and net photosynthesis also decreased in *B. juncea* plants treated with Cd 50 μM; however, applying MeJA at 10 μM enhanced chlorophyll content by 74% and net photosynthesis by 81% (Per et al., 2016). In plants of *G. max*, Cd exposure (500 μM) reduced total chlorophyll content by 57%, and MeJA treatment alone (0.01 and 0.1 mM) slightly decreased total chlorophyll content compared with control plants, but MeJA application after Cd treatment was effective in reducing the damage of Cd stress to chlorophyll content by 20% (Keramat et al., 2010). Similar results were observed in *O. sativa* by Singh and Shah (2014), where Cd at 50 μM decreased chlorophyll content by 50% in leaves of 10-day-old seedlings, and MeJA application increased chlorophyll concentrations by almost 25% after ten days of Cd treatment (Table 12.1). Yan et al. (2013) showed that the application of MeJA at 0.01 μM to Cd-exposed plants of *C. frutescens* increased total chlorophyll content, whereas exogenous MeJA at 1000 μM significantly decreased it.

To protect the photosynthetic apparatus, JAs can decrease Cd uptake and accumulation in plant shoots and leaves, resulting from stomatal closure and decreased transpiration induced by the exogenous hormone application (Ahmad et al., 2017; Chen et al., 2014). A decreased Cd uptake and translocation to the leaves can also result from the downregulation of transporter genes (Lei et al., 2020). It was noticed that in *V. faba* plants exposed to Cd, high concentrations of this metal were found in roots, shoots, and leaves, nevertheless, exogenous JA 0.01 mM was effective in reducing Cd concentrations to 73.3% in leaves, 60.6% in shoots, and 57.5% in roots relative to Cd-stressed plants alone (Ahmad et al., 2017). Plants of *O. sativa* treated with Cd (50 µM) and MeJA (5 µM) showed a decrease in Cd uptake and accumulation on the shoot by ~0.5 times and 2.5 times in roots compared with plants treated with Cd alone after ten days of treatment (Singh and Shah, 2014; Table 12.1). In *K. obovata* plants, after nine days of 200 µM Cd treatment, the leaves and roots accumulated 248.5 and 384 µg g^{-1} dry weight (DW), respectively. By contrast, combined Cd and MeJA treatments significantly inhibited the translocation of Cd to the leaves, being reduced to 22.6 µg g^{-1} DW with the application of MeJA 0.1 µM and 98.4 µg g^{-1} DW with MeJA 10 µM; however, the addition of MeJA did not affect the root uptake of Cd (Chen et al., 2014). Other authors showed that in plants of *S. nigrum* Cd accumulation increases with the application of 40 mg L^{-1} $CdCl_2$, but MeJA treatment at low concentration (0.1 µM) inhibited Cd uptake, whereas MeJA at higher concentrations (> 1 µM) did not significantly reduce Cd accumulation in shoots and roots (Yan et al., 2015).

Also, in Cd-stressed plants, an increase in the levels of malondialdehyde (MDA) and H_2O_2 is usually observed as a consequence of the oxidative damage induced by Cd. A significant increase in MDA levels, considered an indicator of lipid peroxidation, was observed in *G. max* plants treated with 500 µM of Cd (Keramat et al., 2010; Table 12.1) and in *K. obovata* plants with 200 µM of Cd (Chen et al., 2014; Table 12.1). In both studies, the application of MeJA up to 10 µM was effective in alleviating the increase in MDA Cd stress induced. Another study in *G. max* showed that pretreatment with 20 µM JA helped prevent lipid peroxidation in plants (Noriega et al., 2012). MDA and H_2O_2 levels almost doubled in roots and shoots of *O. sativa* subjected to 50 µM Cd, but the application of 5 µM MeJA decreased the content of both compounds (Singh and Shah, 2014). In *B. juncea* plants exposed to 50 µM Cd, the MDA and H_2O_2 contents in leaves were significantly higher than in control plants, but exogenous MeJA (10 µM) helped to reduce the increment of MDA and H_2O_2 levels by about 60%. However, the application of MeJA 20 µM significantly increased H_2O_2 content in leaves compared with Cd-stressed plants (Per et al., 2016). In *C. frutescens* plants subjected to 50 mg L^{-1} $CdCl_2$, MeJA 0.1–1000 µM treatment increased the levels of H_2O_2 in leaves. In the same study, MeJA application (at 10 and 1000 µM) to Cd-stressed plants increased MDA content in leaves, while lower concentrations of the hormone (0.1 and 1 µM) were able to reduce the lipid peroxidation (Yan et al., 2013). Otherwise, Ahmad et al. (2017) showed that applying JA 0.01 mM to Cd treated plants of *V. faba* with $CdSO_4$*$8H_2O$ 150 mg L^{-1} reduced concentrations of H_2O_2 and MDA by around 1.5-fold.

To tolerate Cd-induced oxidative stress, plants need to enhance their antioxidant activity related to enzymatic and non-enzymatic compounds (Chen et al., 2014; Singh and Shah, 2014; Yan et al., 2015). Glutathione (GSH) and ascorbic acid (AsA) are non-enzymatic antioxidants that can react with ROS to prevent oxidative damage to the cell (Keramat et al., 2010; Yan et al., 2013). Cd stress in *O. sativa* plants can oxidize GSH producing glutathione disulfide (GSSG), decreasing GSH/GSSG ratio; however, the application of MeJA can increase the levels of GSH in the plant, improving GSH/GSSG ratio (Singh and Shah, 2014). Otherwise, in *S. nigrum* plants subjected to Cd stress, exogenous MeJA applied at low concentrations (less than 0.1 µM) induced an increase in GSH contents, but higher amounts of the hormone caused a decrease in GSH levels in roots and leaves (Yan et al., 2015; Table 12.1). AsA concentrations in the plant decreased as a consequence of Cd-stress in *K. obovata* (Chen et al., 2014) but increased in Cd-stressed plants of *G. max* (Keramat et al., 2010), in both studies, it was shown that MeJA improves AsA production in plants. In plants of *O. sativa* subjected to 50 µM Cd, improvement of enzymatic activity of ROS scavengers such as superoxide dismutase (SOD), catalase (CAT), and peroxidase (POD) after 5 µM MeJA treatment was observed (Singh and Shah, 2014). Similar studies were performed in *V. faba* plants by Ahmad et al. (2017), who demonstrated that exogenous MeJA 0.01 mM improved levels of SOD, CAT, ascorbate peroxidase (APx), and glutathione reductase (GR) by 87.5, 130.5, 132.6, and 37.8%, respectively. Furthermore, in *C. frutescens* combined treatment of Cd and MeJA enhances enzymatic activity of SOD, POD, and glutathione peroxidase (GPx), whereas application of MeJA at concentrations above 1 µM significantly decreased the activity of CAT and APx (Yan et al., 2013; Table 12.1). Although lower doses of JAs improve the damage of Cd stress in plants, it is species and dose dependent. More studies are needed to elucidate the JAs mechanisms involved in improving Cd stress in plants.

12.5.2 Exogenous Jasmonates and Arsenic Stress

As we mentioned earlier, As toxicity induces detrimental effects on growth and plant development, decreasing crop yields (Ronzan et al., 2019; Figure 12.1). Exogenous plant hormones such as salicylic acid, abscisic acid, and jasmonates are currently used as a tool to improve plant tolerance against metal stress (Bartwal et al., 2013; Ali et al., 2015; Farooq et al., 2016; Figure 12.1). Among these plant hormones, JAs, including JA and its aromatic ester MeJA, have been reported to improve morphological, physiological, and biochemical traits in plants under metal stress such as As (Farmer, 2007; Yoon et al., 2009; Dar et al., 2015; Farooq et al., 2018a; Figure 12.1). At the morphological level, Farooq et al. (2016) reported that As stress (200 µM) significantly reduced root and shoot growth in *Brassica napus* plants (Table 12.1). Meanwhile, exogenous MeJA at 1 µM recovered plant growth in *B. napus* under this stress. Ronzan et al. (2019) showed that in *Oryza sativa* plants, As stress reduced lateral root density, while MeJA increased it. Recently, Mousavi et al. (2020) reported that As stress (50 µM) negatively affected plant growth, crop yield, and yield components such as tiller/hill, grains/panicle, filled grain percentage, 1000-grain weight, and harvest index in *O. sativa* plants. However, the authors showed that 1 µM of MeJA recovered yield components in *O. sativa* under As stress improved crop yield. Similarly, Verma et al. (2020) showed that exogenous 0.25 μM of MeJA increased root and shoot growth in *O. sativa* plants, which were inhibited by 25 μM of As stress (Table 12.1). Likewise, the authors reported that exogenous MeJA significantly decreased As accumulation (about 40%) in the root and shoot of *O. sativa* plants was subjected to As stress compared with non-MeJA treated plants (Table 12.1). Interestingly, transmission electron microscopy analysis showed that As treatment-induced abnormal shaped swollen chloroplasts with loose thylakoid membranes in leaf ultrastructure of *B. napus* plants subjected to As stress (200 μM; Farooq et al., 2018b). In contrast, the authors showed that 1 μM MeJA mitigated the chloroplast damage in *B. napus* under As stress. In roots, the disappearance of the nucleolus and disruption of the nucleus and nuclear membrane were observed in plants exposed to As stress, which were alleviated by 1 μM of MeJA application (Farooq et al., 2018a). At the physiological level, Verma et al. (2020) showed that CO_2 assimilation, stomatal conductance, and chlorophyll content were significantly reduced by about 45% in *O. sativa* plants under As stress, which, in turn, decreased plant growth. Thus, 0.25 μM MeJA was capable of recovering gas exchange and plant growth in *O. sativa* plants subjected to As stress. Some studies have reported that MeJA can scavenge ROS produced by As stress in plants at biochemical levels, inducing the antioxidant defense systems (Meharg and Hartley-Whitaker, 2002; Dave et al., 2013; Petrov et al., 2015). Thus, Farooq et al. (2018a), Mousavi et al. (2020), and Gomes et al. (2020) have reported that MeJA enhanced the activities of SOD, CAT, ascorbate APX, and POD enzymes, which reduced H_2O_2 and $O_2^{\cdot-}$ contents in *B. napus*, *O. sativa*, and *Lemna valdiviana* plants subjected to As stress. At the molecular level, Mousavi et al. (2020) showed that 1 μM MeJA reduced As accumulation in the root and shoot of *O. sativa* plants, decreasing the gene expression of the *Low silicon 1* (*Lsi1*), *Low silicon 2* (*Lsi2*), and *Low silicon 6* (*Lsi6*), which are reported encoding arsenic transporters in plants. According to Ma et al. (2008), Lsi1 and Lsi2 transporters mediate arsenic uptake in the roots of *O. sativa* plants, while Lsi6 transporter mediates arsenic transport from root to shoot in *O. sativa* plants. Indeed, Verma et al. (2020) reported that 0.25 μM MeJA decreased *Lsi1*, *Lsi2*, and *Lsi6* gene expression in *O. sativa* plants subjected to As toxicity, which, in turn, increased plant growth compared with non-MeJA treated plants. Therefore, all these morphological, physiological, biochemical, and molecular studies show that MeJA could be an important tool for improving As toxicity tolerance in plants, although the effects depend on species and dose applied.

12.5.3 Exogenous Jasmonates and Aluminum Stress

As mentioned previously, the acidity of the soil (pH < 5.5) solubilizes the Al complex to phytotoxic aluminum (Al^{3+}), which is harmful to plant crops (Delhaize et al., 2012; Kochian, et al., 2015). At lower concentrations, Al^{3+} negatively affects physiological, biochemical, and morphological processes, depending on species, genotypes, and degree of tolerance (Barceló and Poschenrieder, 2002). It has been reported that the application of JAs neutralizes the toxic effects of metals (Figure 12.1), enhancing the accumulation of osmolytes and antioxidants, among other mechanisms of plants (Poonam et al., 2013). However, whether or not the effects of the jasmonates application enhance the defense mechanisms in the plants is species and dose dependent. A study performed in mutants of *Arabidopsis thaliana* defective in JA biosynthesis indicated that exogenous application of JA enhanced the Al-induced root growth inhibition (Yang et al., 2017). However, the MeJA application to *Brugmansia candida* plants exposed to toxic Al increased H_2O_2 accumulation, peroxidase activity, lignin accumulation, and NADH activity, increasing the oxidative damage (Spollansky, 2000). A similar response was also reported for *Cassia tora* roots exposed to toxic Al, where increased concentrations of MeJA were applied, decreasing root elongation and increasing oxidative damage (Xue et al., 2008). Contrarily, low doses of MeJA applied exogenously and simultaneously with toxic Al improve the antioxidant performance of *Vaccinium corymbosum* plants under greenhouse conditions (Ulloa-Inostroza et al., 2017; Table 12.1). The authors concluded that with MeJA application, the Al-resistant cultivar increased mainly non-enzymatic compounds to counteract Al toxicity. In contrast, the Al-sensitive cultivar increased the SOD activity, indicating that applying a low dose of MeJA could be a good strategy for reducing the negative effects of Al toxicity in *V. corymbosum*, decreasing Al concentration in tissues and strengthening the antioxidant mechanisms. Therefore, the role of MeJA on resistance mechanisms against abiotic stresses as Al toxicity is still controversial and more studies are needed to determine the physiological, biochemical, and molecular mechanisms involved in the protective effect of MeJA against metal stresses.

12.5.4 Exogenous Jasmonates and Other Heavy Metals

Although many metal ions are important nutrients in excess, they can be toxic for plants, provoking oxidative damage that negatively affects diverse physiological processes (Siddiqi and Husen, 2019). Nonetheless, JA application increased carotenoid accumulation and improved the antioxidant system defense to counteract damage by the excess Cu ions in *Cajanus cajan* plants (Sharma et al., 2013). Recent studies showed results suggesting a lack of a clear pattern for MeJA in *Phaseolus coccineus* plants treated with Cu, where observed slight modifications in chlorophyll fluorescence and decreased photosynthetic pigments treated with MeJA compared with the control. Nevertheless, the authors related these results to the plant growth stage (Hanaka et al., 2015). Otherwise, Sirhindi et al. (2015) using JA in *Cajanus cajan* plants under Cu stress, showed that JA increased the protein, sugar, and vitamin content indicating the potential of this phytohormone. Recent studies suggest that the application of JA and MeJA to seeds before germination in *Brassica oleracea* modulates growth, photosynthetic activity, and expression of photosystem II and influenced the assembly, stability, and repair of PSII improved photosynthetic system efficiency, leading to increasing the photosynthetic productivity in terms of saccharides accumulation (Sirhindi et al., 2020). Metals like Cd can generate ROS, which caused oxidative stress in plants (Siddiqi and Husen, 2019; Ali and Baek, 2020). Jasmonic acid natural accumulation in *Phaseolus coccineus* and *Arabidopsis thaliana* was investigated using a combined assay with Cu and Cd using GC-MS measurements. The results showed that these authors suggest that JA is connected with the mechanism of toxic action of both metals, presenting a differential response depending on plant species and growth stage (Maksymiec et al., 2005). Otherwise, it has been described that exogenous application of MeJA can induce anatomical and chemical changes that are components of defense responses in plants (Li al., 2018). In this sense, studies were carried out in species such as *Helianthus annuus*, *Solanum lycopersicum*, and *Glycine max* on the absorption and translocation of Zn, Mn, and Fe using the foliar application. They concluded that changes in leaf properties due to the application of MeJA increased foliar absorption of Zn, Mn, and Fe 3- to 5-fold in sunflower; however, in tomato, it decreased around 0.5 to 0.9-fold, while in soybean, it was unaffected (Li et al., 2017). These differences in the absorption of nutrients could be related to the thickness of the cuticle and epidermal cell wall provoked by MeJA application. Therefore, the exogenous application of MeJA or other jasmonates forms effectively alleviates metal damage by increasing the levels of antioxidative enzyme activity and secondary metabolites that explain the positive effects of JAs to cope with metal stress in plants.

12.6 Conclusion

Currently, many efforts have been performed to understand the complexity of stress signal transduction pathways of jasmonates, which is still not yet fully understood, and deeper research is needed. On the other hand, studies about metal stresses and the signal pathway, transport, and storage of jasmonates are scarce, despite the advancement of molecular technologies. Thus, understanding stress tolerance in plants and the possible strategies to improve their mechanisms are crucial to managing crops subjected to metal stresses. Therefore, it is necessary to study the role of jasmonates in the resistance mechanisms of plants under metal stresses and the possible use of these hormones as a new tool in order to improve stress tolerance in plants, especially crops.

Acknowledgments

We are thankful to the Fondecyt N° 1171286 project and PostDoc Fondecyt N° 3200594 from Agencia Nacional de Investigación y Desarrollo (ANID, ex-CONICYT) of the Government of Chile, and Profesores Patrocinantes DI20-2014 from Dirección de Investigación-UFRO.

REFERENCES

Ahmad et al. *Front. Plant Sci.* 2016, 7:813.
Ahmad et al. *Arch. Agron. Soil Sci.* 2017, 63(1):1–11.
Aldridge et al. *J. Chem. Soc. C Org.* 1971, 1623–1627. The Royal Society of Chemistry.
Ali and Baek. *Int. J. Mol. Sci.* 2020, 21(2):621.
Ali et al. *Biol. Plant.* 2014, 58:131–138.
Ali et al. *PLoS ONE* 2015, 10(4):e0123328.
Amari et al. *S. Afr. J. Bot.* 2017, 111:99–110.
Anjum et al. *Water Air Soil Pollut.* 2017, 228(1):13.
Armendariz et al. *Plant Physiol. Biochem.* 2016, 103:45–52.
Azeem. *Russ. Agric. Sci.* 2018, 44(3):209–215.
Babar and Tariq (eds). *Mechanisms of Arsenic Toxicity and Tolerance in Plants.* 2018. Springer, Singapore.
Barceló and Poschenrieder. *Environ. Exp. Bot.* 2002, 48(1):75–92.
Bartwal et al. *J. Plant Growth Regul.* 2013, 32(1):216–232.
Briat et al. *Curr. Opin. Plant Biol.* 2007, 10(3):276–282.
Çelik et al. (eds). 2020, pp. 137–155. Springer International Publishing.
Chen et al. *Environ. Safe.* 2014, 104(1):349–356.
Choudhury et al. *Plant J.* 2017, 90(5):856–867.
Corbineau et al. *Plant Growth Regul.* 1988, 7(3):157–169.
Dar et al. *Environ. Exp. Bot.* 2015, 115:49–57.
Dave et al. *J. Hazard. Mater.* 2013, 15:1123–1131.
Davies. In: *Plant Hormones.* 2010, pp. 1–15. Springer, Dordrecht.
Delhaize and Ryan. *Plant Physiol.* 1995, 107(2):315–321.
Delhaize et al. *New Phytologist.* 2012, 195:609–619.
Delker et al. *Plant Biol. (Stuttg)* 2006, 8(3):297–306.
Demole et al. *Helv. Chim. Acta* 1962, 45(2):675–685. John Wiley & Sons, Ltd.
Dhankar and Solanki. *Int. J. Pharm. Bio. Sci.* 2011, 2:553–565.
Ding et al. *Plant J.* 2013, 76(5):825–835.
Epstein et al. *Plant Mol. Biol.* 1999, 50:641–664.
Farias et al. *Food Energy Sec.* 2013, 2:49–69.
Farmer. *Nature* 2007, 448(7154):659–660.
Farmer and Ryan. *PNAS* 1990, 87(19):7713–7716.
Farmer et al. *Curr. Opin. Plant Biol.* 2003, 6(4):372–378.
Farooq et al. *Front. Plant Sci.* 2016, 7:768.
Farooq et al. *Plant Growth Regul.* 2018a, 84(1):135–148.
Farooq et al. *Proteomics* 2018b, 18(10):1700290.
Feussner and Wasternack. *Ann. Rev. Plant Biol.* 2002, 53(1):275–297.
Fonseca et al. *Nat. Chem. Biol.* 2009a, 5(5):344–350.
Fonseca et al. *Curr. Opin. Plant Biol.* 2009b, 12(5):539–547.
Footitt et al. *Plant Physiol.* 2007, 144(3):1467–1480.
Gill and Tuteja. *Plant Physiol. Biochem.* 2010, 48(12):909–930.
Gfeller et al. *Sci. Signal.* 2010, 3:4.
Gomes et al. *Acta Physiol. Plant.* 2020, 42:97.
Gratão et al. *Funct. Plant Biol.* 2005, 32(6):481–494.
Hall et al. *Funct. Ecol.* 2020, 34:993–1002.
Han. *J. Exp. Bot.* 2017, 68(6):1323–1331.
Hanaka et al. *Plant Growth Regul.* 2015, 77(2):167–177.
Hassan et al. *Int. J. Phytoremediation* 2017, 19(6):522–529.
Hou et al. *Plant Cell Environ.* 2016, 39(5):1029–1048.
Howe. *J. Plant Growth Regul.* 2004, 23(3):223–237.
Huang et al. *AoB Plants* 2017, 9(6):plx064.
Huang et al. *J. Exp. Bot.* 2017, 68(6):1349–1359.
Ishimaru et al. *Biosci. Biotechnol. Biochem.* 2017, 81(2):249–255.
Islam et al. *Plant Growth Regul.* 2016, 80(1):23–36.
Jin et al. *Afr. J. Biotech.* 2010, 9:1619–1627.
Keramat et al. *J. Plant Nutr.* 2010, 33(7):1016–1025.
Kochian et al. *Annu. Rev. Plant Biol.* 2004, 55:459–493.
Kochian et al. *Annu. Rev. Plant Biol.* 2015, 66:571–598.
Koetje (ed.), 2003, pp. 523–528. Elsevier Science.
Lei et al. *J. Integr. Plant Biol.* 2020, 62(2):218–227.
Li et al. *PNAS* 2002, 99(9):6416–6421.

Li et al. *Ann. Bot.* 2017, 120(3):405–415.
Li et al. *J. Plant Nutr. Soil Sci.* 2018, 181(3):409–418.
Lindsay. *Chemical Equilibria in Soils.* 1979. John Wiley & Sons, New York.
Ma et al. *Proc. Natl. Acad. Sci. U. S. A.* 2008, 105(29):9931–9935.
Maksymiec et al. *J. Plant Physiol.* 2005, 162(12):1338–1346.
Maksymiec et al. *Chemosphere* 2007, 66(3):421–427.
Marschner. *Mineral Nutrition of Higher Plants.* 1995. Academic Press, Boston, MA.
Maynard et al. *J. Exp. Bot.* 2018, 69(22):5341–5354.
Meharg and Hartley-Whitaker. *New Phytol.* 2002, 154:29–43.
Millaleo et al. *J. Exp. Bot.* 2013, 64(1):343–354.
Miller et al. *Plant Cell Environ.* 2010, 33(4):453–467.
Mousavi et al. *Plant Cell Rep.* 2020, 39(8):1041–1060.
Muñoz and Munné-Bosch. *Trends Plant Sci.* 2019, 24(11):1040–1051.
Nascimento et al. *Acta Sci.* 2019, 41:e36483.
Nath et al. *Plant Physiol. Biochem.* 2014, 80:203–210.
Niazi et al. *Int. J. Phytoremed* 2017, 19:670–678.
Norastehnia et al. *Gen. Appl. Plant Physiol.* 2007, 33:13–23.
Noriega et al. *J. Plant Growth Regul.* 2012, 31(1):79–89.
Panuccio et al. *J. Environ. Manag.* 2009, 90(1):364–374.
Paezi et al. *Plant Physiol. Biochem.* 2018, 132:623–632.
Per et al. *Front. Plant Sci.* 2016, 1933.
Per et al. *Environ. Exp. Bot.* 2018, 145:104–120.
Petrov et al. *Front. Plant Sci.* 2015, 6:69.
Pierart et al. *J. Hazard. Mater.*, 289:219–234.
Piotrowska et al. *Environ. Exp. Bot.* 2009, 66(3):507–513.
Poonam et al. *Am. J. Plant Sci.* 2013, 4:817–823.
Raraba et al. *Res. Commun* 2013, 431:409–414.
Raza et al. *Plant Cell Rep.* 2020, https://doi.org/10.1007/s00299-020-02614-z.
Rengel. *New Phytol.* 1996, 134(3):389–406.
Reyes-Díaz et al. In: *Aluminum Stress Adaptation in Plants.* Eds. S. K. Panda and F. Baluška, 2015, pp. 187–202.
Reyes-Díaz et al. *Molecules* 2016, 21(6):567.
Ricachenevsky et al. *Front. Plant Sci.* 2013, 4:144.
Rodriguez-Serrano et al. *Plant Cell Environ.* 2006, 29(8):1532–1544.
Ronzan et al. *Environ. Exp. Bot.* 2019, 165:59–69.
Roselló et al. *J. Inorg. Biochem.* 2015, 152:160–166.
Ruan et al. *Int. J. Mol. Sci.* 2019, 20(10):2479.
Ryan and Pearce. *Ann. Rev. Cell Develop. Biol.* 1998, 14(1):1–17.
Ryu and Cho. *J. Plant Biol.* 2015, 58(3):147–155.
Saito et al. *Plant Cell Environ.* 2010, 33(2):174–187.
Salla et al. *Microchem. J.* 2011, 97:207–212.
Sanglard et al. *Environ. Exp. Bot.* 2016, 123:22–36.
Schaller and Stintzi. *Phytochemistry* 2009, 70(13–14):1532–1538.
Schneider et al. *Polyhedron* 2004, 23(18):3185–3191.
Senesil et al. *Chemosphere* 1999, 39(2):343–377.
Shahid et al. In: *Crop Production and Global Environmental Issues.* Ed. K. R. Hakeem, 2015, pp. 1–25. Springer International Publishing, Cham.
Sharma et al. *Am. J. Plant Sci.* 2013, 4:817–823.
Shri et al. *Metallomics* 2013, 11(3):519–532.
Siddiqi and Husen. *Bull. Natl. Res. Cent.* 2019, 43(1):153.
Singh and Shah. *Phytochem.* 2014; 108:57–66.
Singh et al. *Front. Plant Sci.* 2016, 6:1143.
Sirhindi et al. *Inter. J. Plant Physiol. Biochem.* 2015, 7:30–39.
Sirhindi et al. *Sci. Rep.* 2020, 10:9322.
Skalicky et al. *Int. J. Mol. Sci.* 2018, 19(10):E3115.
Spollansky. *Elec. J. Biot.* 2000, 3:0717–3458.
Srivastava et al. *Biol. Trace Elem. Res.* 2012, 149(1):86–96.
Srivastava et al. *Biol. Plant.* 2013, 57:385–389.
Sytar et al. *J. Plant Growth Regul.* 2019, 38(2):739–752.
Sun et al. *Mol. Plant* 2011, 4(4):607–615.
Tahara et al. *Plant Soil* 2008, 307(1–2):167–178.
Takahashi et al. *Plant Cell Physiol.* 2004, 45(6):803–809.
Tamás et al. *J. Plant Physiol.* 2006, 163(7):781–784.
Tamogami et al. *Plant Signal. Behav.* 2012, 7(11):1378–1381.
Theodoulou et al. *Plant Physiol.* 2005, 137(3):835–840.
Thines et al. *Methods Mol. Biol.* 2013, 1011:13–23.
Thomma et al. *PNAS* 1998, 95(25):15107–15111.
Ueda and Kato. *Plant Physiol.* 1980, 66(2):246–249.
Ueda and Saniewski. *J. Fruit Ornamental Plant Res.* 2006, 14:199–201.
Ulloa-Inostroza et al. *Plant Soil* 2017, 412(1–2):81–96.
Vareda et al. *J. Environ. Manag.* 2019, 246:101–118.
Verma et al. *Ecotoxicol. Environm. Saf.* 2020, 201:110735.
Vick and Zimmerman. *Plant Physiol.* 1984, 75(2):458–461.
Wang et al. *Front. Plant Sci.* 2019, 10:390.
Wang et al. *Int. J. Mol. Sci.* 2020, 21(4):1446.
Wasternack and Hause. *Ann. Bot.* 2013, 111(6):1021–1058.
Wasternack and Strnad. *Int. J. Mol. Sci.* 2018, 19(9):2539.
Wasternack and Xie. *Plant Signal. Behav.* 2010, 5(4):337–40.
Wiszniewska et al. *Ecotoxicol. Environ. Safe.* 2019, 180:12–22.
Xiang and Oliver. *Plant Cell* 1998, 10(9):1539–1550.
Xiang and Oliver. *Plant Cell* 1999, 10:1539–1550.
Xiong et al. *J. Environ.* 2014, 43:1593–1600.
Xu et al. *Plant Signal. Behav.* 2010, 5(6):649–654.
Xu et al. *J. Exp. Bot.* 2016, 67(11):3497–3508.
Xue et al. *J. Agric. Food Chem.* 2008, 56(20):9676–9684.
Yan et al. *Environ. Saf.* 2013, 98:203–209.
Yan et al. *Biol. Plant.* 2015, 59:373–381.
Yang et al. *Plant Physiol.* 2017, 173(2):1420–1433.
Ye et al. *Proc. Natl. Acad. Sci. U. S. A.* 2013, 110:E3631–E3639.
Ye et al. *Plant Cell* 2017, 29(9):2249–2268.
Yildiz et al. *Seed Sci. Technol.* 2008, 36(3):569–574.
Yoon et al. *J. Crop Sci. Biotechnol.* 2009, 12(2):63–68.
Yoshida et al. *Development* 2009, 136(6):1039–1048.
Yu et al. *Funct. Plant Biol.* 2019, 46(3):197–212.
Zamani et al. *Acta Agrobot.* 2019, 72(3):1782.
Zhang et al. *J. Environ. Manag.* 2018, 223:132–139.
Zhao et al. *Annu. Rev. Plant Biol.* 2010, 61:535–559.
Zhao et al. *J. Plant Growth Regul.* 2016, 35(3):603–610.
Zimmerman and Feng. *Lipids* 1978, 13(5):313.

13

Evidence for the Integrative Roles of Jasmonic Acid and Neurotransmitters in Plant Signaling and Communication: An Emerging Field for Future Investigations

Soumya Mukherjee and Ramakrishna Akula

CONTENTS

13.1 Introduction

Among the various plant growth regulators/hormones functioning in plant growth and development, jasmonic acids (JAs) have important roles in plant growth, development, and stress tolerance. JAs exert promotive effects during various processes like vegetative growth, reproductive maturation, trichome development, cell senescence, regulation of cell division, pigment metabolism, fruit ripening, and cell cycle regulation (Yoshida et al., 2009; Reinbothe et al., 2009; Pauwels and Goossens, 2011; Yan et al., 2012). Furthermore, JAs have specific roles in plant biotic defense during pathogen attacks, wound response, and other injuries during environmental stresses (Wasternack, 2007; Howe and Jander, 2008; Browse, 2009*a*; Pauwels and Goossens, 2011). JAs play a pivotal role in sex determination and initiation of buds and can also trigger defense responses in monocots (Wasternack and Hause, 2013).

The last decade has experienced a prolific surge in research papers pertaining to the myriad roles of neurotransmitters (NTs) in plants. Serotonin, melatonin, GABA, dopamine, noradrenaline, and epinephrine in plants exhibit conserved signaling mechanisms that are largely similar to that in mammals (Akula and Mukherjee, 2020). NTs are known to exert a plethora of effects in various plant systems that mainly involve plant growth, development, organogenesis, flowering, embryogenesis, pollen recognition, and stress tolerance. Certain NTs also bear eco-physiological relevance for their allelopathic effect, which might regulate seed germination in the rhizosphere (Akula and Mukherjee, 2020).

Although JA and NTs have been investigated for their diverse roles in plants, few reports substantiate their associative roles or crosstalk mechanisms in plants. In this context, serotonin, melatonin, GABA, and dopamine have been shown to be involved in crosstalk with JA in plants. This chapter provides a comprehensive update on the interactive roles of JA and plant-based neurotransmitters such as serotonin, melatonin, GABA, and dopamine in regulating various facets of plant development and signaling (Table 13.1).

13.2 Regulation of JA Biosynthesis in Plants

The biosynthesis of JA initiates from the substrate α-linolenic acid (18:3) (α-LeA) and is catalyzed by the enzymes present in the plastid, peroxisome, and cytoplasm. This is followed by reduction reactions and serial steps of β-oxidations (Vick and Zimmerman, 1983). In certain flowers, α-LeA is known to be released from the precursor compounds called galactolipids present in chloroplast membranes and is catalyzed by the activity of phospholipase A1 (PLA_1). Analysis of a mutant, namely *DEFECTIVE IN ANTHER DEHISCENCE 1 (DAD1)*, a flower-specific PLA_1, reveals the essential role of JA in the process of filament elongation (Ishiguro et al., 2001). In *Nicotiana attenuate,* it has been known that galactolipase A1 (GLA1) activity is prevalent in roots and leaves, which results in the formation of JA in roots and leaves (Bonaventure et al., 2011). The formation of JA in plant organs is mostly an external stimuli-specific response (Wasternack, 2007; Howe and Jander, 2008; Browse, 2009a; Pauwels and Goossens, 2011).

Oxylipins are produced by the activity of a type of dioxygenases, lipoxygenases (LOXs), which are involved in the addition of oxygen at C-9/13 position, thus resulting in the

DOI: 10.1201/9781003110651-13

TABLE 13.1

Instances of JA-NT Crosstalk in Plants

Plant system	Signaling event	References
Common purslane *Portulaca* sp	JA (100–200 μM) induced noradrenalin production in hairy roots	Pirian and Piri (2012)
Portulaca sp	JA and Si induce dopamine accumulation	Azni et al. (2021)
Grapevine: *Vitis* sp.	JA and dopamine regulate tendril coiling	Malabarba et al. 2018
Watermelon	JA and melatonin induce cold tolerance in rootstock (self-amplifying feedback loop)	Li et al. (2021)
Tomato	Melatonin activates the JA pathway for resistance to *Botrytis cinerea*	Liu et al. (2019)
Arabidopsis	Melatonin and JA regulate anthocyanin biosynthesis	Yu and Zhu (2018)
	Melatonin acts upstream to JA and SA in biotic defense	Zhu and Lee (2015)
Oryza sativa	MeJA promotes serotonin biosynthesis	Hayashi et al. (2016)
Oryza sativa	Serotonin promotes JA accumulation	Yang et al. (2018)
Arabidopsis	Serotonin inhibits root growth through the COI1 and JAR 1 pathways of JA signaling	Pelagio-Flores et al. (2016)

formation of (13S)-hydroperoxyoctadecatrienoic acid (13-HPOT) or (9S)-hydroperoxyoctadecatrienoic acid (9-HPOT; Yan et al., 2013). These steps represent the divergence points within the LOX pathway. Investigations reveal that in the *Arabidopsis LOX* family there are six members of LOX family genes (*AtLOX2, AtLOX3, AtLOX4*, and *AtLOX6*) that are responsible for JA formation during wound responses, reproductive maturation, and senescence (Ding et al., 2016, Ozalvo et al., 2014) The *lox3lox4* double mutant results in a male sterile plant that possesses defective dehiscence, sterile pollen, and impaired anther development. The 13-LOXs have been reported to be involved in both abiotic and biotic stress tolerance signaling in plant organs. For instance, *AtLOX3* functions during salinity stress (Ding et al., 2016) and *AtLOX3* and *AtLOX4* regulate the defense response to root nematodes (Ozalvo et al., 2014). Tomato plants contain six 13-LOXs (TomLoxA-F), among which TomLoxC and TomLoxD are localized in the chloroplast. TomLoxD is essentially expressed during defense against herbivores (Yan et al., 2013). TomloxC functions in formation with the C5 flavor volatiles in plants (Shen et al., 2014).

The foremost step of JA biosynthesis is catalyzed by the activity of 13-ALLENE OXIDE SYNTHASE (13-AOS), which results in the formation of allene oxide. AOS is an enzyme that belongs to the CYP74A enzyme family. Along with reports from *Arabidopsis* (Park et al., 2002), AOS are also known to be present in various plant species. ALLENE OXIDE CYCLASE (AOC) is localized in the chloroplast, which can covert unstable allene oxide into cis (+)-2-oxophytodienoic acid (OPDA) in the plastid. *Arabidopsis* possesses four AOC genes with differential expression according to detection by the promoter::GUS activity (Stenzel et al., 2012). These AOCs exhibit higher expression primarily in the meristematic tissues, root elongation zone, leaves, and lateral root primordia. The JA biosynthesis enzymes, which are localized in the plastids, are capable of catalyzing hexadecatrienonic acid (C16:3) to dinor-OPDA (dnOPDA).

Followed by this step, the cyclopentanone rings of OPDA and dnOPDA are reduced by the activity of peroxisomal OPDA reductase (OPR) (Strassner et al., 2002, Ruan et al., 2019). OPDA and dnOPDA are produced in the plastid and are then transported to the peroxisomal membrane facilitated by COMATOSE. In the membranes, they are further converted to oxophytoenic acid (OPC-8) and OPC-6, respectively. According to the recent report by Guan et al. (2019), the protein JASSY localized in the external membrane of the chloroplast facilitates in the export of OPDA from inside to the exterior of the organelle The mutant *opr3-1* has been known to exhibit defects in stamen development and is involved in inhibition of JA-induced root growth and occurrence of male sterility. Maize mutants of *opr7opr8* show defects in development resulting due to lower JA levels in most of the plant organs (Yan et al., 2012). The *opr7opr8* mutants were also observed to be susceptible to root-rotting oomycetes (*Pythium* spp.). AtOPR3 has been recently associated with its role during perception of Pi deficiency (Zheng et al., 2016). It is involved in the formation of OPC-6/OPC-4 derivatives which then result in JA formation catalyzed by the various enzymes of the ß-oxidation pathway like acyl-CoA oxidase (ACX; Li et al., 2005; Schilmiller et al., 2007), multifunctional protein (MFP; Richmond and Bleecker, 1999), L-3-ketoacyl CoA thiolase (KAT; Castillo et al., 2004), and 4-coumarate: CoA ligase-like enzymes (Schneider et al., 2005). This step is followed by the cleavage of Jasmonoyl-CoA by THIOESTERASE (TE), which results in forming cis-7-iso-jasmonic acid [(+)-7-iso-JA].

In plant cells there exists a crucial balance between production, distribution, and metabolism of JA in response to various developmental or environmental cues. Various elements are known to modify JA levels and modulate the steps of JA biosynthesis and metabolism. Various factors like availability of substrate and spatio-temporal expression function as a feedback loop, thus regulating the expression of JA biosynthesis genes. There are also certain endogenous factors responsible for determining JA levels in cells (Wasternack, 2007; Browse, 2009a,b). Various external stimuli trigger the formation of α-LeA, which is an important substrate in the JA biosynthesis pathway. *Arabidopsis* leaves contain LOX, AOS, and AOC proteins in larger amounts. JA formation, however, mostly occurs in response to external stimuli. Furthermore, stimuli-induced accumulation of JA is transient in nature and occurs as a downstream response to *LOX, AOS*, and *AOC* expression in cells (Stenzel et al., 2003). JA biosynthesis is also reported to be regulated by a feedback loop where all genes for the JA biosynthesis enzymes are induced by the presence of JA in cells. The JA biosynthesis genes exhibit tissue-specific spatial gradient in their distribution and expression in response to external stimuli. Tomato *AOC* is localized to vascular bundles and sieve elements (Hause et al., 2000). There are also reports of post-translational regulation of JA biosynthesis genes. For instance, the activity of OPR3 is regulated by its dimerization and phosphorylation reactions (Breithaupt et al., 2006).

13.3 JA Metabolism and JA Signaling in Plant Cells

Various JAs and their derivatives in plants are known to undergo biochemical modifications. Among these, a majority have been identified and isolated in lower plant groups, including fungi and also in gymnosperms and angiosperms. JA homeostasis is a regulatory mechanism in response to various external stimuli in plants. Various reactions result in the conversion of JAs into active or inactive forms such as amino acid conjugation, hydroxylation, sulfation, O-glycosylation, and lactone formation. Certain derivatives of JA are specific to stress responses mediated by external stimuli to the plant. Investigations pertaining to JA signal transduction have progressed from the discovery of the JA receptor followed by the identification of various molecular pathways related to JA signaling in plants. Accumulating evidence in the last decade has been obtained from transcriptomic, proteomic, and metabolomic approaches that have deciphered the molecular mechanisms of JA perception, signaling, and transport in plants

Xie et al. (1998) have reported the presence of the JA receptor by screening the JA-insensitive mutants after mutagenesis and thus identified plants with reduced or impaired JA signaling. The *coi1-1* mutant exhibited disruption in the JA response associated with male sterility and anthocyanin formation (Xie et al., 1998). MeJA has been known to inhibit root growth and regulates susceptibility to pathogens, which indicates a suggestion of a central role in JA signal transduction. Information obtained from map-based cloning strategy reveals that COI1 is an F-box motif representing leucine-rich repeats (LRRs). Investigations also revealed COIl is similar to human SkpZ and *Arabidopsis* TIRl, having 34% identity (Xie et al., 1998). However, experiments later deciphered that COI1 did not fulfill a putative receptor function for JA signaling.

Investigations performed in 2007 led to the simultaneous identification of JAZ proteins (Chini et al., 2007; Thines et al., 2007; Yan et al., 2007) that belong to the TIFY family of proteins and contain a highly consensus TIF[F/Y]XG motif present within the zinc finger named ZIM (initially named in floral meristem; Vanholme et al., 2007; Bai et al., 2011). The 27 amino acid, JAs motif present near the C terminus helps in interaction with various transcription factors (TFs; Chung et al., 2010). The *Arabidopsis* genome is known to encode around 13 JAZ proteins (JAZ1–JAZ13) which are classified into five phylogenetic groups. Recent investigations have led to the current advancements in deciphering the mechanism of JA biosynthesis and its signaling in plants. Recent reports suggest that the JA biosynthesis pathway does not show OPR3 dependency (Wasternack and Hause, 2018). Furthermore, the mechanism of JA signaling is regulated by its homeostasis, which maintains the balance between the active and inactive forms of JA derivatives. Although various JA derivatives have been reported, further investigations are required to clarify the role of various derivatives and their involvement in systemic responses. There are many questions that still remain unanswered as to how these derivatives function in the downstream signaling response. Apart from its role in biotic/abiotic stress signaling and herbivory defense, JA is also widely known to regulate root formation and its physiological parameters. However, the activity of JAs is largely regulated in association with the combination of various phytohormones. Evidence clarifies that JAs involved in signaling pathways are related to the modulation of auxin homeostasis (Sun et al., 2009; Ishimaru et al., 2018). This crosstalk occurs through various downstream effector molecules and feedback loop mechanisms.

13.4 Regulation of Serotonin and Melatonin Biosynthesis

The biosynthesis of serotonin and melatonin initiates from their common precursor L-tryptophan. This pathway in plants has similar features to that in mammals and prokaryotes in terms of enzyme activity and modulation of the metabolic intermediates. The common precursor of serotonin and melatonin is L-tryptophan, which is also the precursor for auxin plants. Auxin and serotonin biosynthesis originates from a common intermediate precursor of the pathway: tryptamine (Murch and Saxena 2002). Investigations have been performed by using the C14-tryptophan tracer method in the plant St. John's wort (*Hypericum perforatum* L.), which has deciphered the formation of 5-hydroxytryptophan as a main precursor of serotonin. Serotonin biosynthesis occurs through a two-step enzymatic process (Posmyk and Janas 2009). Tryptophan decarboxylase catalyzes the turnover of tryptophan into tryptamine, which is followed by the formation of serotonin. The enzyme- tryptamine hydroxylase catalyzes this conversion step. Followed by serotonin formation, melatonin is produced in this pathway by the conversion of serotonin to *N*-acetylserotonin (precursor for melatonin), which involves an acetylation reaction. In the final reaction step of melatonin biosynthesis, N-acetyl serotonin is metabolized by the activity of hydroxyindole O-methyl transferase (HIOMT). However, according to certain reports, melatonin may also be synthesized from the precursor 5-methoxytryptamine, which involves an acetylation reaction catalyzed by serotonin N-acetyltransferase (Byeon et al., 2015, Zuo et al., 2014). Tryptophan decarboxylase (TDC) is an important rate-limiting and regulatory enzyme in the pathway, which has a higher Km (690 μM) for its substrate, tryptophan (Kang et al., 2008). Tryptophan biosynthesis is important for the regulation of TDC activity, which also modulates the amount of serotonin and melatonin accumulation in different plant organs.

Melatonin was first discovered in vascular plants and dates back to 1995; later detailed investigations have revealed its metabolism and turnover in cells (Arnao and Hernández-Ruiz 2006). The characterization of N-acetylserotonin O-methyltransferase (ASMT) has been performed in *Arabidopsis* (Byeon et al., 2016) where the enzyme participates in melatonin production from its immediate precursor, N-acetylserotonin. The enzyme is mainly present in the cytoplasm. The enzyme caffeic acid O-methyl transferase (COMT) in *Arabidopsis* undergoes methylation activity to its substrate N-acetyl serotonin (Byeon et al., 2014). Furthermore, while considering the reaction kinetics, the Km and Vmax values for ASMT are lower than COMT activity. It is important to note that the two pathways of melatonin biosynthesis in plants (N-acetyl serotonin and via 5-methoxy tryptamine—mediated) operate both during normal and senescent conditions in plants (Back et al., 2016). Senescent organs may produce a higher amount of serotonin which results in enhanced accumulation of 5-methoxytryptamine and melatonin. Thus, according to the accumulating evidence, the biosynthetic regulation of serotonin and melatonin might depend on plant age, light-dependent response, and senescence in various plant organs.

13.5 Jasmonate–Serotonin Crosstalk Crucially Regulates ROS Distribution during Root Growth Regulation

Investigations in *Arabidopsis* reveal the presence of multiple signaling routes to be associated with jasmonic acid, serotonin, ABA, and ethylene-mediated responses. Exogenous serotonin treatment induces inhibition of primary root growth that is known to be mediated by the downstream signaling cascade participated by jasmonic acid–ethylene crosstalk, which, in turn, regulates ROS levels in cells (Pelagio-Flores et al., 2016). Extra-physiological concentrations of exogenous serotonin exert inhibitory effects to root growth. Interestingly, higher concentrations of serotonin also instigate jasmonic acid accumulation in tissues. This finding is in congruence with the involvement of serotonin in the amelioration of biotic stress. Since JA biosynthesis is activated during biotic stress in plants, it is likely that serotonin accumulation may also exert similar effects to JA accumulation. Functional validation has been obtained by using *Arabidopsis* mutants for jasmonic acid and ethylene signaling pathways, which revealed poor sensitivity of serotonin that was measured for its effect on root growth inhibition (Pelagio-Flores et al., 2016). Moreover, the effect of root growth inhibition was brought about by a higher exogenous concentration of serotonin (300 μM), which operated through the jasmonic acid pathway. In line with earlier investigations on the perception of JA signaling, mutant analysis revealed that COI 1 and JAR 1 elements of the JA pathway were crucial for serotonin-induced regulation of root growth. Evidence, therefore, reveals that serotonin–JA crosstalk negatively regulates root growth in *Arabidopsis*. The regulation of root system architecture occurs by the balance between cell proliferation and expansion from the root apical meristem. Serotonin cascades a downstream signal that also involves ROS homeostasis, thus leading to consequent crosstalk to JA signaling (Figure 13.1). Serotonin treatment in *Arabidopsis* triggers the redistribution of ROS in the primary roots which results in growth retardation. Serotonin-induced ethylene signaling has also been found to negatively affect primary root growth. Functional analysis using mutants (etrl) revealed impaired sensitivity to serotonin in the plants that was manifested by recovery in root growth inhibition (Pelagio-Flores et al., 2016). Thus, ROS balance is a mediator of serotonin-JA and ethylene crosstalk during root growth in *Arabidopsis*.

Investigations pertaining to the metabolic control of serotonin biosynthesis in the high free lysine (HFL) phenotype of rice have been observed to be related to the instigation of JA accumulation (Yang et al., 2018). Serotonin biosynthesis (triggered by TDC expression) was also accompanied by the triggering of JA biosynthesis in the endosperm of rice grains. This finding of JA–serotonin crosstalk provides insights into amino acid metabolism and can be further helpful in engineering nutritional quality in rice grains. Serotonin is known to induce a hypersensitive response during fungal infection in rice which protects the adjacent non-infected regions by confinement of ROS accumulation (Hayashi et al., 2016). In this case, serotonin biosynthesis is enhanced by exogenous methyl-jasmonate application. The effect of serotonin in the modulation of plant defense is largely regulated by its turnover from tryptamine.

Information obtained from the RiceXPro database suggests that JA and ABA regulate the expression of genes involved in the serotonin biosynthesis pathway (Sato et al., 2013).

13.6 Melatonin–JA Crosstalk during Biotic and Abiotic Stress Tolerance

Integrative roles of melatonin and jasmonic acid signaling pathways in triggering biotic stress defense have been suggested by Wasternack and Hause (2013) and Shyu and Brutnell (2015). During induction of disease resistance, melatonin is known to act upstream to JA and salicylic acid (SA) signaling (Zhu and Lee, 2015). In addition to biotic stress defense, regulation of the anthocyanin pathway in *Arabidopsis* has also been reported to exhibit involvement of JA–melatonin crosstalk (Yu et al., 2018). In the investigation, it was reported that melatonin inhibits the action of JA-mediated suppression of anthocyanin biosynthesis in leaves. However, on the other hand, melatonin-induced anthocyanin biosynthesis was found to be independent of JA signaling. Further investigations are required to unravel the mechanistic understanding of the regulation of flavonoid biosynthesis by melatonin and jasmonic acid in plants. Regulation of anthocyanin accumulation in plant organs in response to nutrient stress can be investigated in relation to melatonin–JA crosstalk. Melatonin has been reported to trigger tomato resistance to *Botrytis cinerea* by activating the JA signaling pathway (Liu et al 2019).

In a recent investigation by Li et al. (2021), the authors reported the interactive roles of melatonin, JA, and H_2O_2 in improving cold tolerance in rootstocks of watermelon. Interestingly, the application of melatonin/JA mutually influenced the biosynthesis of the two molecules in roots and leaves, which was also accompanied by an increase in H_2O_2 production. Furthermore, analysis by qPCR revealed that methyl-jasmonate (MeJA) gene expression in roots was effective in triggering long-distance cold tolerance in the leaves of watermelon plants. Similarly, the application of melatonin also induced MeJA biosynthesis and subsequent H_2O_2 accumulation, thus resulting in cold tolerance. The authors stated that

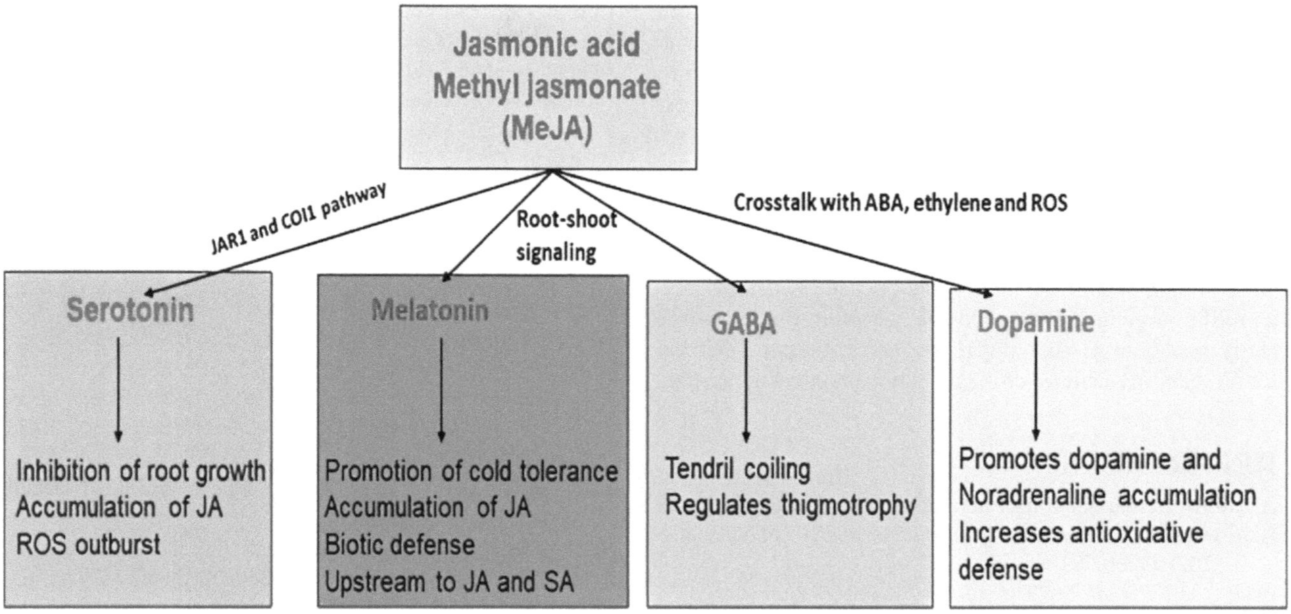

FIGURE 13.1 Associative roles of JA and NTs such as serotonin, melatonin, GABA, and dopamine in plant signaling. Serotonin-mediated root growth inhibition involves the JAR1 and COI1 pathways of JA signaling, which also cause ROS outbursts in cells. Melatonin regulates JA accumulation in plants and is involved in cold tolerance and biotic defense responses. JA and GABA have associative roles in thigmotropism in tendril coiling. Dopamine and noradrenaline accumulation are influenced by MeJA.

melatonin and MeJA interactively exhibit a self-amplifying feedback loop that further enhances cold tolerance in the rootstocks (Figure 13.1). Thus, melatonin–JA crosstalk essentially promotes cold tolerance through long-distance communication from rootstock–scion (leaves). This investigation exemplifies the horticultural importance of exogenous melatonin and jasmonic acid which might encourage future investigations to decipher their roles in detail.

13.7 JA Associates with GABA and Dopamine Crosstalk

Few recent investigations have affirmed the role of JA in regulating the biosynthesis and activity of some other neurotransmitters like GABA, dopamine, and noradrenaline in plants. γ-aminobutyric acid (GABA) is a non-proteinogenic amino acid that has been reported to function in association with JA during tendril coiling in grapevine (Malabarba et al 2019). Along with JA, GABA also assisted in tendril coiling in the plants, where JA did not exhibit any additive effect along with GABA. However, treatment with various phytohormones increased both JA and GABA accumulation in the coiled shoots. This suggests the associative roles of JA and GABA in promoting thigmotropic response in plants. The process of tendril coiling is a mechano-sensitive thigmotropic response, where inhibition of GABA accumulation resulted in decreased coiling that, however, did not affect JA-induced coiling.

In another recent investigation by Azni et al (2021) common purslane (*Portulaca* sp.) plants exposed to exogenous silicon and methyl jasmonate (foliar application) exhibited significantly increased dopamine accumulation. This finding provides substantial information for future investigations related to the regulatory roles of JA in eliciting dopamine biosynthesis. The increased phenol, antioxidant, and nutrient content in common purslane plants was also enhanced in response to JA and Si treatment (Figure 13.1). Thus, JA–dopamine crosstalk requires concrete evidence from future investigations that can provide fruitful findings related to the improvement of nutrient status and stress management in crops.

Similar to dopamine, JA has also been known to enhance noradrenalin production in Agrobacterium rhizogenes-infected hairy roots of *Portulaca* sp. Interestingly, 100–200 μM JA application was optimum and it resulted in an around eight-fold increase in noradrenalin production in the roots (Pirian and Piri, 2012). Although persuasive at present, future investigations should reveal the mechanistic understanding of JA interplay with GABA, dopamine, and noradrenalin signaling in plants.

13.8 Conclusion

Not much information is available to explain the nature of the association between JA and plant neurotransmitters during development and signaling. Although the putative receptor for melatonin has been reported in plants, future investigations are necessary to elucidate the exact mechanisms of receptor-mediated signaling for GABA, dopamine, and noradrenaline interplay in plant communications. Apart from pharmacological studies, it is important to imply various mutants of the neurotransmitter biosynthesis pathways that decipher their roles in plant development exhibit altered sensitivity to JA, if any. Although serotonin and melatonin have been reported to be associated with root growth regulation, biotic and abiotic stress tolerance in relation to JA crosstalk, investigations at the omics level are necessary to detect the transcriptome or proteome arising from JA-Ser/mel crosstalk in plants. Characterization of the major proteins can then obtain a holistic picture of how JA–serotonin or JA–melatonin crosstalk share various regulatory networks in plants. Apart from ROS, MAPK, Ca^{2+}, and cGMP are also some of the common secondary signaling components in plant cells. Thus, it is a requirement to show whether JA–neurotransmitter crosstalk implies any of these components during plant development and communication. Although the last decade has experienced a prolific surge in investigations on neurotransmitters—especially melatonin, GABA, dopamine, and serotonin—evidence for JA–neurotransmitter crosstalk is not abundant at present.

REFERENCES

Arnao and Hernández-Ruiz. *Plant Signal Behav* 2006; 1(3):89–95..
Azni et al. *J Plant Nutr* 2021. https://doi.org/10.1080/01904167.2021.1936027.
Back et al. *J Pineal Res* 2016; 61(4):426–437.
Bai et al. *Genomics* 2011:128–136.
Bonaventure et al. *Plant Cell Environ* 2011; 34(9):1507–1520.
Breithaupt et al. *Proc Natl Acad Sci U S A* 2006; 103(39):14337–14342.
Browse. *Annu Rev Plant Biol* 2009a; 60(1):183–205.
Browse. *Phytochemistry* 2009b, 13: 1539–1546.
Byeon and Back. *J Pineal Res* 2015; 58(3):343–351.
Byeon et al. *J. Pineal Res* 2014; 57(2). https://doi.org/10.1111/jpi.12160.
Byeon et al. *J Pineal Res* 2016; 60(1):65–73.
Castillo et al. *Plant Physiol* 2004; 135(1):85–94.
Chini et al. A. *Nature* 2007; 448(7154):666–671.
Chung et al. *Plant J* 2010; 4:613–622.
Ding et al. *Plant Sci* 2016; 244:1–7.
Guan et al. *Proc Natl Acad Sci* 2019 https://doi.org/10.1073/pnas.1900482116. 201900482.
Hause et al. *Plant J* 2000; 24(1):113–126.
Hayashi et al. *Plant J* 2016; 85(1):46–56.
Howe and Jander. *Annu Rev Plant Biol* 2008; 59(1):41–66.
Ishiguro et al. *Plant Cell* 2001; 13(10):2191–2209.
Ishimaru et al. *Plant Physiol* 2018; 177(4):1704–1716.
Kang et al. *Plant Signal Behav* 2008; 3(6):389–390.
Li et al. *Plant Cell* 2005; 17(3):971–986.
Li et al. *Hortic Res* 2021; 8(1):57.
Liu et al. *J Agric Food Chem* 2019; 67(22):6116–6124.
Malabarba et al. *J Plant Growth Regul* 2019; 38(1):39–45.
Murch and Saxena. *In Vitro Cell Dev Biol Plant* 2002; 38(6):531–536.
Park et al. *Plant J* 2002; 31(1):1–12.
Pauwels and Goossens. *Plant Cell* 2011:3089–3100.
Pelagio-Flores et al. *Physiol Plant* 2016; 158(1):92–105.
Pirian and Piri. *International Research Journal of Applied and Basic Sciences* 2012; 3:213–218.
Posmyk and Janas. *Plant* 2009; 31:1–11.
Ramakrishna and Mukherjee. *Plant Signal Behav* 2020; 15:6.
Reinbothe et al. *FEBS Journal* 2009; 57:219–227.
Richmond and Bleecker. *Plant Cell* 1999; 11(10):1911–1924.
Ruan et al. *Int J Mol Sci* 2019; 20(10):2479. https://doi.org/10.3390/ijms20102479.
Sato et al. *Nucleic Acids Res* 2013; 41(Database issue):D1206–D1213.
Schilmiller et al. *Plant Physiology* 2007; 143(2):812–824.
Schneider et al. *J Biol Chem* 2005; 280(14):13962–13972.
Shen et al. *J Exp Bot* 2014; 65(2):419–428.
Shyu and Brutnell. *J Exp Bot* 2015; 66(14):4165–4176.
Stenzel et al. *Plant J* 2003; 33(3):577–589.
Stenzel et al. *J Exp Bot* 2012; 31(1):1–12.
Strassner et al. *Plant J* 2002; 32(4):585–601.
Sun et al. *Plant Cell* 2009; 21(5):1495–1511.
Thines et al. *Nature* 2007; 448(7154):661–665.
Vanholme et al. *Trends Plant Sci* 2007; 19:2470–2483.
Vick and Zimmerman. *Biochem Biophys Res Commun* 1983; 111(2):470–477.
Wasternack. *Ann Bot* 2007; 100(4):681–697.
Wasternack and Hause. *Ann Bot* 2013; 111(6):1021–1058.
Wasternack and Hause. *Trends Plant Sci.* 2018; 23(4):276–279.
Xie et al. *Science* 1998; 5:1091–1094.
Yan et al. *Plant Cell* 2007.
Yan et al. *Plant Cell* 2012:1420–1436.
Yan et al. *PLoS Genet* 2013; 9(12):e1003964.
Yang et al. *Plant Physiol* 2018 https://doi.org/10.1104/pp.17.01283.
Yoshida et al. *Development* 2009; 136(6):1039–1048.
Yu Ai and Ziqiang Zhu. *J Agric Food Chem* 2018. https://doi.org/10.1021/acs.jafc.8b01795.
Zheng et al. *Scie Rep* 2016; 6(*1*):24778.
Zhu and Lee. *Plant Cell Physiol* 2015; 56(3):414–420.
Zuo et al. *J Pineal Res* 2014; 57(4):408–417.

14

Regulation of Jasmonic Acid and Brassinosteroid Signaling Pathways by MicroRNA

A. Thilagavathy and V. R. Devaraj

CONTENTS

14.1 Introduction

Plant cells use evolved signaling pathways to communicate and coordinate developmental activities and respond to environmental factors. Signaling pathways orchestrate biochemical and physiological responses such as root development, shoot formation, flowering, fruit ripening, germination, and photosynthetic regulation (Mulligan et al., 1997; Huber and Baurele, 2016; Deepika et al., 2020). Signaling pathways utilize a complex network of interacting molecules ultimately leading to the regulation of gene expression that facilitates cells to sense, integrate, and respond to intricate environmental signals (Diaz 2011; Chen and Heidari, 2020). Plant genomes possess an enormous number of functional protein-coding and non-coding RNA genes, which exhibit spatio-temporal expression during growth, development, and fecundity (Wu et al., 2020).

Commonly, genes coding for functional proteins are controlled by a vast array of *trans*-acting factors, known as transcription factors (TFs). There is ample evidence confirming the fact that TFs are the major coordinators in plant growth, development, stress responses, and the crosstalk in different signal transduction pathways (Rushton et al., 2010; Fan et al., 2014; Javed et al., 2006). Recently, microRNAs (miRNAs) have proved to be a crucial regulator in post-transcriptional gene regulation through translational repression or guiding degradation of their mRNA targets (Zhang and Wang, 2015). Single miRNA frequently exhibits multiple regulatory roles in plant growth, organ development, and stresses responses (Li and Zhang, 2016; Chaudhary et al., 2021). Thus, miRNA perhaps acts as a node molecule to coordinate the balance between plant development and environmental cues (Sunkar et al., 2012; Nova-Franco et al., 2015). Studies have demonstrated that plant miRNAs prefer targeting transcription factors, thus, alteration of miRNA expression levels usually results in significant changes in plant growth and development (Chuck et al., 2009; Kamthan et al., 2015). Thorough investigations into the role of miRNAs are expected to pave the way for the discovery of key regulators or regulatory pathways revealing the mysterious molecular network in plants. This information on miRNAs and their targets may offer better tools for crop improvement through plant genetic engineering.

14.2 Post-Transcriptional Regulation of Gene Expression by miRNA

Regulation of gene expression by miRNAs occurs at the post-transcriptional level. From the plant biochemistry perspective, miRNAs are expressed as long primary miRNA (pri-miRNA) transcribed from MIR genes by the activity of RNA polymerase II (Lee et al., 2003). Further, pri-miRNAs are processed in the nucleus by Dicer-like 1 (DCL1), and hyponastic leaves *HYL1* protein to form stem-loop intermediates called miRNA precursors or pre-miRNAs (Bartel, 2004; Kurihara and Watanabe, 2004). In plants, pre-miRNAs are processed into mature miRNA, which are duplexes consisting of two strands: miRNA and miRNA* with a two nucleotide overhang at the 3' end, which is methylated by a methyltransferase, HEN1 (Hua Enhancer 1). The duplex so formed is exported from the nucleus to the cytoplasm by *HASTY* (Park et al., 2002). Upon entry into the cytoplasm, a helicase unwinds the miRNA–miRNA* duplex, exposing the mature miRNA that is incorporated into the RNA-induced silencing complex (RISC) containing *ARGONAUTE1*, which directs the RISC to the target mRNA leading to the mRNA cleavage, or translational repression (Bartel, 2004). While perfect base-pairing between miRNA and the target mRNA causes the latter's cleavage, imperfect base-pairing leads to translational repression. Most

DOI: 10.1201/9781003110651-14

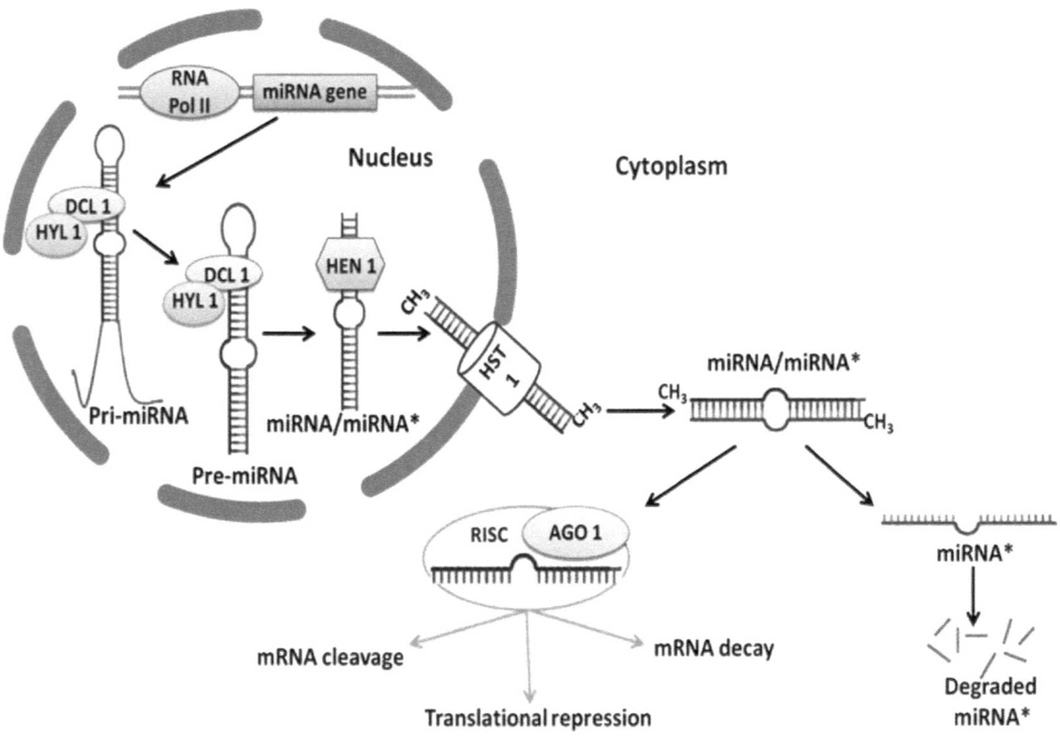

FIGURE 14.1 Biogenesis of miRNAs and their mode of action in plants. Source: Yang and Li (2011).

plant miRNA-mediated regulation of expressions results in the cleavage of target mRNA (Voinnet, 2009). miRNA-mediated decay, which involves the removal of the poly (A) tail of the target mRNA, making it unstable, has been reported in a few cases (Ruegger and Grobhans, 2012) (Figure 14.1).

14.3 Transcription Factors of the Signaling Pathway as Targets of miRNA

Evidence indicates that miRNAs do not directly regulate plant growth and development, but indirectly control plant development by participating in a miRNA-target gene network. Considerable progress has been made in identifying the targets of plant miRNAs (Sunkar et al., 2012; Nova-Franco et al., 2015; Zhang, 2015). In regulatory hierarchy, plant miRNAs have evolved to target TFs, thereby repressing them and indirectly regulating the expression of functional genes (Li and Zhang 2016). When a miRNA is overexpressed in plants, its target TFs are suppressed and exhibit decreased activity. On the other hand, when a miRNA is suppressed, the target TFs are upregulated and exhibit increased activity (Sunkar et al., 2012). The expression level of miRNAs and TFs together mediate the upregulation or downregulation of the genes, ultimately determining the expression of a specific gene to be turned "on/off" (Istrail et al., 2007; Chow et al., 2016). Since both regulators demonstrate a greater impact on the plant genetic system, the circuiting of miRNAs-TFs allows the orchestration of numerous biological processes with high reliability, and their study will help in understanding the organization of several biological pathways.

14.4 Role of miRNA 319 in the Regulation of the Jasmonic Acid Signaling Pathway

Jasmonic acid (JA) and its derivatives, collectively known as jasmonates, play an important role in plant growth and developmental activities that include fertility, flowering, fruit development, root growth, wound response, senescence, and secondary metabolism (Wasternack and Hause, 2013; Ruan et al., 2019). Also, jasmonates are essential signals in the defense system of plants against biotic stress responses (Robert-Seilaniantz et al., 2011; Ali and Baek, 2020).

JA acts as a key player in the control of senescence in plants (Ruan et al., 2019). Senescence, the cessation of growth in cells, organs, or the whole plant, is a highly regulated developmental process that most often terminates with programmed cell death (Lim et al., 2007; Guo and Gan, 2012). Studies in *Arabidopsis* manifest that JA positively affects the timing of leaf senescence in conjunction with other age-related cues. In naturally senescing leaves, a gradual increase in the transcript levels of JA biosynthesis genes was observed (He et al., 2002; Kim et al., 2015). A combination of genetic and biochemical studies discovered that a regulatory network controlled by the miR319-regulated clade of TEOSINTE BRANCHED/CYCLOIDEA/PROLIFERATING CELL FACTOR (TCP) TFs was responsible for JA biosynthesis in *Arabidopsis*. TCPs are plant-specific TFs identified as negative regulators of plant growth and development (Almeida et al., 1997; Luo et al., 1999). miR319 targets and downregulates the mRNA levels of TCP TFs in plants. Schommer et al. (2008) showed that suppression of the expression of class II TCP2/4/10 by miR319 decreased the expression of the JA biosynthesis gene and the endogenous levels of JA. More precisely, the group elucidated the miR319-TCP regulatory module regulating JA biosynthesis by repressing the expression of Lipoxygenase 2 (LOX2), a key enzyme of JA synthesis. They also showed that the reduced levels of JA resulted in a delay in leaf senescence, and the delayed senescence phenotype was reversed by the application of JA. It is therefore predicted that the network of miR319-TCP-*LOX2* represents one of the early regulatory modules that coordinates and balances different mechanisms in leaf development which alters leaf senescence.

Biotic stresses, including diseases caused by viruses, nematodes, insects, and fungus, are a threat to almost all crops (Iqbal et al., 2021). Plant disease resistance and susceptibility are governed by the combined genotypes of host–pathogen interaction and depend on a complex exchange of signals and responses occurring under given environmental conditions (Peck and Mittler, 2020). To date, many miRNAs have been identified that are responsive to disease and insect invasion. Fungus *Magnaporthe oryzae* causes rice blast disease in rice (*Oryza sativa*). It was observed that infection by the *M. oryzae* strain Guy11 specifically induced the expression of rice miR319 and, consequently, suppressed its target gene TCP21, which encodes a TF. In wild-type and miR319 overexpressing transgenic rice, infection by Guy11 caused suppression of JA synthesis, indicating suppression of JA signaling by Guy 11via induction of miR319. In particular, LOX2 and Lipoxygenase 5 (LOX5), key enzymes of JA synthesis, were specifically suppressed by miR319 overexpression, or by Guy11 infection. Thus, it is evident that a circuit of miR319/TCP21 modulates the interaction of *M. oryzae* and rice, wherein Guy11 infection induces miR319 expression, which diminishes the stimulatory roles of TCP21 on key JA synthesis. Decrease in JA signaling results in reduced defense responses in infected rice. Further, in transgenic rice overexpressing TCP21-Res (which is resistant to miR319-mediated silencing), TCP21 was found to be a positive regulator of the rice defense response against blast disease. Therefore, the suppression of miR319-mediated regulation in JA signaling may be a counterdefense mechanism in plant systems (Zhang et al., 2018).

In an attempt to elucidate the role of miRNAs in biotic stress responses in rice plants, Zhang et al. (2016) showed that rice ragged stunt virus (RRSV) infection causes increased accumulation of miR319, but decreased expression of miR319-regulated TCP TF genes, especially TCP21. Transgenic rice plants overexpressing miR319 and plants downregulating TCP21 exhibited disease-like phenotypes and were highly susceptible to RRSV compared with wild-type plants. In contrast, transgenic plants overexpressing TCP21 showed mild disease symptoms. Both RRSV infection and overexpression of miR319 caused decreased endogenous JA levels along with downregulated expression of JA biosynthesis genes. However, the application of JA to rice plants infected by RRSV alleviated disease symptoms and reduced virus accumulation. Collectively, these results suggest that the induction of miR319 by RRSV infection in rice downregulates target TCP genes, which, in turn, suppresses the JA-mediated defense to facilitate virus infection and symptom development.

JA as a systemic signaling molecule is effective against tomato root-knot nematode (RKN) disease and can reduce the number of root knots and defend plants (Cooper et al., 2005; Fan et al., 2014). miR319 is reported to target TCP genes negatively and regulates RKN resistance by altering the JA level in tomato plants. miR319-overexpressing plants have revealed a clear decrease in the levels of all miR319-targeted TCP genes, which positively regulate the JA level in plants. Taken together, these results suggested that overexpressing miR319 inhibited TCP4, which in turn, negatively regulated JA biosynthetic genes, resulting in a lower endogenous JA level. Consequently, the resistance to RKN infection was affected in miR319 overexpressing tomatoes (Zhao et al., 2015).

All this evidence strongly supports viral, fungal, and nematode infection inducing the differential expression of specific miRNAs, which in turn, regulates gene expression, allowing plant–pathogen interaction. This gives a clue that the alteration of miRNA-mediated responses will work as an effective strategy for improving crops for disease resistance.

14.5 Role of miRNA 397 in the Regulation of the Brassinosteroid Signaling Pathway

Brassinosteroids (BRs) are steroidal phytohormones that transduce signals to the nucleus, resulting in the regulation of expression of numerous genes involved in diverse physiological functions, including seed germination, cell elongation, cell division, senescence, vascular differentiation, reproduction, and root development (Nakamura et al., 2017). BRs also play key roles in various biotic and abiotic stress responses (Clouse and Sasse, 1998; Planas-Riverola et al., 2019). Recently, there has been a profound increase in their application in agricultural processes to improve crop productivity under stress and native conditions (Lin, 2020).

Increasing grain yield is an imperative focus of crop breeders around the world. Zhang and colleagues (2013) found that overexpressing miR397 increased the overall grain yield by up to 25% in rice. This is the first study for directly linking crop yield improvement to miRNA-based biotechnology. Overexpression of miR397, which is naturally expressed at higher levels in young panicles and grains of rice, enlarges grain size and promotes panicle branching, leading to an increase in overall grain yield. Molecular and genetic evidence suggests that in rice OsmiR397 increases grain yield by downregulating its target, the LAC gene, whose product is a laccase-like protein found to be involved in the suppression of sensitivity of plants to BR signaling (Zhang et al., 2013).

To understand the role of miR397 in BR signaling, RNA-sequencing of the young panicles of the wild-type, OXmiR397b (transgenic plants overexpressing miR397), and OXLAC (transgenic plants overexpressing *OsLAC*) were performed by Zhang et al. (2013), which showed that numerous brassinosteroid-related genes were differentially expressed between the three samples. Further, they compared the OXmiR397 and OXLAC plants to previously reported BR-related mutants and found that the phenotypic traits of the OXLAC plants were similar to those of the BR-deficient mutants, including erect leaf, shortened internodes, reduced vascular number, and grain size. In contrast, OXmiR397 plants have a phenotype consistent with enhanced BR signaling. These observations supported the involvement of OsmiR397 and *OsLAC* in

BR-regulated plant growth and development. Regulatory interaction between miR397 and LAC mRNA has been predicted to be conserved in many species, including *Arabidopsis*, *Populus trichocarpa*, and tobacco. Therefore, manipulation of miR397 for increasing grain yield can be exploited in rice and other crops (Jones-Rhoades and Bartel, 2004).

14.6 Conclusion

The burgeoning world population demands increased agricultural production compounded by the challenges of climate change and urban development for arable land. This increasing demand, coupled with water shortages, depleting soil fertility, and other abiotic stresses have fostered the scientific community to unravel the signal response mechanisms in plants to exploit these strategies for sustainable agricultural development. Novel strategies are being developed for bringing significant contributions to crop yield quality as well as tolerance to environmental stresses. Although identification and cloning of functional genes are always the prime steps for engineering crops, the emergence of post-transcriptional regulation of gene expression by miRNAs has caused a paradigm shift in crop productivity research. The discovery of miRNAs related to signaling in plants probably heralds the start of investigations into a very important, but previously unsuspected, part of gene regulation in signal transduction. Thus, a consensus has been formed in demonstrating that manipulating miRNA-mediated regulation is an effective strategy for improving agronomic characteristics such as crop yield and stress tolerance.

REFERENCES

Ali and Baek. *Int. J. Mol. Sci.* 2020; 21(2): 621.
Almeida et al. *Development* 1997; 124(7): 1387–1392.
Bartel. *Cell* 2004; 116(2): 281–297.
Chaudhary et al. *Life* 2021; 11(4): 289.
Chen and Heidari. *Int. J. Mol. Sci.* 2020; 21(17): 6062.
Chow et al. *Nucleic Acids Res.* 2016; 44(D1): D1154–D1160.
Chuck et al. *Curr. Opin. Plant Biol.* 2009; 12(1): 81–86.
Clouse and Sasse. *Annu.* Rev. *Plant Physiol. Mol. Biol.*1998; 49: 427–451.
Cooper et al. *J. Chem. Ecol.* 2005; 31(9): 1953–1967.
Deepika et al. *Front. Plant Sci.* 2020; https://doi.org/10.3389/fpls.2020.581666.
Díaz. *Plant Signal. Behav.* 2011; 6(3): 339–343.
Fan et al. *Plant Cell* 2014; 26(2): 828–841.
Fan et al. *J. Plant Growth Regul.* 2014; 34: 196–205.
Guo and Gan. *Plant Cell Environ.* 2012; 35(3): 644–655.
He et al. *Plant Physiol.* 2002; 128(3): 876–884.
Huber and Bauerle. *J. Exp. Bot.* 2016; 67(7): 2063–2079.
Iqbal et al. *Front. Plant Sci.* 2021; 12: 631810.
Istrail et al. *Dev. Biol.* 2007; 310(2): 187–195.
Javed et al. *Annu. Rev. Plant Biol.* 2006; 57(1): 19–53.
Jones-Rhoades and Bartel. *Mol. Cell* 2004; 14(6): 787–799.
Kamthan et al. *Front. Plant Sci.* 2015; 6: 208.
Kim et al. *Front. Plant Sci.* 2015; 6: 1–7.
Kurihara and Watanabe. *Proc. Natl. Acad. Sci. U. S. A.* 2004; 101(34): 12753–12758.
Lee et al. *Nature* 2003; 425(6956): 415–419.
Li and Zhang. *J. Cell. Physiol.* 2016; 231(2): 303–313.
Lim et al. *Annu. Rev. Plant Biol.* 2007; 58: 115–136.
Lin et al. *Front. Plant Sci.* 2020; https://doi.org/10.3389/fpls.2020.00854.
Luo et al. *Cell* 1999; 99(4): 367–376.
Mulligan et al. *Proc. Natl. Acad. Sci. U. S. A.* 1997; 94(7): 2793–2795.
Nakamura et al. *PLOS ONE* 2017; 12(4): e0174015.
Nova-Franco et al. *Plant Physiol.* 2015; 168(1): 273–291.
Park et al. *Curr. Biol.* 2002; 12(17): 1484–1495.
Peck and Mittler. *J. Exp. Bot.* 2020; 71(5): 1649–1651.
Planas-Riverola et al. *Development (Cambridge)* 2019; 146(5): 1–11.
Robert-Seilaniantz et al. *Annu. Rev. Phytopathol.* 2011; 49: 317–343.
Ruan et al. *Int. J. Mol. Sci.* 2019; 20(10): 2479.
Ruegger and Grobhans. *Trends Biochem. Sci.* 2012; 37(10): 436–446.
Rushton et al. *Trends Plant Sci.* 2010; 15(5): 247–258.
Schommer et al. *PLoS Biol.* 2008; 6(9): 1991–2001.
Sunkar et al. *Trends Plant Sci.* 2012; 17(4): 196–203.
Voinnet. *Cell* 2009; 136(4): 669–687.
Wasternack and Hause. *Ann. Bot.* 2013; 111(6): 1021–1058.
Wu et al. *Plants* 2020; 9(4): 408.
Yang and Li. *Bioinformatics* 2011; 27(18): 2614–2615.
Zhang. *J. Exp. Bot.* 2015; 66(7): 1749–1761.
Zhang and Wang. *J. Cell. Physiol.* 2015; 230(1): 1–15.
Zhang et al. *Nat. Biotechnol.* 2013; 31(9): 848–852.
Zhang et al. *Mol. Plant* 2016; 9(9): 1302–1314.
Zhang et al. *Plant Physiol.* 2018; 177(1): 352–368.
Zhao et al. *J. Exp. Bot.* 2015; 66(15): 4653–4667.

15

Brassinosteroids: Potential Agrochemicals

Gurvarinder Kaur, Neha Dogra, Isha Madaan, Renu Bhardwaj, and Geetika Sirhindi

CONTENTS

15.1 Introduction: Multifunctional Roles of Brassinosteroids

The agriculture sector holds an indispensable share in the global gross domestic product. It is a source of income for a substantial proportion of the population, specifically in rural areas. Therefore, agriculture and its allied sectors focus on enhancing the yield of crops that solve problems of food security and offer nutritional and economic benefits. However, growing crops in their natural habitat makes them prone to several challenges of dynamic environmental conditions. These variant conditions pose several threats to plant survival, thereby influencing overall productivity. Plants may face serious consequences in terms of productivity when exposed to sudden changes in temperature, water deficient or excess water conditions, contaminated soil with heavy metals, saline soils, or any attack of biological agents leading to diseases. In order to provide food as well as economic security to the population across the globe, an upsurge in crop production is required along with tolerating harsh environmental conditions (Wik et al., 2008). Relocation of plants from their natural habitat is impractical, therefore, plants have to endure these stress conditions with a quick perception of signals and the generation of a favorable response (Fahad et al., 2015; Planas-Riverola et al., 2019). The potential application of plant hormones and scrutinizing their efficacy in inculcating stress tolerance in economically and nutritionally beneficial crops have been of great interest to researchers across the globe (Ciura and Kruk, 2018).

Phytohormones have been known to regulate several signaling cascades that trigger a response to enhance a plant's survival under environmental cues. Brassinosteroids (BRs), the polyhydroxylated steroid plant hormones, are well known for augmenting plant protection against harsh conditions of the environment, thereby maintaining yield attributes (Ahanger et al., 2018; Ahammed et al., 2020). Out of the various known BRs, 24-epibrassinolide (24-EBL), 28-homobrassinolide (28-HBL), and brassinolide (BL) are the most active forms (Bajguz and Hayat, 2009). These ubiquitously found steroid hormones interact with other plant hormones to trigger a physiological response. BRs regulate a vast array of physiological responses throughout the life span of the plant. Their role has been identified in seed germination, rhizogenesis, reproductive development, and cell senescence (Clouse and Sasse, 1998). BRs also display their role in the process of cell elongation, cell division, and differentiation (Janeczko, 2019). Crop yield is directly associated with biomass. An increase in biomass occurs with effective cell expansion and proliferation (Evans and Fischer, 1999), which is regulated by BRs (Xie et al., 2011). Therefore, BRs determine the yield of a crop. BRs also elucidate responses for anther and pollen development (Clouse, 2015). Besides this, BRs also regulate vascular differentiation and photo-morphogenesis (Nolan et al., 2020). In a nutshell, all the developmental processes that a plant undergoes throughout its life span progress under the control of BRs.

The conditions of stress that a plant encounters pose adverse effects on the overall growth and development of plants, leading to a huge loss in crop yield (Francini and Sebastiani, 2019). This is a grave concern for agriculture-dependent economies as any unprecedented changes in the environment lead to economic loss along with facing challenges regarding food scarcity. Abiotic stress conditions, which may be related to water, temperature, contamination of soils with heavy metals, or other chemicals like pesticides, insecticides, nutrient deficit soils, salinity, and irradiance, affect major processes in crops leading to oxidative burst in plants that leads to a rise in the levels of reactive oxygen species (ROS; Choudhury et al., 2017; Hasanuzzaman et al., 2020). These ROS in turn interfere in the normal functioning of all metabolic processes ongoing in the

DOI: 10.1201/9781003110651-15

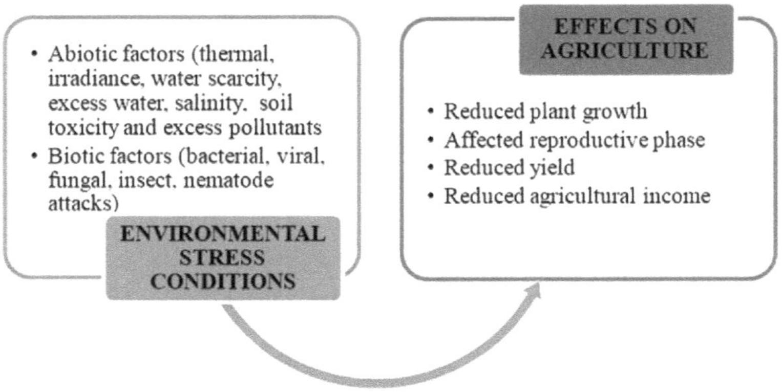

FIGURE 15.1 Effects of environmental constraints on agriculture.

cells. These environmental constraints cause a decline in the content of photosynthetic pigments, reduce the efficiency of photosynthesis, carbon assimilation, respiration, and water use in plants, and disturb the source-sink relationship (Macedo, 2012; Chen et al., 2013). The disrupted aforementioned physiological processes affect the biomass of the crop, vegetative and reproductive characteristics such as number and time of flowering, yield attributes like number of grains, grain filling, and grain nutrition quality of horticultural crops, and post-harvest losses (Figure 15.1).

In order to scavenge these ROS, upliftment in the antioxidant defense system is mandatory. BRs help in the modulation of the components of ADS that monitor the levels of ROS (Zhong et al., 2020). The improved efficiency of ADS molecules confers stress tolerance and supports plants' endurance in harsh environments (Ahammed et al., 2020). BRs are also known to play pivotal roles in improving relative water content, activity of the Rubisco enzyme, net photosynthetic efficiency of plants, and carbon assimilation and enhanced PSII (photo-system) activity that leads to improved growth attributes of crop plants (Xia et al., 2009; Lima and Lobato, 2017; Siddiqui et al., 2018). They also regulate stress-responsive genes that mediate various stress-related responses (Zhao et al., 2016). BRs have also displayed their role in flower and fruit development along with overcoming post-harvest losses (Kang and Guo, 2011). However, the mode of action of BRs and the response produced by the plant may vary with plant species, mode of application of BR (foliar spray, inflorescence spray, root application, seed priming, or addition to growth medium), concentration applied, intensity of stress, and stage of the plant during application (Ahanger et al., 2018). The implication of BRs in improving plant tolerance against environmental cues and maintaining the quality and quantity of productivity emphasizes the use of BRs as agrochemicals in enhancing productivity along with extending plants' endurance to conditions of stress. BRs have already been essential components of commercially available plant growth stimulants like growth "Epin2" (Moiseev, 1998) and Tianfengsu (Ikekawa and Zhao, 1991). The implication of such eco-friendly natural substances that display promising potential in improving agriculture may be of greater interest to plant breeders, researchers, industrialists, and agriculturists. This chapter focuses on deciphering the efficacy of BRs in increasing the productivity and nutritional aspects of cereals, oil crops, legumes, horticultural crops like fruits and vegetables, and beauty and ornamental plants that hold significant economic importance.

15.2 Practical Aspects of BRs in Cereals

Cereals (Poaceae) are important in a healthy and nutritious diet. They form the staple food of the global population. They possess diverse nutritional components and act as a major contributing source of carbohydrates if consumed in the diet (Pal and Molnár, 2021). The cereals contain a perfect blend of carbohydrates, vitamins, specifically vitamin B, minerals, antioxidants, proteins, and fiber (Esfandi et al., 2019). These rich sources of energy are sufficient for providing food and nutrition security to people in poverty. Cereal production holds a major position in the national as well as the international economy. Therefore, the growth and yield, as well as the nutritional entities, of cereal crops must be maintained in order to achieve maximum economic benefits. BRs have played a key role in this regard. They influence the grain size, grain yield, and various other agronomic traits of cereals. BRs are known to promote weight, number of kernels, and length and number of ears in cereals (Ali et al., 2008) To ensure that cereal crops provide all the economic and nutritional benefits to the population across the globe, they must develop tolerance against all conditions of environmental stress.

The first reports of the presence of BRs in cereals were recorded in *Oryza sativa*, which were dolichosterone and castasterone (Abe et al., 1984). Thussagunpanit et al. (2015) studied the effect of 24-EBL and 7,8-dihydro-8α-20-hydroxyecdysone (DHECD), which is a BR-mimic on *Oryza sativa* under heat stress. It was observed that both the compounds enhanced stomatal conductance, PSII activity, and improved water use efficiency in rice. All these parameters affected the yield, which is dependent on grain filling. The number of filled seeds and the amount of starch in seeds of heat stressed plants treated with 24-EBL and DHECD were similar to the non-stressed plants. Conditions of pollen sterility arise when rice plants are exposed to high temperature during the anthesis period, which results in a lower seed setting and reduced grain production due to oxidative burst, reduced carbon assimilation capacity, and reduced fertility of the plants (Jagadish et al., 2010; Chen et al., 2020). However, it was observed by Chen et al. (2021) that BRs (24-EBL and 28-HBL) mitigated heat stress in plants, reduced the levels of hydrogen peroxide (H_2O_2), and enhanced the contents of catalase and ascorbic acid (AsA) in the pistil. This further resulted in enhanced fertility and greater seed setting even under the conditions of heat stress. The number of spikelets per panicle is considered an important agronomical trait in rice and the final panicle size is dependent on spikelet differentiation as well as degeneration (Tan et al., 2011). Studies were conducted to decipher the role of BRs in spikelet differentiation and degeneration (Ashikari et al., 2005). The experiment was conducted with rice which was exposed to five different treatments of nitrogen (N). The application of 24-epicastasterone (CS) and 28-HBL helped in promoting pollen fertility and reducing spikelet degeneration, and BR forms promoted spikelet differentiation and enhanced the grain yield (Zhang et al., 2019). Zhang et al. (2020) studied the effect of soil drying on rice under BR treatment with undersoil drying conditions, which showed that the levels of BRs increased and protected the spikelet from degeneration. Furthermore, it was observed that if any BR-synthesis inhibitor is added to a young panicle or knockdown of BR synthesis gene is done, it leads to enhanced contents of H_2O_2, reduced adenosine triphosphate (ATP) and spikelet degeneration. However, exogenous application of the BR led to an ameliorative effect and change in the energy status as well as uplifted antioxidant defense system (ADS) of the plant.

The effect of 24-EBL was studied by Hniličkа et al. (2007) in wheat (*Triticum aestivum*), which was exposed to conditions of high temperature and drought stress. The results demonstrated increased dry matter, grain yield, and straw under both stressed and control conditions. Similar results were obtained under saline irrigated water, where 28-HBL was reported to enhance all the yield components in wheat (Eleiwa et al., 2011). The enhanced grain yield is associated with enhanced photosynthetic efficiency, enhanced CO_2 assimilation, source-sink relationships, and translocation of more photo-assimilates toward grains, which are under the influence of BRs (Ali and Ashraf, 2008). Fuji and Saka (2001) reported that BRs lead to more accumulation of starch in the grains leading to enhanced grain yield; sucrose uptake is also enhanced in the presence of BRs due to the enhanced activity of invertase (Goetz et al., 2000). The application of BRs as agrochemicals was suggested by Janeckzo et al. (2019). BRs were found to protect wheat plants from frost and it was suggested that BRs and progesterone enter into the membrane lipid during stress conditions in order to stabilize it and also enhance the ADS for protection against frost. The elevation in endogenous BR content was also reported. Heavy metals pose serious effects on the grain in cereals, making them unsuitable for consumption. It was observed by Kroutil et al. (2010) that application of BRs (24-EBL and CS) protected the plant from heavy metal toxicity and the content of lead was found to be significantly lower as compared with that of untreated plants with BRs. High temperature severely affects grain filling and grain number in wheat (Barlow et al., 2015). It was observed that 24-EBL significantly enhanced the efficacy of wheat exposed to high temperature (Hussain et al., 2019). All the phenotypic patterns and net photosynthetic rate of wheat treated with 24-EBL were found to be improved as compared with non-treated plants. Temperature susceptible and temperature tolerant cultivars of wheat were supplemented with 24-EBL and calcium chloride and the supplementation led to protection against high temperature in temperature susceptible varieties and also progressed recovery from damage caused by temperature stress (Hairat and Khurana, 2015). BRs protect plants from heat stress and water-deficient conditions and allow cereals to increase productivity in terms of grains. BR signaling components play a crucial role in this regard, which can be of interest to breeders (Kothari and Lachowiec, 2021). Upregulation of brassinazole-resistant 2 in *Triticum aestivum* (*taBZR2*) leads to protection of the plant against water-deficient stress (Cui et al., 2019), thus promoting and maintaining the yield of wheat in drought conditions. The exogenous application of the phytosterol, β-sitosterol, in wheat in water-stressed conditions ameliorated grain yield, photosynthetic efficiency, and harvest index suggesting the importance of plant steroids in inculcating stress tolerance and maintaining yield (Elkeilsh et al., 2019).

Maize (*Zea mays*) is another major cereal consumed worldwide (Shiferaw et al., 2011). In salt-stressed conditions, BRs (28-HBL and 24-EBL) protected maize from severe damages that would result in the loss of yield. This ameliorative action of BRs can be attributed to the BR-mediated regulation of ADS (Rattan et al., 2020). Similar results have been reported by Kaya et al. (2018), where foliar application and seed treatment maintained the nutrient status of maize in a saline environment. Liu et al. (2019) studied the growth-promoting effects of BR-mimics bikinin and brazide in maize in nicosulfuron toxicity; 24-EBL was used as a positive control in the study. The results reflected the positive role of BR-mimics on maize, which was significant in 24-EBL. They maintained the growth potential of maize along with maintaining ROS and ADS regulation. N-deficiency can act as a major growth limiting factor in maize productivity (Rogers and Benfey, 2015). The role of BRs has also been scrutinized in N-uptake in maize; 24-EBL regulated the nitrate transporter genes (NRT) which in turn maintained N-uptake (Xing et al., 2021). Furthermore, 24-EBL supplementation increased the plant biomass of the plant along with various morphological parameters. Transcriptomic analyses in maize reveal that BRs mitigate biotic and abiotic stress and modulate secondary metabolite production along with controlling lipid signaling and metabolism, which is essential for generating stress responses (Trevisan et al., 2020). The action of BRs is highly significant even at a low concentration (Pérez Quiñones et al., 2018). Some recent studies reflected the role of BRs in regulating kernel size in maize where the BR transcription factors BES1/BZR1 are involved in regulating kernel size (Sun et al., 2021).

Besides abiotic stress, BRs help in protecting plants against biotic attacks, thereby maintaining productivity. In barley (*Hordeum vulgare*), application of 24-EBL to barley plants protected them from *Fusarium* pathogens (*Fusarium culmorum*). These pathogens lead to head blight disease and affect grain quality as myotoxins accumulate in the grain, which has a detrimental effect if consumed. The application of 24-EBL upregulated the expression of pathogenesis-related genes in order to protect the barley from disease damage. It also improved the photosynthetic efficiency of the plants (Ali et al., 2013). Barley semi-dwarf varieties with impaired BR synthesis were used to study the effects of drought where it was observed that BR-mutants showed some defects in the growth parameters and did not satisfactorily produce osmoprotectants as compared with normal BR synthesis. BR-mutant barley plants also displayed reduced efficiency of the PSII system along with reduced carbon assimilation. The BR-mutant plants that were subjected to water-deficient conditions did not upregulate in the transcript levels of the stress-responsive genes like heat shock proteins (*HSP90*; Janeckzo et al., 2016). Similar results were recorded in studies of barley exposed to temperature variations where BR-mutants showed decreased accumulation of HSPs (Sadura et al., 2020). Change in the fatty acid composition of membranes was also considered a response to stress conditions (Horváth et al., 2012). Increased saturation of membrane lipids is essential for sustaining growth during heat stress (Pearcy, 1978), whereas unsaturation is crucial for low-temperature stress (Skoczowski and Filek, 1994). BRs are reported to affect the fatty acid composition of membranes that are essential for various membrane-dependent metabolic processes in plants. BR-mutants were used in barley to study the effects of BR on lipids and their metabolism (Rudolphi-Szydło et al., 2020; Figure 15.2).

Millets have been a major part of the diet of Asian and African populations but they seem to have lost their importance after the introduction of wheat, maize, and other cereals. These small grains are nutritionally richer as compared with wheat and rice, constituting zinc, iron, calcium, dietary fiber, polyphenols, and many more bioactive compounds (Ramakrishnan et al., 2019). Owing to these dietary compounds, gluten-free millets prevent diseases and promote several health benefits. They possess anti-carcinogenic, anti-diabetic, hypocholesterolemic, and anti-inflammatory benefits (Ramakrishnan et al., 2019). Muthamilarasan and Prasad (2020) put forth the idea of the intensification of millet farming amidst the COVID-19 pandemic in order to meet nutritional and economic benefits. Many studies on the action of BRs in millets in various environments suggest the yield promoting action of BRs for future exploitation in agriculture.

Ameliorative action of 28-HBL and 24-EBL on *Panicum miliaceum* L. subjected to heat and drought stress has been acknowledged where the ADS was uplifted and growth parameters that affected the yield were maintained by exogenously applied hormones (Vayner et al., 2014). Studies on finger millet (*Eleusine coracana*) suggested that the application of BR improved all yield parameters, net photosynthetic rate, total chlorophyll, and total soluble protein in plants (Prabha et al., 2016). Basic leucine zipper (bZIP) transcription factors (TFs) are involved in the regulation of stress-responsive genes in conditions of stress. It has been affirmed that *EcbZIP17* in *Eleusine coracana* confers stress tolerance by upregulation of BR responsive genes and signaling pathways in order to maintain growth and yield parameters (Ramakrishna et al., 2018). BR seed treatment (0.1 ppm) enhanced grain yield and total protein in the grain in finger millet as compared with the control. The grain yield was 21.64% compared with the control (Nithila et al., 2007). Mohanabharathi et al. (2019) studied the deleterious effect of drought in finger millet where it was reported that BR helped in mitigating the effect of water deficiency. BRs improved the yield attributes such as number of productive tillers, number, length, and weight of the ears, and number and

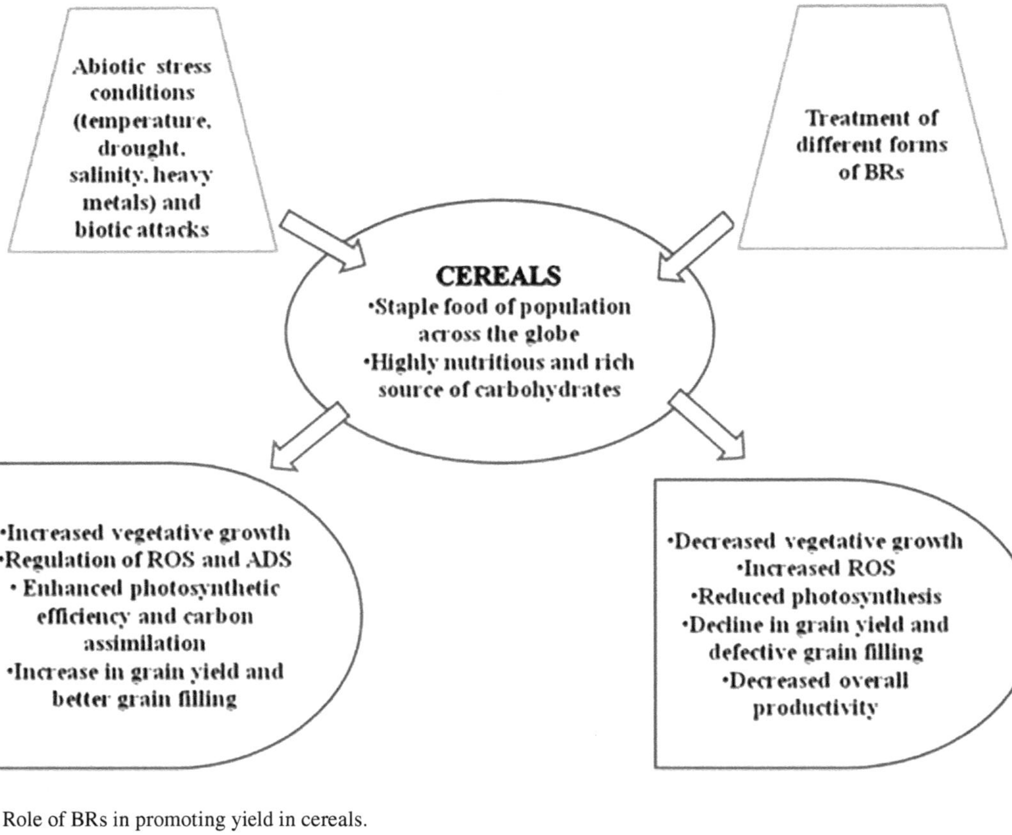

FIGURE 15.2 Role of BRs in promoting yield in cereals.

weight of the grains. All the improved yield parameters can be attributed to the improved physiological status of the plant under BR action. Similar results were obtained in pearl millet (*Pennisetum glaucum*), where with BR foliar supplementation, the grain yield, grain protein, and total sugar content in the grain were significantly maintained under drought stress (Sivakumar et al., 2002). The grain yield was 3591.86 kg ha^{-1} with BR supplementation whereas it was 3018.55 kg ha^{-1} in the control. Under the foliar application of 0.1 ppm BR in pearl millet, the yield enhancement was seen by 19% as compared with the control (Rajavel and Vincent, 2009).

Studies by Vardhini et al. (2011) suggest that application of 24-EBL and 28-HBL alleviated water stress conditions in sorghum with the upliftment of ADS. The tiller number and height of the sorghum were found to be reduced in plants treated with BR-inhibitor propiconazole, which reflects the role of BR in sorghum development (Sutherlin and Weber, 2012). *Secale cereale* L. (rye grass) is another important cereal and is protected from the harsh conditions of winter with the supplementation of 24-EBL. This cold acclimation process is achieved by maintaining photosynthetic efficiency as well as carbon assimilation. EBL conferred stress tolerance against the cold and snow mold pathogen *Microdochium nivale* (Pociecha et al., 2016). The studies of Mantilla Perez (2018) suggest that leaf angle variations play a crucial role in deciding the productivity of sorghum as it influences photosynthetic rates. BRs influence the leaf angle in order to regulate photosynthesis, and hence, productivity.

15.3 Versatility of BRs in Oil Crops

Agriculture faces huge challenges in significantly increasing productivity, diminishing the environmental footprint, and carrying renewable substitutes. These are being addressed by developing new oil crops for the future (Ortiz et al., 2020). Ongoing efforts include the application of BRs to fulfill these needs.

BR regulates various features of plant development and stress responses. In the oilseed crop *Brassica napus*, Sahni et al. (2016) overexpressed the *Arabidopsis* BR biosynthetic gene *AtDWF4*, which attained growth and stress response phenotypes. These transgenic *B. napus* plants displayed amplified seed yield, leading to augmented oil content per plant, significantly improved tolerance to dehydration and heat stress, higher root length and biomass, and enhanced resistance to the fungal pathogens *Sclerotinia sclerotiorum* and *Leptosphaeria maculans* as compared with the wild type. The transcriptome analysis also supported the integrated responses of BRs on growth, development, yield, and stress responses (Ziegler et al., 2019). Supplementation of BR enhanced the growth of *B. napus* microspore-derived embryos (Belmonte et al., 2011). Ma et al. (2009) observed enhancement in photosynthesis, growth, and yield of rapeseed (*Brassica napus* L.) under freezing stress conditions with exogenous application of BRs. Application of EBL (10^{-7}, 10^{-9}, and 10^{-11} M) to three *Brassica* species (*B. juncea, B. napus* and *B. carinata*) under high temperature (34°C and 44°C) stress mitigated the negative result and improved plant development (Pradhan et al., 2013).

BR lightened the negative impact of cold stress in *Brassica juncea* L. and increased seedling growth (Kumar et al., 2010). HBL critically worked on drought stress-induced changes in the photosynthesis of *Brassica juncea* L. (Fariduddin et al., 2009). Foliar application of BRs (0.4 ppm) at pre-flowering and pod development stages augmented the oil quality and yield of mustard (*Brassica juncea*) under drought stress conditions (Kumawat et al., 1997). Further, the improvement in salt tolerance after the treatment of BR was confirmed in *B. juncea* (Ali et al., 2008) and in oil rape seed (Efimova et al., 2014). HBL improved photosynthetic efficiency in *B. juncea* plants subjected to saline stress (Hayat et al., 2007). Foliar application of 28-homoBL mitigated salinity stress by increasing the efficiency of photosynthesis in *B. juncea* (Alyemeni et al., 2013). The appliance of EBL to *B. juncea* L. cv PBR to 91 seedlings inhibited the negative effects of heavy metals like Mn, Zn, Co, and Ni. Furthermore, EBL also increased the growth and abridged the accumulation and uptake of the metals (Sharma and Bhardwaj, 2007a). Exogenous appliance of 0.01 μM of homoBL to Cd (100 or 150 μM) treated *B. juncea* seedlings showed increased plant growth, photosynthesis, content of proline, and antioxidative enzymes (Hayat et al., 2007). Supplementation of HBL (Bhardwaj et al., 2008) and BRs (Bhardwaj et al., 2011) regulated Mn uptake and the growth of *B. juncea* L.

In addition to BR effects associated with growth, a major plant defense signature, expectedly refereed by BES1/BZR1, was noticeable in the transgenic plants. All these results and outcomes established that BR can corporately and simultaneously develop biotic and abiotic stress tolerance and plant productivity. The capability to present pleiotropic favorable effects allied with diverse agronomic traits put forward that BR-related genes are imperative for increasing plant productivity and performance in different stress conditions. From an agricultural perspective, these BR-related genes could serve as reproduction targets for crop improvement (Sahni et al., 2016). *B. juncea* (Indian mustard) L. plants were exposed to arsenic(V) (0.0, 0.1, 0.2, and 0.3 mM concentrations), resulting in the abridged growth and modulation of several biochemical stress markers. The synthesis of BRs was the novel observation made in this study, a good example of either heavy metal induced stress or a stress-defending response of the *B. juncea* plant to assuage the ill effects of stress. These findings show molecular mechanisms that work during interactions between BR synthesis and stress indicators in handling the stress produced by As(V) in *B. juncea* crops (Kanwar et al., 2015). Furthermore, the foliar sprays of HBL significantly improved the seed yields of mustard. The season did not influence the mustard yield responses to HBR treatments. The treatment of HBL was given up to two seasons and it was observed that yields were extensively elevated from the control plots. Quality of crop with HBL foliar sprays given on the 45th day after sowing was considerably lower than that obtained when HBL sprays were given at 30 days after sowing (Ramraj et al., 1997).

In recent years, due to the appropriate composition of fatty acids and oil quality, safflower (*Carthamus tinctorius*) has been established as a major oilseed crop in both arid and semiarid regions. The consequences of EBL on biochemical properties and yield of safflower under stress designated that the steroid hormone is capable of assuaging stress by regulating various physiological and morphological mechanisms of plants. It is hypothesized that EBL can influence safflower biochemical properties, oil content, and crop yield in water-deficit conditions (Zafari et al., 2020). Stress reduced Chl-a, Chl-b, total chlorophyll, carotenoids, seed yield, relative water content (RWC), and oil percentage. Genotypes proceeded significantly differently under drought stress. EBL extensively amplified plant chlorophyll contents and oil percentage, and it reduced the malondialdehyde (MDA) content by raising the proline and carotenoid contents significantly under stress. This is undoubtedly of agricultural as well as economic significance because it is a crucial medicinal plant grown in several parts of the world under stress conditions. In view of that, these results, for the first time, indicated the lightening effects of EBL on safflower physiology, development growth, and quality and yield components under drought stress (Zafari et al., 2020).

Soybean (Glycine max L.) is a globally important crop, providing oil and protein. Drought stress reduced water-use efficiency and efficient quantum yield of PSII photochemistry in soybean. The exogenous appliance of EBL (100 nM) mitigated the negative effects, via increasing these variables. EBL abridged the membrane damages (malondialdehyde and electrolyte leakage) and oxidant compounds such as superoxide and hydrogen peroxide in stressed plants. This study proved that EBL increased electron transport rate (ETR) and net photosynthetic rate (Pn) in stressed plants revealed that ETR had a well-built relationship with Pn. These findings also recommend that soybean plants with elevated values of ETR are more proficient in relation to Pn. The water-deficit conditions promoted the implications on photosynthetic apparatus by reducing the quantum yield of PSII photochemistry, ETR, Pn, and water-use efficiency, however, overall, EBL treatment alleviated these negative effects (Pereira et al., 2019).

Furthermore, Ribeiro et al. in (2019) observed that *G. max* seedlings subjected to water deficiency presented diminution of root endodermis (RDT), root epidermis (RET), root cortex (RCT), root metaxylem (RMD), and vascular cylinder (VCD); however, the application of BR with a concentration of 100 nM encouraged an enhancement in these values in comparison with water deficit+0 nM BR. The treatment with 100 nM BR also induced a raise in the activity of all of the enzymes estimated. Here, BR also enhanced seed germination, length, and dry matter of soybean seedlings. Also, reductions in the content of superoxide and hydrogen peroxide, malondialdehyde, and electrolyte leakage were essentially related to the improved activities of antioxidant enzymes such as catalase, ascorbate peroxidase, and superoxide dismutase, verifying the benefits of BR on the antioxidant system (Ribeiro et al., 2019). Lu et al. (2006) observed that application of BR improved the seedling growth and water-logging resistance of *G. max*. Further, Dong et al. (2008a) reported that supplementation of BR enhanced the photosynthetic characteristics of soybean under Al stress. Furthermore, Ariyoshi et al. (2016) observed that BRs regulated pod growth positively to cell hypertrophy in soybean plants.

The effect of BR application on the growth parameters of *Arachis hypogaea* L. (groundnut) was reported in the 1990s. Spray application of 28-HBL in groundnut during flowering and pegging enhanced pod yields in a dose-dependent manner (Ramraj, 1997). The growth-promoting effect of BRs has also been explored under environmental stress. Verma et al. (2012) studied the effects of BRs on organogenesis from the cotyledonary node on the growth and antioxidant activity of groundnut. The study also used market available BR formulation. It was observed that BR promoted rhizogenesis in groundnut. Further, enhancement in the physiological parameters was also observed along with improved efficiency of ADS to shield the plant. BRs protect groundnut from any environmental constraints that affect the yield. Dong et al. (2008b) stated that application of a BR analog, BRTS303, improved the resistance of peanut plants subjected to chilling stress. The conditions of drought pose serious threats to the productivity of the crop. It was observed that BR supplementation in groundnut exposed to conditions of drought maintained the yield parameters of the crop and also induced several differentially expressed genes (DEGs) that regulate stress responses for drought tolerance which may protect a quick response in later drought conditions (Huang et al., 2020).

Linum usitatissimum, commonly known as flax, has great economic importance owing to its fiber and oil quality and nutritional benefits (Jhala and Hall, 2010). Several reports have unraveled the role of BRs in flax, which can potentially be used for enhancing fiber and oil quality as well as productivity. Water-deficit conditions in fields may pose serious threats to the overall productivity of plants like flax. The role of BRs has been explored in flax in drought conditions. The conditions of drought were created with the treatment of polyethylene glycol (PEG) and the seed priming with 24-EBL led to the development of the plants' immunity to conditions of water scarcity without compensating the yield. It was observed that application of 24-EBL modulated the contents of osmolytes like proline and uplifted the ADS, which helped in mitigating oxidative stress and lipid peroxidation (Aghaee and Rahmani, 2019). The application of 24-EBL also upregulated several stress-responsive transcription factors. Molecular analysis revealed modulation of WRKY 40, KRP2, ERF, Mn-SOD, POD1, and POD3. The enhanced ADS activity, reduced lipid peroxidation, and induced activity of stress-related genes under the influence of BRs put forth strong evidence of the implication of BRs in drought stress conditions in order to maintain the yield of the flax. Amraee et al. (2019) reported that seed treatment of 24-EBL had growth promoting and stress protective effects on flax exposed to salt stress, which is a growth limiting environmental condition. The involvement of BRs in transcription and translation level modulation reflects their importance in generating stress-related responses. The study reported the influence of 24-EBL on the methylation of genes and the generation of an epigenetic response in order to provide stress tolerance and regulate various essential physiological processes to achieve greater yields. The results revealed a reduction in demethylation and an increase in fully methylated bands in 24-EBL treatment. Thus, BR treatment can induce an epigenetic response through cytosine methylation to develop the tolerance of flax in a saline environment. Khodyankov et al. (2013) reported the yield enhancement role of BRs where the foliar application of BRs produced more yield as compared with the non-treated flax plants.

Salt stress caused a decline in the growth and biomass of soybean plants due to oxidative damage and osmotic stress conditions but EBL priming prompted the accretion of osmolytes including glycine betaine, proline, and sugars, ensuing improved photo-protection through the protection of tissue water content. The treatment of EBL under normal as well as salt stress conditions significantly increased antioxidant activity and osmolyte accumulation. However, soybean performance was improved by seed priming with EBL (Soliman et al., 2020).

15.4 Multiple Roles of BRs in Leguminous Crops

Leguminous crops are cultivated widely by farmers owing to their nutritional as well as economic benefits. As far as the nutritional aspect is considered, legumes are a rich source of protein, one of the primary constituents of the major building blocks of the human body. In addition to proteins, legumes also consist of a significant amount of carbohydrates, minerals, fibers, and vitamins. All of these must be a part of a balanced human diet in order to maintain the healthy functioning of body systems. Also, the cultivation of legumes is not a laborious process as they can easily germinate and flourish in dynamic climatic conditions. But devastating environmental fluctuations due to human interference have even posed a threat to legume cultivation, which has resulted in lowered yields and nutritional value. The use of plant growth regulators for crop improvement is now common practice as they do not contaminate the environment and also do not cause any serious ailments in humans. Among the various PGRs, BRs are being explored on a large scale for their application to legumes at various levels of growth, especially seed priming, in order to enhance productivity as well as the qualitative value of the crop (Figure 15.1). BRs have also been reported to reciprocate the effects of dynamic environmental conditions by enhancing the ADS of plants (Bajguz and Hayat, 2009). BRs enhance the vegetative growth of legumes by increasing root and shoot length, number of leaves, leaf area, number of shoots, and branches per plant. All these factors positively add to the overall productivity of the legume crop (Pinol and Simon, 2009). The overall increase in the yield of snap beans grown under conditions of temperature stress was observed by El-Bassiony et al. (2012) due to the elevation of photosynthetic activity, pod size, and number under the influence of 24-EBL. Foliar spray of *Pisum sativum* and lupine with 24-EBL resulted in the enhancement of protein, fat, tocopherol, and β-carotene content, thereby adding to the nutritional value as well as the overall productivity of the crop (Biesaga-Kościelniak et al., 2014). Studies by Janeczko et al. (2009) also revealed that 24-EBL application led to a significant increase in the content of important antioxidants, that is, β-carotene and AsA in soybean. These antioxidants not only act as free radical scavengers but also add to the preserving properties of legume seeds. Foliar application of BL and 24-EBL stimulated activity of DNA and RNA polymerases resulting in the elevation of total protein content in seeds of groundnut along with fat content (Vardhini and Rao, 1998). It has been reported in several research studies that BR application restores the growth of legumes to normal levels under the conditions of biotic as well as abiotic stress. Studies by Yusuf et al. (2014) on the effect of 28-HBL on the growth and yield of mung bean crops grown in a Ni-rich environment revealed that the hormone was very much effective in counteracting the harmful effects of Ni toxicity by enhancing the free amino acid content and antioxidants and creating a positive N-balance overall, thereby making the seeds suitable for human utilization. BRs have also been found to reciprocate the detrimental effects of herbicide and metal contaminated soils in leguminous crops, maintaining their survival and normal chemical composition under such conditions (Mota, 2020). One of the research studies by Ali (2004) also suggested the commercial use of 28-HBL in ameliorating the effects of salinity stress in black gram where the hormone was found to regulate metabolic processes in the crop in such a way that the overall protein content of leaves, as well as seeds, seed size, and number, was found to be significantly higher. Also, seeds of soybean primed with 24-EBL grown under conditions of salinity stress were found to be more tolerant than seeds without the respective hormone treatment. The hormone was found to be effective in creating a positive N-balance in the

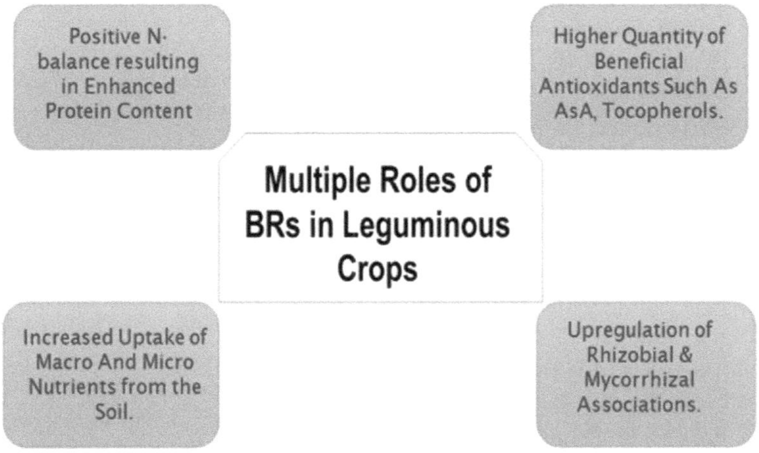

FIGURE 15.3 Multiple roles of BRs in leguminous crops.

legume, maintaining its nutritional properties (Soliman et al., 2020). BRs have also been reported to enhance root growth and the number of root hairs, which helps in retrieving more nutrients in the soil and hence enhancing crop productivity. Jan et al. (2018) found that the detrimental effects of Cd toxicity in pea plants were overcome by 24-EBL, which enhanced the absorption of macro- as well as micronutrients from the soil. These macro- and micronutrients are very much essential for normal metabolic activities of the crop, which result in the augmentation of the crop yield. Legumes are not only being cultivated as a staple food crop but also as an underplant crop as they add to the N content of soil due to their ability to stabilize the symbiotic relationship with N-fixing bacteria, which is called nodulation. Nodulation is beneficial for both the host plant as well as the soil environment. N being an important macronutrient for the growth of plants is not available in the soil for absorption by plants. Therefore, this symbiotic association enables plants to fetch N from the environment to be made available, thereby creating a positive N-balance within the crop and enhancing the overall crop yield. BRs have been reported to enhance the process of nodule formation by stimulating certain pathways that are involved in the biogenesis of nodulation proteins. BRs have been found to control both the processes of nodule formation as well as nitrogen fixation in the root nodules. This has been supported in studies by Cheng et al. (2017) where pea mutants deficient in BR receptors consisted of a smaller number of nodules in the roots with an impairment in the process of nitrogen fixation. BRs upregulate the process of nodulation through a shoot-derived signal irrespective of BR production in roots.

Ferguson et al. (2005) revealed in their studies that wild-type plants with normal BR production in shoots consisted of more nodules in comparison with those with decreased BR biogenesis in shoots. Certain research studies also support the role of ethylene in BR-regulated nodule formation. BRs promote nodulation by suppressing ethylene biosynthesis, which has been shown in studies on both BR and ethylene mutants, where the former showed defects in nodule formation while the latter was found to show normal nodulation (Foo et al., 2016). Sometimes, certain legume crops such as pea may also develop mycorrhizal arbuscular association with some fungi, which, of course, increases crop potential to uptake various vital nutrients present in the soil required for better crop development. Such associations are also upregulated by BR application (Wei et al., 2016; Mc Guiness et al., 2019). In addition to this, BRs also enhance the activity of nitrogenase, nitrate reductase, leghemoglobin content, and so on. All of these factors contribute to enhanced amino acid and protein content of the crop, adding to its alimentary value (Figure 15.3).

Upregulation of nodule formation and N-fixation by BRs in legumes is beneficial to humankind from both a nutritional as well as an economical viewpoint. Nutritionally, it will enhance the protein content of the crop, which is, of course, valuable for human health. Economically, it will not only add to the yield of the growing legume crop but also to the productivity of the next crop to be sown in the same soil. Thus, BRs may be used as potent agrochemicals for increasing the productivity of legumes in different parts of the world.

15.5 BRs as a Vegetable Growth Promoter

Food is one of the vital and basic needs for the survival of life on this planet. Among the various forms of food being consumed, vegetables act as a major food source, being taken in various forms whether raw or cooked. The primary source of vegetable produce is farming. But due to the dynamic climate as well as other environmental discrepancies as a result of human interference in nature, vegetable farming faces a number of challenges and problems. These challenges affect the quality of the vegetables in terms of nutritional value and yield for which they are mainly cultivated by farmers. Therefore, to maintain harmony, such agricultural practices need to prove beneficial to farmers to increase the yield and provide them with economic benefits but also to enhance the qualitative value of food crops to provide health benefits to the people who are consuming them, and, at the same time, these practices should not deteriorate the environment. The use of fertilizers and other chemicals to achieve this aim harms soil, water, and community health as it enters the food chain. Hence, the usage of eco-friendly phytohormones can be promoted in order to accomplish the aim of nutritive as well as economic vegetable farming. Among the various phytohormones, polyhydroxy steroids, that is, BRs, are being widely explored for their roles in controlling various plant processes, which adds to the quality as well as quantity of the crop.

A wide range of physiological processes from seed germination to fruit development are known to be regulated by application of BRs to various plant parts. Studies by Sasse in 1999 revealed the pleiotropic effects of BRs on seed germination, growth, maturation, flowering, abscission, and senescence in different groups of vegetable crops. Research studies on BR-mutant lines also unveiled their indispensable role in controlling the basic growth of a number of vegetable crops such as pea and tomato, thereby affecting their yield and productivity. Foliar application of BRs to the lkb mutant, a BR-deficient dwarf pea mutant, restored its normal growth and resulted in an increase in the overall yield and productivity of the crop (Nomura et al., 1997). Similarly, foliar application of BRs was found to restore the growth of leaves as well shoots in the BR-deficient tomato mutant *dpy* (Bishop and Yokota, 2001). Both natural and synthetic BRs have been reported to enhance the nutritional value of vegetable crops. A very common bioactive BR analog, that is, 24-epibrassinolide, has been found to enhance the protein content of Chinese cabbage, which is a common vegetable consumed in a raw as well as a cooked form (Wu and Zhao, 1993).

Vegetable crops are cultivated as per their growing season but are consumed throughout the year. The problem faced by farmers is not only confined to their cultivation but also extends after harvesting for long-term storage and transport in a healthy condition. In addition to promoting the quality and quantity of vegetable crops, BRs have also been reported to delay senescence and fruit ripening so that they can be stored for longer periods during post-harvest. Apart from this, BRs rescue crops from the damaging effects of various kinds of biotic as well as abiotic stresses such as pathogen attacks,

TABLE 15.1

Effect of BRs on Various Parameters of Oil Crops under the Influence of Different Environmental Conditions

Sr. No.	Oil crops	Remarks	Reference
1.	*Brassica juncea*	Exogenous application of HBL increased crop yield.	Ramraj et al., 1997
2.	*Arachis hypogea*	In the presence of BR, antioxidant enzyme activities were augmented and MDA content progressively declined.	Verma et al., 2012
3.	*Brassica juncea*	EBL significantly reduced the negative effects of stress conditions and defended the cell membrane by ameliorating proline content and activity of antioxidant enzymes.	Pradhan et al., 2013
4.	*Brassica juncea*	BRs played a central role in coping with the stress produced by As(V) by enhancing protein content and antioxidative enzymes and reducing lipid peroxidation.	Kanwar et al., 2015
5.	*Brassica napus*	BRs augmented the seed yield leading to amplified overall oil content per plant.	Sahni et al., 2016
6.	*Brassica juncea*	HBL regulated gene expression and antioxidant enzyme activities in response to temperature-induced oxidative stress.	Kaur et al., 2018
7.	*Linum usitatissimum*	BRs application developed the fiber yield and quality for flax.	Yuan et al., 2019
8.	*Glycine max*	Application of EBL diminished oxidant compounds (superoxide and hydrogen peroxide) as well as membrane damages (MDA and electrolyte leakage).	Pereira et al., 2019
9.	*Linum usitatissimum*	BRs ameliorated the negative impacts of water deficit by reducing membrane lipid peroxidation with enrichment in the activity of APOX, CAT, SOD, and POD enzymes.	Aghaee and Rahmani 2019
10.	*Glycine max*	BRs endorsed a rise in the germination, length, and dry matter of soybean seedlings, and further, these results were clarified via growth in root anatomy.	Ribeiro et al., 2019
11.	*Linum usitatissimum*	BRs supported epigenetic modification to cytosine methylation.	Amraee et al., 2019
12.	*Arachis hypogea*	BR priming influenced many differentially expressed genes (DEGs) that play a role as kinase or transcription factors for signal cascade under drought stress.	Huang et al., 2020
13.	*Glycine max*	BRs triggered the accretion of osmolytes including glycine betaine, proline, and sugars resulting in improved photo-protection through the defense of tissue water content.	Soliman et al., 2020
14.	*Carthamus tinctorius*	EBL significantly amplified chlorophyll contents and oil percentage with extensive reduction in the malondialdehyde (MDA) content via raising the proline and carotenoid content in stress conditions.	Zafari et al., 2020

fungal diseases, salt, drought, temperature, heavy metal, and sodicity stress, all of which are very much common in particular regions of the world. The beneficiary role of BRs among various vegetable crops in different growing conditions has been summarized in Table 15.1 as per the research studies available.

Since the discovery of BRs from rape pollen in rape mustard by Mitchell et al. (1970), extensive research has been carried out across the globe, especially in Russia and Japan, which concluded that the hormone is suitable for the elevation of crop productivity and commercial application. BRs and their various analogs are not just eco-friendly but do not pose any kind of threat to human health owing to their xenobiotic nature (Vardhini et al., 2006). A large number of countries have adopted the commercial use of BR analogs for the improvement of different vegetable crops such as potato, onion, and others as shown in Table 15.2. They have proven to be reasonable for cropping by small farmers as well as for large-scale crop improvement and their application at seed priming or foliar level is not a laborious or harmful process. It has been found in studies by Bhushan Mandava, one of the pioneers in this field of research, that the cost of one gram of a synthetic BR is \$5–\$10 and that a gram of BR may be used for as much crop cultivation as five acres (Maugh, 1981). However, in many countries and many regions, BR application to vegetable crops needs to be introduced and promoted on a large scale for future improvements in the agriculture sector (Table 15.3).

15.6 Fruit Quality Enhancement by BRs

Fresh fruits are well known as horticultural plants and include several annual and perennial herbaceous plants. They show ample diversity and variation in their cultivated varieties and differences in the flower or fruit color and shape. Horticulture, as well as associated green industries, is speedily developing certified fields with rising significance to society. However, the production of horticultural plants is also characterized by robust management, environmental control, high management costs, use of extensive technology, and high risk. A large number of ways have been investigated to improve the productivity of horticultural plants including fertilizers, pesticides, and insecticides. However, their regular and repetitive utilization is proving counterproductive. Furthermore, some of them are harmful to the environment and ecosystems. The application of plant growth regulators is eco-friendly where, in particular, BRs are non-toxic hormones proving to be highly favorable for plant productivity. In order to develop our horticulture industry, the focal point of this section is to explain the role of BRs on horticultural crops. Moreover, progress in the practical application of BRs in fruit plants for enhancing quality, yield, and stress tolerance is a promising vision for the near future.

Fruit in reality is a ripened ovary, suggesting that the yield of crops depends on the occurrence of female flowers. The application of EBL to cucumber plants accelerated and augmented female flower production where the consequences may

TABLE 15.2

Applications of BRs as Vegetable Growth Promoters

Sr. No.	Vegetable crop	Positive effects of BR application	Reference
1.	Radish	Enhancement of yield and storage time of the root crop by application of BL, 24-EBL, and 28-HBL.	Sujatha, 2001
2.	Bush bean	BL application increased the yield.	Meudt et al., 1983
3.	Tomato	Increased fruit yield, i.e., fruit size as well as number by 24-EBL. Early fruit ripening and shortening of life cycle by 28-HBL Enhanced soluble sugars, respiration rate lycopene, and ethylene synthesis, and lower chlorophyll content by BL.	Vardhini and Rao 2002 Zhu et al., 2015
4.	Potato	Increased tuber yield by BL. Longer dormancy and delayed sprouting in tubers enable long-term storage under application of 24-EBL. Stimulation of tuber development and resistance against late blight and dry rot diseases.	Nunez et al., 1995 Korableva et al., 2002 Kazakova et al., 1991
5.	Onion	Enhanced bulb yield by Biobras 6 and 16.	Nunez et al., 1998
6.	Pepper	24-EBL improved flower and fruit number and overall yield per plant. Increase in total antioxidant activity and phenolic content by DI-100.	Samira et al., 2012 Serna et al., 2012
7.	Spinach	Enhanced seed germination by 24-EBL.	Ikekawa and Akutsu, 1987
8.	Cabbage	Increased foliage of inflorescence heads by 24-EBL. Sugar content and overall yield by BL.	Asatova, 1991. Schilling et al., 1991
9.	Cucumber	Augmentation of production of female flowers and hence fruit yield by EBL. Early fruit ripening by induction of ethylene production. Production of parthenocarpic fruits.	Papadopoulou and Grumet, 2005 Fu et al., 2008 Manzano et al., 2011
10.	Coffee	Improved seed number and bean maturation by EBL.	Ramos et al., 1987
11.	Mushroom	Lessens browning, phenolic contents, and polyphenol oxidase activity.	Ding et al., 2016

TABLE 15.3

Brassinosteroids Employed for Commercial Applications

Sr. No.	Country	Traded name of the BR synthetic analog being used	Crops recommended
1.	Cuba	Biobras-6, Biobras-16, DAA-6, DI-31, MH-5, ME	Potato, onion, rice, coffee
2.	India	Bountee, Combine, Double	Rice, wheat, potato, onion, sugarcane
3.	USA	Brassinosteroid-1105	Soybean
4.	Japan	TS-303	Beetroot, soybean, rice
5.	Belarus	Tksamim, Yaponiya	Barley

be attributed, at least in part, to EBL-encouraged production of ethylene (Papadopoulou and Grumet, 2005). External treatment of BRs was given to cucumber and zucchini. Cucumber was found to be more susceptible to BRs than zucchini and it was noticeable in the initial phase of development by a decline in the number of male flowers and the initiation of female flowers in the main shoot. In another study, Fu et al. (2008) demonstrated that BRs play a central role in cucumber during early fruit development. Two different cultivars of cucumber with diverse parthenocarpic capacity illustrated that parthenocarpic growth was induced with exogenous appliance of BRs but repressed by the loss of BRs synthesis. The BRs prompted active cell division associated with amplified transcripts of the cell cycle-related genes, mainly that of cyclin D3 genes. These observations strongly recommend that BRs play a significant role during fruit development in cucumber in the early hours (Fu et al., 2008; Manzano et al., 2011).

BR treatment also amplified the production and quality of fruits of numerous important plants such as apple, strawberry, cranberry, gooseberry, cherry, peach, and *Citrus*. The enhancement in quality and yield of these plants was mediated by avoidance of premature fall of immature fruits, increase in the fruit set, delayed senescence, and several other factors implicated in fruit yield and quality (Khripach et al., 1999). Additionally, molecular biotechnology has shown the participation of BRs in the fruit ripening of strawberry. A collection of expressed sequence tags (ESTs) from the cDNA libraries arranged from various strawberry fruit components and the developmental stages illustrated that FaBRI1 receptor expression was improved in receptacles and evidently increased during ripening (Bombarely et al., 2010). Chai et al. (2013) described the probable pathway of this action of BRs by providing the physiological as well as molecular evidence to show that BRs contribute to strawberry fruit ripening. They examined BR content and BR receptor gene FaBRI1 expression for the period of "Akihime" strawberry fruit development and it was found that BR levels improved during later developmental stages while the mRNA expression levels of FaBRI1 quickly increased from white to early red stages, signifying the association of BRs in fruit ripening.

Current evidence recommends that BRs also play role in the ripening of grapes (a non-climacteric fruit). The expression analysis of genes encoding BR biosynthesis receptors and enzymes during the development of grape berry exposed the transcript accumulation patterns that are dependent on the dramatic increase in endogenous BR levels observed at the onset of fruit ripening. These results support the fact that a rise in endogenous BR levels is linked with grape ripening, and thus the application of EBL to grape berries extensively encouraged ripening. Whereas brassinazole (Brz), a well-known inhibitor of BS biosynthesis, notably delayed fruit ripening (Symons et al., 2006). With subsequent advances in model plant systems, Lisso et al. (2006) found that exogenous appliance of BRs to grape berries appreciably promoted ripening.

Mango is a climacteric fruit in which maturity or ripening shows a pre-climacteric minimum, climacteric rise, climacteric peak, and finally post-climacteric phases (Watada et al., 1984). The ripening procedure of mango fruit includes various biochemical changes, including amplified respiration, fruit softening, ethylene production, carotenoids synthesis, chlorophyll degradation, and numerous other metabolic activities leading to changes in organic acids, carbohydrates, phenolics, and lipids (Singh and Singh, 2011). The treatment of EBL to mango endorsed the fruit ripening and the ripening was associated with the accretion of trace amounts of castasterone and brassinolide, well known as important BR analogs. The level of castasterone was somewhat higher than that of EBL (0.13 ng g^{-1} FW). Furthermore, the exogenous appliance of EBL treatments (45 and 60 ng g^{-1} FW) notably advanced the beginning of the climacteric peak of ethylene production and respiration rate by two days and one day, respectively. Furthermore, the application of EBL promoted fruit softening between days three and seven days of ripening (Zaharah et al., 2012).

Foliar application of EBL improved the yield of melon by 10–20% (Ikekawa and Nagai, 1987). The improved yield was mediated by enhancement in the number of flowers and fruit setting and delayed senescence where fruit setting is an imperative factor for determining crop productivity.

In some fruit species, such as strawberry and mango, endogenous BRs occur in trace amounts and may or may not be essential for fruit ripening (Zaharah et al., 2012). In comparison with some climacteric fruits, the ripening of post-harvest papaya fruit displays fascinating characteristics. Changes related to ripening turn out rapidly in papaya (Fabi et al., 2007; Oliveira and Victoria, 2011). Although respiration is increased, it also sometimes decreases when the fruit is fully ripened. Besides, fruit can show improved respiration without a noteworthy ethylene peak (Bron and Jacomino, 2006). So, in order to make sure whether these phytohormones precede either autonomously or have an interactive effect on fruit ripening, a biochemical approach that merges the use of BRs and ethylene-related inhibitors illuminates the possible interaction between these hormones in papaya post-harvest ripening. When applied, BR biosynthesis inhibitors particularly inhibit BR biosynthetic cytochrome P-450 enzymes and further decrease the number of endogenous BRs in several plant tissues (Asami et al., 2003).

Further, the exogenous application of EBL amplified the alternative oxidase (AOX) capacity. Also, treatment with Brz2001 (a specific inhibitor of BR synthesis) improved AOX capacity but these effects were lacking in fruits treated concurrently with EBL and Brz. The change in EBL level showed no effect on ethylene emission rate in the first 24 hours, but the decline in ethylene production was observed only in Brz-treated fruit on the fifth day. Collectively, Brz and EBL augmented ethylene production on the fifth day following the day in which treatment was carried out. All these observations suggest that BRs and ethylene proceed antagonistically, regulating AOX capacity during fruit ripening of papaya, directly or indirectly (Mazorra et al., 2013).

The effects of BRs were evaluated on commercial yellow passion fruit (*Passiflora edulis*). The treatment with BS analog BB-16 for three weeks after flowering resulted in improvement in the yield of passion fruit plants by approximately 65%. The number of fruit plants and the mean mass of each fruit were considered, which led to an approximate production of 20.1 t ha^{-1} in comparison with the estimated yield of the control, which was 12.6 t ha^{-1}. The brassinosteroid analog was found to be more proficient with appliance for three weeks consecutively after the emergence of the first flower due to the enormous increase in the yield and soluble solids content in passion fruit (Gomes et al., 2006). The application of brassinolide improved the fruit setting in orange trees during flowering. However, when it was applied during fruit development, it diminished the physiological drop of fruits, resulting in an increased number of fruits per plant, accompanied by an enhancement in the average fruit weight. In *Citrus*, treatment of BRs increased juice production and elevated brix/acidity ratio was observed (Kuraishi et al., 1991); however, in *Citrus madurensis*, brassinolide retarded fruit abscission (Iwahori et al., 1990; Figure 15.4).

Besides its effects on the ripening of fruits, BRs also develop fruit quality and yield. Foliar spray of BL on litchi leaves sooner than blossom augmented the activities of polygalacturonase and pectin methylesterase. The content of calcium, water-soluble pectin, and protopectin was amplified in the fruit pericarp, and a reduction in the rate of fruit cracking took place, suggesting the vital role of BRs in raising the commercial value of litchi fruits (Peng et al., 2004). Moreover, the application of BRs (5 M concentration) abridged the post-harvest decay in jujube fruit caused by *Penicillium expansum* and also delayed fruit senescence by reducing ethylene production and respiration rate, increasing the life span of the fruit (Zhu et al., 2010).

15.7 Ornamental Flowers and BRs

Roses are one of the most beautiful, attractive, and widely used flowers for various purposes and they are the majority of flowers used in the cut flower industry (Liorzou et al., 2016). However, they are highly susceptible to the pathogen *Botrytis cinerea*, which causes gray mold disease. The necrotrophic mode of infection severely affects the growth stages of the plant and also causes post-harvest loss. Transcriptomic analysis of the petals was done while they were infested with the pathogen, which revealed the upregulation of BR-responsive genes was more significant compared with other phytohormones. Further to affirm the result, 5 mM BR was exogenously applied to rose petals prior to infection with *B. cinerea*. This application resulted in reduced lesion diameter as compared with the control (Liu et al., 2018). *Lavandula intermedia* is a

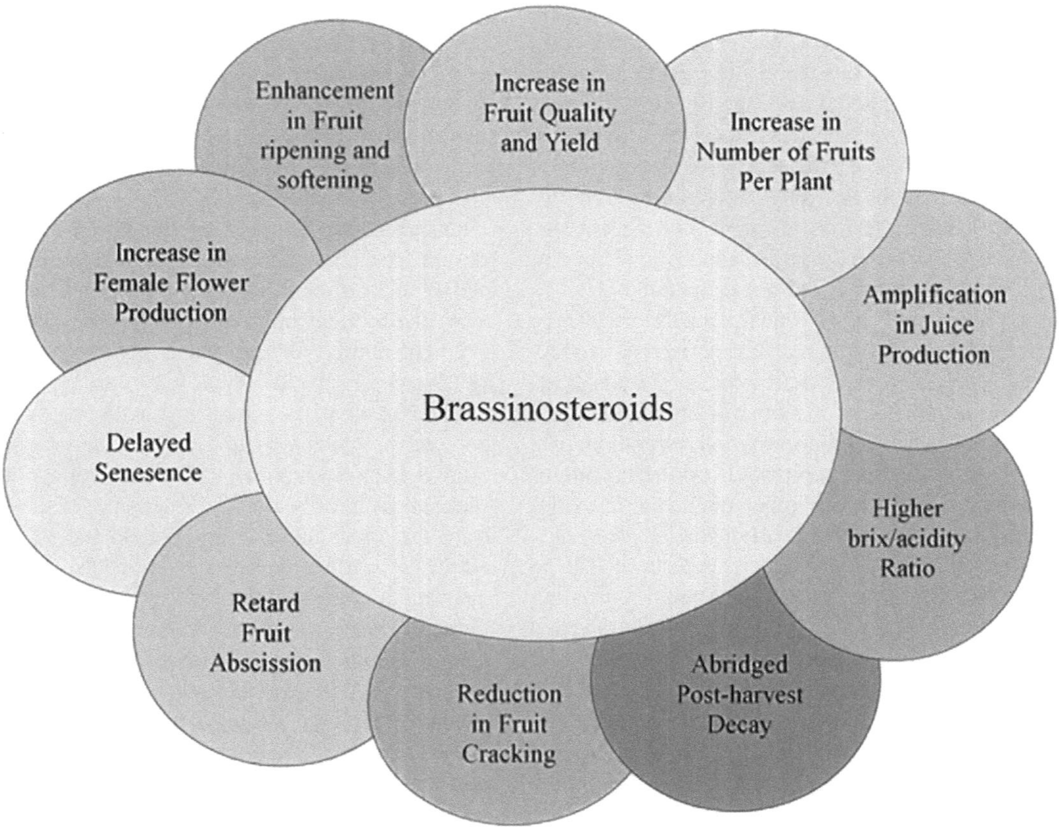

FIGURE 15.4 Effect of BRs on fruits under the influence of various environmental conditions.

hybrid between *L. latifolia* (spike lavender) and L. *angustifolia* (English lavender) where the essential oil lavandin is obtained. It holds medicinal and cosmetic importance. Asciet al. (2019) reported that foliar application of 24-EBL on *Lavandula intermedia* enhanced the essential oil content, which has uses in perfumes and cosmetics and has medicinal importance. The best results were obtained at 1.5 mg/L 24-EBL. Along with increasing the lavandin content, 24-EBL application also promoted total phenolic content and overall growth in the plant. *Zinnia elegans* (zinnia) was subjected to lead (Pb) toxicity that adversely affected the growth, vegetative development, and flower development and led to Pb accumulation in all the parts of the plant. However, foliar spray of 24-EBL ameliorated Pb toxicity and promoted flower production (Badawy et al., 2017). *Solidago Canadensis,* commonly known as Canadian golden rod, is a member of the family Asteraceae. It is used for ornamental purposes along with therapeutic use owing to its antimicrobial, anti-mutagenic, and antioxidant properties. It was observed that foliar application of BR promoted the growth of the herb. The quality, as well as the quantity of the production, was enhanced significantly as compared with the control along with an increase in the content of bioactive compounds like polyphenols (El-Sayed et al., 2020). Application of 50 μg/L of BR promoted the chlorophyll content in the leaves of *Chrysanthemum*. It also led to an increase in the content of anthocyanin in the petals of the flower which led to an increase in the quality of the flower (Liu et al., 2014). Another study in *Chrysanthemum* as reported by Choudhari et al. (2018) scrutinized the role of spray application of BR (0.5 ppm), which enhanced vegetative and floral characteristics of the plant. The application of 24-EBL (0.1 mg L^{-1}) in tulips (*Tulipa greigii*) enhanced pollen viability and the rate of pollen germination (Akçal et al., 2016). The *Gerbera hybrida* of the family Asteraceae is an important cut flower crop with large economic importance. The transcriptome analysis in the *Gerbera* hybrid reveals the major role of BRs in petal formation and their growth. The average length of petal in plants treated with BL was 43.55 mm whereas it was 35.27 mm in the control and 28.64 mm in the plants treated with BR inhibitor BZR (Huang et al., 2014). The growth promotion of daisy (*Aster amellus* L.) was observed with BR supplementation that improved vegetative characters and enhanced vase life (Raveendra et al., 2013; Table 15.4).

15.8 Conclusion

Ever since the discovery of BRs, they have been of great interest to researchers across the globe. BRs regulate various physiological reactions in plants that are vital for maintaining yield and drawing economic benefits. The productivity of crops like cereals, legumes, vegetables, fruits, and oilseeds must be maintained as they offer solutions to food challenges that arise due to a growing population. A significant contribution has been to explore the role of BRs in enhancing the productivity of essential crop plants in their natural environment where they face challenges from both biotic and abiotic factors. BRs and their signaling components are also known to be involved

TABLE 15.4
Effect of BR Treatment on Ornamental Flowers

Sr. No.	Name	Treatment	Action	Reference
1.	*Aster amellus*	BR treatment	Enhanced vegetative growth and vase life	Raveendra et al., 2013
2.	*Chrysanthemum*	Spray of BR	Enhanced chlorophyll content in leaves and anthocyanin in petals	Liu et al., 2014
3.	*Tulipa greigii*	24-EBL	Enhanced pollen viability and pollen germination	Akçal et al., 2016
4.	*Zinnia elegans*	24-EBL (foliar application)	Protected against Pb toxicity and promoted flower production	Badawy et al., 2017
5.	*Gerbera* hybrid	Inflorescence spray of BL	Petal growth and expansion of basal and middle regions of petals	Huang et al., 2014
6.	*Chrysanthemum*	BR (foliar spray)	Enhanced vegetative and floral characters	Choudhari et al., 2018
7.	*Rosa hybrid* (rose)	Upregulated BR-signaling genes	Protection against *Botrytis cinerea*	Liu et al., 2018
8.	*Lavandula intermedia*	24-EBL	Enhanced lavandin content, growth parameters, and total phenolics	Asci et al., 2019
9.	*Solidago canadensis*	BR (foliar spray)	Growth and production promotion	El-Sayed et al., 2020

in regulating gene expression in order to trigger responses in favor of the plant's growth and survival. Therefore, these signaling components should be of great interest to plant breeders for manipulations that may lead to prosperity in agriculture. Also, owing to its yield promoting and defensive properties, this chapter puts forth the idea of the commercialization of plant steroids like BRs. Although many leading industries have already started the cost-effective production of plant growth chemicals that have BRs as essential components, such industries must be encouraged and promoted to produce low-cost chemicals for easy availability to farmers so that crops are not lost as a result of unsuitable environmental conditions.

REFERENCES

Abe et al. *Agr. Biol. Chem.* 1984; 48(4): 1103–1104.
Aghaee & Rahmani. *J. Plant Interact.* 2019; 14(1): 242–253.
Ahammed et al. *J. Plant Growth Regul.* 2020; 39(4): 1–14.
Ahanger et al. *J. Plant Growth Regul.* 2018; 37(4): 1007–1024.
Akçal et al. *III International Symposium on Horticulture in Europe-SHE2016 1242.* 2016; 549–552.
Ali et al. *Chemosphere.* 2008; 72(9): 1387–1392.
Ali & Ashraf. *Plant Growth Regul.* 2008; 56(2): 107–116.
Ali et al. *Phytopathology.* 2013; 103(12): 1260–1267.
Alyemeni et al. *Acta Bot. Bras.* 2013; 27(3): 502–505.
Amraee et al. *Plant Physiol. Biochem.* 2019; 139: 478–484.
Ariyoshi et al. *Plant Growth Regul.* 2016; 80(3): 391–395.
Asami et al. *J. Plant Growth Regul.* 2003; 22(4): 336–349.
Asatova. *Conference on Brassinosteroids.* 1991.
Asci et al. *J. Essent. Oil Bear. Plants.* 2019; 22(1): 254–263.
Ashikari et al. *Science.* 2005; 309(5735): 741–745.
Badawy et al. *J. Environ. Sci. Technol.* 2017; 10(4): 157–174.
Bajguz & Hayat. *Plant Physiol. Biochem.* 2009; 47(1): 1–8.
Barlow et al. *Field Crops Res.* 2015; 171: 109–119.
Belmonte et al. *Planta.* 2011; 233(1): 95–107.
Bhardwaj et al. *J. Pure Appl. Sci.* 2008; 2(1): 149–154.
Bhardwaj et al. *Int. J. Environ. Waste Manag.* 2011; 8(1/2): 114–122.
Biesaga-Koscielniak et al. *Aust. J. Crop Sci.* 2014; 8(3): 378.
Bishop & Yokota. *Plant Cell Physiol.* 2001; 42(2): 114–120.
Bombarely et al. *BMC Genomics.* 2010; 11: 503.
Bron & Jacomino. *Braz. J. Plant Physiol.* 2006; 18(3): 389–396.
Chai et al. *J. Plant Growth Regul.* 2013; 69(1): 63–69.
Chen et al. *Crop J.* 2021; 9(1): 109–119.
Chen et al. *Sci. Rep.* 2020; 10(1):1–13.
Chen et al. *Sci. Hortic.* 2013; 160: 222–229.
Cheng et al. *Sci. Rep.* 2017; 7(1):1–12.
Choudhari et al. *Int. J. Chem. Stud.* 2018; 6(1): 1458–1460.
Choudhury et al. *Plant J.* 2017; 90(5): 856–867.
Ciura & Kruk. *J. Plant Physiol.* 2018; 229: 32–40.
Clouse & Sasse. *Annu. Rev. Plant Biol.* 1998; 49(1): 427–451.
Clouse. *J. Plant Growth Regul.* 2015; 34(4): 828–844.
Ding et al. *Food Bioproc. Tech.* 2016; 9(8): 1327–1334.
Cui et al. *Plant physiol.* 2019; 180(1): 605–620.
Dong et al. *Acta Agron. Sin.* 2008a; 34: 1673–1678.
Dong et al. *Guangxi Zhiwu.* 2008b; 28: 675–680.
dos Santos Ribeiro et al. *Acta Physiol. Plant.* 2019; 41(6): 1–11.
Efimova et al. *Russ. J. Plant Physiol.* 2014; 61(6): 733–743.
El-Bassiony et al. *Gesunde Pflanzen.* 2012; 64(4): 175–182.
Eleiwa et al. *Aust. J. Basic Appl. Sci.* 2011; 5(5): 58–65.
Elkeilsh et al. *J. Plant Res.* 2019; 132(6): 881–901.
El-Sayed et al. *Horticulturae.* 2020; 6(4): 100.
Esfandi et al. *Heliyon.* 2019; 5(4): 01538.
Evans & Fischer. *Crop Sci.* 1999; 39(6): 1544–1551.
Fabi et al. *J. Agric. Food Chem.* 2007; 55(15): 6118–6123.
Fahad et al. *Plant Growth Regul.* 2015; 75(2): 391–404.
Fariduddin et al. *Acta Physiol. Plant.* 2009; 31(5): 889–897.
Ferguson et al. *Plant Physiol.* 2005; 138(4): 2396–2405.
Foo et al. *J. Exp. Bot.* 2016; 67(8): 2413–2424.
Francini & Sebastiani. *Abiotic Stress Effects on Performance of Horticultural Crops.* 2019.
Fu et al. *J. Exp. Bot.* 2008; 59(9): 2299–2308.
Fujii & Saka. *Plant Prod. Sci. (Japan)* 2001; 4(2): 136–144.
Goetz et al. *Plant J.* 2000; 22(6): 515–522.
Gomes et al. *Sci. Horticult.* 2006; 110: 235–240.
Hairat & Khurana. *Am. J. Plant Sci.* 2015; 6(11): 1827.
Hasanuzzaman et al. *Antioxidants* 2020; 9(8): 681.
Hayat et al. *Environ. Exp. Bot.* 2007; 60(1): 33–41.
Hniličk̆a et al. *Cereal Res. Commun.* 2007; 35(2): 457–460.
Horváth et al. *Prog. Lipid Res.* 2012; 51(3): 208–220.
Huang et al. *Peer J.* 2014; 5: 3382.
Huang et al. *Genes.* 2020; 11(8): 919.
Hussain et al. *Environ. Sci. Pollut. Res.* 2019; 26(17): 17163–17172.

Ikekawa & Zhao. *Application of 24-Epibrassinolide in Agriculture.* 1991.
Ikekawa & Akutsu. Jpn. Kokai Tokkyo Koho. JP; 1987 63,239,201 [88,239,201] [C.A. 111, 52465].
Iwahori et al. *Plant Growth Regul.* 1990; 9: 119–125.
Jagadish et al. *J. Exp. Bot.* 2010; 61(1): 143–156.
Jan et al. *BMC Plant Biol.* 2018; 18(1): 146.
Janeczko. *Brassinosteroids: Plant Growth Development.* 2019; 59–88.
Janeczko et al. *Seed Sci. Technol.* 2009; 37(3): 625–639.
Janeczko et al. *Plant Physiol. Biochem.* 2016; 99: 126–141.
Janeczko et al. *Plant Physiol. Biochem.* 2019; 139: 215–228.
Jhala & Hall. *Aust. J. Basic Appl. Sci.* 2010; 4(9): 4304–4312.
Kang & Guo. In: *Brassinosteroids: A Class of Plant Hormone.* 2011; 269–288. Springer, Dordrecht.
Kanwar & Bhardwaj. *Ecotoxicol. Environ. Saf.* 2015; 115: 119–125.
Kaya et al. *J. Plant Growth Regul.* 2018; 37(4): 1244–1257.
Kaur et al. *Sci. Rep.* 2018; 8(1): 1–13.
Kazakova et al. *IzvestiyaTSKhA (Russian Federation).* 1991.
Khodyankov et al. *Agriculture and Plant Protection: Scientific-Practical Journal.* 2013.
Khripach et al. *Brassinosteroids: A New Class of Plant Hormones.* Academic Press, London, UK. 1999; 325345.
Korableva et al. In: *Doklady Botanical Sciences* (Vol. 361, pp. 25–27). Consultants Bureau, New York, 1964; 2002.
Kothari & Lachowiec. *Int. J. Mol. Sci.* 2021; 22(5): 2706.
Kroutil et al. *Plant Soil Environ.* 2010; 56(1): 43–50.
Kumar et al. *Indian J. Biochem. Biophys.* 2010; 47(6): 378–382.
Kumawat et al. *Ann. Biol. – Ludhiana.* 1997; 13: 91–93.
Kuraishi et al. *Brassinosteroids: Chemistry, Bioactivity and Application.* ACS Symposium Series 474. American Chemical Society, Washington, DC. 1991; 312–319.
Lima & Lobato. *Physiol. Mol. Biol. Plants.* 2017; 23(1): 59–72.
Liorzou et al. *J. Exp. Bot.* 2016; 67(15): 4711–4725.
Lisso et al. *Phytochemistry.* 2006; 67(20): 2232–2238.
Liu et al. *Hubei Agric. Sci.* 2014; 53: 97–99.
Liu et al. *J. Plant Growth Regul.* 2019; 38(2): 701–712.
Liu et al. *BMC Genet.* 2018; 19(1): 1–10.
Lu et al. *Chin. Agric. Sci. Bull.* 2006; 23: 37–38.
Ma et al. *Zuowu Xuebao (Acta Agronomica Sinica).* 2009; 35: 1336–1343.
Macedo. *Abiotic Stress Responses in Plants.* 2012; 41–61.
Mantilla Perez. *Genetic and Brassinosteroid Control Underlying Leaf Angle Variation across the Canopy in Sorghum.* 2018.
Manzano et al. *Plant Growth Regul.* 2011; 65(2): 213–221.
Maugh. *Science.* 1981; 212(4490): 33–34.
Mazorra et al. *Theor. Exp. Plant Physiol.* 2013; 25: 223–230.
McGuiness et al. *Front. Plant Sci.* 2019; 10: 269.
Meudt et al. *Proceedings Annual Meeting Plant Growth Regulator Society of America.* 1983.
Mitchell et al. *Nature.* 1970; 225(5237): 1065–1066.
Mohanabharathi et al. *J. Pharmacogn. Phytochem.* 2019; 8(3): 3308–3312.
Moiseev. *Eureka.* 1998; 68: 22–25.
Mota. *The Potential of Brassinosteroids to Alleviate the Effect of Mesotrione Residue on Three Legume Crops* (Doctoral dissertation, University of the Free State). 2020.
Muthamilarasan & Prasad. *Trends Plant Sci.* 2020; 26(1): 33–40.
Nithila et al. *J. Biosci. Agric.* 2007; 3(4): 252–254.
Nolan et al. *Plant Cell.* 2020; 32(2): 295–318.
Nomura et al. *Plant Physiol.* 1997; 113(1): 31–37.
Nunez et al. *Cultivos Trop. (Cuba)* 1998.
Núñez et al. *Cultivostropicales.* 1995; 16(1): 26–27.
Oliveira & Victoria. *Food Res. Int.* 2011; 44(5): 1306–1313.
Ortiz et al. *Curr. Opin. Plant Biol.* 2020; 56: 181–189.
Pal & Molnár. *Growing Importance of Cereals in Nutrition and Healthy Life.* 2021.
Papadopoulou & Grumet. *Hortic. Sci.* 2005; 40: 1763–1767.
Pearcy. *Plant Physiol.* 1978; 61(4): 484–486.
Peng et al. *Sci. Hortic.* 2004; 101: 407–416.
Pereira et al. *LWT.* 2019; 107: 1–8.
Pérez Quiñones et al. *J. Agric. Food Chem.* 2018; 66(7): 1612–1619.
Pinol & Simón. *J. Plant Growth Regul.* 2009; 28(2): 97–105.
Planas-Riverola et al. *Brassinosteroid Signaling in Plant Development and Adaptation to Stress Development.* 2019.
Pociecha et al. *Plant Growth Regul.* 2016; 80(2): 127–135.
Prabha et al. *Res. Crops.* 2016; 17(3): 483–488.
Pradhan et al. *J. Stress Physiol. Biochem.* 2013; 9: 376–384.
Rajavel & Vincent. *Ecobiology.* 2009; 24(3): 237–243.
Ramakrishna et al. *Sci. Rep.* 2018; 8(1): 1–14.
Ramakrishnan et al. *Millets.* In: *Whole Grains.* CRC Press. 2019; 103–127.
Ramos et al. Ministério da Indústria e do Comércio, Rio de Janeiro (Brasil) Instituto Brasileiro do Café, Rio de Janeiro (Brasil). 1987.
Ramraj et al. *J. Agri. Sci.* 1997; 128(4): 405–413.
Rattan et al. *J. Plant Growth Regul.* 2020; 39(4): 1465–1475.
Raveendra et al. *J. Hortic. Sci.* 2013; 8(2): 276–277.
Rogers & Benfey. *Curr. Opin. Biotechnol.* 2015; 32: 93–98.
Rudolphi-Szydło et al. *Cells.* 2020; 9(5): 1125.
Sadura et al. *J. Mol. Sci.* 2020; 21(5): 1889.
Sahni et al. *Sci. Rep.* 2016; 6(1): 1–14.
Samira et al. *J. Stress Physiol. Biochem.* 2012; 8(3): 224–233.
Sasse. *Brassinosteroids: Steroidal Plant Hormones.* 1999; 137–161.
Schilling et al.. *Influence of Brassinosteroids on Organ Relations and Enzyme Activities of Sugar-Beet Plants.* 1991.
Serna et al. *Plant Growth Regul.* 2012; 68(3): 333–342.
Sharma & Bhardwaj. *Acta Physiol. Plant.* 2007; 29(3): 259–263.
Shiferaw et al. *Food Sec.* 2011; 3(3): 307–327.
Siddiqui et al. *Acta Physiol. Plant.* 2018; 40(3): 1–15.
Singh et al. *Crop Post-Harvest: Science and Technology: Perishables*, vol. 3. Blackwell Publishing Ltd, Oxford. 2011; 108–142.
Sivakumar et al. *Indian J. Plant Physiol.* 2002; 7(1): 79–82.
Skoczowski & Filek. *Plant Sci.* 1994; 98(2): 127–133.
Soliman et al. *Physiol. Mol. Biol. Plants.* 2020; 26(3): 501–511.
Sujatha. Ph. D. thesis, Osmania University, Hyderabad, India. 2001.
Sun et al. *J. Exp. Bot.* 2021; 72(5): 1714–1726.
Sutherlin & Weber. *J. Purdue Undergrad. Res.* 2012; 2(1): 15.
Symons et al. *Plant Physiol.* 2006; 140(1): 150–158.
Tan et al. *Plant Breed.* 2011; 130(2): 177–184.
Thussagunpanit et al. *J. Plant Growth Regul.* 2015; 34(2): 320–331.
Trevisan et al. *Plant Growth Regul.* 2020; 92(1): 117.
Vardhin & Rao. *Phytochemistry.* 1998; 48(6): 927–930.
Vardhin & Rao. *Phytochemistry.* 2002; 61(7): 843–847.
Vardhini et al. *Indian J. Plant Physiol.* 2006; 11(1): 1.
Vardhini et al. *J. Phytol.* 2011; 3(10): 38–43.

Vayner et al. *Biotechnologia.* 2014; 7(5): 77–84.
Verma et al. *Int. Sch. Res. Not.* 2012.
Watada et al. *Hortic. Sci.* 1984; 19: 20–21.
Wei & Li. *Mol. Plant.* 2016; 9(1): 86–100.
Wik et al. *Global Agricultural Performance: Past Trends and Future Prospects.* 2008.
Wu & Zhao. *J. Plant Physiol Mol Biol.* 1993; *19*(1):49–52
Xia et al. *Planta.* 2009; 230(6): 1185–1196.
Xie et al. *J. Exp. Bot.* 2011; 62(13): 4495–4506.
Xing et al. *Crop J.* 2021.
Yuan et al. *Industrial Crop Prod.* 2019; 128: 270–281.
Yusuf et al. *Physiol. Mol. Biol. Plants.* 2014; 20(4): 449–460.
Zafari et al. *J. Agric. Food Chem.* 2020; 68(22): 6040–6047.
Zhang et al. *J. Integr. Plant Biol.* 2019; 61(8): 943–963.
Zhang et al. *Environ. Exp. Bot.* 2020; 169: 103887.
Zaharah et al. *J. Plant Growth Regul.* 2012; 31(3): 363–372.
Zhao et al. *Biol. Plant.* 2016; 60(3): 563–571.
Zhong et al. *Ecotoxicol. Environ. Saf.* 2020; 187:109831.
Zhu et al. *Postharvest Biol. Technol.* 2015; 100: 196–204.
Zhu et al. *Postharvest Biol. Technol.* 2010; 56(1): 50–55.
Ziegler et al. *J. Integr. Plant Boil.* 2019; 61(5): 639–650.

16

Exploiting the Recuperative Potential of Brassinosteroids in Agriculture

Barket Ali

CONTENTS

16.1 Introduction

Mitchell et al. (1970) reported brassinosteroids (BS) from *Brassica napus* for the first time and named them "Brassins." This group is now well established and is universally accepted as the sixth class of phytohormones, under the name "brassinosteroids" (Peres et al., 2019). They are of ubiquitous occurrence in the plant kingdom and are reported from the plant groups ranging from algae to angiosperms. Moreover, they are represented by 69 compounds and have been isolated and characterized from more than 100 plant species (Bajguz and Tretyn, 2003; Zullo, 2018; Peres et al., 2019). Among the plant parts, the maximum concentration of BS has been reported in pollen grains (1–100 micro grams per kg fresh weight). Moreover, their concentration varies from species to species. The younger tissue possesses brassinosteroid quantity more than that of the older (Bajguz and Tretyn, 2003). Out of these BS analogs, the most active and stable ones include brassinolide, castasterone, 24-epicastasterone, 28-homocatasterone, 24-epibrassinolide, and 28-homobrassinolide (Figure 16.1).

Brassinosteroids are indispensable for the normal growth and development of plants. They influence diverse physiological responses that range from seed germination to seed production and senescence (Clouse and Sasse, 1998). The involvement of BS in plant growth and development has been demonstrated in two important bioassays: (a) internode bioassay and (b) rice lamina inclination test (Clouse and Sasse, 1998). Other fundamental processes influenced by BS include cell division and elongation, vascular differentiation, reproductive development, vascular differentiation, pollen tube formation and elongation, pollen grain development, elongation of hypocotyls, epicotyls, and peduncles of dicots as well as coleoptiles and mesocotyls of monocots, and the modulation of gene expression (Clouse and Sasse, 1998; Peres et al., 2019). BS also enhance seed germination (Ali et al., 2005), control sex determination (Hartwig et al., 2011), and regulate root foraging responses during nutrient deficiency by modulating the spatial distribution of root system architecture (Pandey et al., 2020). Brassinosteroids also affect the attributes that lead to the increase in the rate of photosynthesis (Bhatia and Kour, 1997; Hayat et al., 2000, 2001). They exhibit the capability to counter abiotic and biotic stresses by rescuing plants from these stressful situations. Exogenous application of BS has been reported to mitigate the adverse effect generated by NaCl (Ali et al., 2007), water stress (Fariduddin et al., 2009), nickel (Ali et al., 2008), cadmium (Hayat et al., 2007, Hasan et al., 2008), aluminum (Ali., et al 2008a), lead (Dalyan et al., 2018), antimony (Wu et al., 2019), and chromium (Jan et al., 2020).

Brassinosteroids improve the quantity and quality of different crops of horticultural importance (Ali, 2017; 2019). They also enhance the concentration of different secondary metabolites and the active constituents of plants (Ali et al., 2021). They also exhibit a great potential to increase the productivity of various crops—cereals, legumes, oil seeds, fruits, and vegetables. This recuperative potential of BS has been exploited in a large number of studies all over the world. These studies are comprehensively presented in this chapter to strengthen the recommendation that brassinosteroids are the efficient alternative to chemical fertilizers as well as other agrochemicals. The continuous and excessive use of chemical fertilizers and other agrochemicals has a hazardous impact on living organisms as well as on the environment. Moreover, some studies have also identified some key genes regulated by BS, which

DOI: 10.1201/9781003110651-16

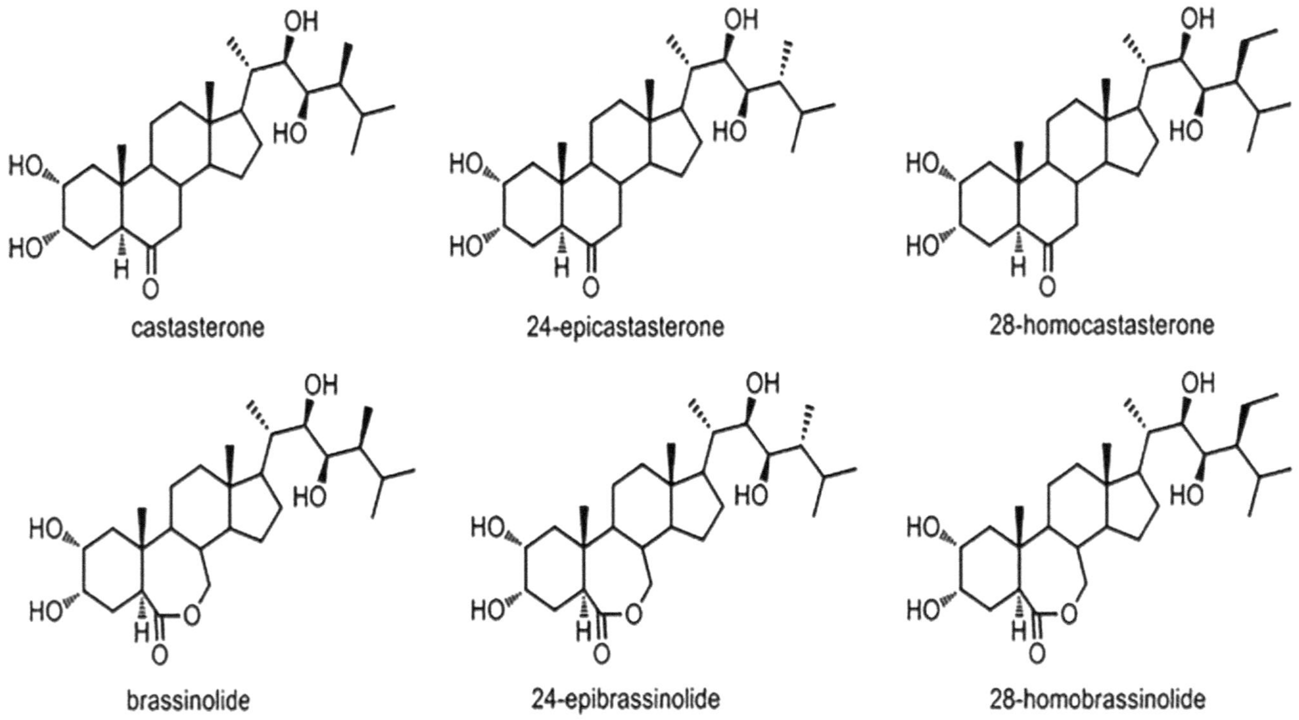

FIGURE 16.1 Active and stable analogs of brassinosteroids.

lead to increased productivity of a crop. These studies are also presented briefly to explain the recuperative potential of BS in agriculture at a molecular level.

16.2 Role of Brassinosteroids in Agriculture

The productivity or yield of any crop is determined by a large number of regulating factors such as environment, genetic constitution, photosynthetic efficiency, mineral nutrition, hormonal balance, flowering, successful pollination and fertilization, fruiting, prolonged attachment of the fruit to the mother plant, delayed senescence, and an overall healthy metabolic state. Brassinosteroids are known to regulate most of these processes efficiently (Clouse and Sasse, 1998). Therefore, they have been exploited to enhance the productivity of different crops; cereals, legumes, oilseed crops, horticultural/fruit crops, and vegetables. Brassinosteroids have been utilized in crop productivity for the following reasons: (a) natural occurrence of brassinosteroids and their eco-friendly nature; (b) efficiency of generating a physiological response at a very low (μM) concentration; (c) involvement in diverse physiological responses; (d) capability to induce resistance against various pathogens and environmental hazards; and (e) free of side effects because of being the non-xenobiotics. Since their discovery and the unveiling of their efficient bioactivity, extensive field trials were started worldwide in countries like the US, Russia, Belarus, Cuba, Japan, and India (Khripach et al., 2003). Some of the common conclusions drawn in these studies suggest that the efficiency of BS depends on (a) time/stage of BS application, (b) frequency of BS application, (c) duration of exposure to BS, (d) type and concentration of BS analog, and (e) existing environmental conditions (Ikekawa and Zhao, 1991; Nunez et al., 2003; Divi and Krishna, 2009; Hola et al., 2010).

One of the most important objectives of agriculture research is the production of the maximum quantity of seed grains (grain yield or seed yield) that can be helpful in feeding the ever-increasing population of the world. The plant part useful for humans is different for different plants such as seeds in the case of cereals and legumes, fruits from horticultural crops, and fruits and leaves from vegetables. The most important seed/fruit producing crops of agricultural interest can be broadly grouped as below, which will be followed in the forthcoming discussion.

a) Cereal crops: wheat (*Triticum aestivum*), maize (*Zea mays*), rice (*Oryza sativa*), barley (*Hordeum vulgare*)
b) Legume crops: chickpea (*Cicer arietinum*), mung bean (*Vigna radiata*), pea (*Pisum sativum*), bean (*Phaseolus vulgaris*), soybean (*Glycine max*)
c) Oil seed crops: Indian mustard (*Brassica juncea*), ground nut (*Arachis hypogea*)
d) Fruit crops: watermelon, strawberry, mango, papaya, passion fruit, orange, litchi, jujube, grapes, etc.
e) Vegetable crops: tomato, pepper, lettuce, sugarcane, sugar beet, potato, cucumber

16.2.1 Effect on Cereal Crops

Wheat, maize, and rice constitute a staple food for many people around the world. There is an immediate need to increase the grain productivity of these crops. One of the traditional methods followed to increase the productivity of crops is the use of chemical fertilizers. However, continuous and excessive use of fertilizers is polluting the ecosystem and exerting a hazardous impact on living organisms. On the other hand, brassinosteroids are non-toxic, non-xenobiotic, and eco-friendly biochemicals, which have been exploited in crop productivity. Several experiments have been conducted to explore the possible utility of these hormones in crop productivity. Jin et al. (1988) sprayed wheat plants with 24-epibrassinolide (EBL) and observed a significant improvement in the percent grain setting, number of caryopsis per ear, and 1000 grain weight. These authors co-related the increase in yield parameters to the mitigation of stress by the hormone. In another study, two cultivars of wheat—C306 (drought tolerant) and HD2329 (drought susceptible)—were given 28-homobrassinolide (HBL) treatment via pre-sowing seed soaking and foliar spray. HBL treatment improved different physiological parameters as well as grain yield, grain number per ear, 1000 grain weight, and harvest index in both cultivars; 0.05 ppm concentration increased the seed yield by 39% in C306 and by 23.8% in HD2329 under irrigated conditions (Sairam, 1994; Figure 16.2A). Similarly, Braun and Wild (1984) and Ramraj (1997) also observed a favorable effect of BL and EBL on the grain productivity of wheat, respectively. Takematsu et al. (1988) also reported an increase in the panicle weight and seed weight by 25–33% and 4–37%, respectively, in response to EBL treatments. These authors also reported a decrease in the sterile portion of the ear. In all these studies, several other parameters were also studied, which show a great synergism with the grain yield of the crop. Janeczko and Swaczynova (2010) established an increase in the endogenous BS level in wheat seedlings in response to the exogenous application of BS via seed soaking, root drenching, and foliar spray. This enhancement in the level of endogenous BS seems the possible reason behind the improvement in overall physiological traits and yield characteristics of wheat.

Maize is another economically important cereal crop. Numerous experiments have been conducted to assess the impact of BS on maize productivity (Ramraj et al., 1997; Rao et al., 2002; Sairam,1994). Hola et al. (2010) conducted an experiment on two inbred lines with exogenous application of EBL at an earlier (V3/4) and later (V6/7) stage of development. The treated plants exhibited a significant improvement in ear length, number of fertile ears, rows per ear, and shelling percentage. Moreover, the response generated was dependent on the stage of BS application, concentration of BS, and cultivar tested. Likewise, the foliar spray of EBL on maize plants before tassel emergence enhanced the yield parameters, total grain yield, number of grains per ear, and 1000 seed weight (Zhang et al., 1987). Moreover, Ju and Ji-Chu (2003) projected an increase of about 10% in grain yield in response to foliar application of 0.01 mg l^{-1} of EBL. Ikekawa and Zhao (1991) also reported an increase in the growth and yield of maize. Likewise, Nunez (2000) also reported an increase of about 10% in grain yield in maize in response to BIOBRAS-16 and BIOBRAS-6, the commercial analogs of BS. However, the former was more effective than the latter. Furthermore, BS treatment also increased the fresh weight and dry weight of the Kwangok cultivar of maize by 7% and 11–14%, respectively (Lim and Han, 1988).

Rice is a staple food of a large population of the world. Therefore, exploring the ways and means to enhance its grain productivity is an immense need so that the increasing population can be provided with sufficient food. Several experiments have been conducted to exploit BS analogs to enhance the yield of rice. Treatment of Nippon cultivars of rice by BS resulted in an increase in the panicle weight and grain weight and also hastened panicle ripening. In this experiment, plants were applied with 2.1×10^{-9} and 2.1×10^{-8} M concentrations twice; ten days before heading and on the day of heading in field conditions (Saka et al, 2003). Yokota and Takahashi (1985) also reported an increase in rice yield in response to BL application. Similarly, BL treatment of rice plantlets with 5 ppm concentration increased the fresh weight and dry weight by 22% and 31.5%, respectively (Lim, 1987). Krishnan et al. (1999) treated the rice plants with BL and recorded a significant increase in the yield. Similarly, exogenous application of

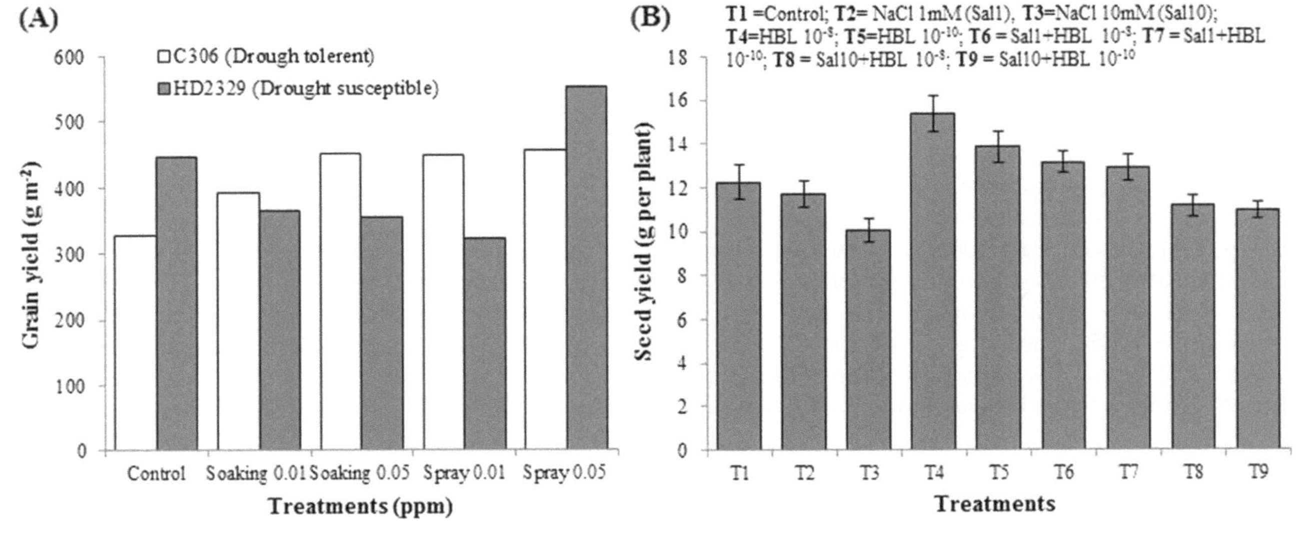

FIGURE 16.2 Effect of 28-homobrassinoloid (HBL) on the grain yield of wheat (A; data adapted from Sairam [1994] and presented in graph form) and chickpea (B; personal data, Ali [2004]).

EBL (Ramraj et al., 1997) and that of HBL and BIOBRAS (Franco et al., 2002) also favored the yield percentage of rice plants. Moreover, BS treatment also increased the percentage of ripe grains cultivated at low temperature (Hirai et al., 1991).

Like natural analogs, the synthetic analogs BIOBRAS-6 and BIOBRAS-16 show a promising impact on the productivity of rice (Vazquez et al 2003). Foliar application of BIOBRAS-6 enhanced the yield in a dose-dependent manner. Besides the concentration, time of BS application also affected the crop yield (panicle number per square meter and filled grain number per panicle) that was increased between 0.83 and 2.48 tons ha^{-1} (Franco et al., 2002) and between 26.1 and 82.9% (Diaz et al., 2002). Similarly, BIOBRAS-6 or BIOBRAS-16 also increased the crop yield by 25.3% under saline conditions (Marina et al., 2002). The increase in the yield characteristics—number of ears, grain number, weight and length per ear—has also been reported in a number of other studies (Rao et al., 2002).

In spring barley, EBL and HBL treatment enhanced the seed yield by 13–21%. The plants in this study exhibited a variation in the relative distribution of Ca and K in the plant organs and also affected the stem stability index by 4–9% (Ageeva et al., 2001). Foliar spray of EBL also enhanced the stem strength and resistance to logeability when the plants were treated at flowering or booting stage (Prusakova et al., 1995). Lapa et al. (1998) also reported an increase in the yield of barley plants in response to EBL treatment. Similarly, BIOBRAS-16 also favored the yield characteristics of this crop (Almenares et al., 1999).

16.2.2 Effect on Leguminous Crops

Legumes are a rich source of protein and cater to a large population of the world for protein requirements. Brassinosteroids have been exploited to enhance the productivity of leguminous crops at various research centers. The legumes treated with BS include chickpea, mung bean, lentil, grass pea, and azuki bean. Root drenching of grass pea (*Lathyrus sativus*) with EBL enhanced its dry weight and yield parameters: filled pod number, grain number, and grain dry weight (Xiong et al., 2016). Similarly, the number of pods per plant and total seed yield also exhibited an increase in response to HBL in lentil (*Lens culinaris*; Hayat and Ahmad, 2003). The foliar spray of BB-6, a commercial BS analog, also showed a favorable impact on soybean (*Glycine max*). These plants were treated with BB-6 three times: once at the flowering stage and two spells of sprays at 24 and 58 days after sowing. Here, 10 mg ha^{-1} dose generated the most prominent response (Cobera and Nunez, 1998). Nunez et al. (2003) also observed a favorable impact of BB-6 (20 mg ha^{-1}) to an extent of 39% in the seed yield. Azuki bean (*Vigna angularis*) also exhibited a positive response to HBL (Mori et al., 1986). Lupin (*Lupinus* sp.), which serves dual purposes (food and ornamental) also exhibited an increase in the yield in response to EBL (Zabolotnyi et al., 2003) and Epin (Tzyganov et al., 2001). Moreover, treatment of bean plants with EBL or HBL at the seedling stage (6–7 true leaves) also generated better yields at harvest (Dulin et al., 2002).

The impact of BS was also explored in chickpea (*Cicer arietinum*), both under normal and stressful situations. Three cultivars of chickpea—IAC India-4, IAC Mexico, and IAC Marrocos—were sprayed with EBL at the flowering stage. Seed yield (kg ha^{-1}) of these cultivars increased by 86%, 76%, and 61%, respectively. Other parameters like crop index and 100 seed weight were also favored by BS application (Ramos et al., 1997). Singh et al. (1993) also reported an increase of about 25% in seed yield in the Pusa256 cultivar of chickpea in response to EBL treatment. Seed priming with HBL favored different yield characteristics in chickpea grown under normal growth conditions and salt stress. A 10^{-8} M dose of HBL increased the seed yield by 26%—improving the seed yield deteriorated by salt stress (Ali et al., 2007; Figure 16.2B). Similarly, in another study, HBL application prior to seed sowing also favored the number of pods per plant, seed yield per plant, and 100 seed yield grown under normal and cadmium stress. Here, HBL also mitigated the toxic impact of cadmium (Hasan et al., 2008).

To compare the impact of HBL with that of kinetin (KIN), Fariduddin et al. (2004) sprayed mung bean (*Vigna radiata*) plants with 0.001, 0.01, and 0.1 μM concentrations of HBL or KIN and observed a high-level impact of HBL on different growth and physiological parameters as well as yield characteristics. The same team of workers compared different modes of HBL application: pre-sowing seed soaking alone, foliar spray alone, and combined seed soaking and foliar spray in mung bean. Pod number and seed yield were significantly enhanced by HBL treatment irrespective of the mode of application. However, BS treatment via combined seed soaking and foliar spray increased pod number and seed yield per plant by 38% and 32%, respectively (Fariduddin et al., 2008).

16.2.3 Effect on Oil Seed Crops

As in cereals and legumes, brassinosteroids have also been used to enhance the productivity of oil seed crops, these include mustard, ground nut, and safflower under different growth conditions. Hayat et al. (2000) sprayed 30-day old plants of Indian mustard (*Brassica juncea*) with varying concentrations of HBL and observed a significant increase in physiological parameters as well as yield parameters—number of pods and seed yield per plant. In another study, the impact of HBL was compared with that of IAA, GA_3, KIN, and ABA. HBL treatment was the most effective, which increased the number of pods and seed yield per plant by 25% and 29%, respectively (Hayat et al., 2001). Application of BL and EBL to the foliage of groundnut (*Arachis hypogea*) at 0.5, 1.0, and 3.0 μM concentrations resulted in a substantial increase in the yield characteristics—number of pods per plant, weight of pods, seed yield per plant, and 1000 seed weight; 3.0 μM concentration was the best treatment dose. Moreover, fat content was also elevated by BS, irrespective of the analog and concentration used (Vardhini and Rao, 1998). Application of HBL (Hayat et al., 2007a; Fig. 16.3 A) and EBL (Wani et al., 2019) also rescued Indian mustard plants from salt stress and improved the physiological and yield attributes impaired by the stress. Safflower (*Carthamus tinctorius*) is also a source of edible oil (Soleymani et al., 2019). Foliar spray of three genotypes (Goldasht, Faraman, and Sina) of this plant with EBL favorably affected the oil yield as well as the oil composition in

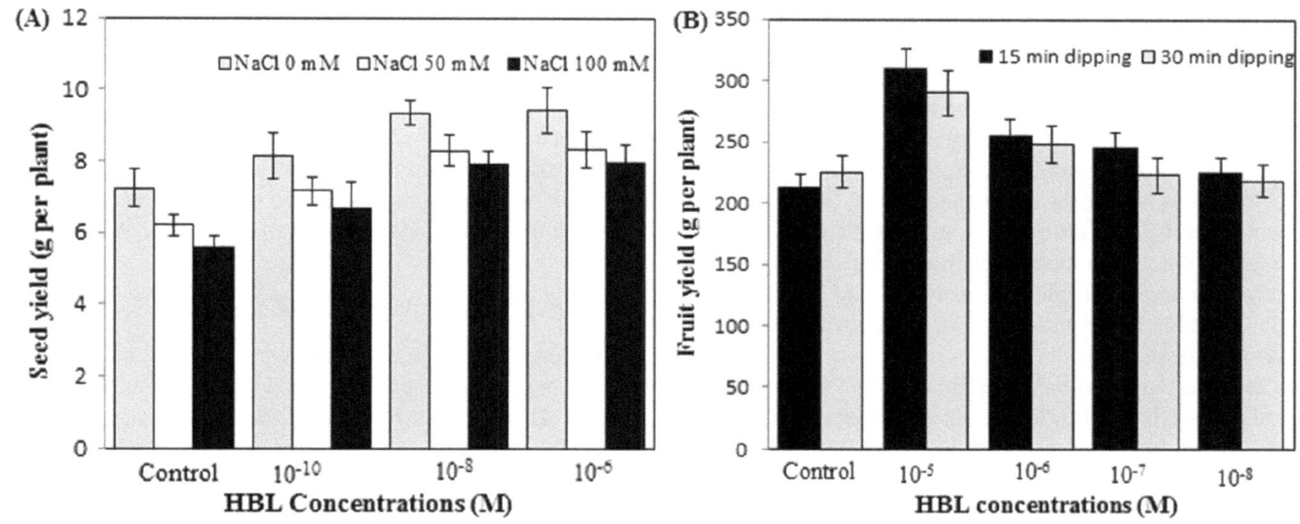

FIGURE 16.3 Effect of 28-homobrassinoloid (HBL) on the seed yield of Indian mustard (A) (data adopted from Hayat et al. [2007] and presented in figure form) and fruit yield per plant in tomato (adapted from Ali et al. [2006]).

the resulting seeds at harvest. EBL treatment enhanced oleic acid (15.95%), linolenic acid (22.5%), linoleic acid (4.26%), and stearic acid (29.6%) compared with the control. Here, EBL also improved the performance of the plants under drought stress (Zafari et al., 2020).

16.2.4 Effect on Fruit Crops

Brassinosteroids play a substantial role in the regulation and enhancement of the pre- and post-harvest attributes in various horticultural crops (Ali, 2017; 2019). Exogenous application of BS (0.1, 0.5, and 1.0 mg L^{-1}) to grape (*Vitis vinifera* L.) vines cv. Flame seedless (at pea stage and at veraison) significantly increased the length, breadth, and weight of the clusters as well as the berries. Treatment of the plants with 0.5 mg L^{-1} concentration effectively reduced the rate of berry softening, maintained external color, stabilized anthocyanins, increased total phenols, and reduced rate of degradation in TSS and TA during low-temperature storage. Pre-harvest foliar spray of BS also exerted a protective impact against decay development and extended the post-harvest life of the grapes (Champa et al., 2015). Vergara et al. (2018) also sprayed grape berries cv Redglobe with three BS analogs (24-epibrassinolide, Triol, or Lactone) (0.4 or 0.8 mg L^{-1}) and a commercial formulation, B-2000 (0.06 mg L^{-1}) at the onset of veraison. The tested BS analogs significantly improved berry color, levels of soluble solids, and anthocyanins, as well as other quality and yield parameters. Similarly, foliar application of Cabernet Sauvignon cultivar of grape berries with EBL and brassinazol (BRz) at the onset of veraison sharply increased the soluble sugars content in the berries but decreased it in the skins. The EBL and BRz+EBL treatments significantly promoted the activities of both invertases (acidic and neutral) and sucrose synthase (sucrolytic) at various stages of ripening. The mRNA levels of genes encoding sucrose metabolic invertase and monosaccharide and disaccharide transporters were increased by the EBL and/or BRz+EBL treatments suggesting an important role of BS in controlling sugar unloading in grape berries during veraison (Xu et al., 2015). The treatment of the grape vines (cv. Alphonse Lavallée) with EBL (0.2 mg L^{-1}) in three spells (seven days after berry set + veraison + 30 days after veraison) substantially enhanced the yield and quality properties of the fruits such as cluster weight, berry weight, and specific gravity. However, the authors recommended the application of a 0.8 mg L^{-1} dose at the veraison stage for the maximum production of total phenolics, β-carotene, ascorbic acid, and transresveratrol in the fruits (Babalık et al., 2020).

Brassinosteroids also play a significant role in the yield and quality attributes of strawberry. Application of BS (0, 0.2, 0.4, 0.6 mg L^{-1}) at five stages (bud break, pre-bloom, middle of bloom, post-bloom, and veraison) enhance the yield and berry quality of the "Thompson Seedless" cultivar of strawberry. The results showed that the BS application has positive effects on bunch morphology and berry quantity and quality. Among the BS treatments, 0.6 mg L^{-1} application at the post-bloom and veraison stages had the greatest effect on berry quality in terms of total phenol and antioxidant capacity (Ghorbani et al., 2017). Application of EBL and two commercial analogs, DI-31 and BB-16, also increased marketable strawberry fruit yield between 9% and 34% (Salazar et al., 2016). BS application also improved the concentration in vitamin C in the "Camiño Real" cultivar of strawberry and reduced the phenolic compounds content, mainly in the red stage, thus helping in fruit conservation by the reduction of oxidative stress and increasing antioxidant activity (Ayub et al., 2018). Sun et al. (2020) also demonstrated the EBL at 1 μM concentration, which showed a good potential for enhancing strawberry fruit nutritional and overall quality, fruit total antioxidant activity, total phenolics, total anthocyanins contents, phytochemical compounds, and post-harvest life. Moreover, EBL (1 and/or 4 μ mol L^{-1}) treatment of strawberry (*Fragaria* × *ananassa* cv. Sabrosa) twice, 15 days after anthesis and at the end of maturation significantly modulated the yield and quality attributes, growth pattern (size, shape, weight), water, color, decreased the fruit growth duration, and enhanced the precocity rate, total phenolics and anthocyanin contents, and carbohydrate metabolism. Higher concentration (4 μ mol L^{-1}) was more effective than the lower one (1.0 μ mol L^{-1}) (Zahedipour-Sheshglani and Asghari, 2020). Furthermore, exogenous applications of BS (EBL and DI-31) encountered the fungal pathogen *Colletotrichum acutatum*, the causal agent of anthracnose disease in strawberry plants cv. Pájaro. Lower concentration of both steroids (0.1 mg L^{-1}) showed a stronger effect than higher (1.0 mg L^{-1}) concentration. BS also protected strawberry plants from *Botrytis cinerea*, the causal fungus of gray mold disease, and also played an important role in growth and fruit quality (Furio et al., 2019a,b). These analogs serve as efficient alternatives to agrochemicals.

In sugar apple (*Annona squamosa* L.), brassinolide (BL) and gibberellic acid (GA) promoted parthenocarpic fruit formation and fruit quality. GA_3 (1000 mg L^{-1})+BL (0.5 mg L^{-1}) combined treatment was the most effective in enhancing the yield-related parameters—fruit set percentage, fruit retention, number of fruits per tree, fruit length, diameter, weight, pulp weight, and peel weight. These improved parameters enhanced the cumulative fruit yield. Moreover, BL treatment improved quality attributes of the fruit, increased the total sugar content, and reduced the acidity (Mostafa and Kotb, 2018). Likewise, foliar spray of EBL to sweet cherry (*Prunus avium* L.) at pre- and post-harvest stages significantly favored the quantity and quality of the produce. BS treatment increased fruit weight, diameter, and length, and also improved fruit color by increasing fruit anthocyanin content, organic acids, ascorbic acid, and phenol content. BS treatment also favored fruit firmness and shelf life after harvest (Roghabadi and Pakkish, 2014). Persimmon (*Diospyros kaki* L.), a typical climacteric fruit, also responds positively to BS treatment. In this plant, EBL application upregulated the enzymes related to fruit ripening and softening, which were mediated by ethylene biosynthesis and cell wall degrading enzymes, respectively, boosted by BS treatment (He et al, 2018).

Wang et al. (1994) and Ikekawa and Nagi (1987) reported a significant (10–20%) increase in fruit production in watermelon in response to EBL treatment. Brassinosteroids also facilitate the ethylene-mediated ripening in mango (Zaharah et al., 2012). EBL treatment regulated the ripening of papaya fruits mediated by the regulation of the alternative oxidative pathway (Mazorra et al., 2013). In passion fruit (*Passiflora edulis-flavicarpa*), foliar spray of a commercial BS (BB-6) increased fruit yield by 65% (Gomes et al., 2006). Similarly, BL treatment enhanced the number of fruits per plant and average fruit weight in orange (*Citrus unshiu*; Kuraishi et al., 1991). BS application also increased the qualitative characteristics of litchi, thereby increasing its market value (Peng et al., 2004). Delay in fruit senescence has also been reported in jujube plants in response to BS application (Zhu et al., 2010). Similarly, BB-6 and BB-16 (0.00001 to 10 mg L^{-1}) were applied to the cladodes of cactus pear (*Opuntia ficus-indica* L Mill. var. Lutea). The treated plants exhibited an increase in precocity and yield, and the number and weight of harvested cladodes (Cortes et al., 2003). BS analog, 28-homocastasterone (10^{-6} M) applied after full bloom increased TSS, color attributes of the skin, total antioxidant activity of the skin and flesh, and peroxidase activity compared with control fruits in "Galaxy'" apples (Ramos et al., 2019). Similarly, the applications of Milagrow (a commercial BL analog) (2.5, 5.0, and 7.5 g L^{-1}) at three growth stages—swollen bud stage, full bloom, and beginning of fruit set—to avocado (*Persea americana* Mill.) tree cv. Fuerte substantially increased all fruit parameters such as fruit set and fruit retention (%). Plants treated with 5 g per 100 L exhibited maximum fruit yield. Moreover, treatments at swollen bud stage showed the most promising effects in terms of both fruit yield and fruit quality (Eid et al., 2016).

16.2.5 Effect on Vegetable Crops

Vegetables constitute an important component of the human diet. These are rich sources of minerals, vitamins, and dietary fibers. Therefore, an increase in the quantity and quality of vegetables becomes an important goal of plant science/agricultural research. Eco-friendly and non-xenobiotic brassinosteroids play an important role in the improvement of various vegetable crops. Application of EBL to tomato plants enhanced their yield by 10–18% (Savelieva et al., 1997). Root dipping of tomato seedlings at the time of transplantation in graded concentrations of HBL for varied durations (15, 30, and 45 minutes) enhanced the number of fruits and fruit yield per plant together with the improvement in the quality attributes—lycopene, β-carotene, and vitamin C concentrations (Ali et al., 2006; Figure 16.3). Ascorbic acid accumulation has also been reported in tomato leaves by Mazorra et al. (2014). Aghdam et al. (2012) also proposed that the application of BS prevents tomato fruit damage generated by oxidative stress.

Exogenous application of BS hastened flowering in cucumber (Fu et al., 2008). Pepper plants also exhibited an increase in fruit number and yield per plant in response to the application of EBL (Samira et al., 2012). DI-31 and DI-100 (commercial BS analogs) enhanced the yield of lettuce (Serna et al., 2012). BS treatment also enhanced the yield of sugar beet (Genma, 1987) and cabbage (Asatova, 1991). Potato tuber productivity has also been reported to be enhanced by BS treatment (Savelieva et al., 1997). BS also enhanced the resistance of potato to various fungal infections (Kazakova et al., 1991) and prevented damping-off disease caused by *Phytophthora melonis* (Ren et al., 2020).

16.2.6 Brassinosteroids and Crop Productivity: Molecular Aspect

There are a large number of studies that demonstrate the involvement of BS in the modulation of genes that determine the productivity of a crop. Zhang et al. (2014) attributed the BS-mediated enhancement in crop yield to the improvement in agronomic traits—plant height, biomass, leaf angle, tiller number, and grain size. Plant height (semi dwarf) is considered an important reason behind the green revolution. Brassinosteroids play a key role in the maintenance of plant height that favors crop yield, particularly, in the members of the Poaceae family. It is pertinent to mention that brassinosteroids regulate this attribute in collaboration or crosstalk with GA, another important hormone determining plant height (Castorina and Consonni, 2020). In crosstalk with IAA, brassinosteroids also play a crucial role in regulating the leaf angle (Nakamura et al., 2009), which facilitates the exposure of most of the plant leaves to sunlight, opening an option of dense planting to favor high productivity. This leaf inclination is regulated by the modulated expression of genes related to BS homeostasis (Zhang et al., 2014) and that of IAA (Zhao et al., 2012). brassinosteroids also regulate the tillering in rice by affecting specific genes. Moreover, brassinosteroids also improve the source-sink relationship efficiency, which plays an important role in the improvement in the size, number, and weight of the granules, leading to enhancement in the yield of the crop (Lin, 2020). Likewise, in soybean (*Glycine max*) plant, brassinosteroids increase the pod length by influencing cellular hypertrophy. Comparing two cultivars differing in pod size revealed the expression of genes related to BS biosynthesis and response in the cultivar with longer pods. This BS-mediated pod elongation is considered one of the reasons responsible for increasing seed yield (Ariyoshi et al., 2016). Rice plants grown under varying moisture levels in soil exhibited an increase in BS biosynthesis in young panicles at moderate soil drying. BS biosynthesis also showed synergism with the key rice inflorescence development genes (*OsAPO2* and *OsTAW1*) and other related attributes. The elevated level of BS promotes spikelet development by enhancing inflorescence activity, the ascorbic acid cycle, and non-structural carbohydrate partitioning of growing panicle (Zhang et al., 2019). Brassinosteroids also upregulate the positive regulators of seed size and repress those which regulate it negatively. Moreover, the development of integuments, endosperm, and embryo are also affected positively by BS, thus determining the shape, size, and weight of the seed (Jiang et al., 2013). These studies clearly reveal that the improvement in crop productivity by brassinosteroids regulated gene expression is mediated via the modulation of plant architecture. However, according to Divi and Krishna (2009), it is not possible to point out the optimal BS effects for all the agronomic traits. Therefore, they suggested an alternate and viable option of the modulation of BS biosynthesis and signal transduction, which can improve plant metabolism as well as make the plants resistant to various stress situations. This kind of genetic manipulation in BS activity has been helpful in increasing the yield by 20–60% (Divi and Krishna, 2009). In addition to the modification of the functions of important components in BS signal transduction, identification and modulation of the regulators of downstream BR signaling or the specific processes with which they are in crosstalk may also be helpful to increase crop yield (Lin, 2020).

16.3 Conclusion

Brassinosteroids are a class of naturally occurring phytohormones. They are eco-friendly and non-toxic, which can serve as an efficient alternative to agrochemicals. The efficient recuperative role of BS has been explored in a large number of studies. The BS analogs that have been used in these studies are brassinolide, 24-epibrassinolide, 28-homobrassinolide, and 28-homocastasterone. Moreover, some commercial analogs of BS (such as BB-6, BB-16, DI-31, DI-100, Milagrow, and Epin) available on the market are also used in some studies. These BS analogs show a promising impact on the productivity of different cereal, legume, oil seed, fruit, and vegetable crops. They elevate the yield of these crops significantly without any side effects on human health and the environment. However, the response generated by BS depends on the type, dose, and duration of exposure to the hormone. In addition to this, the effect of BS also varied with respect to the species and cultivar of the crop as well as environmental or growth conditions. It is pertinent to mention that brassinosteroids enhance the yield of different crops by the modulation of the genes responsible for the regulation of important agronomic traits—plant height, biomass, leaf angle, tiller number, and grain size. Besides this, brassinosteroids also affect the efficiency of the source-sink relation of the crop, which facilitates the improvement in the quantity and quality of the produce. There are also studies that recommend the modulation of BS biosynthesis and signal transduction or their intermediate that might be helpful in the improvement of crop yield, specifically the transgenic crops.

The role of brassinosteroids in crops facing stressful growth conditions has been explored in hundreds of studies. Although it is well established that BS can mitigate various biotic and abiotic stresses. However, the majority of these studies have been restricted to physiological, molecular, or metabolomic levels in laboratory conditions. Only a few studies have been extended to field conditions up to the harvest level. There is an immense need to extend these studies until the harvest of the crop. This extension may be helpful in understanding the role of BS in crop yield under normal and adverse growth conditions. Additionally, more crops are required to be assessed to explore the most effective analog, its optimal dose, appropriate stage, mode of BS application, and overall impact of the hormone on the metabolism and yield of the crop. Identification and understanding of the genes responsible for elevated yield together with the manipulation of BS biosynthesis and signaling, and the subsequent development of transgenic plants, may serve as additional tools to restore the productivity of many crops. It may help in serving food to a growing population and save the environment from the consistent and excessive use of different agrochemicals.

REFERENCES

Ageeva et al. *Agrokchimiya*. 2001;6:49–55.
Aghdam et al. *Sci Hortic*. 2012;144:116–120.
Ali. M. Phil. Thesis, Aligarh Muslim University, Aligarh, 2004.
Ali. *Sci Hortic*. 2017;225:15–21. doi: 10.1016/j.scienta.2017.06.051.
Ali. eds. S. Hayat, M. Yusuf, R. Bhardwaj, and A. Bajguz, Springer, 2019; pp. 349–365. doi: 10.1007/978-981-13-6058-9_12.
Ali et al. *Gen Appl Plant Physiol*. 2005;31:55–63.
Ali et al. *Sci Hortic*. 2006;110:267–273. doi: 10.1016/j.scienta.2006.07.015.
Ali et al. *Environ Exp Bot*. 2007;59:217–223. doi: 10.1016/j.envexpbot.2005.12.002.
Ali et al. *Chemosphere*. 2008;72:1387–1392. doi: 10.1016/j.chemosphere.2008.04.012.
Ali et al. *Environ Exp Bot*. 2008a;62:153–159. doi: 10.1016/j.envexpbot.2007.07.014.
Ali et al. ed. M. Shahnawaz. *Biotechnological Approaches to Enhance Plant Secondary Metabolites*. 2021:125–132.
Almenares et al. *Cultivos Tropicales*. 1999;20:77–81.

Ariyoshi et al. *Plant Growth Regul.* 2016;80:391–395. doi: 10.1007/s10725-016-0176-9.

Asatova. In *2nd. Conference on Brassinosteroids*, Minsk, 1991.

Ayub et al. *Revista Brasileira de Fruticultura.* 2018;40:e-544. doi: 10.1590/0100-29452018544.

Babalık et al. *J Plant Growth Regul.* 2020;39:147–156. doi: 10.1007/s00344-019-09970-5.

Bajguz and Tretyn. *Phytochemistry.* 2003;62:1027–1046.

Bhatia and Kaur. *Phytomorphology.* 1997;47: 421–426.

Braun and Wild. *J Plant Physiol.* 1984;116:189–196.

Castorina and Consonni. *Int J Mol Sci.* 2020;21:1191. doi: 10.3390/ijms21041191.

Champa et al. *Trop Agric Res.* 2015;26:368–379.

Clouse and Sasse. *Ann Rev Plant Physiol Plant Mol Biol.* 1998;49: 427–451.

Corbera and Núñez. *Programa y Resúmenes XI Seminario Científico, Instituto Nacional de Ciencias Agrícolas*, La Habana, 1998; p. 131.

Cortes et al. *Sci Hortic.* 2003;97:65–73.

Dalyan et al. *Plant Physiol.* 2018;65:570–578. doi: 10.1134/S1021443718040118.

Diaz et al. *Memorias 2do*, 2002; pp. 266–272. Encuentro Int de Arroz, La Habana.

Divi and Krishna. *New Biotechnol.* 2009;26:131–136. doi: 10.1016/j.nbt.2009.07.006.

Dulin et al. *Agrokhimiya.* 2002;756–760.

Eid et al. *J Plant Prod Mansoura Uni.* 2016;7:1495–1499.

Fariduddin et al. *Biol Plant.* 2004;48:465–468.

Fariduddin et al. *Turk J Biol.* 2008;32:17–21.

Fariduddin et al. *Acta Physiol Plant.* 2009;31:889–897. doi: 10.1007/s10535-013-0374-5.

Franco et al. *Memorias2do*, 2002; pp. 273–277. Encuentro Internacional de Arroz, La Habana.

Fu et al. *J Exp Bot.* 2008;59:2299–2308. doi: 10.1093/jxb/ern093.

Furio et al. *Eur J Plant Pathol.* 2019a;153:167–181. doi: 10.1007/s10658-018-1551-3.

Furio et al. *Eur J Plant Pathol.* 2019b;154: 801–810. doi: 10.1007/s10658-019-01704-3.

Genma. PCT *Int Appl.* WO 88 04, 1987;890 [C.a. 110, 187813].

Ghorbani et al. *J Grapevine Res.* 2017;56:113–117. doi: 10.5073/vitis.2017.56.113-117.

Gomes et al. *Sci Hortic.* 2006;110:235–240.

Hartwig et al. *PNAS.* 2011;108:19814–19819. doi: 10.1073/pnas.1108359108

Hasan et al. *Environ Poll.* 2008;151:60–66.

Hayat and Ahmad. *Ann App Biol.* 2003;143:121–124.

Hayat et al. *Photosynthetica.* 2000;38:469–471.

Hayat et al. *Photosynthetica.* 2001;39:27–30.

Hayat et al. *Environ Exp Bot.* 2007;60:33–41.

Hayat et al. *Turk J Biol.* 2007a;31:141–146.

He et al. *J Agric Food Chem.* 2018;66:2637–2644. doi:10.1021/acs.jafc.7b06117.

Hirai et al. *Jap J Crop Sci.* 1991;60:29–35.

Hola et al. *Plant Growth Regul.* 2010;61:29–43. doi: 10.1007/s10725-010-9446-0.

Ikekawa and Nagi. Jpn. Kokai Tokkyo Koho. JP 63,243,001 [88,243,0201] [C.A.111, 129007], 1987.

Ikekawa and Zhao. ACS Symposium Series 474, eds. H. G. Cuttler, T. Yokota, and G. Adam, 1991; pp. 280–291, American Chemical Society, Washington.

Jan et al. *J Plant Growth Regul.* 2020;39:1587–1604. doi: 10.1007/s00344-020-10169-2.

Janeczko. *Swaczynova J. Biol Plant.* 2010;54:477–482.

Jiang et al. *Plant Physiol.* 2013;162:1965–1977. doi: 10.1104/pp.113.217703.

Jin et al. *Acta Agric Bor-Sin.* 1988;3:18–22.

Ju et al. In: *Brassinosteroids*, eds. S. Hayat and A. Ahmad, 2003; pp. 159–170, Kluwer Academic Publishers, Netherlands.

Kazakova et al. *Izvestiia Timiryazevskoi seĺskokhoziaistvennoi Akademii.* 1991;82–88 [apud Biological 94(8):85021].

Khripach et al. In: *Brassinosteroids*, eds. S. Hayat and A. Ahmad, 2003; pp. 189–230, Kluwer Academic Publishers, Netherlands.

Krishnan et al. *Curr. Sci.* 1999;76:145–147.

Kuraishi et al. In: *Brassinosteroids*, eds. H.G. Cuttler, T. Yokota, G. Adam, 1991; pp. 312–319, American Chemical Society, Washington.

Lapa et al. *Pochvovedenie Agrokhimiya* 1998;30:96–103.

Lim. *Agric Res Seoul Nat Uni.* 1987;12:9–14.

Lim and Han. *Agric Res Seoul Nat Uni.* 1988;13:1–14.

Lin. *Front Plant Sci.* 2020;11:854. doi: 10.3389/fpls.2020.00854.

Marina et al. *Una alternativa para elevar el rendimiento agrícola. Memorias 2do*, 2002; pp. 277, Encuentro Internacional de Arroz, La Habana.

Mazorra et al. *Theo Exp Plant Physiol.* 2013;25:223–230. doi: 10.1590/S2197-00252013005000003.

Mazorra et al. *Plant Physiol Biochem.* 2014;74:315–322. doi: 10.1016/j.plaphy.2013.11.021.

Mitchell et al. *Nature.* 1970;225:1065–1066.

Mori et al. US Patent 1986;4,604,240.

Mostafa et al. *Mid East J Agric Res.* 2018;7:1341–1351.

Nakamura et al. *Plant and Cell Physiol.* 2009;50:1627–1635.

Núñez et al. Instituto Nacional de Ciencias Agrícolas, La Habana, 2000; pp. 88.

Núñez et al. *Biol. Plant.* 2003;47:67–70.

Pandey et al. *Plant Sci.* 2020;296:110474. doi: 10.1016/j.plantsci.2020.110474.

Peng et al. *Sci Hortic.* 2004;101:407–416.

Peres et al. *Int J Mol Sci.* 2019;20:331. doi: 10.3390/ijms20020331.

Prusakova et al. *Sel'skochosyaistvennaya biologiya 1995;93–97.*

Ramos et al. In: *Abstracts of the II Simpósio Latino Americano de Ciência de Alimentos*, Campinas, Brazil, 1997; pp. 108.

Ramos et al. *Ciência Rural.* 2019;49:04. doi: 10.1590/0103-8478cr20180311.

Ramraj et al. *J Agric Sci.* 1997;128:405–413.

Rao et al. *Curr Sci.* 2002;82:1239–1245.

Ren et al. *Protoplasma*, 2020;257:1519–1529. doi: 10.1007/s00709-020-01523-y.

Roghabadi and Pakkish. *Agric Commun.* 2014;2:49–56.

Sairam. *Plant Growth Regul.* 1994;14:173–181.

Saka et al. *Plant Prod. Sci.* 2003;6:36–42.

Salazar et al. *Revista Agronomica del Noroeste Argentino.* 2016;36:37–41.

Samira et al. *J Stress Physiol Biochem.* 2012;8:224–233.

Savelieva et al. *Technical Report of Gomel Agricultural Station*, Belarus, 1997.

Serna et al. *Sci Hortic.* 2012;143:29–37. doi: 10.1016/j.scienta.2012.05.019.

Singh et al. *Ind J Agric Sci.* 1993;63:395–397.

Soleymani. *Ind Crops Prod.* 2019;131:378–386.

Sun et al. *J Plant Growth Regul.* 2020;39:920–929. doi: 10.1007/s00344-019-10033-y.
Takematsu et al. US Patent 1988;4,767,442.
Tzyganov et al. *6th Conference on Regulators of Plant Growth and Development in Biotechnology*, Moscow, 2001; pp. 290–291.
Vardhini and Rao. *Phytochem.* 1998;48:927–930.
Vazquez et al. (eds.) S. Hayat and A. Ahmad. Kluwer Academic Publishers, Netherlands, 2003; pp. 87–117.
Vergara et al. *Front Plant Sci.* 2018;9:663. doi: 10.3389/fpls.2018.00363.
Wang et al. *Zhiwu Shenglixue Tangxun.* 1994;30:423–425.
Wani et al. *Plant Physiol Biochem.* 2019;135:385–394. doi: 10.1016/j.plaphy.2019.01.002.
Wu et al. *Plant Physiol Biochem.* 2019;141:51–59. doi: 10.1016/j.plaphy.2019.05.011.
Xiong et al. *Plant Growth Regul.* 2016;78:217–231. doi: 10.1007/s10725-015-0087-1.
Xu et al. *Plant Physiol Biochem.* 2015;94:197–208. doi: 10.1016/j.plaphy.2015.06.005.
Yokota and Takahashi. (ed.) M. Bopp. Springer–Verlag, Berlin/Heildberg, 1985; pp. 129–138.
Zabolotnyi et al. *Prikladnaya Biokhimiyai Mikrobiol.* 2003;39:99–104.
Zafari et al. *J Food Comp Analysis* 2020;92:103544. doi: 10.1016/j.jfca.2020.103544.
Zaharah et al. *J Plant Growth Regul.* 2012;31:363–372. doi: 10.1007/s00344-011-9245-5.
Zahedipour-Sheshglani and Asghari. *Sci Hortic.* 2020;268:109376. doi: 10.1016/j.scienta.2020.109376.
Zhang et al. *Acta Agric Uni Henan.* 1987;21(1):56–64.
Zhang et al. *Plant Cell Rep.* 2014;33:683–696. doi: 10.1007/s00299-014-1578-7.
Zhang et al. *BMC Plant Biol.* 2019;19:409. doi: 10.1186/s12870-019-2025-2.
Zhao et al. *Mol Plant.* 2012;6:174–187.
Zhu et al. *Postharvest Biol Technol.* 2010;56:50–55.
Zullo. 1st ed.; Lambert Academic Publishing, Beau Bassin, Mauritius, 2018.

17

Application of Jasmonates in the Sustainable Development of Agriculture and Horticulture Crops

Anmol Sidhu, Shruti Kaushik, Anil Kumar Singh, and Geetika Sirhindi

CONTENTS

17.1 Introduction

Growing industrialization and urbanization in the 21st century has raised the structure of the global economy at the expense of natural resources. This rampant and conspicuous use of natural resources has ultimately accelerated environmental degradation. As a result, it is urgent to underscore the importance of eliminating these negative externalities that are responsible for environmental degradation and the depletion of natural resources. Thus, a more balanced and nuanced view of development is required where we can sustainably ensure development without environmental degradation.

Agriculture has been the main livelihood for human societies for thousands of years, as it caters to our food demands. But, in this industrial paradise, agricultural produce has developed a more mechanistic approach where fossil fuels are being replaced by industrially produced chemical fertilizers and pesticides (Smil, 2004). Although continuous inputs of fertilizers are essential to sustain food production, their constant use will eventually degrade the uptake of plant nutrients. Notable losses are nitrogen and phosphorous uptake, which, in turn, negatively influence the concentration of greenhouse gases and water quality (Trenkel, 1997; Haygarth et al., 2013). Also, many researchers have attributed the continued and intensive use of agrochemicals as the prime cause of loss of soil organic matter and the further depletion of soil reserves (Lal, 2004; Lipper et al., 2014; Baishya, 2015). Therefore, it is highly uncertain as to how long we can ensure food security using chemicals for crop improvement. Thus, a more sustainable approach toward agriculture is required.

Sustainable agriculture is a holistic approach toward sustainable development that concedes the finite nature of natural resources, acknowledges the check on economic growth, and encourages resource allocation equity. Sustainable agriculture can be defined as a form of agriculture that attempts to produce sufficient amounts of food for the present generation without exhausting soil fertility and irreversibly degrading the environment (Horrigan et al., 2002). Among various practices of sustainable agriculture, the potential use of phytohormones to enhance crop yield and productivity without compromising environmental degradation is being explored for commercial availability (Mohanta et al., 2020).

Phytohormones are well known to exhibit a vital role in plant growth and development as well as acting as major stress regulators in plants. Their controlled synthesis and regulated homeostasis are known to orchestrate the entire life cycle of plants by regulating cellular metabolism. Therefore, for regulating various aspects of plant productivity, phytohormones are being actively used in agriculture and horticulture (Sakamoto, 2006; Kurepin et al., 2014). Considerable success has been gained in recent years in the use of phytohormones in many crop improvement processes like flowering and fruiting, ripening, harvesting, and post-harvesting of fruits and vegetables (Tadeu Dias, 2019). The exogenous application of phytohormones like methyl jasmonate (MeJA) and brassinosteroids are

DOI: 10.1201/9781003110651-17

reported to increase thermal stability and therefore increase rice productivity under high temperature (Fahad et al., 2016). Thus, phytohormones could play an active role in the sustainable development of the agriculture sector, presenting practical advantages and perspectives for future use.

Jasmonates, a class of lipid-derived cyclopentanones, are known to elicit plant growth and physiology, and thus, used in the plant response against various environmental stresses. Jasmonates are reported to actively participate in plant defense responses, for instance, pest or insect attacks (Rohwer and Erwin, 2008). Also, MeJA, an active derivative of JA, is involved in a myriad of plant responses including crop growth, development, and harvesting. Meyer et al. (2003) observed that post-harvest application of gaseous MeJA can actively prevent harvest losses by activating varied defense responses in a number of horticulture crops. Thus, jasmonates can be further explored for future prospects of sustainable development in agricultural and horticultural crops.

17.2 Definition and Distribution of Jasmonates in Plants

Jasmonates are a group of lipid-derived cyclopentanones belonging to the family of oxylipins, which are synthesized from α-linolenic acid (α-LeA) through a series of reactions and occur ubiquitously and exclusively in the plant kingdom (Creelman and Mullet, 1997; Wasternack and Song, 2017). Jasmonate was first discovered as a methyl ester of jasmonic acid (MeJA) in 1962 as a fragrant constituent of the ethereal oil of *Jasminum grandiflorum* (Demole et al., 1962). Later, in 1971, free jasmonic acid was first isolated from the fungus *Lasiodiplodia theobromae* (Aldridge et al., 1971). In addition to higher plants, jasmonates are also actively found in various algae and fungi (Sembdner and Parthier, 1993). The jasmonate group of phytohormones is well established to elicit various cellular responses at low concentration and therefore regulate plant growth and development (Wasternack and Parthier, 1997). This fatty acid metabolized phytohormone is known to play multifarious roles in plant stress responses and also actively participate in various physiological and developmental activities of plants including senescence, fruit ripening, sex determination, root elongation, reproductive development, elicitation of secondary metabolites, accumulation of storage proteins, and so on (Moreno et al., 2009; Avanci et al., 2010; Nafie et al., 2011; De Geyter et al., 2012).

Jasmonic acid and its various derivatives are actively distributed throughout the plant body with the highest level being reported in reproductive tissues and a higher concentration in leaves than in roots, possibly because chloroplasts make the bulk of oxo-phytodienoic acid (OPDA) and JA (Bell et al., 1995). In all higher plants, the concentration of JA is assessed to be <10 μM or in the range of 0.01–3.0 ng g^{-1} fresh weight (Mason et al., 1992) where JA levels vary according to their assigned distribution in tissues, under development, and against external stimuli (Vick and Zimmerman 1987; Sembdner and Parthier 1993) and are transported among various organs in both the liquid and vapor phases (Farmer and Ryan, 1990).

Jasmonates are synthesized via the octadecanoid pathway through the consecutive action of enzymes present in chloroplasts, peroxisomes, and cytosol with (+)-7-epi-JA as the end product (Wasternack, 2006). The biosynthesis of jasmonates is initiated in the chloroplast thylakoid membrane where α-LeA acts as their precursor (Wasternack and Kombrink, 2010), which in the first step is oxidized via the 13-lipoxygenase (LOX) enzyme to form an unstable intermediate 13-hydroperoxy-9,11,15-octadecatrienoic acid (13-HPOT). This 13-HPOT is further acted upon by allene oxide synthase (AOS) to form an oxidized product allene oxide, which is in turn converted to 12-OPDA, the active precursor of JA (Schaller and Stintzi, 2009; Wasternack and Hause, 2013). This 12-OPDA is further transported to cytosol and then to peroxisome through the action of *JASSY* proteins and *COMATOSE* (*CTS*) ABC transporters (Guan et al., 2019) for its subsequent conversion into the active form. Lastly, 12-OPDA undergoes three cycles of β-oxidation in peroxisomes to form (+)-7-epi-JA (Wasternack and Hause, 2013) and is further subjected to diverse modifications to form a series of metabolites. For active homeostasis, jasmonic acid is exposed to six metabolic conversions:

- **Hydroxylation** at C-11 or C-12 of the pentyl side chain to form 12-hydroxy-jasmonic acid (12-OH-JA), commonly known as tuberonic acid (Sembdner and Parthier, 1993), followed by sulphation or *O*-glucosylation (Gidda et al., 2003).
- **Decarboxylation (-CO_2)** of jasmonic acid yields a volatile compound *cis-jasmone* (Koch et al., 1997).
- **Adenylation** at the carboxylic acid side chain by AMP transferase (Staswick et al., 2002), which is in turn followed by **conjugation** with an amino acid (Staswick and Tiryaki, 2004).
- **Methylation** of jasmonic acid at the carboxylic acid side chain catalyzed by S-adenosyl-L-methionine; jasmonic acid carboxyl methyltransferase (JMT) to form volatile derivative MeJA (Seo et al., 2001).
- **Reduction** of the keto-group on C-6 carbon of the cyclopentanone ring of Jasmonic acid forms cucurbic acid (Sembdner and Parthier, 1993).
- **Glycosylation (+glucose)** of JA forms jasmonyl-1-β-glucose, jasmonoyl-1- β-gentiobiose, and hydroxyjasmonoyl-1- β-glucose (Swiatek et al., 2004).

17.3 Derivatives of Jasmonates Used in Agriculture

Jasmonates were labeled as the major plant immunity hormone when first discovered in higher plants and since then the biological activities of jasmonic acid and its various derivatives have been extensively explored and expansively experimented for the growth and development of plants. For the past 35 years, JA and its active metabolites have captivated the attention of plant physiologists worldwide to explore the practical importance of jasmonic acid and its much-valued metabolites (Eng et al., 2021). In recent years, credible evidence has been reported for their active

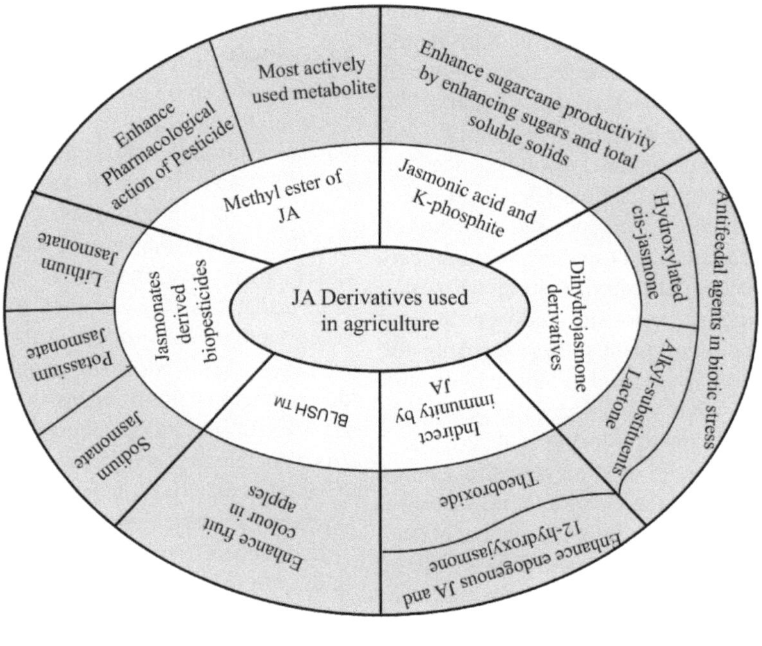

FIGURE 17.1 Derivatives of jasmonates used in agriculture.

application in agriculture and horticulture sectors in regulating plant defense against various biotic agents (Gavin et al., 2012, Wasternack, 2014, Sanches et al., 2017, Moreira et al., 2019). This targeted use of phytohormone jasmonates in agriculture seems to be in accordance with the principles of sustainable agriculture with the least harm to the environment (Figure 17.1).

Since the 1960s, from the time of its discovery, increasing interest has been noted in the jasmonate class of phytohormones, which led to a growing number of patents for commercial production and the use of JA in the last decade (Pirbalouti et al., 2014). The use of jasmonate salts like sodium jasmonate, potassium jasmonate, lithium jasmonate, or ammonium jasmonate in biopesticides has been a suggested method for protection and propagation of plants under fungal pathogen attacks and therefore imparts resistance against biotic stress factors (Gewehr, 2014). Jasmonates, specifically MeJA, are recognized for enhancing the pharmacological action of pesticides. Furthermore, jasmonates are now patented as mixing partners in many biopesticides like chitosan, spermine, inter alia methyl jasmonate, maleic hydrazide, and so on. Some of the jasmonic acid salts or derivatives include without limitation the dimethyl ammonium jasmonate, isopropyl ammonium jasmonate, diol ammonium jasmonate, diethtriethanolammonium jasmonate, jasmonic acid amide, methyl ester of JA, methylamide of jasmonic acid, L-amide linked conjugates of jasmonic acid like L-isoleucine, L-valine, or L-phenylalanine, coronatine, linoleic acid, 12-OPDA, or their other derivatives with combinations of any of the above are reported to provide a pesticidal composition to ameliorate plant health and in particular increase crop yield (Gewehr, 2014).

The use of chemical fertilizers like urea or any other ammonium-containing fertilizer in agriculture is often reported to act as a major limiting factor in crop yield. The use of jasmonate derivatives in many biofertilizers was reported to actively tackle ammonium toxicity by triggering the induced systemic resistance (ISR) in plants by enhancing ethylene production and a further increase of calcium flux in cells to neutralize the ammonium build-up (David Marks, 2012). In the present decade, various jasmonate-containing agricultural formulations have been described to enhance the yield and health of plants where a water-soluble salt of jasmonic acid is prepared as a biofertilizer along with some macro and micronutrients, for example, potassium jasmonate 50 ppm in 100% w/w formulae or specifically manufactured dihydrojasmonic acid magnesium salt 0.200 kg in 1000 kg agricultural formulation (David Marks, 2012). A low concentration dose of jasmonic acid and K-phosphite was observed to be beneficial for improving the quality of sugarcane crop by enhancing sugars and total soluble solids (Moreira et al., 2019). Along with the above-mentioned derivatives, a new set of synthetically produced JA products, namely microbial hydroxylated *cis*-jasmone and alkyl-substituted δ-lactone derivative from dihydrojasmone, were observed to act as antifeedant agents under biotic stress (Gliszczyńska et al., 2015; Gliszczyńska et al., 2016). Thus, the production of modified forms of naturally occurring JA derivatives viz. hydroxy and epoxy cis-jasmone and dihydrojasmone were studied to be more biologically active than other compounds used as aphid repellents (Paprocka et al., 2018). Apart from the direct use of JA metabolites, exogenous application of some naturally derived fungal products like theobroxide (a novel compound isolated from fungus *Lasiodiplodia theobromae*) also enhances the endogenous levels of jasmonates and its derivative tuberonic acid (TA, 12-hydroxyjasmonic acid) and therefore induces tuber formation in potato (Yang et al., 2004). Theobroxide spray is also reported to induce certain defense responsive genes like *PR1a, PR1b, GST*, and so on by triggering an elevated level of jasmonates and thus is known to impart a faster resistance toward disease development (Ahn et al., 2013). Thus, natural products like theobroxide could also be explored for their use in jasmonate-derived agricultural biopesticides.

Among various derivatives of JA, MeJA and prohydrojasmon are the most actively used synthetic derivatives. Prohydrojasmon, marketed under the name Blush™ is reported to enhance the red color of apples when applied over fruits 7–28 days before their harvesting day at the rate of 190–381 mL/100L of water (100-200 ppm) without any notable negative effects on the natural fruit ripening process (Brighenti et al., 2017). Prohydrojasmon is known to enhance the red color of apples by boosting the accumulation of anthocyanin and carotene and speeding up the degradation of chlorophyll (Atay, 2015). Apart from enhancing fruit color and quality, prohydrojasmon is also used as a common plant growth regulator for the cultivation of rice where it is reported to induce biotic and abiotic stress marker genes when applied to roots and seeds and therefore prevent spindly growth (Morino et al., 2020).

Analogs of MeJA and JA can be used on plants in a number of different ways to enhance their physiological activity. MeJA, a methyl ester of jasmonic acid, is the most actively used metabolite of JA in agriculture (Rohwar and Erwin, 2008). MeJA is mostly applied to plants in a gaseous form but many researchers also delineated the liquid hydroponic solution use of it (Baldwin and Schmelz, 1996). Foliar spray of MeJA can be used as a novel method to enhance the yield and quality of pomegranate fruit (Mohammadreza et al., 2019). MeJA is also reported to be actively used in enhancing disease resistance and the color of fruits (Jin et al., 2009; González-Aguilar, 2001). Thus, various synthetically derived forms of jasmonic acid could provide valuable inputs in phytohormone-driven sustainable agriculture.

17.4 Horticulture Applications of Jasmonates

For horticulture functions, plant adaptability in an ecological sense like fecundity or survivorship is not necessary. Although the instigation of defense responses may increase resistance at the cost of increased yield. For example, after applying 10 mM JA instead of 1 mM JA, tomato fruit number, fruit weight, and seed number per plant are reduced (Redman et al., 2001; Thaler et al., 1999). Applying JA (0.5–1 mM) to strawberry plants was reported to accelerate flowering and enhance the total yield per plant by 26–37% (Yilmaz et al., 2003). In a discreetly herbivorous environment, in natural ecosystems, MeJA induces an increase in the lifetime and number of capsules in *Nicotiana attenuata*. However, at small herbivorous levels, the performance of non-induced plants was superior to that of induced plants, while at high herbivorous levels, there was no dissimilarity between MeJA induced and non-induced plants (Baldwin, 1998). Soybean yield improved when seeds infected with *Bradyrhizobium* were cultivated with 50 μM MeJA than when seeds were inoculated with untreated *Bradyrhizobium* (Mabood et al., 2006a). Although the amount of pollen grains in each flower is insignificant for the formation of a cut-flower crop, it may be an essential concern in the process of seed development and/or fruit size. In a potted geranium, the number of fruits per plant is insignificant, but in cucumber production, it is crucial. Thus, further research on crops is necessary to determine the influence of JA on commercially important characteristics of food and decorative plants.

17.4.1 Flowers

Flowers are the most important reproductive units of angiosperms and exhibit great diversity in morphology. The development of a flower is determined by the combined action of various pathways that mainly consist of hormone signaling molecules and floral homeotic genes (Gan et al., 2013). Plant hormones including auxins, cytokinins, abscisic acid, gibberellins, and jasmonic acid play an important role in flower development (Wellmer et al., 2014). Various studies have shown that JAs play a crucial role in the transition of inflorescence meristems to flower meristems. Furthermore, MeJA exogenous application also promotes flowering time and the development of floral organs in oilseed rape (*Brassica napus* L.; Park et al., 2009). Unexpectedly, it was later noted that JAs also help in touch-induced delay in *Arabidopsis* flowering (Chehab et al., 2012). A diversified role of JAs was observed in reproductive development.

In monocots, the function of JA in flower development is diverse from that of dicot species. In the case of rice, the flower spikelet is a condensed branch in which small flowers are enclosed in leaf-like glumes, which open to release wind-spread pollen and pistil pollination. In this system, studies on several mutants, containing extra glumel (*eg1*) encoding rice *DAD1* homologs and *eg2* encoding *OsJAZ1* with distorted Jas domains, point out that JA is active very early during spikelet development (Cai et al., 2014). Even the phenotypes of *eg1* and *eg2* are relatively mild, mainly due to the redundancy of coding gene functions; other genetic and molecular analyses have shown a model in which JA-mediated degradation of O*sJAZ* allows rice *MYC2* homologs to induce expression of *OsMADS1*, *OsMADS7*, and *OsMADS8* transcription factors which start and stop spikelet development (Yuan et al., 2015). The rice MADS box proteins are similar to the SEPALLATA transcription factor family in *Arabidopsis* and are required for the expression of flower development (Wu et al., 2012). Therefore, in rice, the core JA signal module JAZ/JA-Ile/COI1/MYC2 seems to have been redirected to control the basic process at the beginning of spikelet and floret development, rather than being required for the final stages of stamen and pollen maturation.

Among the large number of mutants that alter sex determination in maize are those called *tasselseed* (*ts*). The mutations in *ts1* are recessive and have no effect on the early development of spikelets and florets but lead to the complete feminization of tassels. When *ts1* was cloned, it turned out that it encodes 13-lipoxygenase, which is expected to be located in the chloroplast and catalyzes the peroxidation of polyunsaturated fatty acids (Acosta et al., 2009). Because the synthesis of JA is carried out by the 13-lipoxygenated product 13-hydroperoxylinolenic acid, the clear implication is that the JA signal may be important in controlling the pathway identified by the *tasselseed* mutant. After checking the level of JA of homogenous *ts1-ref* plants, it was found that its concentration was less than 10% of the concentration obtained from the measurement of the wild-type tassels. In addition, when the mutant's tassel was treated with 1 mM JA, the mutant's tassel seed phenotype was partially rescued. It was also found that some other tassel seed mutants were defective in JA signaling in the developing tassel (Wang et al., 2019).

17.4.2 Fruits and Vegetables

Fruits and vegetables are known as major providers of minerals, vitamins, and antioxidant chemicals of phenolic origin such as flavonoids, phenolic acids, and anthocyanins, depending on the crop. As a result, several fruits/vegetables have demonstrated strong radical scavenging activity, making them beneficial in preventing the oxidation of human low-density lipoproteins (Weng et al., 2012). Recent studies have demonstrated that MeJA treatments increase antioxidant activity by enhancing bioactive components in pomegranates and blackberries, enhancing their health-promoting qualities. MeJA enhances the amount of antioxidant chemicals such as anthocyanins and other phenolic metabolites in a variety of fruit crops when applied through vapor, dipping, or spraying methods (Sayyari et al., 2011). MeJA was also reported to increase the resveratrol accumulation in leaves of *Vitis vinifera* L. according to Belhadj et al. (2006), which serves various health benefits for humans (Opie and Lecour, 2007). Jasmonates also enhanced the amount of glucosinolates in cruciferous crops like *Brassica napus*, *Brassica rapa*, *Brassica juncea*, and *Raphanus sativus* (Agrawal et al., 1999). Further isothiocyanates, which serve as chemo-preventive agents, are formed from glucosinolates (Hecht, 1999). Similarly, increasing the anthocyanin content in fruits and vegetables may be feasible. Jasmonates increased anthocyanin accumulation in *V. vinifera* cell cultures (Curtin et al., 2003), and MeJA boosted anthocyanin accumulation in *V. vinifera* cell cultures and anthocyanin and carotenoid contents of apple fruit (Jung, 2004). Fruit of the raspberry (*Rubus idaeus*) treated with 100 mM MeJA vapor when kept at 10°C was found to be darker and redder than untreated fruit, possibly because of increased content of anthocyanin (Wang, 2003). Treatment of young *R. idaeus* and *R. occidentalis* L. fruits with 0.01 mM or 0.1 mM MeJA boosted anthocyanin, antioxidant, phenolic, and flavonoid contents (Wang and Zheng, 2005). This was further demonstrated by the fact that after ten days in storage, MeJA-treated raspberries exhibited a higher oxygen radical absorbance capacity (ORAC) than untreated raspberries (Wang, 2003). Strawberry fruit treated with MeJA had increased anthocyanin and phenolic content, as well as greater ORAC, after 12 days of storage (Ayala-Zavala et al., 2005). MeJA, on the other hand, was found to decrease anthocyanin accumulation in apples in another investigation (Saniewski et al., 1988). Thus, jasmonates may improve the healthfulness of treated goods since certain glucosinolates and anthocyanins have chemo-preventive or other health-related qualities (Wattenberg, 1985). The amount of ascorbic acid (vitamin C, a potent antioxidant in the human diet; Wang, 1999) in drought-stressed strawberry leaves is lower than in normal leaves. However, prior to drought stress, treating strawberry plants with 0.1 mM MeJA kept their ascorbic acid levels stable (Wang, 1999). After two-day drought stress, drought-resistant maize seedlings treated with 112 μM MeJA showed increased ascorbic acid levels. (Li et al., 1998). *A. thaliana* cell cultures were treated with 50 μM MeJA for 21 hours and had 64% more ascorbic acid content as compared with untreated cells, however, cell growth (and ascorbic acid levels) was dependent on the occurrence of cytokinin auxin (Wolucka et al., 2005). Nevertheless, after spraying with a high concentration of MeJA (>4 mM), the total carotenoid content of *S. lycopersicum* and *S. tuberosum* dropped. The carotenoid content of *Pisum sativum* seedlings was lowered by MeJA (Fedina and Tsonev, 1997). MeJA, either as a gas or as a 5% (w/w) solution in lanolin paste, reduced lycopene levels in tomato fruit while boosting carotene levels (Czapski and Saniewski, 1985). According to previous studies, reduced endogenous levels of jasmonates in tomato fruit during the later stages of ripening are necessary for lycopene synthesis (Fan et al., 1999). Significantly, the application of MeJA on fruits increases red color, β-carotene and anthocyanin content increases the content of various phenolic compounds. Ethylene and several ester compounds also congregated in a climacteric fruit (Rudell et al., 2002). In non-climacteric fruits, such as raspberry, blackberry, and strawberry, the soluble solid content (SSC), glucose, and sucrose concentrations were also elevated after the application of MeJA (Wang et al., 2008). MeJA treatment changed the expression levels of various cell well-modifying (CWM) genes in the pre-climacteric peach (Ziosi et al., 2008), indicating a possible function for JAs in fruit cell wall metabolic control. The physiological role of JAs in fruit ripening has been explored, but less is known about their influence on gene expression, particularly in non-climacteric fruits. The impact of MeJA on grape development and ripening was investigated at the transcriptional and physiological levels. The gene functions of *VvAOS* and *VvCOI1* (grape jasmonic acid pathway gene) were examined by utilizing a strawberry gene transient expression system to determine the involvement of JAs in resisting *Botrytis cinerea* effects (Jia et al., 2016a). Thus, to examine the exact impacts of MeJA on healthy components in foods, further research is needed in this area.

17.5 Targeted Use of Jasmonates in Sustainable Agriculture

Environmental health, economic profitability, and social and economic equality are the three fundamental aims of sustainable agriculture. These objectives have been achieved through a range of beliefs, policies, and methods. This vision has been shared and contributed by people in a variety of roles, from farmers to consumers. Sustainability is based on the premise that we must fulfill current demands without compromising future generations' ability to fulfill their own. As a result, the management of both natural and human resources is crucial. Agriculture crops have changed dramatically in recent years as a result of the need for sustainability and food security. Therefore, trade obstacles limiting pesticide residues have emerged. In this context, attempts have been made to limit the usage of inorganic pesticides, and when they are used, organic ones are preferred (Juszkiewicz et al., 2002). As a natural substance, MeJA has no constraints for pre- and post-harvest treatments, and it has thus been studied to extend the life of a variety of agricultural crops (Ghasemnezhad et al., 2008). MeJA is produced in several plant organs as a phytohormone, however, the highest amounts are seen in various vegetables and fruits. It has a significant impact on secondary metabolites in many crops, as well as the development of natural defense against biotic and abiotic stressors (González-Aguilar et al., 2003).

17.5.1 Legumes

In India's rural areas, legumes are regarded as a low-cost source of protein and energy. When compared with the world average, pulses and nuts account for a significant portion of total dietary energy in India, whereas animal-based food products account for the opposite. However, the country's pulse output is concerning because it is quite low when compared with other crop outputs. Furthermore, there is evidence that fertilizer usage does not support long-term production. Phytohormones such as jasmonic acid and its conjugates have a lot of potential in this context to maintain crop fertility at high levels of productivity (Das and Ghosh, 2012). Legume crops are split into two categories based on their capacity to grow in different seasons: cool-season food legumes and warm-season food legumes (Toker and Yadav, 2010). Broad bean (*Vicia faba*), lupin (*Lupinus spp.*), lentil (*Lens culinaris*), dry pea (*Pisum sativum*), chickpea (*Cicer arietinum*), grass pea (*Lathyrus sativus*), and common vetch (*Vicia sativa*) are all cool-season dietary legumes (Andrews and Hodge, 2010). Cowpea (*Vigna unguiculata*), mung bean (*Vigna radiata* var. *radiata*), pigeon pea (*Cajanus cajan*), urd bean (*Vigna mungo*), and common bean (*Phaseolus* spp.) are warm-season food legumes that are mostly cultivated in hot, humid conditions (Singh and Singh, 2011). Exogenous application of jasmonates has been shown to promote leguminous plant development for long-term production (Abdelgawad et al., 2014). For example, exogenous application of MeJA to soybean plants promoted abiotic stress tolerance by modifying the level of endogenous plant hormones, increasing chlorophyll and proline content and photosynthetic rate (Yoon et al., 2009). The use of MeJA as a foliar spray on soybean leaves can improve their ability to withstand water stress and further study showed elevated levels of phenols, flavonoids, and sugars (Mohamed and Latif, 2017). Organs such as the hypocotyl hook, axes, and plumules of the soybean plant have greater levels of JA than the hypocotyl zone of the elongated cells and the non-elongating stems and roots (Creelman and Mullet, 1997). For the first time, JAs have been identified and isolated from the soybean plant, where they control vegetative storage proteins (VSPs). In soybean, VSPs are found in the bundle sheath cells and vacuoles (Franceschi et al., 2002). VSPs are found in developing reproductive organs and pods, but not seeds, suggesting that VSPs are transient storage of amino acids formed by the disassembly of leaf proteins and Rubisco, which become active during seed production. Leguminous plants have a bulbous root structure called a nodule (Creelman and Mullet, 1997), which includes the bacterial enzyme nitrogenase, aiding in the fixation of nitrogen from the atmosphere. According to Stougaard (2000), lipochitooligosaccharide Nod factors are released by rhizobial bacteria and have a role in nodulation initiation. The cytokinin signaling pathway is required for nodule formation to begin (Reid et al., 2016). Hormones like gibberellins, ethylene, abscisic acid, auxins, jasmonates, and so on have been implicated in the development of nodules in several studies (Ferguson et al., 2005). Ethylene regulates nodulation adversely by inhibiting Ca spiking caused by the Nod factor (Ding et al., 2008). Hormones appear to modulate nodule organogenesis and senescence, although the exact mechanism is unknown. Inhibition of nodulation via JA has been found in *Lotus japonicus* and *Medicago truncatula* lately. EST study of *Lotus japonica* revealed that genes related to patriotism were activated during *Rhizobium–Lotus* interaction (Sun et al., 2006). EST study of *Lotus japonica* revealed that genes implicated in pathogen defense responses and enzymes of JA biosynthesis (AOC and OPRs) were elevated and repressed at the time of the late nodule development process during *Rhizobium–Lotus* interaction (Bordenave et al., 2013). According to Kouchi et al. (2004), genes involved in lipoxygenase were mainly downregulated in the JA biosynthesis pathway. MeJA given to the shoots of wild-type and hypernodulation mutant harl plants significantly reduced nodulation, demonstrating that JA adversely affects nodule quantity as possible mobile signaling from shoot to root (Nakagawa and Kawaguchi 2006). MeJA also suppressed the early phases of nodulation, including as NFs-induced calcium spiking, infection thread production, and expression of the nodulin genes such as RIP, NIN, and ENOD11 (Sun et al., 2006). In hypernodulation mutant nts (harl in *L. japonicus*), however, a JA biosynthesis inhibitor applied to soybean leaves drastically decreased root nodulation (Kinkema and Gresshoff 2008). A modest concentration of JA (0.1 μM) promoted root nodule development in *L. japonicus*, indicating a consistent beneficial effect of JA in nodulation. Surprisingly, JA and MeJA may act as a signal to encourage the synthesis and secretion of other proteins (Mabood et al., 2006b). Depending on the legume species and the applied JA concentration, current data suggest that JAs can act as either positive or negative regulators in root nodule symbiosis. Future research should focus on elucidating the underlying processes in more depth.

17.5.2 Oil

In the quest for sustainable agriculture, much emphasis is being placed on renewable feedstocks, and vegetable oils account for a significant amount of the chemical industry's current use. The chemical potential of renewable oils and fats, on the other hand, has yet to be completely realized. Indeed, the majority of oleochemical events have occurred in the fatty acid carboxy group, with just a small number of those involving alkyl chain modifications. Nonetheless, the latter reactions offer a lot of promise for broadening the spectrum of chemicals that oils and fats may produce. The epoxidation of vegetable oils is being investigated in this area due to the existing and future commercial use of the resulting compounds. Oil epoxides are employed as stabilizers and plasticizers, as well as potential intermediates in the synthesis of polyurethanes, polyols, biolubricants, and epoxy resins (Jia et al., 2016b). Furthermore, fatty acids obtained from vegetable oils are pure enough to be employed in other chemical reactions. Their conversion to fatty acid methyl esters (FAMEs) is a well-known application, with biodiesel manufacturing being a major focus. Unsaturated fatty acids and FAMEs can also be epoxidized and employed in commercial chemical and intermediate syntheses. Essential oil components are a category of secondary metabolites found in plants that are important to both the plants and the people who consume herbal plants. Essential oil content in herbal plants is determined by plant species, cultivar, ambient conditions, and agronomic approaches, among other factors (Hussain et al., 2008).

Foliar application of jasmonic acid induced production of basil essential oils. The principal constituents of the essential oils were affected differently by varying concentrations of the tested elicitor. According to Urszula Zlotek et al. (2016), basil treated with 100 µM jasmonic acid generated the greatest amounts of linalool (20.88%) and eugenol (20.88%), while the highest levels of methyl eugenol (40.69%) were found in basil treated with 1 lM jasmonic acid (40.69%). The highest concentrations of 1,8-cineole (20.34%) and b-pinene were found when 0.1 l M jasmonic acid was used (3.43%). Urszula Zlotek et al. (2016) findings were consistent with a prior study that methyl jasmonate treatment might generate eugenol and linalool in sweet basil. In a research conducted by Malekpoor et al. (2015), it was found that basil treated with jasmonic acid produced more linalool and 1,8-cineole, similar to Fard et al. (2012), however, these researchers saw the greatest rise in the concentration of methyl chavicol in basil stimulated with 400 µL of JA. Furthermore, in a study by Ghasemi et al. (2014), the level of certain summer savory essential oil components, such as sabinine, a-phellandrene, b-phellandrene, (Z)-b-ocimene, (E)-b-ocimene, linalool, (Z)-verbenol, citronellol, and bornyl acetate, increased significantly after foliar application of JA at 50 µL, whereas the content of two main components like carvacrol and cterpinene decreased. Exogenous application of JA stimulated biosynthesis of other categories of plant physiologically active compounds like anthocyanins, flavonoids, vitamin C, or carotene, as reported in earlier studies (Złotek et al., 2016). Furthermore, it should also be noted that elicitation with jasmonic acid had a considerable influence on various antioxidant activities. The essential oil of basil induced with 100 µM jasmonic acid has the strongest antiradical potential (against DPPH and ABTS). This is most likely due to the increased quantities of linalool and eugenol in the test sample compared with the control sample. Our findings are consistent with those of Alavi-Samani et al. (2015), who found that foliar application of 400 µL JA increased the antioxidant activity of thyme essential oil. Trevisan et al. (2006) also noted a connection between a high amount of eugenol and essential oil antioxidant potential. In comparison with a concentration of 1 mM, a lower concentration of MeJA (0.1 mM) was more efficient for inducing essential oil in *Agastache foeniculum*. Other researchers have shown that using MeJA at a dose of 0.1 mM enhances secondary metabolite synthesis (Mangas et al., 2006). Furthermore, MeJA treatment for 48 hours at 0.1 mM concentration obtained the maximum quantity of methyl chavicol (97.6%). Similarly, Deschamps et al. (2006) discovered that eliciting basil leaves with methyl jasmonate increased total essential oil and methyl chavicol synthesis. Methyl chavicol was found to be produced via the phenylpropanoid pathway, which was started by phenylalanine ammonialyase (Gang et al., 2001). Stress, elicitors, JA, and MeJA can all cause phenolic compounds to be generated via this pathway (Kim et al., 2006). As a result, the induction of methyl chavicol in *A. foeniculum* treated with MeJA has been linked to phenylalanine ammonialyase. Treatment of JAs in soybean seeds improved oil quality by lowering oleic acid and increasing linoleic acid and linolenic acid.

17.5.3 Cereals

Cereals are the most important agricultural staple food in the world, accounting for more than half of the daily calorie consumption. Plants have distinct and more complicated processes for responding to and tolerating environmental stress (Qin et al., 2011). Traditional breeding strategies have had little success with stress tolerance characteristics because of their complexity. They need practical solutions to meet the world's expanding food needs. Novel and effective ways should be explored in this direction. Phytohormone engineering might be a viable option for producing climate-resilient crops with high yields. Phytohormones are molecules that are generated in very small amounts yet have the ability to control a range of cellular processes in plants. They function as chemical messengers in higher plants, allowing cellular activity to be communicated (Vob et al., 2014). Phytohormones play an important role in the abiotic stress response, coordinating numerous signal transduction pathways. They control both external and internal stimuli (Kazan et al., 2015). Phytohormones, such as ABA, have been discovered as stress hormones. Apart from modulating abiotic and biotic stress responses, ABA plays an important function in plant growth: seed dormancy maintenance, inhibition of germination, stomatal closure, fruit abscission, and regulation of growth (Lin et al., 2010). Under salinity stress, endogenous levels of JA were elevated in rice roots, which was found to offset the negative effects of salinity stress (Wang et al., 2001). There have been reports of jasmonic acid's role in grain development under stress conditions (Tiwari et al., 2016). The usage of jasmonic acid has been shown to boost grain production in previous experiments conducted by Anjum et al. (2011).

Vahabi et al. (2017) discovered that increased stress affected leaf area, number of grains per spike, spike length, biological yield, and wheat grain production. On the other hand, using growth regulators boosted grain yield components and ultimately increased grain yield. However, because the impact of chlormequat chloride was greater than that of jasmonic acid, thus spraying this regulator boosted grain production by 7.20, 8.13, and 24.7% in the greenhouse and 3.18, 2.12, and 1.8% in the field, respectively. Growth regulators are more convenient and cost-effective than breeding procedures, which are typically time-consuming and costly (Bartels and Sunkar, 2005). The use of plant growth regulators like MeJA can be a useful strategy to decrease stress-related damage. Plant growth regulators, sometimes known as plant hormones, have been increasingly used to increase plant development in both stress and non-stress situations in recent years. In both cultivars of rice, 100 M MeJA prolonged spike emergence. Under stress conditions, MeJA could increase photosynthesis, spike emergence time, and days to maturity (Sio-Se Mardeh et al., 2006). Many crop plants, such as rice, maize, and wheat have been discovered to increase drought tolerance by gathering organic solutes and growth hormones, as well as increasing antioxidant activity (Abdelgawad et al., 2014). On the other hand, rice yields increased significantly after foliar treatment of MeJA. The effects of salt stress on endogenous JA levels in rice plants were examined. JA concentrations in salt-sensitive cultivar plants were lower than in salt-tolerant cultivar plants, according to the reports conducted by Kang et al. (2005). Furthermore, at 200 mM NaCl, MeJA levels in rice roots elevated considerably. As a result, large levels of JA accumulated following salt treatments in salt-tolerant plants can provide significant protection against excessive salinity. However, nothing is known about how salt affects the amounts of endogenous JA in plants. According to Kang et al. (2005), post-application with exogenous JA can alleviate salt-stressed rice seedlings, particularly salt-sensitive rather than salt-resistant varieties.

17.6 Pre- and Post-Harvest Applications of Jasmonates

Some research on the horticulture use of MeJA has focused on pre- and post-harvest treatments to prevent microbial development from occurring on harvested tissues. Treatment of MeJA reduces pathogen growth on pepper (*Capsicum annuum* L.) and celery (*Apium graveolens* L.; Buta and Moline, 1998); 100 mM MeJA vapor was more effective to protect and maintain the quality of strawberries at 7.5°C (Ayala-Zavala et al., 2005). Wang (2003) reported that 100 µM MeJA vapor eliminated decay in raspberry (*Rubus idaeus* L.) fruit that were stored at 10°C for ten days. In some Rosaceae species, such as apple (*Malus domestica* "Fuji"), pre-harvest application of MeJA can also increase red color during fruit development (Rudell et al., 2005). Previous reports of exogenous MeJA application on strawberry in an *in vitro* ripening system displayed a promising role in promoting the red color of commercial strawberry fruits (Delgado et al., 2018). All these prerequisites indicate that MeJA can be used as a color promoter agent for strawberries, including field applications. MeJA is also used to cut pineapples (*Ananas comosus* Merr.) as a dip or a vapor for 6-12 days after cutting, reducing softening and microbial load (Martínez-Ferrer and Harper, 2005). MeJA applied to peach fruits showed slower decay and better quality than untreated fruit (Jin et al., 2006). Meir et al. (2005) reported the 200–600 µM MeJA application to cut roses (*Rosa hybrida* L.) by pulsed solution or spray to reduce the incidence and severity of *Botrytis* and increase the color persistence of flowers. Various other ornamental crops can also benefit from the application of MeJA pre- or post-harvest to block post-harvest *Botrytis* development. Current reports of chemical and physical and biological introduction of post-harvest disease resistance show that jasmonates have not attained much more interest than the other means of induced post-harvest defense (Terry and Joyce, 2004). In some research fields prior to chilling, MeJA has also been used as a gas ($\leq$ 100 µM) or as a dip ($\leq$ 25 µM) to grapefruit, avocado (*Persea americana* Mill.), or pepper fruit to reduce chilling injury (Fung et al., 2004). González Aguilar et al. (2004) also found that MeJA treatment reduces chilling injury in papaya (*Carica papaya* L.) fruit as well as in two cultivars of guava (*Psidium guajava* L.) and resulted in the least fruit decay. The application of 1 mM of MeJA had eminent polygalacturonase activity in peaches that helped to reduce woolliness in chilling fruits (Budde et al., 2005); 0.01 mM concentration of MeJA applied as a gas in tomato fruit was reported to reduce chilling injury and improve ripening (Ding et al., 2002). Volatile MeJA treatment enhanced the total soluble sugar content and slowed down the process of loss of organic acids in raspberries after storage (Wang, 2003). Hence, MeJA can optimistically modify post-harvest physiology for horticultural gain in various crops.

Pre-harvest treatments of jasmonic acid on *Pharbitis nil* (Japanese morning glory) produced an impact similar to the application of abscisic acid, which reduces the growth of shoots, roots, leaves, and buds; still, these impacts were partially upturned by gibberellic acid (GA3) application (Maciejewska et al., 2002). Furthermore, pre-treatment application of MeJA improved superoxide dismutase (SOD), catalase (CAT), and peroxidase (Px) activities, anthocyanin content, and glutathione reductase (GR) to counter the oxidative stress produced due to decreased photosynthetic activity (Jung, 2004). Although there is little research on the application of MeJA before harvest, this is the stage where fruits are most receptive to agrochemical applications. For example, it has been shown in sweet cherries that MeJA application in the early stages of vegetative growth produces an improved post-harvest response to the *Monilinia fructicola* pathogenic fungus (brown rot, 0.2 mM). The mode of action takes place by improving the activity of PAL and β-1,3-glucanase that inhibit fungal mycelia growth and spore formation. Rudell et al. (2005) noted that the application of 0.5 mM MeJA to apples can enhance the biosynthesis of β-carotene by adapting to low temperature, thereby reducing temperature fluctuations in the orchard and giving fruits a photoprotective effect. Besides this, MeJA induced the expression of the *CYP71A2* gene, which encodes the cytochrome P450s. This cytochrome appears to be important for the ripening of avocado fruit (Li et al., 2010). Lately, the application of MeJA in the pre-harvest stage of raspberry plants has led to a significant increase in related health promoting compounds like quercetin, ellagic acid, and myricetin. The author concluded that this enhancement was attributed to the promotion of PAL enzyme activity by MeJA (Flores et al., 2014). In fact, Wang et al. (2009) have previously confirmed this conclusion, where the enhancement in PAL activity in Chinese bayberry is due to the response to the application of MeJA. Similarly, Kucuker et al. (2014) noted that *Prunus salicina* (plums) treated with MeJA had a better yield and maintained higher flesh firmness than the control. Nevertheless, the author pointed out that pre-harvest MeJA application during plum ripening may be considered an effective tool for maintaining the firmness of the pulp during commercial harvesting. Therefore, most studies have shown that the application of MeJA before harvest may be a promising tool to improve fruit quality and extend shelf life, but the optimal concentration of this hormone depends on the species and variety.

On the other hand, in the post-harvesting application of MeJA (0.01 mM) on loquat fruit, the ratio of saturated/unsaturated fatty acid is higher, which increases the ability to resist chilling injury (Cao et al., 2009). When stored fruits of loquat (0.1 mM), mango (0.1 mM), pineapple (0.01 mM), and peach (0.1 mM) are stored at a very low temperature, MeJA treatment reduces the symptoms of chilling injury. This harmful consequence is mostly caused by the disintegration of the cell wall caused by physiological processes such as ion leakage, polysaccharide solubilization, and ethylene biosynthesis. (Meng et al., 2009). Some research works have shown that 0.1–10 mM MeJA application can promote color alteration in apples and mangoes by reducing the content of chlorophyll and improving carotene accretion by endorsing the biosynthesis of ethylene (González-Aguilar et al., 2001). The application of exogenous MeJA enhanced the disease resistance of the fruit after harvest and reduced fungal invasion, thereby allowing a longer and improved post-harvest life (Osorio et al., 2013). Since the plant's own defense mechanism has broad-spectrum antibacterial properties, inducing fruit resistance after harvest seems to be an important strategy to reduce the incidence of disease (Meyer, 1984). MeJA induces the expression of plant defense genes in tomato, *Vitis labrusca* L. cv. "Kyoho" (grape berry), *Vitis vinifera* L., and loquat, particularly β-1,3-glucanase and chitinase, both encode disease-related (PR) proteins (Wang et al., 2015). According to the study, the genes for chitinase and β-1,3-glucanase encoded by MeJA can hydrolyze the chitin polymer of the fungal cell walls, showing that these genes are implicated in the defense mechanism of plants against fungal infections. Usually, small doses are used to control post-harvest disease. The application of MeJA in sweet cherries can control the fungus *M. fructicola*. However, the MeJA vapor method is not effectual as spraying or immersing the fruits in pesticides because the peel is very thick (Yao et al., 2005). Although many studies have been conducted on the effects of MeJA on fruit disease, further studies are still needed to clarify the molecular mechanisms of MeJA-induced post-harvest fruit defense responses.

17.7 Summary

Sustainable agriculture is a holistic approach toward sustainable development that concedes the finite nature of natural resources, acknowledges the check on economic growth, and encourages resource allocation equity. Among various practices of sustainable agriculture, the potential use of phytohormones to enhance crop yield and productivity without compromising environmental degradation has been investigated since their discovery and commercial availability. Considerable success has been attained in recent years in the use of phytohormones in many crop improvement processes like flowering and fruiting, ripening, and harvesting and post-harvesting of fruits and vegetables. Since the 1960s, from the time of its discovery, increasing interest has been noted in the jasmonate class of phytohormones, which has led to the growing number of patents for commercial production and use of JA in the last decade. Presently, jasmonic acid is being synthetically metabolized into various derivatives like sodium jasmonate, potassium jasmonate, lithium jasmonate, ammonium jasmonate, and so on, which are actively used in biopesticides against fungal and various other biotic stress factors. Among various derivatives of JA, MeJA and prohydrojasmon are the most actively used synthetic derivatives in fruit harvesting and post-harvesting processes. As a natural substance, MeJA has no constraints for pre- and post-harvest treatments, and it has thus been studied to extend the life of a variety of agricultural crops. MeJA is an acclaimed phytohormone to control post-harvest diseases in fruits. It is also reported to have a significant impact on secondary metabolites in many crops, as well as the development of a natural defense against biotic and abiotic stressors. Treatment of jasmonic acid on soybean seeds is widely used to improve their oil quality by lowering oleic acid and increasing linoleic acid and linolenic acid. Apart from these, a great many other applications of jasmonates and their various derivatives have been extensively studied in numerous cereals, vegetables, fruits, legumes, and so on. Thus, more research on jasmonic acid and its mechanism of action could prove to be valuable in phytohormone-driven sustainable agriculture.

17.8 Future Prospects

Due to their striking involvement in plant defense and developmental processes, jasmonates have been extensively researched in recent years. To explore natural plant immunity responses for their possible exogenous use has long been a goal of modern biodynamic agriculture. In order to further explore the agricultural applications of jasmonates, a thorough understanding of JA-regulated mechanistic level and enzymatic activities are big gaps in the understanding of JA biochemistry for agricultural exploitation. Although many derivatives of JA are reported to have a positive influence on plant growth and pest control, their efficacy and further research are obligatory before commercially using them as control strategies. Therefore, the work on JA-regulated agricultural produce is still in its infancy and requires further research on the perception and signal transduction of diverse biotic and abiotic signals that would undoubtedly result in valuable uses of jasmonates in agriculture and horticulture sectors.

REFERENCES

Abdelgawad et al. *J. Agric. Sci.* 2014:1077–1088.

Acosta et al. *Science* 2009; 323(5911):262–265.

Agrawal et al. *Evolution* 1999; 53(4):1093–1104.

Ahn et al. *Plant Pathol. J.* 2013; 29(1):110.

Alavi-Samani et al. *Hortic. Environ. Biotechnol.* 2015; 56(4):411–420.

Aldridge et al. *J. Chem. Soc. C Org.* 1971:1623–1627.

Andrews et al. *Climate Change and Management of Cool Season Grain Legume Crops*, eds S. Yadav and R. Redden. Springer, Dordrecht. 2010; 1–9.

Anjum et al. *Int. J. Agric. Biol.* 2011; 13:251–256.

Atay. *Bulg. J. Agric. Sci.* 2015; 21(6):1242–1247.

Avanci et al. *Genet. Mol. Res.* 2010; 9(1):484–505.

Ayala-zavala et al. *Eur. Food Res. Technol.* 2005; 221(6):731–738.

Baishya. *International Conference on Science Technology and Management.* University of Delhi, New Delhi. 2015; 778–786.

Baldwin. *Proc. Natl. Acad. Sci. USA* 1998; 95(14):8113–8118.

Baldwin and Schmelz. *Ecology* 1996; 77(1):236–246.

Bartels and Sunkar. *Crit. Rev. Plant Sci.* 2005; 24(1):23–58.

Belhadj et al. *J. Agric. Food Chem.* 2006; 54(24):9119–9125.
Bell et al. *Proc. Natl. Acad. Sci. USA* 1995; 92(19):8675–8679.
Bordenave et al. *PLOS ONE* 2013; 8(12):e83199.
Brighenti et al. *Pesquisa Agropecuária Brasileira.* 2017; 52(11):1118–1122.
Budde et al. *Acta Hortic.* 2005; 682:363–368.
Buta and Moline. *J. Agric. Food Chem.* 1998; 46(4):1253–1256.
Cai et al. *Nat. Commun.* 2014; 5:3476.
Cao et al. *Food Chem.* 2009; 115(4):1458–1463.
Chehab et al. *Curr. Biol.* 2012; 22(8):701–706.
Creelman and Mullet. *Annu. Rev. Plant Biol.* 1997; 48(1):355–381.
Curtin et al. *Biotechnol. Lett.* 2003; 25(14):1131–1135.
Czapski and Saniewski. *Gartenbauwissenschaft* 1985; 50:35–37.
Das and Ghosh. *Outlook Agric.* 2012; 41(4):279–284.
David Marks. *Agricultural Composition.* Patent no. US8137429B2. 2012.
De Geyter et al. *Trends Plant Sci.* 2012; 17(6):349–359.
Delgado et al. *Molecules* 2018; 23(6):1433.
Demole et al. *Helv. Chim. Acta* 1962; 45(2):675–685.
Deschamps and Simon. *J. Essent. Oil Res.* 2006; 18(6):618–621.
Ding et al. *Planta* 2002; 214(6):895–901.
Ding et al. *Plant Cell* 2008; 20(10):2681–2695.
Eng et al. *PeerJ* 2021; 9:e10873.
Fahad et al. *J. Agron. Crop Sci.* 2016; 202(2):139–150.
Fan et al. *Planta* 1999; 204:444–449.
Fard et al. *J. Med. Plant Res.* 2012; 6(45):5701–5705.
Farmer and Ryan. *Proc. Natl. Acad. Sci. USA* 1990; 87(19):7713–7716.
Fedina and Tsonev. *J. Plant Physiol.* 1997; 151(6):735–740.
Ferguson et al. *Plant Physiol.* 2005; 138(4):2396–2405.
Flores and Ruizdel Castillo. *Postharvest Biol. Technol.* 2014; 97:77–82.
Franceschi et al. *Am. J. Bot.* 2002; 89(4):578–586.
Fung et al. *Plant Sci.* 2004; 166(3):711–719.
Gan et al. *Int. Rev. Cell Mol. Biol.* 2013; 305:115–161.
Gang et al. *Plant Physiol.* 2001; 125(2):539–555.
Gavin et al. *Nat. Prod. Res.* 2012; 10:1–8.
Gewehr. Patent no. PCT/EP2013/070149. 2014.
Ghasemi Pirbalouti et al. *J. Essent. Oil Res.* 2014; 26(5):342–347.
Ghasemnezhad and Javaherdashti. *J. Environ. Sci.* 2008; 6:73–78.
Gidda et al. *J. Biol. Chem.* 2003; 278(20):17895–17900.
Gliszczyńska et al. *Curr. Microbiol.* 2015; 71(1):83–94.
Gliszczyńska et al. *Molecules* 2016; 21(9):1226.
González-Aguilar et al. *Int. J. Food Sci. Technol.* 2001; 36:767–773.
González-Aguilar et al. *Postharvest Biol. Technol.* 2003; 28(3):361–370.
Gonzalez-Aguilar et al. *Biochem. Biophys. Res. Commun.* 2004; 313(3):694–701.
Guan et al. *Proc. Natl. Acad. Sci. USA* 2019; 116(21):10568–10575.
Haygarth et al. Wiley-Blackwell, West Sussex, UK. 2013; 132–159.
Hecht. *J. Nutr.* 1999; 129:768–774.
Horrigan et al. *Environ. Health Perspect.* 2002; 110(5).
Hussain et al. *Food Chem.* 2008; 108(3):986–995.
Jia et al. *Funct. Integr. Genomics* 2016a; 16(1):79–94.
Jia et al. *Korean J. Chem. Eng.* 2016b; 33:1080–1087.
Jin et al. *Acta Hortic.* 2006; 712:711–716.
Jin et al. *J. Sci. Food Agric.* 2009; 89(5):802–808.
Jung. *Plant Physiol. Biochem.* 2004; 42(3):225–231.
Kang et al. *J. Agron. Crop Sci.* 2005; 191(4):273–282.
Kazan. *Trends Plant Sci.* 2015; 20(4):219–229.
Kim et al. *J. Agric. Food Chem.* 2006; 54(6):2327–2332.
Kim et al. *Plant Physiol.* 2009; 149(4):1751–1760.
Kinkema and Gresshoff. *Mol. Plant Microbe Interact.* 2008; 21(10):1337–1348.
Koch et al. *Helv. Chim. Acta* 1997; 80(3):838–850.
Kouchi et al. *DNA Res.* 2004; 11(4):263–274.
Kucuker et al. *Sci. Hortic.* 2014; 176:162–169.
Kurepin et al. *J. Sci. Food Agric.* 2014; 94(9):1715–1722.
Lal. *Geoderma* 2004; 123(1–2):1–22.
Li et al. *Plant Growth Regul.* 1998; 25(2):81–87.
Li et al. *Proteomics* 2010; 1804:929–940.
Lin et al. *PLOS ONE* 2010; 5(8):e11928.
Lipper et al. *Nat. Clim. Chang.* 2014; 4:1068–1072.
Mabood et al. *Plant Physiol. Biochem.* 2006a; 44(11–12):759–765.
Mabood et al. *Field Crops Res.* 2006b; 95:412–419.
Maciejewska and Kopcewicz. *J. Plant Growth Regul.* 2002; 21(3):216–223.
Malekpoor et al. *J. Herb. Drugs* 2015; 6(1):13–21.
Mangas et al. *Phytochemistry* 2006; 67(18):2041–2049.
Martinez-Ferrar and Harper. *J. Food Qual.* 2005; 28:3–12.
Mason et al. *Plant Physiol.* 1992; 98(3):859–867.
Meir et al. *Acta Hort.* 2005; 669:91–98.
Meng et al. *Food Chem.* 2009; 114(3):1028–1035.
Meyer et al. *J. Plant Growth Regul.* 1984; 3:1–8.
Meyer et al. *Phytochem. Anal.* 2003; 14(3):155–159.
Mohamed and Latif. *Physiol. Mol. Biol. Plants* 2017; 23(3):545–556.
Mohanta et al. *Glob. Trends Phytohorm. Res.* 2020; 9(9):1248.
Moreira et al. *J. Agric. Sci.* 2019; 11(14):254–264.
Moreno et al. *Proc. Natl. Acad. Sci. USA* 2009; 106(12):4935–4940.
Morino et al. *Plant Prod. Sci.* 2020; 23(3):350–359.
Nafie et al. *Braz. J. Plant Physiol.* 2011; 23(2):161–174.
Nakagawa and Kawaguchi. *Plant Cell Physiol.* 2006; 47(1):176–180.
Opie and Lecour. *Eur. Heart J.* 2007; 28(14):1683–1693.
Osorio et al. *Plant Pathol.* 2013; 38:63–67.
Pak et al. *Planta* 2009; 231(1):79–91.
Paprocka et al. *Molecules* 2018; 23(9):2362.
Qin et al. *Plant Cell Physiol.* 2011; 52(9):1569–1582.
Redman et al. *Oecologia* 2001; 126(3):380–385.
Reid et al. *Plant Physiol.* 2016; 170(2):1060–1074.
Rudell et al. *J. Am. Soc. Hortic. Sci.* 2002; 127:435–441.
Rudell et al. *Hort.Sci.* 2005; 40(6):1760–1762.
Sakamoto. *Transgen. Res.* 2006; 15:399–404.
Sanches et al. *Bull. Entomol. Res.* 2017; 107(6):828–838.
Sayyari et al. *Food Chem.* 2011; 124(3):964–970.
Schaller and Stintzi. *Phytochemistry* 2009; 70(13–14): 1532–1538.
Sembdner and Parthier. *Plant Mol. Biol.* 1993; 44:569–589.
Seo et al. *Proc. Natl. Acad. Sci. USA* 2001; 98(8):4788–4793.
Singh and Singh. *J. Food Legumes* 2011; 24:83–90.
Sio-Se Mardeh et al. *Field Crops Res.* 2006; 98(2–3):222–229.
Smil. The MIT Press, Cambridge, MA. 2004.
Staswick and Tiryaki. *Plant Cell* 2004; 16(8):2117–2127.
Staswick et al. *Plant Cell* 2002; 14(6):1405–1415.
Stougaard. *Plant Physiol.* 2000; 124(2):531–540.
Sun et al. *Plant J.* 2006; 46(6):961–970.
Swiatek et al. *Plant Physiol.* 2004; 135(1):161–172.
Tadeu Dias. *Biotecnol. Veg.* 2019; 19(1):3–14.
Terry and Joyce. *Postharvest Biol. Technol.* 2004; 32(1):1–13.

Tiwari et al. *Plant Physiol. Biochem.* 2016; 99:108–117.
Toker and Yadav. Eds S.S. Yadav, R. Redden. 2010; 351–376.
Trenkel. The International Fertilizer Industry Association (IFA), Paris, France. 1997; 151.
Trevisan et al. *J. Agric. Food Chem.* 2006; 54(12):4378–4382.
Vahabi et al. *Iran. J. Field Crops Res.* 2017; 15:124–135.
Vick and Zimmerman. Academic Press. 1987; 53–90.
Vob et al. *Trends Plant Sci.* 2014; 19(5):311–319.
Wang. *J. Plant Growth Regul.* 1999; 18(3):127–134.
Wang. *Int. J. Food Sci. Technol.* 2003; 38:869–875.
Wang and Zheng. *Int. J. Food Sci. Technol.* 2005; 40:187–195.
Wang et al. *J. Chem. Ecol.* 2001; 27(2):327–342.
Wang et al. *Food Chem.* 2008; 107:1261–1269.
Wang et al. *J. Agric. Food Chem.* 2009; 57(13):5809–5815.
Wang et al. *Int. J. Food Microbiol.* 2015; 194:32–39.
Wang et al. *J. Integr. Plant Biol.* 2019.
Wasternack. Blackwell Publishing, Oxford, UK. 2006; 185–228.
Wasternack. *Biotechnol. Adv.* 2014; 32(1):31–39.
Wasternack and Hause. *Ann. Bot.* 2013; 111(6):1021–1058.
Wasternack and Kombrink. *ACS Chem. Biol.* 2010; 5(1):63–77.
Wasternack and Parthier. *Trends Plant Sci.* 1997; 2(8):302–307.
Wasternack and Song. *J. Exp. Bot.* 2017; 68(6):1303–1321.
Wattenberg. *Cancer Res.* 1985; 45(1):1–8.
Wellmer et al. *J. Exp. Bot.* 2014; 65(1):1–9.
Weng and Yen. *Cancer Treat. Rev.* 2012; 38(1):76–87.
Wolucka et al. *J. Exp. Bot.* 2005; 56(419):2527–2538.
Wu et al. *Proc. Natl. Acad. Sci. USA* 2012; 109(9):3576–3581.
Yang et al. *Z. Naturforsch. C* 2004; 59(11–12):828–834.
Yao and Tian. *J. Appl. Microbiol.* 2005; 98(4):941–950.
Yilmaz et al. *J. Am. Pomol. Soc.* 2003; 57:32–35.
Yoon et al. *J. Crop Sci. Biotechnol.* 2009; 12(2):63–68.
Yuan and Zhang. *Curr. Opin. Plant Biol.* 2015; 27:44–51.
Ziosi et al. *J. Exp. Bot.* 2008; 59(3):563–573.
Złotek et al. *Food Chem.* 2016; 213:1–7.

18

Post-Harvest Physiology of Cut Flowers: Use of Methyl Jasmonate as a Quality-Retention Agent

Takanori Horibe

CONTENTS

18.1 Introduction

Cut-flower quality depends on the external characteristics of the plant, such as color, length, volume, freshness, fragrance, and perishability, which is determined by the duration of these aforementioned conditions. Vase life is a particularly critical factor in determining the market value of cut flowers. Most cut flowers will no longer be suitable for ornamental purposes within a week or so, making it difficult to enjoy them for an extended period of time. Flower senescence is the biggest cause of the depreciating ornamental value of cut flowers. There are mainly two types of senescence: petal wilting and detachment (Woltering and van Doorn, 1988). Specifically, ethylene is a major factor in the aging of several cut flowers. The ornamental value may also be lost due to non-flowering or yellowing of the leaves (van Doorn et al., 1992). In addition to the depreciating ornamental value due to petal senescence, severe water stress may cause petals to wither (van Doorn and Suiro, 199). Also, if the stem is soft, it will lose its ornamental value due to stem bending or breaking. For cut-flower quality improvement, quality preservatives are commonly used. Treatment with ethylene inhibitors can make the flowers last longer than the original life expectancy and using quality-preserving agents is the most important quality-preserving technology for several cut flowers (Ichimura and Shimizu-Yumoto, 2007). Recently, several studies have indicated that methyl jasmonate (MeJA) has potential as a quality-retention agent for cut flowers (Mizuno et al., 2017; Horibe et al., 2019). It is also important to understand flowering mechanisms and establish methods to control such mechanisms, as well as focusing on slowing the aging process, in order to achieve improvement of cut-flower quality since ornamental value also lies in the process of blooming from buds in several cut flowers, including roses. In this review, the authors describe recent progress in the research of post-harvest physiology of cut flowers and the utilization of MeJA to improve cut-flower quality.

18.2 Petal Growth and Flower Opening

The "flower opening" phenomenon is a reaction caused by the growth of petals due to an increase in the size and quantity of petal cells (Evans and Reid, 1986, 1988). The adaxial and abaxial epidermal cells in cut roses show marked expansion during flower opening, and adaxial epidermal cells become corn-shaped when the flower is fully open (Yamada et al., 2009c). In several flowers, the differences in the cell expansion patterns among cell types and locations, including adaxial and abaxial epidermal cells, are believed to cause petal reflection during rose flower opening. Also, in rose petals, the changes in cell structure and strength were thought to cause differences in epidermal cell growth on the adaxial and abaxial sides, resulting in the reflection of petals (Horibe et al., 2013). In carnations, the amount of DNA contained in petal cells stops increasing at the time when petals emerge from the calyx, suggesting that cell division stops at a relatively early stage of flower development (Kenis et al., 1985). In *Gaillardia grandiflora*, the division of petal cells stops at an early stage of flowering (Koning 1984). Also, in roses, petal cell division stops almost completely while the petals are still covered by the calyx since rose flowering is mainly caused by petal cell enlargement (Roberts et al., 2003; Yamada et al., 2009c). These data show that flowering is mainly caused by the enlargement of petal cells. It is thought to be caused by water influx into the cells due to the increase in osmotic pressure caused by the

DOI: 10.1201/9781003110651-18

accumulation of solutes, such as sugar in the vacuole, and the increase in cell wall extensibility (Horibe and Yamada, 2017).

18.3 Function of Carbohydrates in Flower Opening

Carbohydrates are essential substances for plant growth as respiratory substrates, cell wall synthetic substrates, and osmolytes. During petal growth, a large amount of soluble sugar accumulates in the petals (van Doorn et al., 1991; Ichimura et al., 2003). Sugars, namely glucose and fructose, accumulate during petal development of most flowers (Ichimura et al., 1998, 2000b; Yamada et al., 2009a). Glucose and sucrose accumulate in plant species that do not accumulate fructose in petals, such as in *Eustoma* (Shimizu and Ichimura, 2005) and sweet pea (Ichimura et al., 1999).

It has been reported that treating cut flowers with sugar is also effective in improving cut-flower quality, and sucrose is widely used to extend the vase life of cut carnations and roses (Paulin and Jamain, 1982; Ichimura et al., 2003). This suggests that soluble sugar plays an important role in regulating the osmotic pressure of petal cells. Carbohydrate accumulation in petal cells is thought to reduce the water potential of the petals, thereby causing the petal cells to flower by water influx (Ho and Nichols, 1977). In sink organs, such as tomato fruits, the soluble sugars accumulate in the apoplast, in addition to the vacuole (Damon et al., 1988). The flower is also regarded as a sink organ, and glucose and fructose mainly accumulate in vacuoles in the tulip perianth (Wagner, 1979). In roses, sugar tends to accumulate in the apoplast of petal cells (Yamada et al., 2009b). Osmotic pressure needs to be maintained at higher levels in the symplast than in the apoplast in order to promote water influx into cells.

In addition, other sugars may accumulate in petal cells depending on the flower type. Gentiobiose is the main carbohydrate constituent of gentian (Ichimura et al., 2005), whereas sugar alcohol is the major constituent in several flowers. Mannitol is the main carbohydrate in the petals of delphinium and snapdragon (Norikoshi et al., 2015; Ichimura et al., 2016), and 2-C-methylerythritol accumulates in the petals of phlox (Enomoto et al., 2004). The bud growth of cut delphinium can be promoted by mannitol treatment, but it is suppressed by 3-O-methylglucose treatment, a non-metabolizable carbohydrate (Ichimura et al., 2000). Therefore, mannitol is suggested to not only function as an osmolyte but also as an important respiratory substrate.

18.4 Petal Growth-Related Proteins

Sucrose synthase and acid invertase (β-fructofuranosidase) are enzymes that metabolize sucrose translocated from leaves to sink tissues and that are present in the cell wall (insoluble form) and vacuole (soluble form) in several higher plants. Cell wall acid invertase converts sucrose into hexoses after sucrose is translocated from the phloem to the apoplast (Roitsch and Gonzalez-Garcia, 2004), enabling petals to uptake more sucrose from the phloem. Sucrose synthase activity in rose petals is much lower than that of acid invertase (Kumar et al., 2008). These results suggest that acid invertase plays a major role in sucrose degradation in rose petals and that the translocation of sugar from the phloem to the petal is partially via the apoplast. Sugars are partially transported through the apoplast in tomato and apple fruits where the apoplast sugar content is relatively high (Damon et al., 1988; Zhang et al., 2004). In addition, acid invertase activity has been reported to strongly influence the sink strength of sink organs (Tang et al., 1999; Balibrea et al., 2004; Roitsch and Gonzalez-Garcia, 2004). Acidic invertase activity in rose petals increases when flowers bloom on trees but markedly decreases in cut flowers (Yamada et al 2007). The petal size of cut flowers is smaller than that of roses on trees. This may be due to a decrease in acid invertase activity in cut flowers, thus reducing the ability to decompose translocated sucrose and inhibiting water absorption due to increased osmotic pressure. Acid invertase activity is affected by phytohormones (Miyamoto et al., 1993; Trouverie et al., 2004; Pan et al., 2006; González and Cejudo, 2007). Horibe et al. (2013) showed that the acid invertase activity of cut rose flowers was kept high when treated with MeJA, which coincided with the inhibition of petal wilting.

Changes in cell wall extensibility are also important for petal cell growth. When the surrounding cell wall remains rigid, water cannot flow into the cell, thereby limiting expansion. In fact, cell wall loosening has been shown to occur during rose petal development (Yamada et al., 2009b). Cell wall-loosening candidate proteins include endotransglucosylases/hydrolases (XTH) and expansins (Cosgrove, 2001). XTH catalyzes either molecular grafting or the disassembly of xyloglucan crosslinks within the cellulose–xyloglucan framework (Nishitani and Tominaga, 1992; Okazawa et al., 1993), and expansins are believed to disrupt hydrogen bonding between cellulose microfibrils and matrix glucans (McQueen-Mason and Cosgrove, 1994). Expansins reportedly function in carnation petal growth (Harada et al 2011) and are expressed during the flowering process in tomato and petunia (Brummel et al., 1999; Zenoni et al., 2004), whereas XTH activity has been associated with flowering in *Sandersonia aurantiaca* (O 'Donoghue et al., 2002). These cell wall-loosening proteins are also target candidates for cut-flower quality improvement.

18.5 Ethylene and Cut-Flower Senescence

Ethylene is synthesized in plants through the following pathway: L-methionine→ *S*-adenosyl-L-methionine→1-aminocyclopropane-1-carboxylic acid (ACC)→ethylene. The last two reactions are catalyzed by ACC synthase and ACC oxidase (Lin et al., 2009; Shibuya, 2018). Ethylene is strongly involved in the senescence of several flowers and promotion of petal wilting or detachment, although the ethylene sensitivity of flowers varies significantly among plant species (Woltering and van Doorn, 1988). In flowers that are highly sensitive to ethylene, including carnations and *Eustoma*, pollination promotes petal senescence (Larsen et al., 1995; Shimizu-Yumoto and Ichimura, 2006). In these flowers, pollination increases ethylene production first in the style and then in the ovary.

Petal ethylene production eventually increases, leading to petal senescence.

Since the senescence-promoting effect of ethylene is strong in several cut flowers, ethylene-production/action inhibitors are widely used as quality-retention agents. Silver thiosulfate complex (STS) is a typical ethylene-inhibiting agent. Its effects on extending the vase life of cut flowers are very high, especially in ethylene-sensitive cut flowers, such as carnations and sweet peas (Veen, 1979; Mor et al., 1983). On the other hand, 1-methylcyclopropene can irreversibly bind to ethylene receptors but it only blocks newly synthesized receptors. As a result, its quality-retaining effect is limited compared with that of STS (Ichimura et al., 2002).

In cut flowers, such as carnations (Pun et al., 2016) and delphiniums (Ichimura et al., 2000), ethylene production is induced as the sugar content decreases. In these cut flowers, sugar treatment delays the peak of ethylene production and prolongs the vase life, so the decreasing sugar content may indicate increased ethylene production.

18.6 Water Relation

The water status of cut flowers is determined by the difference between water absorption and transpiration. When the amount of transpiration is larger than the amount of absorbed water, the water condition deteriorates and flower wilting proceeds. What directly suppresses water absorption is xylem occlusion, which is mainly caused by air bubbles generated in the stem and by bacterial growth (van Doorn, 1997).

Air enters the xylem when the cut stem is exposed to air. In addition to air entering from the cut end, it has been suggested that air bubbles that are generated inside the stem (cavitation) also impede water transport (van Doorn and Suiro, 1996). In cut rose flowers, using degassed or filtered water promotes water absorption (Durkin, 1979). In contrast, water aeration suppresses its absorption (van Doorn, 1990). These findings suggest that air bubbles generated in the water and stem can cause xylem occlusion.

Bacterial growth in the xylem and vase water is also thought to contribute to xylem occlusion (van Doorn et al., 1989). Zagory and Reid (1986) reported that adding bacteria to vase water promoted xylem occlusion. On the contrary, when antibacterial agents were applied to vase water, the bacterial concentration became low, and xylem occlusion was suppressed (Jones and Hill, 1993). These results strongly support the idea that the presence of bacteria and other microorganisms in the stem and vase water is one of the major causes of xylem occlusion. Thus, antibacterial agents are frequently used as quality-retention agents for cut flowers.

18.7 MeJA as a Potential Quality-Retention Agent

MeJA is a plant-growth regulator that belongs to the jasmonate family and plays a role in various plant functions, from the morphological to the molecular levels (Ueda et al., 2006). These functions include mediating intra- and inter-plant communications and modulating plant defense responses, including antioxidant systems (Wasternack, 2007). MeJA biosynthesis starts in the chloroplast and occurs through the following pathway: linolenic acid→13-hydroperoxylinolenic acid→12-oxophytodienoic acid→(′)-7-iso-JA→MeJA. These reactions are catalyzed by several enzymes, including lipoxygenases, allene oxide synthase, allene oxide cyclase, and JA methyltransferases. The MeJA biosynthesis and signaling pathway details are widely reported and discussed in previous reports (Wasternack, 2007; Wasternack and Hause, 2013; Dar et al., 2015).

As discussed previously, quality-preserving agents for cut flowers often include ethylene inhibitors, sugars, and antibacterial agents. MeJA application is not widely commercially used for cut-flower quality improvement. Recently, however, several studies have suggested that MeJA application can be a promising tool for increasing fruit quality and extending shelf life by enhancing post-harvest disease resistance and reducing fungal attack (Ghasemnezhad and Javaherdashti, 2008; Osorio et al., 2013; Reyes-Díaz et al., 2016). For example, spraying MeJA led to increases in the quality of apples, accumulating phenolic compounds, such as anthocyanins, chlorogenic acid, phloridzin, flavanols, and flavonols in the fruit skin without negative fruit-quality effects (Shafiq et al., 2013). In plums (*Prunus salicina*), MeJA treatment resulted in higher fruit yields and significantly higher fruit-flesh firmness compared with controls, although the diameters of MeJA-treated fruits were lower than those of control fruits (Kuculer et al., 2014). In addition, Guo et al. (2014) reported that 100 μM MeJA treatment of mandarin (*Citrus reticulata*) can lower disease incidence and growth of green mold (*Penicillium digitatum*). Several studies indicated the interaction between MeJA and ethylene. It has been shown that MeJA treatment enhanced the ethylene biosynthetic enzymes ACC oxidase and ACC synthase in tomato and apple fruits, promoting pigmentation and ripening (Fan et al., 1998; Rudell et al., 2005). Furthermore, other studies have shown that MeJA-induced changes in the expression of genes responsible for fruit ripening, pollen production, and foliar bud, shoot, and root hair growth (Creelman and Mullet, 1997). The relationship between MeJA and fruit quality has been reviewed recently by Reyes-Díaz et al. (2016).

We have previously investigated the effects of MeJA on flower opening of three rose cultivars (Horibe and Makita, 2019). In this previous study, 1000 μM MeJA treatment clearly delayed cut-flower opening, resulting in longer vase life (Horibe and Makita, 2019; Figure 18.1). In addition, flower wilting was suppressed in all cultivars, and the petal fresh weight was maintained high at seven days following treatment, compared with the control. However, the physiological mechanisms involved in MeJA-induced delayed flower opening remain to be identified. We have previously shown that MeJA treatment activates invertase in rose petals, promoting the translocation of sucrose from leaves to the petals during petal growth (Horibe et al., 2013). Furthermore, sugar metabolism in leaves may also be affected by MeJA treatment. Ochiai et al. (2013) reported that MeJA accelerates expansin and XTH expression. These cell wall-loosening proteins are also thought to be mainly involved in petal expansion (Takahashi et al., 2007; Yamada et al., 2009b). The functions of cell wall-loosening

FIGURE 18.1 Flower opening pattern of cut rose "Princes Meg" two to six days after MeJA treatment. Flowers were cut to a length of 30 cm, and the stem ends were continuously held in 2% w/v sucrose with 0.02% w/v 8-hydroxyquinoline monohydrate (control), 100 μM MeJA in addition to 2% w/v sucrose with 0.02% w/v 8-hydroxyquinoline monohydrate (100 μM), and 1000 μM MeJA in addition to 2% w/v sucrose with 0.02% w/v 8-hydroxyquinoline monohydrate (1000 μM). Cut flowers were maintained at 25°C, 60% relative humidity, and a 12-hour photoperiod (photosynthetic photon flux density: 20–30 $\mu mol \cdot m^{-2} \cdot s^{-1}$) in a plant growth chamber.

proteins and sugar-metabolizing proteins in rose petals may also be affected by MeJA treatment. Moreover, MeJA treatment may delay senescence by slowing down chlorophyll and carotenoid degradation (Kovač and Ravnikar, 1994; Hamberg and Gardner, 1992). Further research regarding the roles of these proteins in response to MeJA treatment is warranted.

Askari-Khorasgani and Mortazaeinezhad (2016) also showed the effectiveness of MeJA as a pre- and post-harvest quality-retaining agent. They investigated the effect of MeJA alone or in combination with benzyl adenine (BA) on rose qualitative characteristics, including flower diameter, leaf chlorophyll content, petal carotenoid content, and vase life. The results showed that the flower-head diameter and leaf chlorophyll content were improved by applying 100 ppm MeJA + 20 ppm BA, and 150 ppm MeJA treatment resulted in the longest vase life (Askari-Khorasgani and Mortazaeinezhad 2016). In this experiment, MeJA treatment suppressed carotenoid degradation and petal-color fading, resulting in rose quality and vase life improvement. In contrast to cut roses, Ochiai et al. (2013) reported that MeJA accelerated cut *Eustoma* flower opening. The maximum flower diameter at the fully opened stage did not significantly change, and only the timing of the flowering occurred earlier with MeJA treatment compared with the control group. The author attributed early flower opening induced by MeJA treatment to the accelerated cell wall-loosening protein expression (Ochiai et al., 2013) since the expression of the cell wall-loosening protein genes (*EgEXPA2*, *EgEXPA3*, and *EgXTH1*) was accelerated in MeJA-treated flower petals.

Flower color is an important factor affecting cut-flower quality. Several environmental conditions, including high temperature and sugar depletion, must be avoided when storing several ornamental flowers, such as roses, lilies, and carnations, because they inhibit anthocyanin biosynthesis (Maekawa and Nakamura, 1977; Dela et al., 2003; Lai et al., 2011; Zhang et al., 2015). Carbohydrate treatment is known to improve flower coloration in lilies and tree peonies by facilitating anthocyanin biosynthesis and accumulation (Han, 2003; Zhang et al., 2015). Recently, several studies reported that MeJA is also useful for improving the coloration of ornamental flowers and fruits (Saniewski et al., 1998a,b; Rudell et al., 2002; Mizuno et al., 2017). Mizuno et al. (2017) investigated the effects of MeJA on *Eustoma* petal coloration and showed that the non-uniform coloration of petals was reduced by continuous exposure to MeJA vapor. In addition, several studies indicated that shorter MeJA treatments, such as one to two days after harvesting, also reduced non-uniform coloration, similar to continuous MeJA exposure. Our research also showed that MeJA is effective in promoting anthocyanin biosynthesis and inhibiting rose bluing (data not shown). Guo et al. (2016) reported that MeJA treatment affected the flavonoid UDP-glycosyltransferase gene expression pattern in safflower petals, thus changing the amounts of several flavonoids, including rutin, kaempferol-3-O-β-D-rutinoside, kaempferol-3-O-β-D-glucoside, carthamin, luteolin, and quercetin-3-O-β-D-glucoside.

Lastly, MeJA application has also been shown to activate defense-related genes and promote phytoalexin synthesis, which are important for the plant's pathogen-induced defense responses (Wang et al., 2015). Darras et al. (2005) reported that MeJA vapor treatment significantly suppressed *Botrytis cinerea* on cut freesia flowers, delayed flower senescence, and improved vase life and fresh weight of the flowers. Regarding the application method, using MeJA treatment as a post-harvest spray or pulse treatment to freesia flowers also suppressed *B. cinerea*, although the suppressive effect fluctuated depending on MeJA concentration and application method (Darras et al., 2007).

As shown earlier, several studies suggest that MeJA application can also be a powerful tool for increasing cut-flower quality and extending vase life by enhancing post-harvest disease resistance, hastening flower opening, and improving petal coloration.

18.8 Conclusion

Recently, numerous studies have reported the usefulness of MeJA as a pre- and post-harvest quality-retaining agent for fruits and ornamental flowers. The strong requirement for sustainability and food security these days has led to dramatic changes in the horticulture market, including fruit and flowers. In this context, there have been efforts to reduce the use of inorganic pesticides and promote natural substances in accordance with integrated disease management strategies to ward off the harmful effects of chemical compounds, which threaten both environmental and human health. Since MeJA is a natural phytohormone and is present in different plant organs, it is well-suited for an environmentally friendly market. It is of interest to verify the effects of MeJA on flowering using numerous ornamental flowers and different cultivars. Lastly, additional studies are warranted to investigate the mechanisms of MeJA's quality-retaining effect on ornamental flowers.

REFERENCES

Askari-Khorasgani and Mortazaeinezhad. *JCEA* 2016, 17:369–378.
Balibrea et al. *Plant Cell* 2004, 16(5):1276–1287.
Brummel et al. *Plant Mol. Biol.* 1999, 39(1):161–169.
Cosgrove. *Plant Physiol.* 2001, 125(1):131–134.
Creelman and Mullet. *Annu. Rev. Plant Physiol. Plant Mol. Biol.* 1997, 48:355–381.
Damon et al. *Plant Physiol.* 1988, 87(3):731–736.
Dar et al. *Environ. Exp. Bot.* 2015, 115:49–57.
Darras et al. *PBT* 2005, 38:175–182.
Darras et al. *Plant Pathol.* 2007, 36:332–340.
Dela et al. *Plant Sci.* 2003, 164(3):333–340.
Durkin. *J. Amer. Soc. Hort. Sci.* 1979, 104:860–863.
Enomoto et al. *J. Plant Physiol.* 2004, 161(8):977–980.
Evans and Reid. *Acta Hort.* 1986, 181:55–63.
Evans and Reid. *J. Amer. Soc. Hort. Sci.* 1988, 113:884–888.
Fan et al. *Planta* 1998, 204(4):444–449.
Gonzalez-Garcia and Cejudo. *Plant Sci.* 2007, 172(4):839–846.
Guo et al. *PBT* 2014, 88:72–78.
Guo et al. *PLOS ONE* 2016, 11(7):e0158159.
Ghasemnezhad and Javaherdashti. *Casp. J. Environ. Sci.* 2008, 6:73–78.
Hamberg and Gardner. *Biochim. Biophys. Acta Lipids Lipid Metab.* 1992, 1165(1):1–18.
Han. *Hortscience* 2003, 38(3):412–416.
Harada et al. *J. Exp. Bot.* 2011, 62(2):815–823.
Ho and Nichols. *Ann. Bot.* 1977, 41(1):227–242.
Horibe and Makita. *J. Hortic. Res.* 2019, 27(2):1–10.
Horibe and Yamada. *J. Hort. Res.* 2017, 25:5–18.
Horibe et al. *PBT* 2013, 86:195–200.
Ichimura and Shimizu-Yumoto. *Bull. Natl. Inst. Flor. Sci.* 2007, 7:17–27.
Ichimura et al. *J. Plant Physiol.* 1998, 152(4–5):363–367.
Ichimura et al. *Biosci. Biotechnol. Biochem.* 1999, 63(1):189–191.
Ichimura et al. *Physiol. Plant.* 2000, 108(3):307–313.
Ichimura et al. *Bull. Natl. Inst. Flor. Sci.* 2002, 2:1–8.
Ichimura et al. *J. Japan. Soc. Hort. Sci.* 2003, 72:292–298.
Ichimura et al. *J. Hort. Sci. Biotech.* 2005, 80:280–286.
Ichimura et al. *PBT* 2016, 113:20–28.
Jones and Hill. *J. Amer. Soc. Hort. Sci.* 1993, 118(3):350–354.
Kenis et al. *Physiol. Plant.* 1985, 65(4):455–459.
Koning. *Am. J. Bot.* 1984, 71(1):1–8.
Kovač and Ravnikar. *Plant Sci.* 1994, 103(1):11–17.
Kucuker et al. *Sci. Hortic.* 2014, 176:162–169.
Kumar et al. *J. Hortic. Sci. Biotechnol.* 2008, 83(4):520–524.
Lai et al. *Sci. Hortic.* 2011, 132:59–65.
Larsen et al. *Plant Physiol.* 1995, 108(4):1405–1412.
Lin et al. *J. Exp. Bot.* 2009, 60(12):3311–3336.
Maekawa and Nakamura. *J. Jpn. Soc. Hortic. Sci.* 1977, 45(4):375–382.
McQueen-Mason and Cosgrove. *Proc. Natl. Acad. Sci. U. S. A.*, 91(14):6574–6578.
Miyamoto et al. *Physiol. Plant.* 1993, 88:301–306.
Mizuno et al. *The. Hortic. J.* 2017, 86(2):244–251.
Mor et al. *Plant Physiol.* 1983, 71(3):541–546.
Nishitani and Tominaga. *J. Biol. Chem.* 1992, 267(29):21058–21064.
Norikoshi et al. *Planta* 2015, 242(6):1467–1477.
Ochiai et al. *PBT* 2013, 82:1–5.
O'Donoghue et al. *J. Exp. Bot.* 2002, 53(368):513–523.
Okazawa et al. *J. Biol. Chem.* 1993, 268(34):25364–25368.
Osorio et al. *Trop. Plant Pathol.* 2013, 38(1):63–67.
Pan et al. *J. Integr. Plant Biol.* 2006, 48:536–549.
Paulin and Jamain. *J. Am. Soc. Hort. Sci.* 1982, 107:258–261.
Pun et al. *PBT* 2016, 121:151–158.
Reyes-Díaz et al. *Molecules* 2016, 21(6):567.
Roberts et al. *Encyclopedia of Rose Science*, 1st edition. Elsevier Academic Press, Amsterdam, 2003. pp. 1200.
Roitsch and Gonzalez-Garcia. *Trends Plant Sci.* 2004, 9(12):606–613.
Rudell et al. *J. Amer. Soc. Hort. Sci.* 2002, 127:435–441.
Rudell et al. *Hort.Sci.* 2005, 40(6):1760–1762.
Saniewski et al. *J. Plant Growth Regul.* 1998a, 17(1):33–37.
Saniewski et al. *J. Plant Growth Regul.* 1998b, 17(3):121–124.
Shafiq et al. *J. Sci. Food Agric.* 2013, 93(3):611–618.
Shibuya *Breed. Sci.* 2018, 68(1):99–108.
Shimizu and Ichimura. *J. Jpn. Soc. Hort. Sci.* 2005, 74(5):381–385.
Shimizu-Yumoto and Ichimura. *J. Japan. Soc. Hort. Sci.* 2006, 75(1):66–71.
Takahashi et al. *Acta Hortic.* 2007, 755:483–488.
Tang et al. *Plant Cell* 1999, 11(2):177–189.
Trouverie et al. *Planta* 2004, 219(5):894–905.
Ueda and Saniewski. *J. Fruit Ornam Plant Res.* 2006, 14:199–210.
van Doorn. *J. Plant Physiol.* 1990, 137(2):160–164.
van Doorn. *Hort. Rev.* 1997, 18:1–85.
van Doorn and Suiro. *Physiol. Plant.* 1996, 96(2):305–311.
van Doorn et al. *J. Plant Physiol.* 1989, 134(3):375–381.
van Doorn et al. *PBT*, 1991, 1:47–57.
van Doorn et al. *Plant Growth Regul.* 1992, 11:59–62.
Veen. *Planta* 1979, 145(5):467–470.
Wagner. *Plant Physiol.* 1979, 64(1):88–93.
Wang et al. *Int. J. Food Microbiol.* 2015, 194:32–39.
Wasternack. *Ann. Bot.* 2007, 100(4):681–697.

Wasternack and Hause. *Ann. Bot.* 2013, 111(6):1021–1058.
Woltering and Van Doorn. *J. Exp. Bot.* 1988, 39(11):1605–1616.
Yamada et al. *PBT* 2007, 43:174–177.
Yamada et al. *Planta* 2009a, 230(6):1115–1127.
Yamada et al. *J. Japan. Soc. Hort. Sci.* 2009b, 78(2):242–251.
Yamada et al. *J. Japan. Soc. Hort. Sci.* 2009c, 78(3):356–362.
Zagory. and Reid. *J. Am. Soc. Hort. Sci.* 1986, 114:154–158.
Zenoni et al. *Plant Cell* 2004, 16(2):295–308.
Zhang et al. *Plant Physiol.* 2004, 135(1):574–586.
Zhang et al. *PBT* 2015, 101:73–81.

19

Biofortification of Crop Plants with Brassinosteroids in Managing Human Health Issues

Isha Madaan, Manish Kumar, Harpreet Kaur, Renu Bhardwaj, Neha Dogra, Gurvarinder Kaur, and Geetika Sirhindi

CONTENTS

19.1 Introduction: BRs and Their Analogs

Brassinosteroids (BRs) belong to a class of steroidal phytohormones that are crucial regulators of plant growth and development. Several phytohormones have been discovered and isolated from plants that are reported to control various functions in plants at different stages of their life cycle (Bajguz, 2007). BRs have recently been discovered and represent the sixth class of phytohormones after auxins, gibberellins, cytokinin, ethylene, and abscisic acid. Although the imperative role of BRs in plants has still not been unveiled, it is known to play an indispensable role in the mitigation of stressful environmental conditions.

BRs are polyhydroxylated sterols found endogenously in plants and are known to stimulate a variety of physiological and developmental plant processes. However, these steroidal phytohormones were isolated from the pollen grains of *Brassica napus* for the first time by Mitchell et al. in 1970. The organic extracts of the pollens of this rape plant were found to enhance cell division and stem elongation in *Phaseolus vulgaris*, thereby leading to the discovery of this steroidal plant growth regulator. Brassinolide (BL) was the active constituent that was actually isolated from the pollens and was also reported to enhance seed viability and crop yield. Studies on the phytohormone since its discovery revealed its presence in other parts of plants, which include buds, seeds, leaves, roots, and fruits (Bajguz and Piotrowska-Niczyporuk, 2014). However, their quantity in these different parts varies in age, that is, younger parts of the plant, such as pollen grains, have been found to consist of high concentrations of active BRs as compared with mature tissues such as leaves and shoots (Takatsuto, 1994).

BRs have been reported in both free as well conjugated forms in different plant parts. Being steroidal in nature, conjugated forms ensure their stability in a hydrophilic plant environment. They are known to exist in the form of glucosides, sulfates, and in conjugation with fatty acids and sugars (Hayat et al., 2003; Bajguz and Hayat, 2009). These different forms of BRs

DOI: 10.1201/9781003110651-19

regulate numerous growth and developmental activities such as regulation of senescence, sterility, and stress conditions. Until now, more than 70 different forms of BRs with varying functional aspects have been discovered and isolated from different plants (almost 64 different species) in their various parts at different stages of development (Sadura and Janeczko, 2018). Diverse plant species ranging from thallophytes (*Chlorella Vulgaris, Cystoseira myrica*, and *Hydrodictyon reticulatum*), bryophytes, pteridophytes (one), gymnosperms (six) to angiosperms (12 monocots and 41 dicots) are known to abandon these BR analogs. However, among the 74 analogs, about 52 different forms have been reported to be bioactive in controlling numerous plant activities. Research evidence suggested the role of these numerous forms in the regulation of the antioxidant defense system (ADS), stress specific signaling systems, and various interactions with other phytohormones in order to maintain homeostasis in plant growth under the conditions of biotic and abiotic environmental stress (Ye et al., 2017; Kim et al., 2012; Fariduddin et al., 2014; Zou et al., 2018; Rajewska et al., 2016). BRs have been found to be very similar in structure to ecdysteroids, which represent a group of steroidal hormones found in different phyla of animals, that is, arthropods as well as androgens, estrogens, and corticosteroids from vertebrates (Thummel and Chory, 2002).

Recently, a number of advanced molecular and genetic assays paved the way for the study of various analogs, biosynthetic pathways, and mechanisms of physiological action of BRs in plants (Camoni et al., 2018). The basic structure of BR resembles a natural steroidal skeleton and consists of four fused rings, namely, A, B, C, and D. Each ring consists of hydroxyl groups and other alkyl substituents in their side chain. Based on the number of carbon atoms and alkyl substituents in the side chain, all the BRs can be broadly classified into three different categories: C-27, C-28, and C-29, each consisting of 27, 28, and 29 carbon atoms, respectively, in their steroidal skeleton (Fujioka and Yokota, 2003). The different analogs of BRs consist of a basic steroidal skeleton but differ in the nature of conjugation in the hydroxyl group present in the side chain as well as in the number of carbon atoms in the alkyl side chain. The most common naturally occurring form of BR, that is, BL, represents a polyhydroxylated derivative of 5α-cholestan, namely (22*R*, 23*R*, 24S)-2α,3α,22,23-tetrahydroxy-24-methyl-B-homo-7-oxa-5α-cholestan-6-one, consisting of a cis-vicinal hydroxyl group at C-2 and C-3, a lactone functional group at C-6, a hydroxyl group in R configuration at C-22/C-23, and a methyl substitution at C-24 (Figure 19.1).

Other active BR analogs differ from BL at C-2 and C-3 positions in the A ring, hydroxyl group configuration in the side chain (R or S), presence of ketone, de-oxo or lactone group at C-6 of the B ring, and occurrence of a methyl or ethyl group at C-24 (Clause, 2011; Figure 19.2). The hydroxyl group present in ring A or in the side chain is subjected to meristoylation, laurylation, or glycosylation in different BR analogs. The most studied BRs analogs include BL, castasterone (CS), 24-epibrassinolide (24-EBL), 28-homobrassinolide (28-HBL; Figure 19.1), typhasterol (TY), 6-deoxoCS, and teasterone (TE; Bajguz and Tretyn, 2003 a,b). The huge structural diversity in cyclic and side chains of the steroidal skeleton is responsible for significant functional and biologically active conformations of these analogs, allowing them to be eco-friendly and environmentally sustainable tools for crop improvement.

19.2 Applications of BRs in Crop Improvement

The positive and stimulative effects of BRs on the phenotypic, physiological, and metabolic processes of crop plants have been extensively studied by researchers through various modes of their exogenous application as well as in BR mutant varieties of crops. BR mutant crops showed a significant reduction in plant height, fertility, and seed viability. However, these reductive effects were found to be rescued by exogenous application

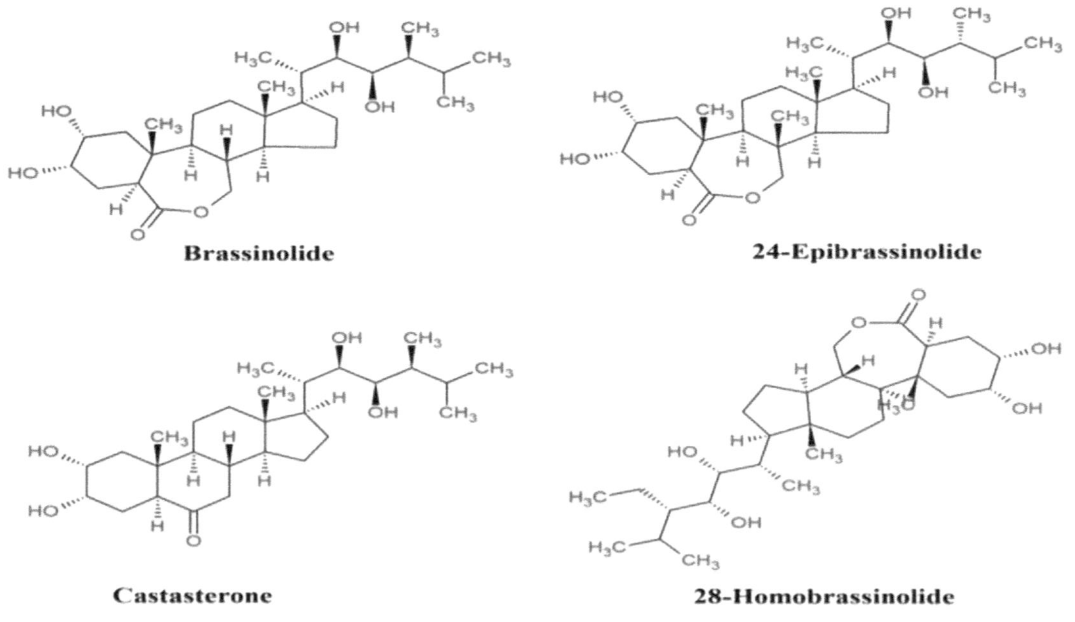

FIGURE 19.1 Chemical structure of common BR analogs found in plants.

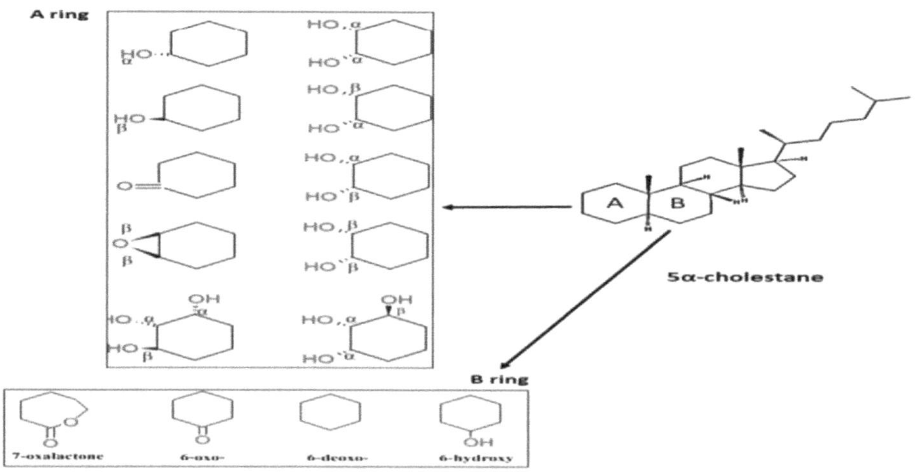

FIGURE 19.2 Various A and B ring substituents present in different BR analogs.

of BRs. Research studies by Farriduddin et al. (2014) revealed the role of BRs in promoting stem elongation, leaf differentiation, cell division, and stomatal activities in different environments. Application of BRs to crop plants can be done at different stages of the plant life span as well as to the fruits after harvesting. The pre-harvest application of BRs has been found to be relatively more beneficial than post-harvest application as the latter only helped in the maintenance of quality, not enhancement (Hayat and Ahmad, 2010). Exogenous application of the bioactive BR analogs may be done in a number of ways such as seed priming, foliar spray, and root treatment. Stimulative effects of BRs were reported in all of these methods of application, whereby the age of the crop plant and the part being exposed were also found to impose significant contributions. Until now, BRs have been utilized for studying their imperative effects on a huge number of agricultural crops, which include basic cereals, oilseed crops, legumes, and horticultural crops and their different varieties. Most studies proved BRs to be excellent non-toxic and eco-friendly compounds for the amelioration of crop health in conditions of abiotic and biotic stresses commonly prevailing due to rapid industrialization.

BRs are potent regulators of crop performance and productivity as they are known to influence a vast array of agronomic traits of crops including plant height, senescence, flowering, adaptations to environmental changes, leaf angle, tillering, and grain size (Divi and Krishna, 2009). BRs enhance seed production in crop plants and cause an increase in the number of seeds per plant. Studies on BR-deficient and insensitive rice plants by Morinaka et al. (2006) revealed their stimulatory effects in the regulation of seed size and weight in respective plants. It has also been found through microarray analysis that enhanced levels of BRs in leaves and stems supplemented photosynthetic efficiency and starch assimilation in rice seeds (Wu et al., 2008). Exogenous application of aqueous BRs to grapes influenced their various physiological attributes such as sugar and phenolic content significantly, and also reduced decay development, which is beneficial for long-term cold storage and transport (Champa et al., 2015). Treatment of another non-climacteric fruit—strawberry—after harvesting with BRs was also found to be advantageous to crop plants as it led to enhanced levels of carotenoids, soluble sugars, and antioxidants (Chai et al., 2013). BRs are also known to impose other diverse benefits to plants such as shortening of vegetative growth, enhanced success of fertilization, fruit quality, number, size, crop, and yield, and adaptation to unfavorable environmental conditions (Khripach et al., 2000). It has been found by Tang et al. (2016) that BRs also influence numerous plant processes such as fruit setting, transition to the reproductive stage, photosynthesis, and seed viability, thereby enhancing crop pomological value. Recently, BRs were also utilized in plant tissue culture techniques, such as micropropagation and somatic embryogenesis proving them as potent compounds for the upregulation of plant development. A fluoro derivative of 28-homocastasterone—5F-HCTS—when applied exogenously to the *in vitro* cultures of apple rootstock stimulated elongation of branches (Pereira-Netto et al., 2006). Similar results were also obtained by Azpeitia et al. (2003) for *in vitro* cultures of coconut where significant enhancement in the growth of callus and the formation of somatic embryos were found by application of BRs. Khripach et al. (1996) unveiled the promoting effects of 24-EBL and 28-HBL on the starch and vitamin C content of potato tubers, thereby enhancing the nutritional as well as economic value. Further, it was also found that these BRs also play a crucial role in the regulation of ion uptake from soil by plants. BR application to crop plants growing in heavy metal contaminated soils was found to be influential to crop plants in not only balancing their normal growth but also in the reduced accumulation of heavy metals and their radioactive counterparts within the plant's root and shoot system. It has been shown by Chaiwanon and Wang (2015) through biochemical and physiological analyses that the treatment of plant *in vitro* cultures with a low concentration of BRs enhanced root growth while the same was inhibited if higher concentrations were used. Similar results were also observed by Kalpan and Gokabayarak (2012) regarding the efficiency of BRs to increase the number of adventitious roots in grape rootstock.

Exogenous application of BRs to tomato and grape plants resulted in increased fruit ripening (Vardhini and Rao, 2002). It has also been reported by Shahid et al. (2011) that treatment of pea plants with 24-EBL augmented seed germination, phenotypic growth, and the elongation of the embryonic axis. At the cellular level, BRs have been reported to promote tracheid formation, resulting in xylem differentiation in the mesophyll cells of leaves of *Zinnia elegans*, thereby contributing to plant overall growth (Zurek et al., 1994). Vast and extensive research studies have been carried out which proved BRs as potent ameliorators of plant growth and development under conditions of various abiotic stresses such as drought, salinity, heavy metal, and temperature stress, as well as biotic stresses, which include infection by crop cellular parasites. BRs help plant survival under such harsh environments through activation of different defense mechanisms such as the primary process of photosynthesis, enhanced content of photosynthetic pigments, and stimulation of enzymatic and non-enzymatic antioxidants and signaling molecules involved in stress tolerance (Mazorra et al., 2011). Rice yield and photosynthetic efficiency were enhanced to significant levels by planting seeds primed with BRs (Sakamoto et al., 2006). It has also been reported by Sayed et al. (2009) that seed priming of *Cucurbita pepo* with low concentrations of BL resulted in an enhanced germination rate as well as metabolic activities. Similar results were also obtained by Symons et al. (2006) in grape fruits where amelioration in endogenous BR levels and fruit ripening were observed with foliar spray with low concentrations of BRs. Further, increased production of ethylene in response to exogenous application of BRs enhanced the ripening of tomato fruits (Vardhini and Rao, 2002). Photosynthesis, being primarily a metabolic process, provides energy to plants to complete their metabolic activities related to development and reproduction, thereby contributing to plant overall vigor and yield. Various research studies have revealed the imperative role of BRs in amplifying photosynthetic efficiency directly by activation of enzymes such as RUBISCO involved in different stages of the process, content of photosynthetic pigments, such as chlorophyll a, b, and also by increasing the D2 protein repair process under the conditions of abiotic stress in different crop plants (Hayat et al., 2011; Swamy and Rao, 2010). BRs also upregulate the nutritional value of the seeds of crop plants by enhancement of efficiency of carbohydrate translocation from source to sink as well as number of grains per plant (Choe et al., 2001).

Research studies by Shi et al. (2006) unveiled the role of BRs in augmenting fiber growth in cotton plants by their exogenous application and it was also demonstrated by treating with an inhibitor of BR which imposed negative effects. Also, Li et al. (2016), through molecular studies, unveiled the functional aspect of BRs in improving the accumulation of carotenoids in tomato fruits, enhancing their shelf life and nutritional value. Foliar spray of watermelon plantlets with even very low doses of BRs escalated their nutritional composition such as total soluble sugars and phenols as well as overall yield (Susila et al., 2012). Similarly, BRs are also known for their contribution to economic markets as they supplement various edible and exported crops with more nutritional value and vigor and yield. These results were observed by Salazar et al. (2016), who found up to a 35% increment in fruit yield of marketable yellow passionfruit and strawberry, thereby adding to the economy. BR-primed Lucerne seeds showed significant enhancement in phenotypic traits such as overall plant length, physiological attributes such as fresh and dry weight of roots and shoots, and root health. BRs are also known to reduce lipid peroxidation and MDA content under stressful environmental conditions, thereby contributing to plant survival under such conditions. However, reactive oxygen species (ROS) produced by plants diminish their growth and BRs have been found as potential scavengers of the same by upregulation of various enzymatic and non-enzymatic antioxidants involved in ROS clearance (Zhang et al., 2007). Xu et al. (2015) reported the escalating effects of BR treatment on the activities of various enzymes involved in fruit ripening and sweetening in grape berries, increasing the edibility of the fruit and hence the market price. Moreover, BRs also show health benefits to the plant itself such as increased disease resistance, delayed senescence, and reduced decay of jujube fruits when treated with even low doses of BRs (Zhu et al., 2010).

The ameliorative effects of BRs on overall plant growth and development have proved to be advantageous not only to plants but also to the life dependent on them for survival. In addition, BRs are non-toxic and environmentally friendly steroids, enhancing their utilization in the agricultural sector. The positive effects of BRs on the nutritional and economic value of the major crop plants on a large scale are discussed in the following sections.

19.2.1 Cereal Crops

Cereals are considered the most important food crop. They provide food as well as economic security to people worldwide. Cereals belong to the family of Poaceae (Gramineae), which comprises the grass members and the starch grains hold economic importance. The members of the family include rice, wheat, maize, millets, rye, sorghum, and barley. Owing to their enormous dietetic benefits, these staple crops are grown globally by humans in order to meet economic and nutritional demands. Major cereals like maize, wheat, and rice contribute 57% of the total energy output produced agriculturally (Ray et al., 2012). The maximum energy obtained from cereals can be attributed to their high starch content along with proteins, fats, vitamins, and minerals. Along with being an excellent food for the human population, cereals can also act as food for livestock, which are sources of meat, milk, and other dairy products. Owing to the large benefits of cereals, any decline in their productivity or nutritional quality cannot be ignored. However, exploitation by synthetic means will eventually deplete the environment. Therefore, eco-friendly approaches such as the use of potential phytohormones like BRs should be encouraged to maintain or improve the value of cereals under normal as well as unfavorable environmental circumstances (Figure 19.3).

19.2.1.1 Photosynthetic Pigments

Plants perform versatile functions of converting the energy of the sun into biologically active molecules, that is, carbohydrates, which are further made available to the organisms on higher trophic levels. Therefore, in order to sustain energy in the ecosystem, the plants need to perform the function of photosynthesis efficiently. Photosynthetic pigments play a major role in the process by absorbing light. However, plants

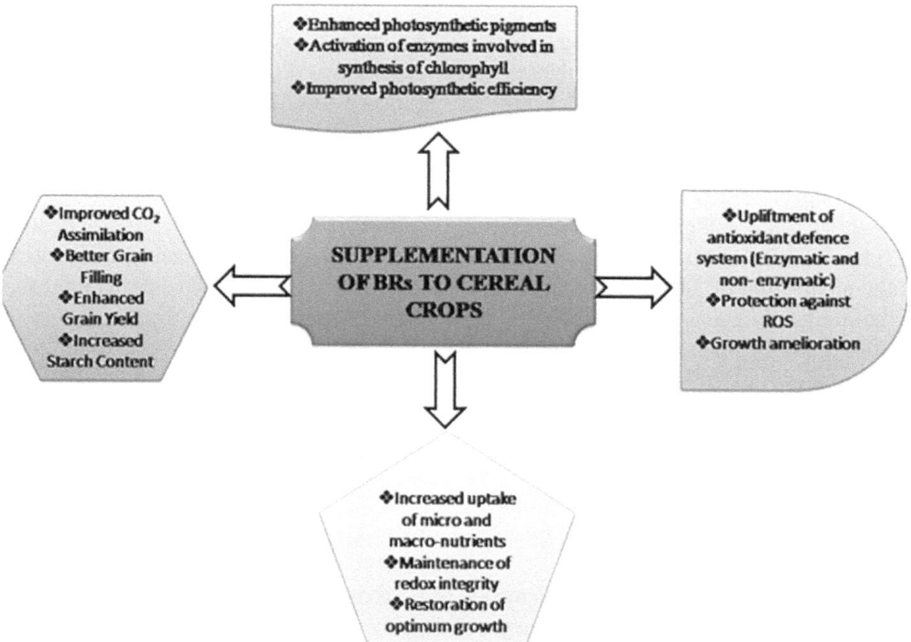

FIGURE 19.3 BRs and the enhancement of the nutritional value of cereal crops.

encountering any of the unfavorable conditions prevailing in the environment result in the degradation of the photosynthetic pigments that further hinder the overall process of photosynthesis (Batra et al., 2014). Therefore, it is necessary to maintain the levels of photosynthetic pigments. BRs are known to act as saviors for plants in terms of protecting the photosynthetic machinery, including photosynthetic pigments. BRs keep the levels of photosynthetic pigments maintained by regulating the enzymes involved in their synthesis (Hola, 2011).

BRs are reported to enhance photosynthetic pigments during favorable environmental conditions as reported in *Triticum aestivum* (Braun and Wild, 1984) and *Oryza sativa* (Wang, 1997). The study conducted by Eleiwa et al. (2011) revealed the protective action of 28-HBL in wheat (*T. aestivum*) against salinity stress, where the foliar sprays of 28-HBL protected photosynthetic pigments along with various parameters of growth, thereby elevating the overall yield. Another report in *T. aestivum* suggested that BRs conferred protection to chlorophyll a and chlorophyll b contents when the plants were exposed to a saline environment. However, such enhancement was not reported under normal circumstances by the author (Shahbaz et al., 2008).

The action of BRs has also been studied against salt stress in *O. sativa* where the overall growth of the plants exposed to salt stress was restored by maintaining the levels of chlorophyll (Anuradha and Rao, 2003). In the case of Pusa Basmati-1, a common rice variety, it was observed that with supplementation of 24-EBL, the contents of chlorophyll retained that were degraded due to saline stress (Sharma et al., 2013). In the case of rice, it was observed that heat stress had an adverse effect on the total chlorophyll content, which displayed an overall decline. However, 24-EBL protected the pigments and an increase in the pigment content was observed as compared with the control. This also affected the photosynthetic efficiency of the rice plants (Hussagunpanit et al., 2015). The study also included an investigation of the effect of a BR-mimic, 7,8-dihydro-8α-20-hydroxyecdysone (DHECD), on growth parameters, including photosynthetic pigment content; although a positive effect was observed, it was not more pronounced than 24-EBL. Maize plants that were exposed to toxic concentrations of nicosulfuron exhibited decreased content of photosynthetic pigments that in turn affected the yield. However, the exogenous application of BRs displayed the elevated contents of photosynthetic pigments in order to execute a rescue operation to restore the growth of the plants (Liu et al., 2019). A study by El-Khallal et al. (2009) showed that reduced contents of chlorophyll a, chlorophyll b, total chlorophyll, and carotenoids were ameliorated with the help of application of BL in cases of maize grown in salt-stressed conditions.

19.2.1.2 Carbohydrates

BRs are known to enhance the photosynthetic efficiency of plants, which, in turn, leads to the regulation of carbohydrate production and their accumulation in the plant body. The process of photosynthesis is distinguished into light and dark reactions that eventually lead to carbohydrate synthesis. It is the dark and light reaction of photosynthesis that is targeted by BRs for enhancing carbohydrates in the plants. BRs regulate major enzymes of the Calvin cycle (Siddiqui et al., 2018). The results reported by the studies of Pociecha et al. (2016) in cases of winter rye (*Secale cereale*), which is the major cereal predominantly grown in Central and Eastern Europe, indicated that the treatment of 24-EBL in winter rye modulated photosynthesis and carbohydrate metabolism in order to inculcate winter tolerance. The total soluble sugars as well as the reducing sugars were found to increase in *O. sativa* straw on the supplementation of 24-EBL and DHECD (BR-mimic compound). It was also observed that this exogenous supplementation resulted in an overall increase in the starch content of seed, husk, and straw (Hussagunpanit et al., 2015). Another study on rice reported that BRs induce grain filling, which is one of the secondary effects that BRs produce, as they do not themselves enter into seeds. BRs regulate the movement of sucrose and some other sugars into the seed from the parts of the plant like leaves and the stem. This loading of grains with sugars is dependent upon CO_2 assimilation and BRs are known to participate in such physiological processes (Wu et al., 2008). Xu (2007) has also reported similar results in cases of sorghum where the application of 24-EBL displayed positive results during the grain filling stage. Furthermore, it has also been reported that carbohydrates function as essential substrates or signals for various essential metabolic reactions in plants (Koch, 1996) and BR-deficient mutants result in the faulty carbohydrate metabolism that can directly be related to the impaired activities of various enzymes and genes (Schlüter et al., 2002). Therefore, it can be interpreted that BRs demonstrate positive impacts on CO_2 assimilation that can be correlated with carbohydrate accumulation in cereal crops.

Another report suggested that when 24-EBL supplementation was given to wheat plants at the heading stage, the grain yield was enhanced by 24-EBL and the content of soluble sugars was found to be enhanced (Janeczko et al., 2010). Fujii and Saka (2001) reported that the content of starch was found to be increased in rice kernels with supplementation with BRs.

19.2.1.3 Antioxidants

Any condition of stress that prevails in the environment, even for a short duration of time, will ultimately disturb the plant's metabolism. This, in turn, leads to the outburst of the ROS that triggers various toxic activities such as lipid peroxidation and DNA degradation that eventually interfere with the normal growth and physiology of plants (Mittler, 2002). However, plants have developed certain mechanisms that give them protection against such phytotoxic ROS. They have the inbuilt antioxidant defense system (ADS) which scavenges and neutralizes the ROS produced, however, the upliftment in ADS capacity by eco-friendly can prove fruitful to plants. BRs are such eco-friendly approaches that help plants improve their ADS efficiency. Plant ADS includes two major components: non-enzymatic and enzymatic. The non-enzymatic components include ascorbic acid (AsA), tocopherols, glutathione (GSH), phenolics, and proline, whereas superoxide dismutase (SOD), EC 1.15.1.1, catalase (CAT), EC 1.11.1.6, glutathione reductase (GR), EC 1.6.4.2, peroxidase (POD), EC 1.11.1.7, ascorbate peroxidase (APOX), EC 1.11.1.11, and guaiacol peroxidase (GPX), EC 1.11.1.7 comprise the enzymatic antioxidants. The greater the efficacy of the ADS in plants, the higher the chance of the plants thriving, even if the environmental conditions do not favor the optimum growth of the plant. When wheat plants were under salt-stressed conditions, 24-EBL supplementation helped in ameliorating the toxic effects of the saline environment and boosted antioxidants (Talaat et al., 2013). The study also reported that there was an increase in the concentrations of various osmoprotectants like proline, free sugars, glycine betaine, and free amino acids followed by reduced levels of hydrogen peroxide and the reduced extent of lipid peroxidation.

The foliar applications of 24-EBL helped in enhancing the activities of antioxidants like SOD, POD, and CAT in the case of two different cultivars of *Triticum aestivum* that were exposed to salinity, however, this increase in the activities of SOD, POD, and CAT was highly in a dose-dependent manner (Shahbaz et al., 2008). GSH is an essential detoxifier in plants with a prominent role in scavenging ROS. When maize plants were exposed to the toxic environment of nicosulfuron, it was found that 24-EBL enhanced the content of GSH to detoxify the ROS produced due to pesticide toxicity (Liu et al., 2019). Along with GSH, an enhancement in the activities of SOD, POD, and CAT was also observed in maize plants. In the case of *Sorghum vulgare*, the application of 28-HBL and 24-EBL helped in enhancing the activities of CAT in different varieties. Along with the CAT, levels of proline were also enhanced that act as an osmolyte under drought stress conditions (Vardhini and Rao, 2003, 2005). Yusuf et al. (2011) reported that 28-HBL helped in uplifting the activities of SOD, POD, and CAT in different cultivars of wheat that were exposed to toxic concentrations of nickel. The heat-sensitive and heat-tolerant cultivars of rice were subjected to high-temperature stress; it was observed that BRs helped in enhancing the expression levels of POD and SOD (Cao and Zhao, 2008). In the case of *Zea mays*, the application of BL helped in alleviating the adversities caused by water-deficient conditions and the activities of enzymatic antioxidants. Total protein and proline contents were found to increase (Anjum et al., 2011). Similar results were also obtained by Li and Van Staden (1998a,b) in their studies in *Zea mays* where the activities of SOD, POD, APX, GR, and CAT were enhanced by BL. The results of Bhardwaj et al. (2007) corroborated with the aforementioned studies that under heavy metal toxicity, 28-HBL uplifted the ADS in maize.

19.2.1.4 Micro- and Macronutrients

Plants require certain essential bioactive molecules for their proper growth and development. They acquire these nutrients from the soil. These nutrients must be obtained in the optimum quantity as any less will result in deficiency diseases and more will result in catastrophic toxicity symptoms. Macronutrients are required by plants in a higher amount as compared with micronutrients or trace elements that are required in lesser quantities. Macronutrients include carbon (C), oxygen (O), hydrogen (H), nitrogen (N), phosphorus (P), potassium (K), sulfur (S), magnesium (Mg), and calcium (Ca), while micronutrients include copper (Cu), cobalt (Co), selenium (Se), boron (B), zinc (Zn), chlorine (Cl), manganese (Mn), iron (Fe), molybdenum (Mo), and nickel (Ni). These elements serve functions to the plant body as they act as structural elements of cells, maintain the osmotic potential of the cell, form essential components in enzymes, and help in the activation of various enzymes and metabolic reactions. Therefore, they are required by the plants in optimum concentrations in order to sustain their growth and development. Sometimes, due to environmental constraints, there may be the possibility of reduced uptake of such elements from the soil, which may hinder the normal processes of growth and development. BRs can be incorporated to help plants tackle such circumstances. They help the plant to uptake these nutrients not only during stressed environmental conditions but also during normal circumstances as better incorporation of such nutrients will result in better yields and quality of crops like cereals in order to meet the nutritional demands of the population.

BRs help in maintaining the integrity of the transport channels for the regulation of the uptake of ions, which, in turn, maintains the integrity of the plasma membrane (Khripach et al., 2003). In *Triticum aestivum*, it was observed that due to saline conditions, there was reduced uptake of elements like N, P, K, Ca^{2+}, and Mg^{2+} as a result of disturbed ionic balance. Under such circumstances, the application of 24-EBL improved the uptake of the aforementioned elements along with protecting the plant from saline adversities (Talaat et al., 2013). Salinity stress also disturbs the ionic balance in the plants as it increases the Na^+ ion concentration that hinders the entrance of K^+ ions into the plant body. It has been reported the 24-EBL supplementation results in maintaining the K^+/Na^+ ratio along with enhancing the Ca^{2+} ion uptake by wheat cultivars (Ali et al., 2006). Reduced uptake of Ca^{2+} and K^+ in a saline environment has also been reported by Shahbaz and Ashraf (2007) in the case of *Triticum aestivum*, but on 24-EBL supplementation, the optimum ratio of various anions and cations were restored in wheat plants. Fe is another essential element that is required by crops like cereals for the operation of various metabolic processes. BRs are known to regulate the concentrations of Fe under both deficient as well as excess circumstances in order to maintain Fe homeostasis (Wang et al., 2015). El-Khallal et al. (2009) reported that the content of essential elements like N, P, K^+, Ca^{2+}, and Mg^{2+} was enhanced during salt-stressed conditions in the case of *Zea mays* when the plants were subjected to treatments of BL in the form of pre-soaking as well as foliar spray. Thus, the maintenance optimum amount of K^+ and exclusion of Na^{2+} from shoots with the help of BRs help the plant to counteract the toxic concentrations of various anions and cations that arise due to salinity.

The entry of essential nutrients is also blocked due to the presence of various heavy metals in the soil that tend to compete with essential elements and do not allow their entry into the plant. BRs are also known to stop the entry of harmful elements and allow the plant to maintain its redox integrity. Kroutil et al. (2010) showed protection in wheat plants against cadmium (Cd), lead (Pb), Zn, and Cu toxicity. Similarly, Archana et al. (2006) studied the effects of BR against Ni toxicity. Another approach that can eventually enhance the levels of minerals in plants is symbiotic association. Almost all plants are involved in a symbiotic association with different micro-organisms, the most prominent example being rice. Various symbiotic relations help plants to acquire minerals from the soil by enhancing the surface area for absorption and allowing them to survive in nutrition deficient soils (Liao et al., 2018). BRs have also been reported to enhance this association among plants in order to facilitate the movement of micro- and macronutrients from the soil to plant parts (Bitterlich et al., 2014). Similar reports have also been documented by McGuiness et al. (2019).

Many reports explore the role of BRs in cereals for protection as well as improving their nutritional status. All the aforementioned reports suggest that BRs can be exploited as promising candidates for enhancing the carbohydrate content of cereal crops by maintaining their photosynthetic efficiency. This is evident from reports that suggest that BRs helped in improving as well as restoring degraded photosynthetic pigments that were once lost due to environmental cues. Furthermore, BRs also regulate various enzymes involved in the Calvin cycle. The activities of BRs have also been explored in maintaining the balance between micro- and macronutrients that help in the continuation of growth and developmental processes in cereals. Therefore, BRs can be utilized as eco-friendly approaches to bio-fortify cereal crops for human consumption and elevate their economic importance.

19.2.2 Oil Crops

Oilseed crops are the backbone of the agricultural economy of India. These crops and their products are the second most sold commodities in world trade. Besides income for farmers, oil crops provide an inexpensive source of high-quality nutrition (Savage and Keenan 1994). Triacylglycerol (TAG) is the major form of oil in plant species. Plant oils have a great use in the agriculture industry as they are principally utilized as edible oil in the processing and preparation of food. For different industrial applications, plant oil also serves as a renewable resource that can be easily extracted and utilized in various industrial applications. Due to its structural similarity with the long chain hydrocarbons, TAG represents an appropriate and competitive replacement for the hydrocarbons created by different products. Therefore, relying on the physiochemical properties contained by fatty acids, plant oils are thus utilized in the production of biodiesel and also as an industrial raw material for the synthesis of a wide array of products such as lubricants, paints, coatings, and inks (Cahoon et al., 2007). The majority of plant oils are produced from four main crops: soybean, oilseed rape, oil palm, and sunflower (Dyer and Mullen, 2008). Principally, in developing countries, the demand for oil in food and feed progressively amplifies due to rapidly upgrading standards of living, the growing population, and changing diets (Marchive et al., 2014). Similarly, its inclusive consumption and requirement for biofuels production have been growing by around 5% every year for the last half century (Samarth and Mahanwar, 2015). By 2050, the worldwide requirement for vegetable oil is expected to be doubled in comparison with current production. Agriculture lands are relatively finite, thus augmenting the production of oilseed crops has turned out to be a challenge (Harwood et al., 2013). Uplifting the content of oil in oilseed crops without expanding the area of cultivation is a suitable way to enhance the oil yield without using the additional crops needed to fulfill imminent global oil scarcity. In order to boost the yield of oilseed crops, several elements on which yield comprises need to be amplified (Bhat, 2010). However, the regulation of oil content is a complicated process that occurs at various oil biosynthesis and growth phases (Guschina et al., 2014). Although India is the world's major producer of edible oils, nevertheless, it imports large quantities of oils (~ 1.2 million tons). The main reason for India's large edible oil deficit is the deprived yield of oilseed crops per hectare, chiefly caused by the accessible cultivars being susceptible to several diseases and different abiotic stress conditions, as well as poor farm practices. For that reason, it is important to assess the potential of these crops in *in vitro* conditions with new combinations of PGRs for regeneration and biochemical manipulation

BRs are plant-produced compounds structurally analogous to animal steroid hormones that function as growth regulators (Khripach et al., 2000; Haubrick and Assmann 2006). Exogenous appliance of BRs affects a wide spectrum of physiological responses such as vascular differentiation, cell expansion, seed germination, reproductive development, flowering, and fruit set in plants (Cao et al., 2005; Yu et al., 2004; Monotoya et al., 2005). Apart from these growth-promoting effects, BRs are also accounted to confer resistance against various abiotic and biotic stress conditions like drought, heavy metals, heat, salt, infections, pesticides, and even viruses (Krishna 2003; Wachsman et al., 2004a; Bhardwaj et al., 2007; Sharma and Bhardwaj 2007). Nunez et al. (2003) and Ozdemir et al. (2004) observed that resistance to these stress conditions involves the regulation of antioxidant enzyme activities under the control of these plant growth regulators. Although different components like antioxidants, secondary metabolites, nutrients, and so on are present in crops, various innumerable works have assured the potential of plant hormones to improve crop performance synergistically (Figure 19.4) with augmentation of these components under normal and abrupt conditions.

19.2.2.1 Unsaturated Fatty Acids

In most plant species, the predominant unsaturated fatty acids (UFAs) are the three 18-carbon (C18) species that are 18:1 (oleate), 18:2 (linoleate), and 18:3 (α-linolenate) (Harwood 1988). These simple compounds play numerous critical roles and are associated with both biotic and abiotic stress conditions. Besides acting as membrane ingredients, the C18 UFAs have a role as intrinsic antioxidants, precursors of different bioactive molecules (typically the stress hormone jasmonic acid and brassinosteroids), and stocks of extracellular barrier components such as cutin and suberin (Ohlrogge and Browse, 1995; Harwood, 1996). Moreover, these UFAs also play regulatory roles in plant defense (Lim et al., 2017). Different changes in the fatty acid composition of membrane lipids in the response to low-temperature stress have been described for some plant species (Willemot 1979; Johnson and Williams 1989; Bakht et al., 2006). Theoretically, unsaturation of the fatty acids ameliorates as a result of cold treatment. However, dependency with a high rate on the species and their environmental requirements may exist (Yoshida and Uemura 1984). In winter oilseed rape, an enhancement in the content of linolenic acid is usually accompanied by a decline in the linoleic and oleic acid content in hypocotyls and in their subcellular fractions (Skoczowski and Filek 1986, 1994). If we do a comparison, in the cold-induced changes in unsaturated fatty acid composition in oilseed rape plants with those obtained in this research for oilseed rape calli, there were generally similar effects observed. Here, the intracellular action of BRs may be functionally allied with the phospholipases that control several plant growth reactions (Wang et al., 2012). These phospholipases hydrolyze the membrane lipids to produce the second messengers that are essential for regulatory signaling. With the application of BRs, phospholipases also contribute to the generic lipid metabolism and lipid turnover (Li et al., 2006) counting TAG biosynthesis (Lee et al., 2011). Additionally, phospholipases are well known to play a part in stress and hormone signaling (Janda et al., 2013). As an example, production of phosphatidic acid (PA) via phospholipase D activity was rapidly induced in salt-stressed conditions with BR treatment (Pokotylo et al., 2012), whereas overexpression of

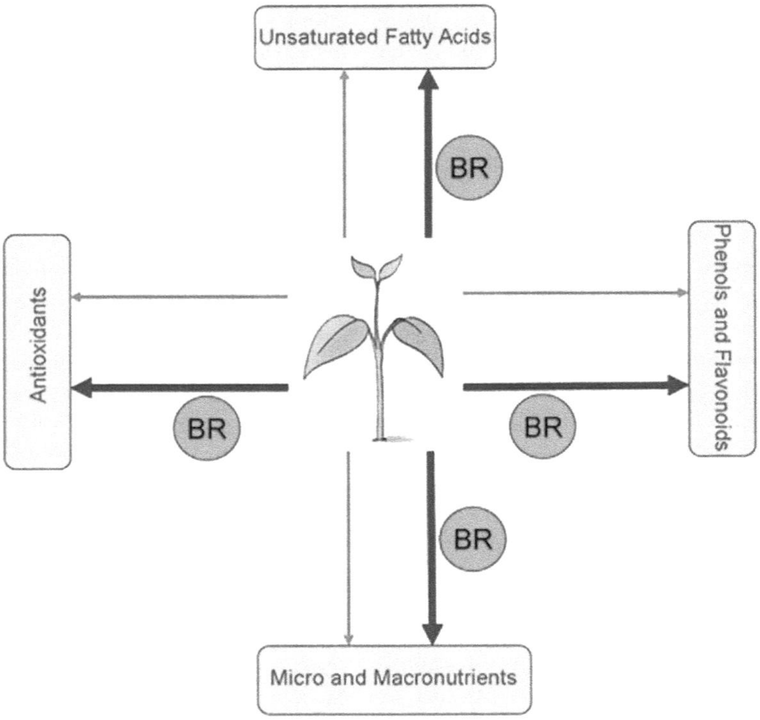

FIGURE 19.4 BRs and the enhancement of the nutritional value of oilseed crops.

phospholipase D resulted in improved drought and salt tolerance of plants (Peng et al., 2010; Zhang et al., 2008). It is interesting to know that plant non-specific phospholipases C (NPC) are able to directly cleave structural membrane phospholipids and produce diacylglycerol (DAG; Pokotylo et al., 2013), which is activated after BR treatment in BY-2 cells (Wimalasekera et al., 2010). BRs are known to enhance UFA quantity and seed oil accumulation. For example, the end-point enhancement in oil content induced via tissue-specific photosynthesis in the silique walls can be imitated by application of BRs (Hua et al., 2012). UFA composition is an influential factor in oil quality and usability both for biodiesel production and food purposes.

19.2.2.2 Phenols and Flavonoids

Secondary metabolites are also referred to as secondary products or natural products. These are organic compounds that are not directly involved in growth, development, or reproduction. These secondary metabolites do not even directly contribute to respiration, protein synthesis, nutrient assimilation, translocation, photosynthesis, and differentiation. But the absence of secondary metabolites may increase long-term harm to the plant's survivability or also immediate death. These compounds are more complex than primary metabolites. Such compounds in general include pigments, effectors of ecological competition, antitumor agents, and molecules of the plant's chemical defense like alkaloids and steroids. Primary metabolites are present in all living dividing cells while secondary metabolites are found infrequently and these are not of much significance for primary plant life. Generally, secondary metabolites present at low concentrations in living organisms, whereas these compounds are extensively distributed in plants and play a crucial role. In recent times, these compounds have developed into the subject of rising interest for their significant practical use for medicinal, nutritive, and cosmetic purposes, and their indisputable role in plant stress physiology. Plants are devoid of motility and an immune system, therefore, they have developed alternative defense strategies involving an enormous variety of secondary metabolites serving as tools to overcome stress constraints, an adaptation to the changing environment that helps in survival in suboptimal conditions (Edreva et al., 2008). There is a very delicate balance between the generation and destruction of the oxidant agents that may be beneficial or deleterious to plants. Under different physiological conditions, the level of ROS is kept low via the coordinated activity of antioxidant defense systems that include secondary plant metabolites and scavenging enzymes. Many *in vitro* studies have shown that various compounds like flavonoids, sesquiterpenes, coumarins, and phenolic acids have significant antioxidant activity. Phenols and flavonoids have been studied broadly for their antioxidant activities (Khan et al., 2010). Transcriptional regulation has long been acknowledged as a major determinant of phenotypic variation (Wray et al., 2003). Using standard molecular biology techniques, the coordinated expression of different secondary metabolite synthesizing enzymes has been described in numerous reports such as in flavonoid synthesis during seedling development (Pelletier et al., 1999) and in stress inducible indole metabolism (Zhao et al., 1998). Similarly, key enzymes of the phenylpropanoid pathway, named phenylalanine ammonia-lyase (PAL), 4-coumaroyl CoA ligase (4CL), and coumarate-4-hydroxylase (C4H) isoforms, are expressed in a very coordinated manner under various stress conditions (Koopmann et al., 1999). However, these results of previous studies on secondary metabolism necessitate being reinterpreted in light of the abundant duplication of underlying genes (Bowers et al., 2003). Numerous different secondary metabolites have been identified in the approximate 500,000 higher plant species found globally. Here, as the name clearly indicates, secondary metabolites are considered the side products of plant metabolism. This approach reflected the lack of knowledge in the scientific community, further ignoring the fact that cellular metabolism is a complicated result of the relationship of a plant with its environment.

The synthesis of phenols and flavonoids in plants was increased in those raised from EBL-primed seeds by maintaining tissue osmotica. This accumulation of secondary metabolites can help plants in several ways for better amelioration of the deleterious effects of stress (Pourcel et al., 2006). These can avert stress-triggered decline in the chlorophyll and help in the maintenance of nitric oxide homeostasis for better stress amelioration (Klein et al., 2015). Therefore, plants demonstrating greater accumulation of secondary metabolites show rapid stress acclimation (Farooq et al., 2019). It was studied that seed priming with EBL extensively improved soybean performance. The external application of EBR enhanced the phenolic compounds and pigments in these plants (Sharma et al., 2013; Sharma et al., 2016a). In leaves of *Brassica juncea*, ten phenolic compounds were detected viz., caffeic acid and umbelliferone (hydroxycinnamic acid), gallic acid (phenolic acid), chlorogenic acid (ester of caffeic acid), tert-butyl hydroquinone (naphthoquinone), ellagic acid (stilbene), rutin (flavonol glycoside), catechin, epicatechin, and kaempferol (flavonoids), and were classified according to Harborne (1980). Here, the total polyphenol content increased significantly in plants grown from seeds soaked with 24-EBL (Sharma et al., 2016b)

19.2.2.3 Antioxidants

In order to diminish the damaging effects of ROS, plants upregulate the activities of enzymatic antioxidants such as SOD, POD, CAT, and enzymes of the ascorbate-glutathione (AsA-GSH) cycle (Ahmad et al., 2018a). SOD catalyzes the translation of superoxide radicals (O^{2-}) into hydrogen peroxide (H_2O_2) and molecular oxygen (O_2). H_2O_2 produced here is also very deleterious for the cell and by the action of CAT this H_2O_2 can be converted to water (H_2O) and molecular oxygen (O_2; Ahmad et al., 2010a, 2010b; Rasool et al., 2013). This enzymatic antioxidant system is further supported by non-enzymatic components including ascorbic acid, glutathione, polyphenols, and osmolytes (Ahanger et al., 2017a; 2017b; Ahmad et al., 2018b). To increase crop production, scientists are working on long-lasting solutions to this problem. One of the sustainable solutions is the external use of plant hormones to enhance crop production without any damage to the plants. Among the plant hormones, BRs are known for their function in growth, development, productivity, and increased resistance against abiotic stress conditions (Sharma et al., 2011; 2013; 2014) Plants with external EBR treatment showed improved growth, pigments, photosynthetic efficiency, and activities of antioxidants (Kanwar et al., 2012; Sharma et al., 2013; Vardhini and Anjum 2015; Xia et al., 2015; Sharma et al., 2016b; Sirhindi et al., 2017). Brassinosteroids also interact with other various cellular molecules to amalgamate the tolerance networks for superior stress mitigation (Ahanger et al., 2018). Exogenous application of EBR has been well established to evade the harmful effects of stress conditions by upregulating the antioxidant system (Ahmad et al., 2018a). Soybean (*Glycine max*) is cultivated for its higher protein source. It is also used as a source of oil and livestock feed. EBR treatment helped to enhance the antioxidants in soybean that boosted growth (Alam et al., 2019). This was further revealed that treatment of 24-EBL significantly amended the activities of antioxidant enzymes and protein content plays a major role in imparting defense to plants against stress conditions. These alterations in the activities of antioxidative enzymes are a common response to ROS produced by several abiotic and biotic factors (Arora et al., 2002). Exogenous application of 24-EBL to the treated plants caused the scavenging of ROS efficiently.

An increase in CAT activity after treatment with BR had been reported earlier in groundnut (Vardhini and Rao 2000) and *B. juncea* (Sirhindi et al., 2009; Arora et al., 2010). It may be concluded from the augmentation in CAT activity that BR application modulated the antioxidant metabolism resulting in better growth of plants (Vardhini and Rao 1998a; 1998b). In response to the application of BRs, a rise in POX activity has also been reported (Sharma et al., 2007; Sirhindi et al., 2009; Arora et al., 2010). This POX has been implicated in the production of lignin and phenolic polymers. The improvement in POX activity might protect cells against destructive concentrations of hydroperoxides. Sirhindi et al. (2009) also practically found a raise in the activity of POX after exogenous appliance of 28-homobrassinolide in the seedlings of *B. juncea* L. The activity of APX was found to be reliable with the former reports in which APX activity was modulated after the application of 28-homobrassinolide (Kartal et al., 2009), 24-epibrassinolide (Mazorra Behnamnia et al., 2009; Arora et al., 2010), and brassinolide analog BR-16 (Nunez et al., 2003). Experimental evidence indicated that an *in vitro* effect of BR was involved in the mitigation of not only antioxidant capacity but also biochemical parameters. Nonetheless, there were some differences in the responsive behavior of two groundnut cultivars in the presence of BR that may be attributed to genetic differences.

19.2.2.4 Micro- and Macronutrients

Micronutrients and macronutrients have proven to be indispensable components for the proper growth, development, and productivity of plants as they play noteworthy roles in controlling various reactions in plants in severe conditions and in the management of diseases. Nevertheless, there is not a specific rule that specific nutrients can reduce the severity of disease or any harm. Despite these facts, the significance of nutrients in controlling plant growth has been widely recognized but accurate nutrient management to control several conditions in sustainable agricultural practices has regrettably received little attention. Nutrients are well known to affect the tolerance and full resistance against various mischief conditions. The macronutrients of plants are known to play a crucial role in the appropriate growth and development of plants because their major functions range from being structural units to redox-sensitive agents. Out of the different macronutrients, potassium is enormously mobile and is essential in plants, and is plentifully found in the younger parts of plants. Likewise, calcium is another fundamental element of a living organism and is specifically required as the calcium ion participates and helps in various cellular processes. Calcium is a vital constituent for plant growth and development and also entails enzyme activation in salt balance and water movements in the cells of plants, thereby activating potassium in order to control the opening as well as the closing of stomata. Further, magnesium is the central metal atom of chlorophyll, which plays a significant role in the photosynthesis of plants. Deficiency allows the degradation of chlorophyll with yellowing of leaves, also known as chlorosis, however, adequate magnesium availability keeps the plant healthier. Here, sulfur is the most beneficial plant element for proper growth, development, and survival of plant life (Tripathi et al., 2014). Some essential components that are necessary for very small quantities by plants are called micronutrients. The quality, as well as yield of agricultural products, are enhanced with amplification in the concentration of micronutrients, thus further animal health and human life are included within the feed of enriched plant materials. All essential plant elements in particular perform their nutritional role in a well-balanced ratio, which is required for the proper growth and development of various species (Tavakoli et al., 2014). It is essential to maintain nutrient availability via fertilizers by changing the soil environment with the use of plant hormones, which significantly influence nutrient availability (Dordas 2008). In recent times, the majority of growers believe that a focus on plant growth regulators like BRs should be utilized in order to control the diseases of plants as well as making plants tolerant against various stress conditions. In addition to the economic consequences, utilization of BRs augments food safety and environmental concerns (Fageria 2016). Fe is indispensable for chlorophyll biosynthesis and functions in the chloroplasts. Fe deficiency aggravates the negative effects on photochemical efficiency and electron transport. Thus, 24-epibrassinolide plays a role as a natural molecule with potential advantages including a natural origin, high plant steroid bioactivity, biodegradability, and improving metabolism. Fe is the micronutrient most absorbed by higher plants (Baker et al., 2003). This metal is implicated in the synthesis of chlorophylls and is crucial for the maintenance of the structure and function of chloroplasts (Rout and Sahoo 2015; Chen et al., 2015) and in catalyzing the enzymatic reactions linked to respiration, synthesis of DNA, and redox metabolism (Barberon et al., 2014; Xiong et al., 2014). But when this micronutrient is not presented to plants, frequent detection of the yellowing of upper leaves, a lower growth rate, and chlorosis in the leaf nervure occur (Jeong and Connolly 2009); 24-epibrassinolide treatment boosted Fe content (Bishop and Koncz 2002; Zhiponova et al., 2013).

19.2.3 Leguminous Crops

Leguminous crops are one of the most important nutritional food sources being utilized by humans after cereals. They serve as a sustainable and standard source of vital nutrients such as proteins, fibers, carbohydrates, and certain vitamins and minerals. Being very low in fats, they are utilized on a large scale worldwide for therapeutic and pharmaceutical benefits. However, large bioactive compounds have been discovered from the various parts of leguminous crops that are used in the treatment as well as prevention of various disorders and syndromes such as osteoporosis, diabetes, cancers, heart diseases, and other degenerative diseases. Malnutrition is one of the biggest dilemmas in densely populated countries of Africa and Asia and legumes, as part of a diet, are very helpful in the eradication of protein deficiency at easily affordable prices.

Legumes belong to the family "Leguminosae" which comprises a large number of species of herbs, shrubs, and trees. But among the huge diversity of legumes, few are consumed as part of food. The common legumes consumed as a part of food include peas, grams, pulses, soybean, lentils, peanuts, and lupins. In different parts of the world, they are eaten as part of the diet in varying forms such as raw salad, sprouted food, and cooked food. Seeds of certain legumes are rich in oil content and hence their oil is utilized for various beneficiary purposes as these possess various pharmaceutical effects. Most of the legumes possess anticarcinogenic, hypocholesterolemia, and hypoglycemic properties and also have proven to be a substantial source of income for farmers.

The cultivation of legumes is now becoming a common practice among the agriculturalists of the world. As every farming practice faces huge challenges, legume cultivation is also not spared from it. The presence of vast abiotic stresses, such as drought, salinity, heavy metal contamination of soils, and temperature fluctuations, in the environment results in a deterioration of crop quality as well as quantity. At the same time, different types of exercises are being adopted by agriculturists to respond to the detrimental effects of these stresses on crops in order to maintain the nutritional health of the crop. The use of various synthetic chemicals has been undertaken for a long time, but such chemicals have proven to deteriorate soil health and hence making the soil unfit for long-term use. Therefore, recent approaches utilize non-toxic and sustainable compounds that may provide benefits to the crop as well as maintaining environmental balance by avoiding polluting it. BRs are also included in the category of such compounds being exploited to uphold crop growth and development under stressful environments. Different bioactive forms of BRs when applied exogenously to the legume seedlings have been found to be advantageous in maintaining their normal survival under stressed conditions and upregulating their nutritional value in terms of enhanced protein composition, antioxidants, and nutrient composition under normal conditions (Figure 19.5).

Various instances of such benefits to the crop by BR treatment are discussed in the following sections.

19.2.3.1 Protein Content

Among the various types of foods being utilized, legumes are the richest source of good quality protein and consist of about 35-45% of proteins. The protein content is approximately double that of cereals and oilseed crops. Most of the proteins present in legumes are rich in lysine amino acid, one of the most abundant essential amino acids in the cellular proteins of human beings. The high protein content of leguminous crops is mainly due to its symbiotic association with certain nitrogen-fixing bacteria called diazotrophs in their roots and stem. These diazotrophs are well known for their potential to convert atmospheric nitrogen into ammonium ions, which are further incorporated as part of amino acids through the process of reductive transamination occurring in different parts of these legumes. However, certain nitrate ions being absorbed by these plants from the soil are also reduced to ammonium ions and hence incorporated into amino acids. These amino acids finally serve as the building blocks of proteins and hence make the crop peculiar for its protein composition.

BRs, steroidal phytohormones, are known for their influential effects on both fresh matter, i.e., protein content, as well as on the process of nodulation in legumes, which indirectly contributes to the high protein value of the crop. Bassiony et al. (2012) observed the ameliorative role of very low doses of BR on free amino acid content, overall protein content, and quality of pods of two cultivars of *Phaseolus vulgaris* namely *Oxzira* and *Paulesta* commonly known as snap or green beans utilized in the form of a sprout salad on a large scale.

Also, a number of research studies indicated the stimulatory effect of exogenous application of BRs on the process of nodulation in legumes (Yusuf et al., 2012). Cheng et al. (2017) and Ferguson et al. (2005) suggested through their independent research studies that the number of nodules in roots were decreased to significant values in the BR receptor and biosynthesis mutant plants of *Medicago truncatula* and *Pisum sativum* (pea), respectively, in comparison with wild-type control plants. The white-colored nodule formation was observed in *M. truncatula* plants, demonstrating them as non-functional. However, such a phenotype was not distinguished in the pea plants. The mechanism of action of BRs on nodulation number is suggested to be through ethylene. BR-deficient mutants of pea plants consisting of low nodulation were found to consist of elevated levels of ethylene and, on the other hand, ethylene receptor mutants consisted of a high nodulation phenotype (Foo et al., 2016). Similar approaches by other researchers also suggested that BRs influence nodule formation at different stages in different parts of the plant. Also, BRs act as a shoot-derived signal to influence nodule number as shown by the higher number of nodules in plants with a wild-type shoot system than those with a shoot system producing low levels of BRs, regardless of the amount of BR production by the roots (Ferguson et al., 2005). A more comprehensive explanation of this was provided in research studies by Symons et al. (2008) that showed BRs are immobile in plants. However, they act through a signal, which is translocated from the shoot to other plant parts. BRs are also found to influence nodule function and formation by their action on the autoregulation of nodulation (AON) pathway in an upstream direction as suggested by a study of double mutants (Foo et al., 2014). This systemic effect is yet to be explored further.

The analog 28-HBL was found to be very effective in reciprocating the toxic effects of Ni on the growth and survival of *Vigna radiata* plant through various mechanisms, leading to upregulation of various enzymes involved in nitrogen metabolism such as nitrate reductase, nitrogenase thereby leading to

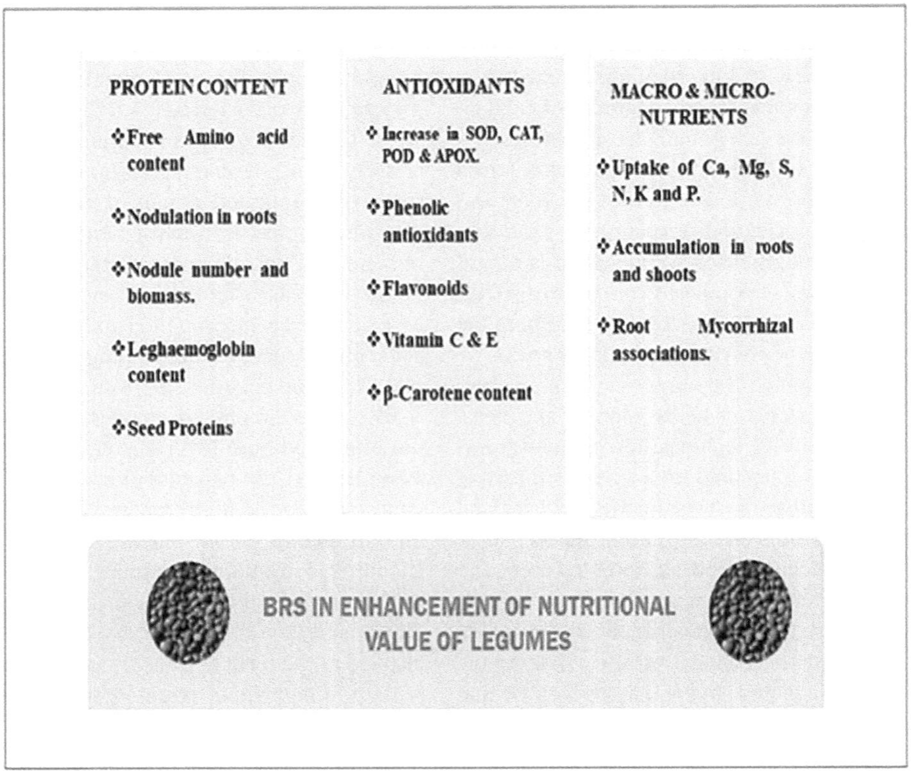

FIGURE 19.5 Role of BRs in the enhancement of the nutritional value of legumes.

amelioration of protein synthesis in the legume crop. Also, the hormone application resulted in enhanced leghemoglobin, nodule formation to significant levels and hence increasing the positive nitrogen balance in the plant and making the crop rich in amino acids and proteins. Thus, BRs not only optimized the stress effects but also ensured the rise in quality and survival of the crop (Yusuf et al., 2014). Similar results were also obtained by Biesaga-Koscielniak (2014) for the rise in protein content in seeds of pea and lupine in the treatment of 24-EBL, making the seeds more effective for nutritional purposes.

Another important legume well known for its high protein content is soybean. BRs are reported to control the growth and developmental processes at different stages in legumes. Major biological processes influenced by BRs include RNA and protein synthesis, root formation, and nodule formation in the plant, which contribute to the overall nutritional value of the crop (Song et al., 2019). Ameliorating the effects of 24-EBL on the protein content of seeds of *P. sativum* was also observed under the conditions of salinity stress (Shahid et al., 2011). Further research studies by Janeczko et al. (2015) demonstrated the influence of the seed priming of 24-EBL in upregulating the protein content and amount of FAA in lupine and pea seeds, thereby enhancing the nutritional value of both the legumes. Foliar spray of BRs to the plants of *Arachis hypogea* resulted in a significant increase in the number of nodules, activity of nitrogenase enzyme, and fresh biomass of the nodules. All of these factors are involved in contributing to the protein content of the crop (Vardhini and Rao, 1999). Similarly, exogenous treatment of plants of *Phaseolus aureus* and *Phaseolus radiatus* with BRs resulted in enhanced levels of RNA and protein in the crops thus adding to their nutritional value (Kalinich et al.1985). Seed priming of *Cicer arietinum* with 28-HBL resulted in an increase in the number of nodules, fresh mass of nodules, nitrogenase activity, leghemoglobin content, activity of nitrogen reductase, and protein content of leaves, as well as seeds under normal as well as salt-stressed conditions (Ali, 2004). BRs have also been reported to mitigate the deleterious effects of herbicide and metal contaminated soils in legumes, ensuring their survival and normal chemical composition under such conditions (Mota, 2020). Hence, it could be established that BR application may prove advantageous to the various leguminous crops in adding to protein content directly or by enhancing the processes that contribute to it, making the crop nutritionally beneficial.

19.2.3.2 Antioxidants

The term "antioxidant" itself explains its meaning: a substance that prevents oxidation. In broad terms, an antioxidant is a substance that can prevent the oxidation of primary metabolites of the body such as carbohydrates, proteins, lipids, and nucleic acids. A large number of ROS are produced as a result of various metabolic activities in both plants and animals. These ROS are responsible for the oxidation of the primary and secondary metabolites that lead to the downregulation of various cellular processes and hence cause degenerative disorders. Therefore, antioxidants play a crucial role in the maintenance of normal body processes by neutralizing these ROS. Under stressful conditions, the accumulation of ROS in both plants and animals may rise, and hence, the functional aspect of antioxidants becomes more prominent under such conditions. Moreover, human beings preferably utilize natural antioxidants in their diet from various food sources obtained from plants as synthetic antioxidants have proved to be toxic to human health. Plants themselves also require antioxidants for their normal functions and also act as a source of the same to animal life. Antioxidants include enzymes as well as non-enzymatic molecules. Enzymatic antioxidants include SOD, CAT, GR, POD, APOX, and other enzymes involved in the DNA repair process. Non-enzymatic antioxidants include vitamin C, vitamin E, β-carotene (precursor of vitamin-A), carotenoids, flavonoids, phenols, and GSH (Sindhi et al., 2013). Legumes being a good source of antioxidants are utilized by humans on a large scale and BRs act as potential candidates for the amelioration of the antioxidant content and activity in the edible parts of legumes. Under stressful environments, BRs have been reported to enhance levels of both enzymatic as well as non-enzymatic antioxidants in legumes, which provide an advantage to humans utilizing the latter in their diet in making them less prone to diseases and also boosting immunity.

One of the research studies demonstrated the rise in levels of non-enzymatic antioxidants in lupine and pea seeds in 24-EBL foliar spray of both plants. These include phenolic antioxidants, that is, tocopherols and tocotrienols, vitamin C, also called ascorbic acid, and β-carotene (Biesaga-Koscielniak et al., 2014). Tocopherols are mainly associated with plant responses to stress conditions and hence ensure their normal growth and development under stressful environmental conditions by hampering the negative effects of ROS on their primary metabolic processes (Szarka et al., 2012). Similarly, tocotrienols and vitamin C are responsible for the scavenging of lipid peroxyl radicals and hence maintaining membrane integrity and normal cellular functions. Moreover, β-carotene is related to the homeostasis of processes of photosynthesis and photoprotection (Munné-Bosch and Alegre, 2002). Hence, these antioxidants not only ensure crop survival but also add to their nutritional value by maintaining themselves in seeds that are consumed by humans in various forms in diets. Similarly, enzymatic antioxidants such as CAT, POD, and SOD were found to enhance to significant levels in *V. radiata* plans in 28-HBL treatment grown in conditions of Ni stress (Yusuf et al., 2014). Further research studies by Lima et al. (2019) demonstrated the supplementary role of 24-EBL on the content of enzymatic antioxidants such as SOD, APX, POD, and CAT in conditions of water stress in cowpea (*V. unguiculata*) plants. It has also been reported that pre-sowing soaking treatments of lupine and pea seeds with 24-EBL augmented the antioxidant and vitamin content in their plants as well as their seeds, making them a more efficient part of the diet to be consumed (Janeczko et al., 2015). The amount of enzymatic as well as non-enzymatic antioxidants has also been upregulated in pea plants and seeds in the treatment of low doses of 28-HBL in water-deficit environments (Shahid et al., 2011). Legumes when consumed in either form by humans provide them with these antioxidants that in the human body serve to perform various beneficiary purposes such as synthesis of erythrocytes, prevention of heart attack and neural spasms, and anti-cancerous, anti-allergic, and anti-inflammatory effects (Sindhi et al., 2013).

19.2.3.3 Micro- and Macronutrients

Micro- and macronutrients are the primary elements required by legumes for their normal survival and reproduction. The former is present in large quantities in fresh mass while the latter occurs in relatively small amounts. Usually, plants obtain these nutrients from their growing environments, that is, from soil, water, or air in different forms. Macronutrients include N, P, K, Ca, S, Mg, C, O, and H, while micronutrients, also called trace elements, include Fe, B, Cl, Mn, Zn, Cu, Mo, and Ni. These nutrients serve to act as constituents of various vital metabolites such as amino acids, monosaccharides, and fatty acids that form the basic machinery of primary plant processes. Yet, deficiency, as well as excessive uptake and accumulation, of these nutrients may lead to toxicity and hence prove to be detrimental to crop survival.

Uptake of legumes by living organisms provides them with the organic form of nutrients required for meeting their metabolic needs and producing energy for their well-being, that is, these nutrients in legumes act as a source of energy in the form of carbohydrates, fats, protein, minerals, and vitamins to the living organisms. BRs play an important role in the uptake of these nutrients and also prevent legumes from experiencing the deleterious effects of their deficiency or toxicosis.

Exogenous application of 24-EBL independently, as well as in combination with silicon (Si), enhanced the absorption of macronutrients—Ca, Mg, S, K, P—and micronutrients—B, Mn, Cu, Fe, Zn—by the roots of pea plants under normal as well as conditions of Cd toxicity. This resulted in their increased accumulation in roots and shoots counteracting the deleterious effects of Cd toxicity (Jan et al., 2018). However, BRs have also been reported to promote arbuscular mycorrhizal associations in the roots of legumes, which enhance their ability in the uptake of vital macronutrients, i.e., N and P (McGuiness et al., 2019). Comparable results were also obtained by Dalio et al. (2013) with regard to the enhanced accumulation of nitrogenous compounds in roots and shoots of pigeon pea plants as a result of treatment with 24-EBL, which increased the uptake of nitrogen in the form of nitrate ions by roots under the condition of salt stress. BRs are also found to be involved in the regulation of root development, formation of root hairs, and development of mycorrhizal and nodule associations, all of which promote the plant's ability to more efficiently absorb the nutrients available in the surrounding environment (Wei et al., 2016). Seed priming of soybean with 24-EBL also resulted in enhanced N uptake, and hence a positive N balance in plant parts in conditions of salt stress, thereby enhancing the tolerance of the legume to salinity (Soliman et al., 2020). A number of research studies demonstrated the stimulatory effects of BRs on the uptake and accumulation of micro- and macronutrients in some leguminous crops. However, the exact molecular and physiological mechanisms by which BRs show their effects need to be further explored in detail.

19.3 BRs as Mediators of Human Welfare

BRs as already discussed are natural polyhydroxy steroids found ubiquitously in the plant kingdom, where they are well known for their regulatory effects on a wide range of metabolic processes, especially in stressful environments. However, they have been found to share many similarities with steroidal hormones present in various groups of animals. Therefore, recently, these plant hormones have been explored for their direct utilization in animal systems as they have been reported to influence certain biological activities.

In animal systems, receptor proteins for BRs are located on the plasma membrane or in the cytosol or nucleus of the target cells. BR-receptor association is reversible and the BR-receptor complex enters the nucleus, undergoes the process of dimerization with another nuclear receptor, and the dimer in turn interacts with specific DNA sequences in the chromatin. This consecutively results in the upregulation or downregulation of transcription of the genes present in those DNA sequences (Kim and Wang, 2010). The messenger RNA (mRNA) being transcribed enters a process of translation after processing and proteins are synthesized. This alters the protein-specific biological activities of animal systems (Figure 19.6). The mechanism of action of BRs is very much complementary to the action of steroid hormones secreted naturally in animals.

It has been established that these naturally occurring plant steroids may have many benefits to humans by imposing anti-cancerous, anti-viral, anti-fungal, and anti-inflammatory effects (Figure 19.7). Therefore, they may now be considered potent molecules for application in clinical and biomedical sciences. Some of the most prominent biological effects of BRs in animal systems are described in the following sections.

19.3.1 Inhibition of Cancerous Growth

Cancer or malignant tumors in humans are characterized by the presence of abnormal mutations resulting in excessive cell proliferation and inhibition of cell apoptosis, thereby increasing their life span and the ability to undergo numerous divisions. In addition, the proteins and enzymes involved in the repair of such mutations are also downregulated in malignancy. BRs containing a basic steroidal structure and oxygenated side chains have been proved to be toxic to mammalian tumor cells and hence these characteristics of BRs can be utilized in the synthesis of new anti-cancerous drugs (Mehtiev et al., 2008; Gupta et al., 2013; Greenwell et al., 2015). Research studies by Kaushik et al. (2014) and Zhabinskii et al. (2015) demonstrated that the various bioactive analogs of BRs with side chains undergoing the process of oxygenation act in a similar manner to anti-cancerous steroids. They are known to stimulate various factors that may act to antagonize the cell cycle and division process and hence inhibit excessive proliferation of cells. BRs have also been found to play an effective role in the stimulation of apoptosis which results in the death of these cancerous cells (Malikova et al., 2008). A large number of research studies have been carried out on the anti-cancerous effect of BRs on lung and cervical carcinoma, multiple myeloma, prostate, and breast cancer cell lines through Calcein AM assay.

The highest cytotoxicity for these cancer cell lines has been reported to consist of 28-homoCS and CS. Moreover, fibroblast cells did not show any loss in their viability with BR treatment, which resulted in different responses in various kinds of cells. Also, the viability of the T-lymphoblastic and

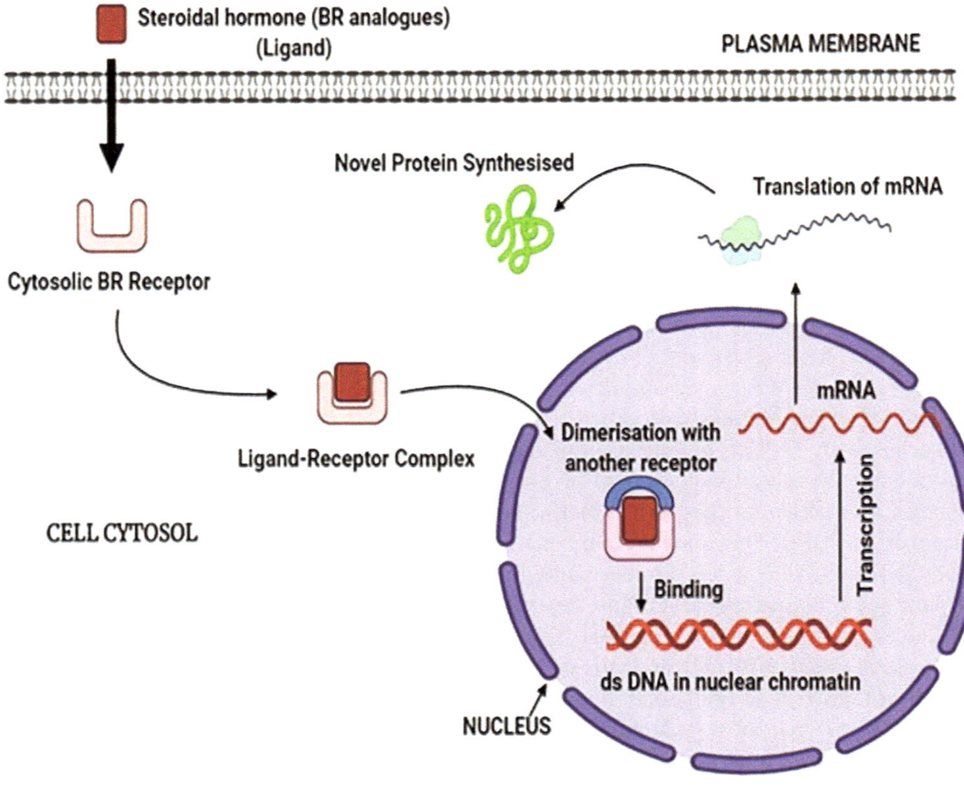

FIGURE 19.6 Mechanism of action of BR analogs in human cells.

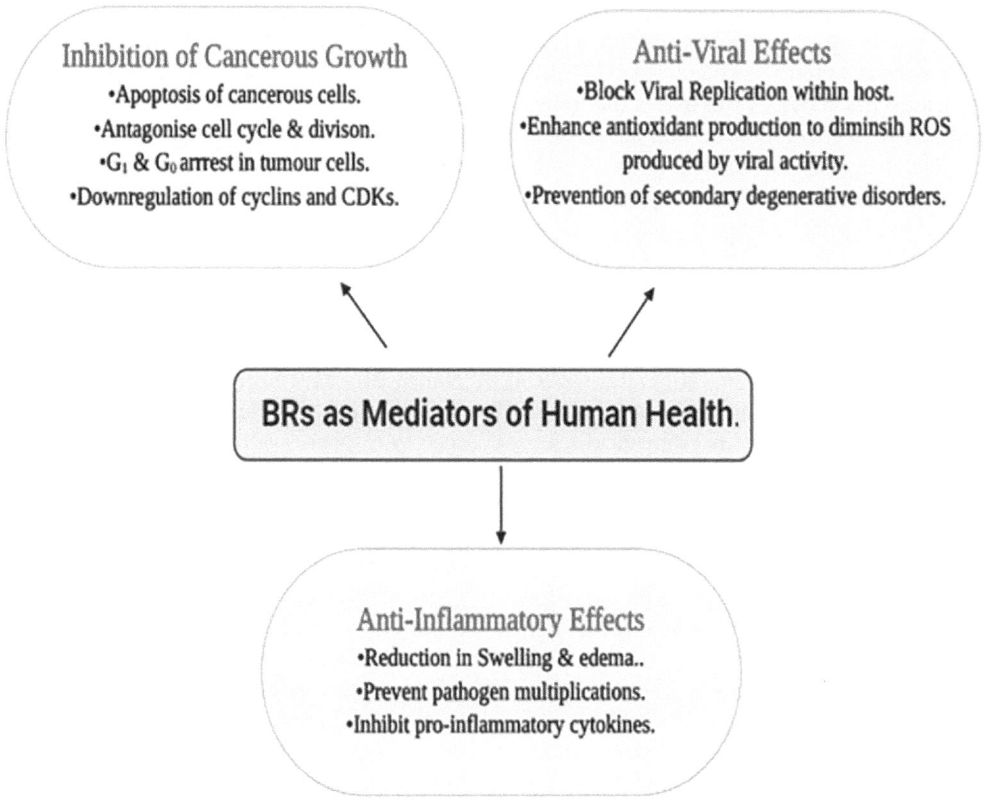

FIGURE 19.7 The benefits of BR applications in humans.

B-lymphoblastic cancer cell lines was found to decrease even in the presence of micromolar concentrations of 28-homoCS and 24-EBL wherein 28-homoCS was found to be more effective than T-lymphoblastic cancer cell lines (Malikova et al., 2008). Estrogen sensitive and insensitive breast cancer cell lines demonstrated diverse behavior in treatment with BRs. Similar results were obtained for androgen sensitive and insensitive prostate cancer cell lines where BRs proved to be potent for the elimination of cancerous growth and its probability of occurrence (Denmeade et al., 1996; Weisburger et al., 1998).

Bioactive BR analogs—28-homoCS and 24-EBL—have been reported to downregulate the cell cycle, proliferation, and survival of breast cancer cell lines, namely DU-145 in a dose-dependent manner. BR anti-cancerous effects were also demonstrated by flow cytometric studies where G_1 arrest was observed in cancer cell lines of breast as well as prostate tissues. This in turn resulted in a fewer number of cells entering the S phase for the process of replication and hence reduced cell proliferation and cell number (Swaczynova et al., 2006). Research studies by eminent biologists also suggested the role of BRs in hampering the proliferation of cancerous cell lines without affecting the cell cycle and division of normal cells. This property of BRs made them a remarkable component to be incorporated into anti-cancer drugs for the treatment of malignancy.

Cancer in different body tissues may result in some other additional disorders such as osteoporosis, excessive secretion of adrenal cortisol, thereby repressing immune responses, sexually differentiation diseases, and asthma. BRs not only suppress the growth of cancer cell lines but also cure these disorders by activation of certain immune responses required in specific tissues of the body (Franek et al., 2003). In addition to flow cytometry, other techniques such as immunofluorescence, western blotting, and the DNA ladder test also supported the arrest of the cell cycle of breast and prostate cancer cell lines at the G_1 phase along with downregulation of CDK4/6, phosphorylated retinoblastoma, and cyclin D1. All of these factors suppress cell proliferation (Steigerova et al., 2010; Steigerova et al., 2012). Moreover, activation of pro-apoptotic proteins by BRs was also observed in these studies.

Further research studies by Swaczynova et al. (2006) also showed the inhibitory effects of BRs on the proteins and factors that enhanced cancerous growth by stimulation of the cell cycle such as cyclins, related CDKs, α-estrogen receptor, p27, p53, and anti-apoptotic proteins that suppress apoptosis and hence enhance the life span of the cell; 24-EBL resulted in the elevation of the membrane potential of mitochondria, arresting cells in G_0 and G_1 phases and a relatively smaller number of cells in the S phase of the cell cycle, thereby proving its anti-cancerous potential. Another molecular technique commonly called SILAC also displayed anti-cancerous properties of 24-EBL through which enhanced apoptosis was observed in the androgen sensitive prostate cancer cell lines; 24-EBL treatment to these cancer cell lines resulted in increased translocation of CHOP proteins from the cytoplasm to the nucleus, which, in turn, activated the prominent enzymes that upregulate apoptosis, i.e., caspase-3 and 9. It was also supported by the stimulation of enzymes such as polyamine oxidase (PAO) and spermine N1-acetyltransferase (SSAT) in prostate cancer cell lines after 24-EBL treatment where these enzymes are also known for their inducing effects on apoptosis (Obakan et al., 2014). Comparable results were also obtained by Wu et al. (2007), who demonstrated the potential of 24-EBL in efficiently upregulating the activity of caspases that promote apoptosis and inhibiting expression of an anti-apoptotic protein, i.e., Bcl-2. Also, exogenous supplementation of 24-EBL to colon cancer cell lines resulted in the activation of enzymes that promote mitochondrial-regulated apoptosis in these cells (Coskun et al., 2015). Another mechanism by which BR analogs have been found to prevent the development of malignancy is inhibition of the process of angiogenesis, which is necessary for the conversion of benign tumors into malignant tumors and hence cancer (Rarova et al., 2012; Rarova et al., 2018). All these properties of BRs and their various analogs enable them to be successfully utilized in biomedical sciences for the preparation of anti-cancer drugs.

19.3.2 Anti-Viral Properties

BRs exhibit potential to protect the host from viral infections by triggering various signaling mechanisms that inhibit the lytic or lysogenic cycle of replication of viruses in host cells. Viral infection in any cell induces an immune response through the interaction of specific pathogen-associated molecular patterns (PAMPs) present on the virus particles with host specific receptor proteins, that is, pattern recognition receptors (PRRs) located on the cell surface (Calil et al., 2016; Macho et al., 2014). The PAMP-PRR complex is in turn activated by receptor-like kinases to start the signal transduction for the elucidation of the immune response. These receptor-like kinases require assistance of a co-receptor for their functioning. BR insensitive-1 related kinase (BAK1) acts as a co-receptor that interacts with the PAMP-PRR complex and hence activates the immune response against the virus (Liebrand et al., 2014). Thus, BRs have been found to display potential applications in the prevention of plants from virus infections of various types and hence control diseases caused to nutritionally important crops by viruses (Yang et al., 2010; Zhang et al., 2015). BRs have also been found to enhance antioxidant activity in order to neutralize the deleterious effects of ROS produced in large amounts in virus infections. This, in turn, helps the host cell to acclimatize and survive viral infections and promote viral clearance from the body.

In humans, various kinds of viruses with different kinds of genetic material are known to cause several diseases. Various synthetic forms of BRs show anti-viral properties against arena virus, herpes simplex virus type-1 and type-2 (HSV-1 and HSV-2), polio virus (PV), vesicular stomatitis virus (VSV), and measles virus (MV; Wachsman et al., 2002; Sasse et al., 2003; Wachsman et al., 2004 (b)). The mechanism by which a virus infects its target host cells remains the same and involves sequential steps of attachment to the host plasma membrane, penetration or injection of its genetic material into the host cytoplasm, and replication and expression of its genetic material in host cells through the lytic or lysogenic cycle and hence formation of a large number of copies of virus particles leading to the lysis of host cells. One of the BR analogs, (22S,23S)-3β-bromo-5α,22,23-trihydroxystigmas tan-6-one, inhibits the late phase of virus replication inside host cells, expression of viral proteins, and packaging of viral capsid and genetic material into viral particles in order to rescue the host cell from the viral multiplication within it. No inhibitory effects of the compound were observed in the initial stages of infection by a virus (Romanutti et al., 2007). This compound showed specificity in hampering infection by the Junín virus by downregulating the early growth stages of the virus inside host cells. In addition to this, the compound targets viral RNA replication and causes incomplete synthesis of the antigenomic RNA. However, attachment of the virus to the host plasma membrane and its internal development was not affected by the compound (Castilla et al., 2005). In certain plants, BRs have been found to stimulate virus resistance by accumulation of nitric oxide that mainly promotes pro-inflammatory responses of the immune system against viruses (Zou et al., 2018). This compound also exhibits anti-viral properties against VSV by inhibiting its replication in Vero cells. This virus is responsible for causing an economically important disease in cattle and horses.

Some of the synthetic analogs of BRs have also been reported to impede the replication of HSV-1 and arena virus *in vitro*. Wachsman et al. (2004a) investigated various methods to prepare synthetic BR derivatives that display anti-viral properties against both RNA as well as DNA viruses; 24-EBL also acts as a potent compound to inhibit replication of HIV in cultures in *in vitro* conditions (Khripach et al., 2000).

19.3.3 Anti-Inflammatory Properties

Inflammation generally represents an immune response occurring in the body under conditions of allergy, injury, lesions, and any pathogenic infection. Common signs of tissue exhibiting inflammation are redness, swelling, bulging out, edema, itching, and pain. Excessive inflammation may result in a variety of disorders such as allergic conjunctivitis, chronic sinusitis, asthma, bowel disorders, and Crohn's disease (Konstantopoulos et al., 2005). BRs and their various synthetic analogs have been widely reported in suppressing these inflammatory disorders and hence are known to be anti-inflammatory in nature (Patel et al., 2015). Common examples of synthetic BR analogs possessing anti-inflammatory properties are stigmastane and androstane, which modulate immune responses in order to reduce tissue inflammation. Other antiviral BR analogs (22S,23S)-3β-bromo-5α-22,23-trihydroxyst igmastan-6-one, (22S,23S)-3β-5α-22,23tetrahydroxystigmas tan-6-one, and (22S,23S)-5α-fluoro-3β-22,23-tetrahydroxystig mastan-6-one suppress replication of HSV in host Vero cells and also reduced inflammation within three successive days after treatment (Wachsman et al., 2004a). Another synthetic derivative of stigmastane, (22S,23S)-22,23-dihydroxystigm ast-4-en-3-one, is very much structurally similar to a commonly used anti-inflammatory drug, dexamethasone (DEX; Lan et al., 2012). This compound is found to be responsible for responding to the deleterious effects of the herpes virus as well as stromal inflammation.

Some other synthetic derivatives of BRs found to possess anti-inflammatory properties are ursolic acid, (23R)-2α,3 β,23,28-tetrahydroxy-14,15-dehydrocampestrol, and oleanolic acid. Also, 24-EBL has been utilized to treat rough and wrinkled skin formed as a result of inflammation of fibroblast and epidermal cells in human beings (Quinones et al., 2010). The anti-inflammatory properties of various BR derivatives have been analyzed through various experiments in mice and rabbits. These properties are also being utilized for reducing inflammation and related disorders in humans as well. Stigmastane derivatives are known to shut down the multiplication of HSV-1 in nerve cells as well as epithelial cells, thereby reducing inflammation. Moreover, these compounds also resulted in reduced amounts of anti-inflammatory cytokines, that is, interleukin-6 and interferon-α in HSV-1 infected cells, suggesting their anti-inflammatory role again (Ferrari et al., 2009). In addition to this, the compounds are also found to be involved in the elevation of anti-inflammatory cytokines in macrophages and thus providing an innate immune response. Other anti-inflammatory agents upregulated by these compounds are IL-10 and suppressor of cytokine signaling (Socs-2), which inhibit the JAK-STAT activation pathway of inflammatory cytokines (Shen et al., 2009). Further research studies also suggested the anti-inflammatory potential of 28-HBL, which inhibits the synthesis of vasodilators such as leukotrienes that increase inflammation and also downregulate the expression of phospholipase and thus decreasing calcium ion concentration in the cellular pool. All these effects finally result in a reduction of inflammation. The anti-inflammatory nature of BRs is shown in the anti-glycemic properties of homoCS, induced synthesis of RBC (red blood corpuscles), WBC (white blood corpuscles), platelets, and hemoglobin, and the reduction of inflammatory disorders such as leukemia, bowel disorders, and neoplasms in living organisms (Athithan et al., 2018). Therefore, it can be established that BRs and their various analogs may prove beneficiary to living beings if utilized in an appropriate manner for the reduction of tissue inflammation.

19.4 Conclusion

BRs are a natural, environmentally friendly, non-poisonous, and biologically safe category of phytohormones that can be utilized in their different forms for the biofortification of crop plants as well as the regulation of human metabolic processes for the treatment or prevention of certain disorders. They have been reported to influence a vast array of agronomic traits effectively, which makes them suitable to be used in agriculture for the augmentation of qualitative (nutritional value) and quantitative (crop biomass and production) traits of commonly cultivated crop plants. A wide range of physiological and molecular attributes are regulated by BRs and their different analogs, therefore, they may be mentioned as pleiotropic hormones. They are known to control a number of processes throughout the life history of plants from seed germination, vegetative growth, flowering, and fruit development to an increase in the overall production yield of the crops. BRs have been proved to be very effective and potent in enhancing the survival of different kinds of crops—cereals, legumes, oilseed, and horticultural—not only in normal growth conditions but also in the harsh conditions of biotic and abiotic stresses in the environment.

By enhancing the qualitative as well as the quantitative value of crops, BRs pose indirect benefits to the community in terms of good health and the economy. In addition, recent screening of BRs for their wide applications in controlling human metabolic activities has made them useful compounds for exploration and utilization. BRs block the replication of viruses in human cell lines and pose toxic effects on tumor cells without affecting normal ones. Many BR analogs have also been found to have anti-inflammatory, anti-genotoxic, anti-hepatic, anti-angiogenic, and anti-bacterial activities. Furthermore, they have also been reported to possess ecdysteroidal and anabolic properties as well. These properties of BRs enable them to be a potent compound for utilization in pharmaceutical and biomedical laboratories to synthesize anti-cancerous drugs.

Despite the considerable number of benefits of BRs to plants and humans, their application is still limited in common agricultural practices as well as in therapeutics for use in living organisms. Therefore, further insights are being carried out to explore molecular mechanisms in the regulation of various plant processes such as symbiosis and its crosstalk with other plant hormones. This will aid in achieving the goal of sustainable crop production, along with enhanced nutritive value, by lessening the harmful effects of prevalent farming practices. Also, further investigations of BR signaling in animal systems will expose their aptitude in clinical applications.

REFERENCES

Ahanger et al. *Plant Physiol. Biochem.* 2017a; 115:449–460.
Ahanger et al. *Protoplasma.* 2017b; 254(4):1471–1486.
Ahanger et al. *J. Plant Growth Regul.* 2018; 37(4):1007–1024.
Ahmad et al. *Crit. Rev. Biotechnol.* 2010a; 30(3):161–175.
Ahmad et al. *Russ. J. Plant Physiol.* 2010b; 57(4):509–517.
Ahmad et al. *Sci. Rep.* 2018a; 8:41598–41618.
Ahmad et al. *J. Plant. Growth Regul.* 2018b; 4:157–159.
Ahmad et al. *J. Plant Interact.* 2018c; 13(1):64–72.
Alam et al. *Biomolecules.* 2019; 9(11):640.
Ali. (Doctoral dissertation: Aligarh Muslim University). 2004.
Ali et al. *Pak. J. Bot.* 2006; 38(5):1621–1632.
Anjum et al. *J. Agron. Crop Sci.* 2011; 197(3):177–185.
Anuradha and Rao. *Plant Growth Regul.* 2003; 40(1):29–32.
Archana et al. *J. Plant Biol.* 2006; 33(3):221.
Arora et al. *Curr. Sci.* 2002; 82:1227–1238.
Arora et al. *Physiol. Mol. Biol. Plants.* 2010; 16(3):285–293.
Athithan et al. *Int. J. Pharm. Pharm. Sci.* 2018; 10:162–167.
Azpeitia et al. *J. Hortic. Sci. Biotechnol.* 2003; 78(5):591–596.
Bajguz. *Plant Physiol. Biochem.* 2007; 45(2):95–107.
Bajguz and Hayat. *Plant Physiol. Biochem.* 2009; 47(1):1–8.
Bajguz and Piotrowska-Niczyporuk. *Plant Physiol. Biochem.* 2014; 80:76–183.
Bajguz and Tretyn. *Phytochemistry.* 2003a; 62(7):1027–1046.
Bajguz et al. In: *Brassinosteroids.* 2003b;1–44.
Baker et al. *Proc. Natl Acad. Sci. USA* 2003; 100(7):3579–3583.
Bakht et al. *J. Exp. Bot.* 2006; 57(14):3707–3715.
Barberon et al. *Proc. Natl Acad. Sci. USA* 2014; 111(22):8293–8298.
Batra et al. *J. Plant Interact.* 2014; 9(1):712–721.
Bhardwaj et al. *Asian J. Plant Sci.* 2007; 6(5):765–772.
Bhat. *J. Plant Biochem. Biotechnol.* 2010; 19(1):1–7.
Biesaga-Koscielniak et al. *Aust. J. Crop Sci.* 2014; 8(3):378.
Bishop et al. *Plant Cell.* 2002; 14:97–110.

Bitterlich et al. *Plant Signal. Behav.* 2014; 9(10):970426.
Bowers et al. *Nature.* 2003; 422(6930):433–438.
Braun and Wild. *J. Plant Physiol.* 1984; 116(3):189–196.
Cahoon et al. *Curr. Opin. Plant Biol.* 2007; 10(3):236–244.
Calil and Fontes. *Ann. Bot.* 2017; 119(5):711–723.
Camoni et al. *Front. Plant Sci.* 2018; 9:297.
Cao and Zhao. *Rice Sci.* 2008; 15(1):63–68.
Cao et al. *Physiol. Plant.* 2005; 123(1):57–66.
Castilla et al. *Antiviral Res.* 2005; 68(2):88–95.
Chai et al. *Plant Growth Regul.* 2013; 69(1):63–69.
Chaiwanon and Wang. *Curr. Biol.* 2015; 25(8):1031–1042.
Champa et al. *J. Sustain. Trop. Agric. Res.* 2015; 26(2):368–379.
Chen et al. *Plant Cell Rep.* 2015; 34(1):83–96.
Cheng et al. *Sci. Rep.* 2017; 7(1):1–12.
Choe et al. *Plant J.* 2001; 26(6):573–582.
Clouse. TAB. *American Society of Plant Biologists.* 2011; 9.
Coskun et al. *Exp. Cell Res.* 2015; 338(1):10–21.
Dalio et al. *Bot. Stud.* 2013; 54(1):9.
Denmeade et al. *Prostate.* 1996; 28(4):251–265.
Divi and Krishna. *New Biotechnol.* 2009; 26(3–4):131–136.
Dordas. C. *Agron. Sustain. Dev.* 2008; 28(1):33–46.
Dyer and Mullen. *Physiol. Plant.* 2008; 132(1):11–22.
Edreva et al. *Gen Appl Plant Physiol.* 2008; 34:67–78.
El-Bassiony et al. *Gesunde Pflanzen.* 2012; 64(4):175–182.
Eleiwa et al. *Agron. Sustain. Dev.* 2011; 5(5):58–65.
El-Khallal et al. *Res. J. Agric. Biol. Sci.* 2009; 5(4):380–390.
Fageria et al. CRC Press. 2016.
Fariduddin et al. *Biol. Plant.* 2014; 58(1):9–17.
Farooq et al. *Plant Physiol. Biochem.* 2019; 141:353–369.
Ferguson et al. *Plant Physiol.* 2005; 138(4):2396–2405.
Ferrari et al. *Investig. Ophthalmol. Vis. Sci.* 2009; 50(13): 3098–3098.
Foo et al. *Ann. Bot.* 2014; 113(6):1037–1045.
Foo et al. *J. Exp. Bot.* 2016; 67(8):2413–2424.
Franěk et al. *Collect. Czechoslov. Chem. Commun.* 2003; 68(11):2190–2200.
Fujii and Saka. *Plant Prod. Sci.* 2001; 4(2):136–144.
Fujioka and Yokota. *Annu. Rev. Plant Biol.* 2003; 54(1):137–164.
Greenwell and Rahman. *Int. J. Pharm. Sci. Res.* 2015; 6(10):4103.
Gupta et al. *J. Steroid Biochem. Mol. Biol.* 2013; 137:242–270.
Guschina et al. *Biochim. Biophys. Acta.* 2014; 1838(6):1488–1500.
Harborne et al. *Encyclopedia of Plant.* 1980; 329–395.
Harwood. *Annu. Rev. Plant Physiol.* 1988; 39:101–138.
Harwood. *Biochim. Biophys. Acta.* 1996; 1301(1–2):7–56.
Harwood et al. *FEBS Lett.* 2013; 587(13):2079–2081.
Haubrick and Assmann. *Plant Cell Environ.* 2006; 29(3):446–457.
Hayat et al. *Brassinosteroids.* 2003; 231–246.
Hayat et al. Springer Science & Business Media. 2010.
Hayat et al. *Photosynthetica.* 2011; 49(3):397.
Hola. *Brassinosteroids: A Class of Plant Hormone.* 2011; 143–192.
Hua et al. *Plant J.* 2012; 69(3):432–444.
Jan et al. *BMC Plant Biol.* 2018; 18(1):146.
Janda et al. *Plant Cell Rep.* 2013; 32(6):839–851.
Janeczko et al. *J. Agron. Crop Sci.* 2010; 196(4):311–321.
Janeczko et al. *Legum. Res.* 2015; 38(2):185–193.
Jeong and Connolly. *Plant Sci.* 2009; 176(6):709–714.
Johnson, and Williams. *Plant Physiol.* 1989; 91(3):924–929.
Kalinich et al. *J. Plant Physiol.* 1985; 120(3):207–214.
Kanwar et al. *Acta Physiol. Plant.* 2012; 35:1351–1362.
Kartal et al. *Plant Growth Regul.* 2009; 58(3):261–267.
Kaushik et al. *Glob. J. Pharm. Educ. Res.* 2014; 3:(1–2).
Khan et al. *Molecules.* 2010; 15(3):1562–1577.
Khripach. *Ann. Bot.* 2000; 86(3):441–447.
Khripach et al. *Pathology RU.* 1996; 2:119–285.
Khripach et al. *Brassinosteroids.* 2003; 189–230.
Kim and Wang. *Annu. Rev. Plant Biol.* 2010; 61:681–704.
Kim et al. *Mol. Cells.* 2012; 34(6):539–548.
Klein et al. *S. Afr. J. Bot.* 2015; 96:13–18.
Koch et al. *Annu. Rev. Plant Biol.* 1996; 47(1):509–540.
Konstantopoulos. *Curr. Drug Targets Inflam. Allergy.* 2005; 4(1):39–39.
Koopmann et al. *Plant Physiol.* 1999; 119(1):49–55.
Krishna. *J. Plant Growth Regul.* 2003; 22(4):289–297.
Kroutil et al. *Plant Soil Environ.* 2010; 56(1):43–50.
Lan et al. *Ocul. Surf.* 2012; 10(3):137–148.
Lee et al. *Plant Biotechnol. J.* 2011; 9(3):359–372.
Li and van Staden. *S. Afr. J. Bot.* 1998a; 64(2):116–120.
Li, and van Staden. *Plant Growth Regul.* 1998b; 24(1):55–66.
Li et al. *Plant Physiol.* 2006; 142(2):750–761.
Li et al. *Plant Biotechnol. J.* 2016; 14(3):1021–1033.
Liao et al. *Int. J. Mol. Sci.* 2018; 19(10):3146.
Liebrand et al. *Trends Plant Sci.* 2014; 19(2):123–132.
Lim et al. *Annu. Rev. Phytopathol.* 2017; 55:505–536.
Liu et al. *J. Plant Growth Regul.* 2019; 38(2):701–712.
Macho et al. *Mol. Cell.* 2014; 54(2):263–272.
Malíková et al. *Phytochemistry.* 2008; 69(2):418–426.
Marchive et al. *Eur. J. Lipid Sci. Technol.* 2014; 116(10):1332–1343.
Mazorra et al. *Plant Physiol. Biochem.* 2011; 49(12):1420–1428.
McGuiness et al. *Front. Plant Sci.* 2019; 10:269.
Mehtiev et al. *Biomed. Chem.* 2008; 2(1):1–17.
Mitchell et al. *Nature.* 1970; 225(5237):1065–1066.
Mittler. *Trends Plant Sci.* 2002; 7(9):405–410.
Montoya et al. *Plant J.* 2005; 42(2):262–269.
Morinaka et al. *Plant Physiol.* 2006; 141(3):924–931.
Mota et al. (Doctoral dissertation: University of the Free State). 2020.
Munné-Bosch and Alegre. *Crit. Rev. Plant Sci.* 2002; 21(1):31–57.
Nunez et al. *Plant Biol.* 2003; 47:67–70.
Ozdemir et al. *Plant Growth Regul.* 2004; 42:203–211.
Obakan et al. *Amino Acids.* 2014; 46(3):553–564.
Ohlrogge and Browse. *Plant Cell.* 1995; 7(7):957–970.
Patel et al. *J. Phytopharmacol.* 2015; 4(2):121–125.
Pelletier et al. *Plant Mol. Biol.* 1999; 40(1):45–54.
Peng et al. *Plant Cell Rep.* 2010; 29(7):793–802.
Pereira-Netto et al. *Trees.* 2006; 20(3):286–291.
Pociecha et al. *Plant Growth Regul.* 2016; 80(2):127–135.
Pokotylo et al. *Cytol. Genet.* 2012; 46(3):131–135.
Pokotylo et al. *Prog. Lipid Res.* 2013; 52(1):62–79.
Pourcel et al. *Trends Plant Sci.* 2006; 12:29–36.
Quiñones et al. *Carbohydr. Polym.* 2010; 80(3):915–921.
Rajewska et al. *Front. Plant Sci.* 2016; 7:629.
Rárová et al. *Steroids.* 2012; 77(13):1502–1509.
Rárová et al. *J. Steroid Biochem. Mol. Biol.* 2018; 178:263–271.
Rasool et al. *Acta Physiol. Plant.* 2013; 35(4):1039–1050.
Ray et al. *Nat. Commun.* 2012; 3(1):1–7.
Romanutti et al. *Int. J. Antimicrob. Agents.* 2007; 29(3):311–316.
Rout and Sahoo. *Rev. Agric. Sci.* 2015; 3:1–24.
Sadura and Janeczko. *Biol. Plant.* 2018; 62(4):601–616.
Sakamoto et al. *Nat. Biotechnol.* 2006; 24(1):105–109.
Salazar et al. *Rev. Agron.* 2016; 36(1):37–41.
Samarth et al. *Open J. Org. Pol. Mat.* 2015; 5:1–22.

Sasse et al. *J. Plant Growth Regul.* 2003; 22(4):276–288.
Savage et al. *The Groundnut Crop—A Scientific Basis for Improvement.* 1994; 173–213.
Sayed et al. *Aust. J. Basic Appl. Sci.* 2009; 3(4):4465–4477.
Schlüter et al. *Plant Cell Environ.* 2002; 25(6):783–791.
Shahbaz et al. *Pak. J. Bot.* 2007; 39(2):513.
Shahbaz et al. *Plant Growth Regul.* 2008; 55(1):51–64.
Shahid et al. *Aust. J. Crop Sci.* 2011; 5(5):500.
Sharma and Bhardwaj. *Acta Physiol. Plant.* 2007; 29(3):259–263.
Sharma et al. *Braz. J. Plant Physiol.* 2007; 19(3):203–210.
Sharma et al. *Ecotoxicology.* 2011; 20(4):862–874.
Sharma et al. *Plant Physiol. Biochem.* 2013; 69:17–26.
Sharma et al. *Toxicologist.* 2014; 21:44.
Sharma et al. *Indian J. Plant Physiol.* 2016a; 21(1):70–75.
Sharma et al. *Front. Plant Sci.* 2016b; 7:1569.
Shen et al. *FEBS Lett.* 2000; 480(2–3):132–136.
Shi et al. *Plant Cell.* 2006; 18(3):651–664.
Siddiqui et al. *Acta Physiol. Plant.* 2018; 40(3):59.
Sindhi et al. *J. Pharm. Res.* 2013; 7(9):828–835.
Sirhindi et al. *Physiol. Mol. Biol. Plants.* 2009; 15(4):335–341.
Sirhindi et al. *Acta Physiol. Plant.* 2017; 39(10):228.
Skoczowski, and Filek. *Plant Sci.* 1994; 98(2):127–133.
Skoczowski et al. *Acta Physiol. Plant.* 1986; 8:203–212.
Soliman et al. *Physiol. Mol. Biol. Plants.* 2020; 26(3):501–511.
Song et al. *Sci. Rep.* 2019; 9(1):1–14.
Steigerová et al. *Chem. Biol. Interact.* 2010; 188(3):487–496.
Steigerová et al. *Food Chem. Toxicol.* 2012; 50(11):4068–4076.
Susila et al. *World J. Agric. Sci.* 2012; 8(3):223–228.
Swaczynová et al. *Pol. J. Che.* 2006; 80(4):629–636.
Swamy et al. *Am. J. Plant Physiol.* 2010; 5(4):212–218.
Symons et al. *Plant Physiol.* 2006; 140(1):150–158.
Symons et al. *J. Exp. Bot.* 2008; 59(1):17–24.
Szarka et al. *Int. J. Mol. Sci.* 2012; 13(4):4458–4483.
Takatsuto. *J. Chromatogr. A.* 1994; 658(1):3–15.
Talaat et al. *Acta Physiol. Plant.* 2013; 35(3):729–740.
Tang et al. *BMC Biol.* 2016; 14(1):1–5.
Tavakoli et al. *Adv. Environ. Biol.* 2014:31–36.
Thummel and Chory. *Genes Dev.* 2002; 16(24):3113–3129.
Thussagunpanit et al. *Photosynthetica.* 2015; 53(2):312–320.
Tripathi et al. *Improvement of Crops in the Era of Climatic Changes.* 2014; 197–216.
Vardhini and Anjum. *Front. Environ. Sci.* 2015; 2:67.
Vardhini and Rao. *Indian J. Plant Physiol.* 1998a; 13:58–60.
Vardhini and Rao. *Phytochemistry.* 1998b; 48(6):927–930.
Vardhini and Rao. *Plant Growth Regul.* 1999; 28(3):165.
Vardhini and Rao. *Indian J. Plant Physiol.* 2000; 5(1):89–92.
Vardhini and Rao. *Phytochemistry.* 2002; 61(7):843–847.
Vardhini and Rao. *Plant Growth Regul.* 2003; 41(1):25–31.
Vardhini and Rao. *Indian J. Plant Physiol.* 2005; 10(4):381–384.
Wachsman et al. *Antivir. Chem. Chemother.* 2002; 13(1):61–66.
Wachsman et al. *Int. J. Antimicrob. Agents.* 2004a; 23(5): 524–526.
Wachsman et al. *Curr. Med. Chem.* 2004b; 3(2):163–179.
Wang et al. *Int. Rice Res. Notes.* 1997; 22(1):20–21.
Wang et al. *Methods Mol. Biol.* 2012; 861:123–137.
Wang et al. *J. Exp. Bot.* 2015; 66(9):2749–2761.
Wei et al. *Mol. Plant.* 2016; 9(1):86–100.
Weisburger et al. *Mutat Res-Fund Mol. M.* 1998; 402(1–2):331–337.
Willemot et al. *Low Temperature Stress in Crop Plants.* 1979; 411–430.
Wimalasekera et al. *Mol. Plant.* 2010; 3(3):610–625.
Wray et al. *Mol. Biol. Evol.* 2003; 20(9):1377–1419.
Wu et al. *Int. J. Pharm. Sci. Res.* 2007; 62(5):392–395.
Wu et al. *Plant Cell.* 2008; 20(8):2130–2145.
Xia et al. *J. Exp. Bot.* 2015; 66(10):2839–2856.
Xiong et al. *Plant Physiol. Biochem.* 2014; 80:83–89.
Xu et al. *Research Signpost.* 2007; 195–202.
Xu et al. *Plant Physiol. Biochem.* 2015; 94:197–208.
Yang et al. *Eur. J. Plant Pathol.* 2010; 127(1):149–156.
Ye et al. *Nat. Commun.* 2017; 8(1):1–13.
Yoshida and Uemura. *Plant Physiol.* 1984; 75(1):31–37.
Yu et al. *J. Exp. Bot.* 2004; 55(399):1135–1143.
Yusuf et al. *Arch. Environ. Contam. Toxicol.* 2011; 60(1):68–76.
Yusuf et al. *Plant Physiol. Biochem.* 2012; 57:143–153.
Yusuf et al. *Physiol. Mol. Biol. Plants.* 2014; 20(4):449–460.
Zhabinskii et al. *Steroids.* 2015; 97:87–97.
Zhang et al. *Aust. J. Agric. Res.* 2007; 58(8):811–815.
Zhang et al. *Chin. Sci. Bull.* 2008; 53(23):3656–3665.
Zhang et al. *Planta.* 2015; 241(4):875–885.
Zhao et al. *Plant Cell.* 1998; 10(3):359–370.
Zhiponova et al. *New Phytol.* 2013; 197(2):490–502.
Zhu et al. *Postharvest Biol. Technol.* 2010; 56(1):50–55.
Zou et al. *Physiol. Plant.* 2018; 163(2):196–210.
Zurek et al. *Plant Physiol.* 1994; 104(2):505–513.

Index